The Asia Pacific Region in the Global Economy: A Canadian Perspective

THE RESEARCH ASSEMBLED IN THIS VOLUME is mainly the product of work undertaken by academic researchers. Industry Canada staff, however, formulated and managed the project and provided constructive feedback throughout the process. Nevertheless, the papers ultimately remain the sole responsibility of the authors and do not necessarily reflect the policies or opinions of Industry Canada or the Government of Canada.

GENERAL EDITOR:
RICHARD G. HARRIS

The Asia Pacific Region in the Global Economy: A Canadian Perspective

The Industry Canada Research Series

University of Calgary Press

© Minister of Supply and Services Canada 1996

ISBN 1-895176-87-5
ISSN 1188-0988

University of Calgary Press
2500 University Dr. N.W.
Calgary, Alberta, Canada T2N 1N4

Canadian Cataloguing in Publication Data
Main entry under title:
The Asia Pacific region in the global economy

(Industry Canada research series, ISSN 1188-0988; v.7)
Also published in French under title: La région de l'Asie-Pacifique et l'économie mondiale.
Includes bibliographical references.
ISBN 1-895176-87-5

Cat. no. Id53-11/7-1996E

1. Canada—Commerce—Pacific Area.
2. Pacific Area—Commerce—Canada.
3. Pacific Area—Foreign economic relations.
I. Harris, Richard G.
II. Series.

HF3228.P3A84 1996   382'.097101823   C96-910592-4

All rights reserved. No part of this work covered by the copyrights hereon may be reproduced or used in any form or by any means – graphic, electronic or mechanical – without the prior permission of the publisher. Any request for photocopying, recording, taping or reproducing in information storage and retrieval systems of any part of this book shall be directed in writing to the Canadian Reprography Collective, 379 Adelaide Street West, Suite M1, Toronto, Ontario M5V 1S5.

The University of Calgary Press appreciates the assistance of the Alberta Foundation for the Arts (a beneficiary of Alberta Lotteries) for its 1996 publishing program.

EDITORIAL & TYPESETTING SERVICES: CIGC Services-Conseils Inc.
COVER & INTERIOR DESIGN: Brant Cowie/ArtPlus Limited

Printed and bound in Canada

∞   This book is printed on acid-free paper.

Table of Contents

PREFACE xiii

1. INTRODUCTION 1
   RICHARD G. HARRIS

   Growth and Trade Issues 3
   Service Exports 8
   Policy Issues 15
   Conclusion 19

2. ECONOMIC GROWTH AND SOCIAL CAPITAL IN ASIA 21
   JOHN F. HELLIWELL

   Asian Growth in a Global Perspective 21
   Comparative Growth in the Asian Economies 25
   How Important Are Institutions? 31
   Social Capital and Cultural Differences 33

   COMMENT 42
   FRANCIS X. COLAÇO

3. CANADA AND THE ASIA PACIFIC REGION:
   VIEWS FROM THE GRAVITY, MONOPOLISTIC
   COMPETITION, AND HECKSCHER-OHLIN MODELS 47
   WALID HEJAZI & DANIEL TREFLER

   Broad Facts About Canada's Trade with East Asia 49
   Labour and Capital Market Implications:
   The Factor Content of Canada's Trade 49

      Formal Modelling 58
      Conclusions 70
      Data Appendix 73

  COMMENT 84
  MARCUS NOLAND

4. RIVALRY FOR JAPANESE INVESTMENT IN NORTH AMERICA 87
  KEITH HEAD & JOHN RIES

      The Stakes of the Rivalry 89
      The Relative Attractiveness of Provinces and States to
         Japanese Investors 92
      The Effects of Investment Promotion Policies 105
      Conclusion 108
      Appendix 1 Notes on Sources 110

  COMMENT 114
  JOHN W. CRAIG

      Assessment of the Paper 115
      Anecdotal Comments 117
      Conclusion 119

5. CANADA'S NATURAL RESOURCE EXPORTS TO THE ASIA PACIFIC REGION 121
  ROBERT N. MCRAE

      Historical Natural Resource Exports 122
      The Relationship Between Natural Resource Consumption and
         Economic Development 126
      The Potential for the Expansion of Natural Resource Exports to
         Asian Countries 128
      Summary 135
      Appendix A 136
      Appendix B 145
      Appendix C 147
      Appendix D 150
      Appendix E 157

COMMENT   162
TIM HAZLEDINE

6. CANADIAN EXPORTS OF BUSINESS AND EDUCATION
   SERVICES TO THE ASIA PACIFIC REGION   165
   LAWRENCE L. SCHEMBRI

   APR Service Imports: An Overview   168
   Canadian Exports of Business Services   172
   Exports of Business Services to China   178
   Canadian Exports of Education Services   183
   Trade Barriers and the General Agreement on Trade in Services   186
   Conclusion   190

   COMMENT   196
   FRANK FLATTERS

   Growth and Potential Export Opportunities in the Asia Pacific Region   197
   Are We Making Enough of the Opportunities?   198
   Conclusion   200

7. FINANCIAL SECTOR OPPORTUNITIES IN THE
   ASIA PACIFIC REGION: THE CASE FOR
   CANADIAN BANKS   203
   JOHN F. CHANT

   Trade in Financial Services   205
   Demand for Financial Services in the Asia Pacific Region   210
   Financial Reform in the Asia Pacific Region   213
   Reform and the Treatment of Foreign Banks   215
   The Competitive Advantages of International Banks   220
   The International Business of the Canadian Banks   225
   The Prospects for Canadian Banks in the Asia Pacific Region   240
   Appendix 1 Sources of Data   241
   Appendix 2 Financial Reform in the Asia Pacific Region   242

   COMMENT   252
   REUVEN GLICK

   Determinants of Canadian Banking Abroad   253
   Future of the Banking Industry   257

8. INTERNATIONAL TOURISM IN THE ASIA PACIFIC REGION
    AND ITS IMPLICATIONS FOR CANADA   259
    RICHARD G. HARRIS & STEPHEN T. EASTON

   International Tourism: A Review of Trends and Facts   260
   The Economic Theory of International Tourism   271
   Tourism Flow Models   279
   The Arrivals of Tourists to Canada in 1972-92:
      A 17-Country Analysis   289
   Chinese Tourism in Canada: Some Illustrative Calculations   292
   Conclusion   297

   COMMENT   300
   TAE H. OUM

9. ASIA PACIFIC IMMIGRATION AND THE
    CANADIAN ECONOMY   303
    MICHAEL BAKER & DWAYNE BENJAMIN

   Immigration Flows to Canada   304
   A Profile of the Immigrants   318
   The Human Capital Contribution   322
   Immigration and Trade Links   333
   Immigration and Foreign Direct Investment   339
   Conclusions   341
   Appendix Data Sources and Explanations of Variables   343

   COMMENT   348
   DONALD DEVORETZ

   Trade and Investment   352
   Immigrant-Induced Growth: Vancouver's Projected
      Expenditure Patterns   353

10. TRADE, ENVIRONMENT AND GROWTH IN THE
    ASIA PACIFIC REGION: IMPLICATIONS FOR CANADA   357
    BRIAN R. COPELAND

   Air Pollution in Selected Asia Pacific Countries   359
   Effects of Growth and Trade on the Environment   362
   Policy   372

Environmental Dumping Examined 377
Conclusion 389

11. CANADIAN SMALL AND MEDIUM-SIZED ENTERPRISES: OPPORTUNITIES AND CHALLENGES IN THE ASIA PACIFIC REGION 395
    SOMESHWAR RAO & ASHFAQ AHMAD

    Canada and the Asia Pacific 397
    The Growing Importance of SME in Canada 405
    Outward Orientation and Economic Performance of Canadian SME 408
    Obstacles to SME Participation in the Asia Pacific Region 420
    Conclusions 430
    *Appendix 1 Main Characteristics of the Canadian and
         American Sample Firms* 434
    Appendix 2 SME Performance Indicators 442
    Appendix 3 Regression Results 444

    COMMENT 452
    JAMES MCRAE

12. PACIFIC TRIANGLES: U.S. ECONOMIC RELATIONSHIPS WITH JAPAN AND CHINA 457
    WENDY DOBSON

    Policy Frameworks 458
    Economic Dimensions of Interdependence 462
    Triangular Relationship: The Canadian-U.S.-Asian Triangle 472
    Implications and Conclusions 477

    COMMENT 484
    MASAO NAKAMURA

13. CANADIAN TRADE AND INVESTMENT POLICIES AND THE ASIA PACIFIC REGION: CONFRONTING AMBIVALENCE? 489
    MURRAY G. SMITH

    Globalization and the Asia Pacific 490
    Canadian Barriers to Trade and Investment with the Asia Pacific Region 495

Asia Pacific Trade and Investment Barriers   500
Canada-Asia Pacific Trade and Investment: An Assessment of
   the Options   503
What Can Be Done in the Asia Pacific?   505
Conclusion   513
Appendix 1   515
Appendix 2   535

COMMENT   573
EDWARD M. GRAHAM

14. SCIENCE AND TECHNOLOGY POLICIES IN
    ASIA PACIFIC COUNTRIES: CHALLENGES AND
    OPPORTUNITIES FOR CANADA   577
    RICHARD G. LIPSEY & RUSSEL M. WILLS

    S&T Policy in Korea   581
    S&T Policy in Singapore   591
    Conclusions and Agenda for Future Research   604

    COMMENT   609
    LOUISE SÉGUIN-DULUDE

15. AUSTRALIAN EXPERIENCE WITH EXPORTING TO ASIA   613
    RICHARD POMFRET

    Australia's Trade Patterns   614
    Australia's Trade Policies   624
    Explaining Australian Export Success in Asia   633
    The Elaborately Transformed Manufactures Debate   636
    The Role of Regional Trading Arrangements   637
    Lessons for Canada   638
    Conclusions   640
    Appendix 1 Selected Three-Digit SITC Export Categories   641

    COMMENT   648
    PETER J. SAGAR

16. THE NEW TRADING SYSTEM: GLOBAL OR REGIONAL?   651
    SYLVIA OSTRY

    Structural Changes   652
    Globalization   652
    The ICT Revolution   654
    U.S. Trade Policy   654
    Globalism, Regionalism and APEC   657

ABOUT THE CONTRIBUTORS   661

Preface

INTERNATIONAL TRADE IS THE LIFEBLOOD OF THE CANADIAN ECONOMY. Together, exports and imports of goods and services currently amount to about 85 percent of real gross domestic product (GDP) in Canada and their importance in the economy has been increasing rapidly. Trade has been the engine of growth and job creation in the 1990s. This growing outward orientation of the Canadian economy bodes well for future productivity and real income growth in Canada.

However, future growth in Canada's international trade depends critically on progress in four key, but interrelated, areas: increased penetration of non-U.S. markets; improved cost position through gains in relative productivity; increased outward orientation of small and medium-sized enterprises (SME), and broadening of the export base. A deepening of Canada's trade and investment linkages with the dynamic and fast growing Asia Pacific markets would be helpful in alleviating these structural weaknesses.

The Asian economies offer immense potential for Canadian exports and investment in the 21st century. From a little over 12 percent in 1965, these economies will account for over one-quarter of world GDP by year 2000. Canada's commercial linkages with the Asian economies have strengthened considerably, although from a small base, and the Asia Pacific region has surpassed the European Union as Canada's second largest trading partner. But Canada needs to do much better. We have been losing ground to other countries in these markets.

To examine fully the Asia Pacific opportunities and challenges for Canada, the Micro-Economic Policy Analysis Branch of Industry Canada organized a two-day conference, held in Vancouver in December 1995, on *The Growing Importance of the Asia Pacific Region in the World Economy: Implications for Canada*. At the sessions, experts presented 14 research papers on a wide spectrum of economic growth issues, ranging from trade and investment to the environment, tourism, and immigration. Other papers looked at the implications of policy developments in the region for Canada's commercial and innovation policies.

The revised drafts of the conference papers are published in this impressive research volume. The research indicates that Canada could benefit a great deal from closer links with the Asia Pacific region. The gains would be realized in trade and investment flows, tourism, human resource development, innovation, specialization, resource utilization and allocation, and real incomes. The research assembled in this volume will also contribute to the development of government policies aimed at taking full advantage of the tremendous opportunities offered by the rising economic growth in the Asia Pacific region.

I would like to take this opportunity to thank all of the authors for their excellent work, and in particular Professor Richard Harris in his capacity as both author and general editor.

JOHN MANLEY
MINISTER OF INDUSTRY

Richard G. Harris
Department of Economics
Simon Fraser University
Canadian Institute for Advanced Research

Introduction

A CASUAL READING OF THE BUSINESS PRESS IN CANADA over the last several years might lead one to conclude that the East Asia region has had an enormous impact on the Canadian economy. A virtual barrage of media reports on the Japanese trade surplus, Asian growth rates, investments in China, Team Canada visitations to the region, Asia Pacific Economic Cooperation (APEC) summits, and the flow of Hong Kong immigrants and money into Canada leaves one with the strong impression that the impact of the Asia Pacific region on Canada has been significant, if not overwhelming.

But what are the facts? In 1994 Canadian exports to the Asia Pacific region represented less than 8 percent of total Canadian exports (Figure 1). On the import side, the same region supplied about 13 percent of total imports – an increase of 6 percentage points since 1980 (Figure 2). Within the region, more than half of both exports and imports is accounted for by Japan. The picture on the direct investment side is about the same. The Asia Pacific region accounts for less than 6 percent of direct inward investment and 7 percent of outward investment. To coin a popular phrase, "Asia is everywhere but in the trade and investment statistics." To put this in context, approximately 37 percent of U.S. trade is with Asia.

This volume is the outcome of a research program organized by Industry Canada, which brings together a group of researchers who attempt to systematically assess whether the popular view of Canada-Asia links is correct or whether the statistical picture referred to above is more accurate. As will be seen, the truth is probably somewhere in between. Current and past data show that the Canadian economy does have links with Asia, but these are not as strong as one is often led to believe. On the other hand, there are very good reasons to think that the links between the Asia Pacific region and Canada will become stronger over time. This has a number of implications for Canadian economic growth, the pattern of industrial development, wage distributions, trade and industrial policy, the future of Canada within the multilateral trading system, and our relationships with the United States and Mexico.

FIGURE 1

DISTRIBUTION OF CANADIAN MERCHANDISE EXPORTS, BY REGION, 1980 AND 1994

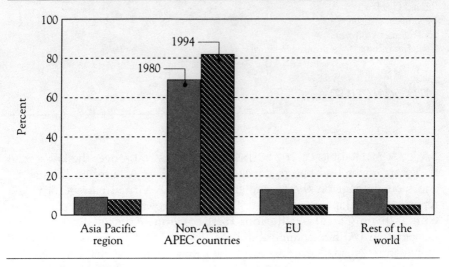

Source: Industry Canada compilations using various sources, including APEC Economic Committee (1995).

FIGURE 2

DISTRIBUTION OF CANADIAN MERCHANDISE IMPORTS, BY REGION, 1980 AND 1994

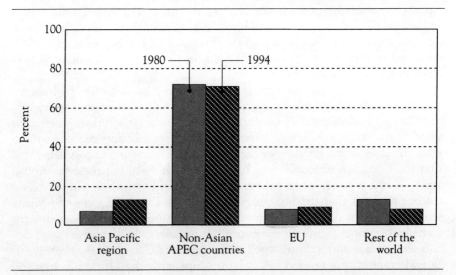

Source: Industry Canada compilations using various sources, including APEC Economic Committee (1995).

The papers in this volume represent the application of a wide range of methodological approaches in economics. This is only natural in this instance. Globalization has blurred the traditional lines of division between specialists in international versus domestic economics. There are papers written by specialists in macroeconomics, labour, environment, trade, public policy, business development, and so forth. They converge in their focus on the Asia Pacific region and on its past and potential impact on the Canadian economy. While there are "region specialists," more often they are focused on the region itself rather than on the external implications of regional developments. The contributors to this book were asked to turn their own discipline's optics, with a Canadian perspective, onto the Asia Pacific region.

A word on terminology: "Asia Pacific" in this book refers to East Asia – a region that includes Japan, China, the newly industrialized countries (the NIC, also referred to as "the Gang of Four" or "the Four Tigers" – Hong Kong, South Korea, Taiwan, Singapore), the other members (Brunei, Indonesia, Malaysia, Philippines, Thailand) of the Association of South-East Asian Nations (ASEAN), and the few remaining developing economies in the region (Vietnam, Cambodia, Papua New Guinea, etc.). It does not include South Asia (India, Pakistan, Sri Lanka), Australia, or New Zealand. The attention in this volume is firmly fixed on the implications for Canada of developments in the Asia Pacific region. A number of subjects related to prospective developments in the region, such as the outlook for further reform in China or in Japanese macroeconomic policy, are not covered, as they do not relate directly to Canada and have been the subject of numerous other specialized volumes.

GROWTH AND TRADE ISSUES

A FIRST SET OF PAPERS DEALS WITH two general issues – the pattern of economic growth in the region, and the pattern of international trade. This is a natural starting point for the volume, since much of the interest in the region on the eastern side of the Pacific derives from the links between the trade and economic growth performances of these countries.

The countries of the Asia Pacific region have had a spectacular growth performance. This includes both Japan (an advanced industrialized country), the NIC, and many of the developing countries in the region, including China. This growth performance is the reason why volumes such as this one are appearing and why Canada's interest in the region is increasing. There are at least two aspects to the East Asian growth debate. First, why did it occur and what are the important explanatory factors? Second, will this growth performance continue in the lower-income countries in the region? The answer to the latter question is obviously of critical importance in looking at the Asia Pacific region's potential impact on Canada. But to answer the latter we clearly need a good answer to the first question.

In the opening paper, John Helliwell provides an overview of the debate on the sources of economic growth in the East Asia region and of obvious dynamism. It is generally agreed that, whatever other factors may have played a role in the growth process experienced by these countries, the openness of the trade regimes is a common and important underlying characteristic. Helliwell finds, using either quantitative or qualitative indicators of openness, that once growth rates are adjusted for the "catch-up" effect, openness is the most important explanatory variable. It is important, however, not to confuse "openness" with the absence of interventionist trade policies. As is well known, many of these countries actively intervene in both export and import markets, a subject that comes up a number of times in the volume.

The growth performance of the Asia Pacific region is notable in a number of respects. First, it reminds us that there is no evidence of aggregate scale economies in growth, as predicted by the new endogenous-growth theories: small countries can grow just as fast as large countries. Second, the history of the region suggests that while catch-up is obviously occurring, convergence in income levels is not, contrary to what has been observed in the OECD region. It is not clear how to interpret this finding. It may well be that the "growth miracle" (World Bank, 1993) is still under way and that we have yet to observe a period in which growth slows down and convergence sets in. As Helliwell reminds us, most of these countries are still a very long way from the advanced industrialized world. GDP per capita is still less than half that of the world as a whole, while the OECD's GDP per capita is eight times the Asian average.

The quest for an explanation of the remarkable Asian growth performance has led to a search for non-traditional explanations, many of which have focused on institutions. It may be that the institutional environment was responsible – a hypothesis held dear by industrial strategy advocates. Helliwell reminds us that the cross-country evidence is quite weak in this respect. Adding a variable that captures bureaucratic efficiency to the growth regression has no effect, but there is a possibility that causation with respect to income levels runs the other way – that richer countries tend to have more efficient bureaucracies. Helliwell expresses surprise at these results, which compound the task of explaining the growth performance of the region.

A number of writers have suggested that Asian growth can be explained in part by a set of cultural institutions that are unique and provide what has been termed "social glue" or "trust." The social sciences have found evidence that trust has been declining in the United States over the past three decades. Helliwell refers to the presence of both trust and engagement as "social capital." Inglehart (1994) has constructed indexes of social values across countries in an attempt to explain growth. He finds a significant correlation between growth performance and such values as thrift, hard work, and determination. Among the 43 countries in his sample, China and South Korea have the highest index of "values." In the remainder of the paper, Helliwell explores the possibility that social capital might explain growth by checking for its significance in a sample of industrialized countries. The results are generally negative and not encouraging for the "social capital" hypothesis. Thus much of the Asian "miracle" awaits an explanation. In the meantime, there are some clouds on the horizon, including population growth, infrastructure bottlenecks, environmental degradation, and human resource constraints. Nevertheless, these concerns have been present for some time and their impact on growth rates have so far been unmeasurable.

Canadian trade patterns have been subject to considerable analysis over the years. There are two major "facts" to acknowledge regarding Canadian trade. First, the bulk of it is with the United States – at around 80 percent of total trade. Second, trade has traditionally been concentrated in automobiles and resource products. The general view is that Canadian imports from East Asia are in the low-skilled manufacturing area, and this is consistent with the view that Canada has a comparative disadvantage in these industries. In the second paper of the volume, Walid Hejazi and Daniel Trefler provide an update on Canadian trade patterns and analyze the implications of growth in trade with East Asia. First, they report on the trend in the net factor content of aggregate Canadian trade. The factor content of trade is a measure of the amount of capital, labour, and resources that is implicitly embedded in the bundle of goods exported to or imported from a country. Hejazi and Trefler report that from 1972 to 1992, there was a 5 percent reduction in the factor content of unskilled labour and capital in total Canadian exports, the implication being that this may have reduced the demand for these factors and thus their market returns.

Hejazi and Trefler then turn their attention to three alternative empirical models of Canadian international trade – the gravity model, which emphasizes transaction costs; the monopolistic-competition approach, focused on increasing returns to scale in firm-level production; and the traditional factor-endowments-based theory of trade. They come to two main conclusions regarding Canadian trade, by controlling for within-region versus between-region effects, with North America and East Asia each defined as regions. First, Canada has unusually high levels of exports in natural resource-abundant products, after controlling for the obvious abundance that Canada has in resource endowments. Second, East Asia has unusually large exports in low-end manufactures and a few high-end manufactures. The authors suggest that the presence of East Asia in the global trading system may be responsible in part for these patterns. One possibility is that the export-oriented industrial policies of East Asia may have resulted in an expansion of exports in sectors beyond those predictable by factors common to all countries. This, in turn, may have led to a reorientation of Canadian exports in favour of natural resource products. Hejazi and Trefler hint that this may be consistent with the deterioration in wages of unskilled workers that has been observed in Canada over the last decade.

Given the particular nature of the integration of the Canada and U.S. economies, there is clearly a problem with Canada's trade patterns that these models do not capture. Integration has occurred within firms located on both sides of the Canada-U.S. border, and Canadian plants may serve as part of an integrated, North American-based production system. Products from that system are exported to Asia as either components or final goods. Many of these are counted in trade statistics as exports from the United States. This alternative explanation of the Canadian trade data would be consistent with the Hejazi-Trefler results in that the missing variable is some index of the degree of North American corporate integration by sector. The "unexplained" trade in resource-intensive products may be a measurement error, as total intra-firm trade is not measured correctly, or a consequence of complementarity between Canadian and U.S. factor inputs within the same firm, which are not included in the type of trade models they use.

The next paper, by Keith Head and John Ries, looks at inward foreign direct investment (FDI) from the largest economy in the Asia Pacific region – Japan. There is little question that Japan has had a major impact on the North American economy. Its large trade surplus, its domination of a number of world industries, and its large-scale direct investment abroad have attracted a great deal of policy attention, particularly in the United States. (Japan-U.S. links are also considered by Wendy Dobson, as discussed later on.) While these trends have not been given the same degree of attention in Canada, there are good reasons to believe that Japan's impact on Canada has been considerable in terms of both trade and investment. The loss of U.S. comparative advantage to the Japanese in a number of industries, most notably automobiles, has clearly had repercussions in Canada. The great success of the Canadian automobile sector does not belie the fact that for much of the past 20 years the North American auto industry has been an import-competing industry rather than an export industry. Only recently has North American deindustrialization begun to slow down, and this has coincided with continued wage growth and appreciation of the yen in Japan. There has also been large-scale direct investment by Japan in North America. The Head and Ries paper looks at how that investment has been allocated within North America and, in particular, at how it has spurred competition among state and provincial governments for Japanese plants locating within North America.

Although the importance of FDI to Canada has not been an issue in recent policy history, Canada has not done as well as the United States in attracting Japanese investment, which accounts for only 4 percent of total FDI in Canada. Canada attracts only one Japanese "greenfield" establishment for every 15 that go to the United States. Head and Ries investigate the relative attractiveness of Canadian provinces versus U.S. states, using a discrete-choice, investment-location model that includes such variables as wages, taxes, unionization, unemployment rates, agglomeration effects, income levels, and subsidies. They then construct an index of competitiveness, intended to measure the estimated probability that a particular province or state will attract a new Japanese "greenfield" investment. Of the 10 provinces, only Ontario ranks in the top 20 locations within North America. Head and Ries suggest that a number of policies could increase the attractiveness of an individual province in this type of locational competition for foreign investment – including the "Newfoundland solution" of offering $2,000 per job created. While foreign investment has well-documented benefits, the possibility that provinces might engage in mutually damaging "subsidy wars" that would actually lower the economic welfare of the winning province or state has been well recognized. The temptation for governments to offer such incentives unilaterally is clearly supported by the Head and Ries analysis. Given the predilection of local governments to engage in this type of subsidy, there seems to be a clear case for a legal or constitutional commitment to avoid such behaviour. While there has been some progress on this issue among the provinces, there is no agreement between Canada and the United States.

Canada is a large net exporter of resource products. Approximately 40 percent of total Canadian exports in 1994 were resource-based. Most of these exports went to the United States, but Asia – and Japan in particular – was a net purchaser of Canadian resource products (fish, forest, energy, and mineral products; cereals). The low volume of Canadian resource exports to East Asia is explained, in part, by the fact that many countries in the region – Indonesia, Thailand, and Malaysia, for example – are also resource exporters and thus compete with Canada. As economic growth proceeds in the region, however, Canada could become a much larger net exporter of resource products to Asia, as argued in Robert McRae's paper.

McRae reviews the evidence on the relationship between economic development and resource consumption, and finds a solid connection between income levels per capita and resource consumption. Using "status quo" growth scenarios, he looks at the aggregate resource consumption in China, India, and Indonesia over the next 25 years. The forecasts are quite instructive. The forecast values for energy and coal consumption in China are compared to 1990 consumption levels in the United States. The forecast for coal is more than four times the 1990 U.S. level; for iron ore, 10 times; for steel, 5 times – and so on across a range of commodities. The forecasts for Indonesia are comparable to the 1990 values for Japan. McRae notes that these estimates ignore supply effects and demand substitution induced by relative-price changes. The growth assumptions, however, are very much in the middle of the range of plausible estimates. On balance, they suggest that there is likely to be some combination of price and volume increases in exports from Canada to the region. This presumes, of course, that world supply conditions do not change dramatically. On this issue, we have even less evidence.

In discussing the McRae paper, Tim Hazledine wonders how plausible these estimates are, since they suggest that world resource depletion may become a very serious issue and raise concerns about political stability as regions fight over dwindling resources. On balance, the discussion suggests that exports of resource-saving technology may be as important, if not more so, than exports of the resources themselves. An Asian demand-driven resource boom over the medium term does not seem out of the realm of possibilities at this time. One implication of these demand projections is that the current trade pattern between Asia and North America is unlikely to change a great deal as imports of resource products into Asia will have to be matched by a surplus of manufactured exports from the region. Furthermore, this could shift the terms of trade adversely against these countries and in favour of the resource exporters.

Service Exports

THE NEXT SET OF PAPERS DEAL WITH Canadian service exports to the Asia Pacific region. Growth in the service sector is a characteristic of virtually all economies, both advanced and developing, and trade in services increases each year, aided by new technology that has reduced the cost of moving either the service itself, the people who provide it, or the customers who receive it. Three papers deal respectively with business and education services, banking services, and international tourism.

Trade in international business services has become an increasingly important feature of the international economy. In the last decade, it has grown at more than 50 percent the rate of growth in international trade. In 1993 Canadian service exports were equivalent to approximately 7.6 percent of merchandise trade exports. While Canada traditionally runs a net deficit on service trade, with the Asia Pacific region it has recorded a surplus. The key services here are engineering, consulting, financial services, telecommunications, R&D services, and postsecondary education. In his paper, Lawrence Schembri provides an overview of this trade and argues that there is potential for Canadian firms to make substantial headway in the export of services to the Asia Pacific region. His argument is threefold. First, Canada has a comparative advantage in these types of services, based on a strong skill base and on the experience of Canadian firms in areas such as resource extraction and construction. Second, the demand for business services is very income-elastic. As economic growth proceeds in the region, there will be strong growth in the demand for these services, which, if not met, may lead to a bottleneck in the development process. Recent data suggest that the demand for services has grown faster in Thailand, China, South Korea, and Hong Kong than in the world as a whole. As Canada gains experience in the region, income and service-demand growth should lead to further exports. Third, the recent completion of the General Agreement on Trade in Services (GATS) under the World Trade Organization (WTO) will provide additional liberalization on market access issues, which should help Canadian firms to enter these markets.

Schembri documents a number of barriers to trade in services and notes that these are expected to be reduced under the GATS. Furthermore, the commitments under the GATS to reducing limitations on market access in the areas of professional services (e.g., accounting, consulting, engineering services, etc.) and advanced telecommunications are significant and will have added importance when China is admitted to the WTO.

Schembri also argues that Canadian exports of education services to the region could grow. In 1992-93 there were 23,000 students from the Asia Pacific region pursuing a postsecondary education in Canada. This amounts to about $200 million or 25 percent of total business service exports to the region. Given low levels of educational attainment in the region and given the strong emphasis in Asian culture on education, Canada could well expand its market opportunities in this area. As secondary education is almost exclusively in the domain of the provinces, this would require either explicit government policies directed to developing these markets or a deregulation of the secondary education sector. Part of Schembri's paper deals with service exports to China, on which unfortunately there are very little data. However, given the large amount of infrastructure spending that is expected to occur in China, he argues that there should be substantial opportunities for Canadian firms in construction and engineering services, based on studies of the experience of American firms in that country.

The growth of the international financial system is a distinguishing characteristic of the global economy. This development has had an impact on the Asia Pacific region but not to the extent that one might expect. John Chant provides an overview of Canadian banking activities in the region and outlines the prospects for future growth in banking service exports. His paper begins with the observation that income growth leads to "financial deepening" and thus to greater demand for banking services. An interesting cross-country regression analysis shows that money/income ratios increase systematically as per capita income rises. This suggests that the Asia Pacific banking sector will experience substantial growth, although its absolute size should not be overstated. A useful benchmark to keep in mind is that in Canada, banking services account for 1.8 percent of GDP in value-added terms.

The main difficulty in most of Asia, as Chant discusses, is the absence of liberalization in banking regulation. Banks provide intermediation services, act as agents for their customers, or trade on their own behalf. When these activities are carried out in a foreign country, the regulatory regime is critical. Throughout the East Asia region, there is considerable variability in regulation in the form of interest rate controls, direct credit allocation, and the degree of government ownership. Singapore and Hong Kong are free from government control while China and Vietnam still have a system based on a single state-run bank that allocates capital to state enterprises according to a central plan.

For foreign banks, freedom of entry is a basic requirement. Another critical element is national treatment, which in most countries is not a characteristic of the regulatory regime. Only Hong Kong, Japan, and South Korea offer unrestricted entry to foreign banks. It is not evident this picture will change soon, as banks have long been a traditional tool for state intervention. It is probably reasonable to expect continued hostility towards foreign banks, even among those countries that are signatories to the GATS. One problem in that respect is that the GATS provides for very weak commitments in the area of banking services.

Chant goes on to assess the potential for the future development of international banking services in the region. Two theories of international banking are reviewed. The first approach, labelled "follow the leader," is based on the idea that one can expect banks to follow their national exporters into a region. In the case of Canada, this does not lead to an optimistic view, given the relatively low volume of exports to the Asia Pacific region. It certainly suggests a scale of entry to the market that is likely to be too small to be successful. The second theory is the "banking advantage approach." In this instance, advantage is based on cost, technology, or experience. The problem with this approach is that culture and technology may not be easily transportable from North America or Europe to Asia. Chant's reading of the explanatory value of either theory is weak in the case of Asia.

What is evident is that Canadian banks are concentrated in Hong Kong, Japan, and Singapore. These are obvious regional financial centres, and so it may be that agglomeration and network effects, and a lack of regulation are dominant factors, at least in the first stage of entry into a new region. The net result is that the Canadian presence in banking services in the Asia Pacific region is very small, with exports of less than 0.25 percent of Canadian GDP in 1993. However, Canada is not unique in that respect. Reuven Glick points out in his discussion of the Chant paper that foreign banking penetration in the Asia Pacific region accounts for less than 10 percent of total deposits. He goes on to argue that the real issue may be non-bank competition and how international banks might move into related activities, such as foreign-exchange trading, securities underwriting, and the provision of trade credit. It is clear that for further gains to be made in the area, further relaxation on the constraints governing foreign banks will be needed. In the meantime, the best predictor of Canadian success in Asian financial service markets may be how well they do elsewhere in global financial markets. If they can succeed in the United States and Europe, then that success may well ultimately carry them into Asia when further deregulation occurs.

A prominent feature of globalization has been the growth of international tourism, which has been substantially faster than that of world income. Tourism, both domestic and international, is a highly income-elastic good whose demand rises rapidly with per capita income. For this reason, the tourism market originating in the Asia Pacific region has attracted a lot of attention. Canada has certain obvious advantages as a tourist destination for the Asian market. In our paper in this volume, Stephen Easton and I examine recent trends in the Asia Pacific tourism market and attempt to assess what the likely impact of Asian growth might be on exports of Canadian tourism services. One should be wary of overstating the case for tourism in Canada. In 1992 Canada's share of world tourism arrivals was only 3.43 percent, and much of this was of U.S. origin. In the balance of payments, tourism receipts account for only about 0.5 percent of total export receipts. Nonetheless, tourist arrivals from Asia have grown considerably over the period 1972 to 1994, rising from 13 percent to just over 31 percent of non-U.S. arrivals. Of total Asian arrivals in Canada, Japan accounts for about 40 percent. The potential is clearly there, but it is difficult to estimate just how large the net flows are likely to be.

There are few systematic data on Asian tourism, except for Japanese tourism. The paper describes a model of Japanese international tourism for the period 1971–91, using a panel of 12 country destinations. The results indicate that four important effects on international Japanese tourism have been identified in the data. One is a very strong per capita income effect, with elasticities in the range of 1.5 to 3. A second effect is that on real exchange rates, as the appreciation of the yen has led to an increase in the price of "at home tourism" relative to international tourism. Third, a demographic effect is noted: long-distance international tourism tends to be concentrated in the under-55 age category. Fourth, the reduction in air transport costs has contributed to the growth in long-distance international tourism from Japan. The fixed-effects coefficient in the regression model indicates that, when other factors are held constant, Australia and Canada are about equal in terms of attractiveness to Japanese international tourists, ranked well ahead of the United States but behind New Zealand.

To what extent will the Japanese experience be replicated in other countries in the region? That is a difficult question to answer. Intra-regional tourism may prove to be more important in East Asia than has been the case in Japan over the past three decades. It is also clear that for the poorer countries in the region, income levels will have to grow considerably before the 1970 levels of Japanese participation in international tourism are approached. Nevertheless, the growth of populations and incomes in the region should mean that growth rates in the international tourism market are potentially large. When the econometric parameters estimated for Japan are applied to China, the predicted growth rates for arrivals from China range between 10 and 20 percent over the next two decades. What is needed to complete the analysis is some identification of supply-side factors across countries, which are potentially important. Supply-side factors might prove to be in Canada's favour simply because of the geographic size of the country. One difference between tourism and other traded services is that tourism is almost unencumbered by regulation. Competitive forces should lead to lower prices and faster growth rates than in the more highly regulated service sectors.

Immigration from Asia is one area where statistics and public perceptions are much closer than is the case for trade and investment. Asian immigration to Canada stands at about 50 percent of total immigration, or about 100,000 persons per year. Within the region, China, Hong Kong, and Taiwan account for the bulk of the immigrants. In their paper, Michael Baker and Dwayne Benjamin attempt to estimate the impact of Asian immigration to Canada on the labour market outcomes of Canadian-born workers and on Canada's international trade. Their analysis of immigrant supply is based on a framework in which Canada competes as a destination country for immigrants. Their results may surprise those who are not familiar with the subject. While the number of immigrants to Canada has been large, they do not have an appreciable effect on labour market outcomes. Asian immigrants assimilate at about the same rate as other immigrants but start from a lower initial earnings level.

The link between trade and immigration has been relatively unexplored. Given that Canada has experienced strong growth in Asian immigration, is there any evidence to suggest that trade or investment follows immigration? Using a gravity model of bilateral trade flows, Baker and Benjamin control for the general effect of globalization by looking at Canada-U.S. differences but controlling for differences in immigrant sources. They find some evidence that imports – and, to a lesser degree, exports – are correlated with immigration. Given the discussion in the Hejazi-Trefler paper on trade patterns, Baker and Benjamin suggest that these results should be interpreted cautiously. Identifying the impact of immigration on trade is problematic when the United States is used as the benchmark against which to judge Canadian trade flows, since some unmeasured Asian trade flows with Canada occur through the United States. In the case of foreign direct investment, Baker and Benjamin find no evidence of a relationship with immigration. In general, the links running from immigration to trade and investment appear to be weak.

In discussing the Baker-Benjamin paper, Donald DeVoretz takes issue with a couple of points. First, he argues that a supply-side approach to the explanation of immigration may be inappropriate. The Canadian data are better explained, he argues, by using a demand-based approach in which the excess demand for immigrant entry is "managed" through the point system. He further argues that aggregate national analysis misses an important effect of immigrants on the economy, in that they tend to locate in the three major Canadian cities – Vancouver, Toronto, and Montreal. One indicator of their impact is measured by their effect on local consumption prices of goods such as housing. In an appropriate cost-benefit analysis of changes in immigration, these effects should be accounted for.

Brian Copeland examines the potential links between trade with Asia and the quality of the environment in the region. Given the increased concern about pollution and environmental degradation in Asia, the question arises whether developed-country demand for pollution-intensive goods makes the situation better or worse, and also whether there has been environmental dumping, in the sense that pollution-intensive activities have been shifted from the high-income countries to Asia. Copeland points out that perceptions in this area can often be wrong. Canada, for example, produces more CO_2 and SO_2 than any Asian country, but Canadian air quality is much higher than that of most Asian countries. He also argues that the past may not be a good predictor of the future and that rapid growth in Asia may make matters worse. Offsetting these adverse trends is the effect of higher incomes on the demand for environmental quality. A number of studies suggest a "humped effect," with pollution first rising with increases in per capita income, then falling. Where the hump is likely to occur is a matter in some dispute, but many of the poorer countries are still on the "wrong" side of the hump: more growth means more pollution. Reviewing the theoretical literature, Copeland notes that international trade has a number of channels through which pollution can help or harm countries engaged in trade; in particular, free trade may be an inappropriate trade policy if environmental externalities are important and unregulated or priced incorrectly. This is another example of the well-known "second-best" dilemma that policymakers face. Liberalizing trade may make matters worse because of the unintended effect on the environment. The problem is compounded by the fact that differences in environmental regulations across countries create not only global efficiency losses but also significant distributive effects that can be intergenerational in scope. For example, a country with weak pollution controls that grows by exporting dirty products may be benefiting its current workers at the expense of its future population.

The policy implications of this analysis for Canada are far from clear. As a member of the multilateral system, Canada can attempt to push the WTO agenda either for or against the "greening of trade," but environmental regulation is viewed by many countries as a matter of national sovereignty. In its relations with the United States, Canada has always strongly rejected extraterritorial application of U.S. laws. The idea that Canada might countervail environmental dumping in an attempt to enforce global environmental standards leaves many people uneasy because most of the countries affected would be the poor countries. The basic question here is, Who is the polluter? If rich-country consumers are the ultimate polluters, then *they* should pay, not the producers. Implementing the international income transfers required to realize such a scheme seems at the moment an institutional impossibility. Finally, there is the important fact that Canada itself is likely to be a target of such actions. Canadian forest practices are routinely the subject of media attention in Europe. One can view this as a good or a bad thing, but in either case it must be realized that because Canada is a large resource exporter – a role that may increase in the future, if McRae is correct – any change in the international regime on trade and the environment is very likely to spill over into the regulation of trade in natural resources.

Small and medium-sized enterprises (SME) account for over 98 percent of all business enterprises and play an increasingly important role in the Canadian economy. They are particularly important in the service sector and thus can be viewed as potential engines of growth in facilitating the business service exports described in the Schembri paper. Someshwar Rao and Ashfaq Ahmad investigate how these enterprises have fared in terms of their contribution to Canadian participation in Asian markets and in relation to larger firms that are also present in Asia. Their analysis is based on a detailed survey of Canadian and American companies, disaggregated by size class (SME, large firms, and very large firms) and by degree of outward orientation. Their major finding is that outward-oriented Canadian SME are competitive vis-à-vis their American counterparts, but the overall participation rate of Canadian SME in Asia is very low relative to larger firms. SME account for only 9 percent of total exports, although they constituted 97 percent of all Canadian export firms in 1992.

In the Rao-Ahmad survey, virtually all firms identify the same factors as motivating their expansion into the Asia Pacific region — market diversification, production for export, favorable regulations, production for the local market, and reduced production costs. The rate-of-return performance of outward-oriented Canadian SME is impressive and significantly better than that of large Canadian firms. On the other hand, SME have had lower labour productivity growth than the larger firms. Accounting for this difference is an important puzzle. It may reflect the service-oriented nature of SME output, in which case there are the well-known downward biases in the measurement of service output growth, or it may reflect the relative capital intensity of the large firms. The paper then turns to a review of the barriers to SME participation in the region. There are a large number of trade, investment, and regulatory barriers in the Asia Pacific region that raise the cost of doing business. These regulations are biased against the smaller firms in that they constitute a fixed cost and, therefore, get spread over a smaller volume of sales. However, three other factors were identified as crucial to the SME attempting to enter Asian markets: a lack of financial resources; a lack of market intelligence; and a lack of managerial experience. This evidence supports those who argue that the presence of these impediments or market failures in export entry provide a rationale for government assistance to SME in export markets.

It is evident that, given the discussion on trade in business services, the prospects for SME will depend heavily on progress in the GATS under the WTO. In this sense, if one believes that SME in the area of consulting and professional services will be the vehicle by which Canada as a human capital-intensive country benefits from the growth in Asian markets, then Canadian trade policy should be focused on these areas. On the domestic "push" side it appears that the reduction of financial and informational barriers to induce entry of small firms into these activities is important. It may be that the Canadian experience in entering the U.S. market after the implementation of the Canada-U.S. Free Trade Agreement will be the first step in the globalization of Canadian small business. If this hypothesis is true, then entering the Asian market might occur faster than recent trends suggest.

Policy Issues

THE NEXT SET OF PAPERS TURNS TO the broader political and policy perspectives. As a small country Canada has limited leverage in the global geopolitical system. This does not mean, however, that there are no important policy decisions to be made.

In examining Asia Pacific and Canadian interests, it is critical to understand what the United States and Japan are doing and how their bilateral relationship is evolving. In her wide-ranging paper, Wendy Dobson reviews well-known irritants in the Japanese-U.S. relationship, including trade and macroeconomic imbalances. In her view, there is bound to be more trouble in the future as the "realists" in both countries gain the upper hand in response to the lack of progress in resolving some of these key disputes. The realists view the world in geopolitical, "us versus them" terms, in which governments are the key players. On the U.S. side, we have seen the Structural Impediments Initiative, which irritated Japan considerably. On the Japanese side, we have seen generations of U.S. trade negotiators throw up their hands in attempting to open the Japanese market to U.S. producers. The implications of this impasse are disturbing for Canada. Both Dobson and her discussant, Masao Nakamura, take the view that the Japanese and other Asians see Canadians as being "like" the Americans and, therefore, as being apt to be treated in similar fashion. Canada has been remarkably silent on the issue of market access to Japan, even though that issue should affect Canadian and American producers in similar ways. This undoubtedly reflects the policy judgment that there is not much that Canada can do about it and that Canada obviously benefits to a certain degree as an exporter of raw materials to Japan. To the extent that the Japan-U.S. relationship gets better or worse, it will affect Canadians similarly.

Dobson discusses a number of uncertainties in these scenarios, including how Japan will interact with the rest of Asia and, in particular, with China. In East Asia there is some evidence of growing interdependence, as measured by intra-regional trade flows. In addition, Japanese investment in the region is more regionally focused than in North America or Europe. Japanese producers are setting up plants that produce for export back to the Japanese market. These data have raised the spectre of a possible Japanese-led trade bloc (from which China would be absent) in the Asia Pacific region. Neither Japan nor the United States have a coherent policy approach towards China. Dobson argues that one can expect bilateralism to dominate in those relationships. There is a distinct possibility, however, that Japanese "realists" who argue that Japan should take a leadership role in Asia will win the day. How the rest of Asia would react to such attempts is not clear, but it could exacerbate the current U.S.-Japan disputes.

Dobson has three main policy recommendations. First, a U.S.-Japanese dispute panel should be set up, along the lines of the Canada-U.S. FTA panels. Second, APEC should be promoted as a channel through which Japan can link its regional thrust into Asia with global governance systems such as the WTO. APEC might also serve as an intermediary in disputes by providing a formalized "dispute mediation service," as suggested by the APEC Eminent Persons Group. Finally, she argues that Canada is invisible in Asia and needs to intensify efforts to raise its profile in the region, as Australia and New Zealand have done, for example.

In his discussion, Nakamura suggests that this is perhaps not as clear as it seems. Canadian economic interests are strongly linked to U.S. interests, and Canada may not be seen as a "neutral" third party in its dealings with Asia. Both Dobson and Nakamura agree that one aspect of the Pacific imbalance has been the absence of Canadian direct investment in Japan, and it might be the right time for Canadian firms to attempt to crack the Japanese market. Clearly, one cannot talk about Asia without talking about Japan.

This brings us to Canadian trade policy. Murray Smith analyzes the trade policy scenarios that could arise as a result of developments in the Asia Pacific region, the completion of the Uruguay Round of GATT, and the implementation of various regional agreements – the North American Free Trade Agreement (NAFTA), in particular. In the Asia Pacific region, there are no formalized regional preferential trade agreements, with the exception of the ASEAN free trade area. However, there has been a great deal of discussion about the role of APEC. Canada has been an active participant in APEC, and some understanding of where APEC might go is critical to an assessment of other trade policy initiatives. The possibility of extending APEC into a formalized free trade agreement has been subject to some discussion but so far no progress has been made in that direction. As Smith points out, there is very little agreement among APEC members about what role APEC should play. The term "open regionalism" is often used by Asian participants to describe the role of APEC – that is, its role as a promoter of trade liberalization in the region but without commitments and timetables. APEC, in this view, would assist in the implementation of WTO obligations agreed to in the Uruguay Round and promote bilateral liberalization in the region. How this fits in with the NAFTA and the European Union remains a matter of considerable dispute. Multilateralists would argue that regionalism and preferential trading agreements have gone much too far; others would claim that achieving liberalization is a pragmatic matter and that any form of liberalization is worth pursuing. Given the lack of any form of concrete arrangements in the Asia Pacific region, APEC serves a useful purpose, if only to promote dialogue and information exchanges within the region. Some have described it as sort of "OECD for the Pacific Rim."

From the Canadian trade policy perspective, Smith reminds us of three important considerations in looking at the Asia Pacific region. First, Canadian trade barriers against the exports of most of the countries in the region are still quite restrictive in a number of areas – in particular, agriculture, textiles, and manufactured goods. In addition to quotas administered under the Multi-Fibre Agreement (MFA), there has been a substantial increase in the use of Canadian anti-dumping and countervail legislation against the exports of the region's developing countries. Second, Canada itself faces substantial export barriers in the region, including high tariffs, non-tariff barriers, and informal barriers to trade. These barriers may explain, in part, Canada's relatively poor export performance in East Asia. Reducing trade barriers should therefore be an important objective for Canadian trade policy in the region, and in this respect Canadian and U.S. interests are convergent. Third, assuming that Canada adopts a more liberal regime towards the import of goods from Asia, there is likely to be some reduction of the trade diversion towards Mexico within North America. Mexico now serves as the low-cost labour platform for labour-intensive activities within North America. In this respect, Mexico competes with Asia directly in both Canadian and U.S. markets. This will inevitably put pressure on NAFTA, and the outcome will affect the direction taken in North America towards a more open form of regionalism or, alternatively, towards a strengthening of preferential arrangements.

The Asian growth experience has affected Canadian economic policy in ways other than prompting concern about its trade and investment consequences. It has, for example, had an enduring impact on the debate, in academic and policy circles, on the role that governments have played in promoting growth. This issue was hinted at in Helliwell's discussion of comparative growth, but it runs deeply through much of the policy literature on the Asia Pacific region. Richard Lipsey and Russell Wills step into this debate in their paper by looking at the role that government science and technology policies have played in South Korea and Singapore. They view the NIC as examples of what Lipsey refers to as "massaged market economies" – with governments doing the massaging. Government policy has facilitated the substitution of knowledge-based inputs for low-skill labour and resource-based inputs with success. They argue that Canada might benefit from studying the details of these policies. In addition, they support greater Canadian involvement in R&D in East Asia via strategic alliances and partnerships with both South Korea and Singapore as a means of keeping open potential windows of opportunity for Canadian business. The reader will have to work through this paper in order to get a sense of their arguments. In this case, the details of individual policies matter, and the authors argue that one cannot summarize by saying that intervention was either all good or all bad.

In the latter part of their paper, they comment on some institutional barriers that exist in Canada to implementing "pick-the-winner" policies, and Louise Séguin-Dulude in her discussion alludes to a number of these points. First, there is no detailed cost/benefit analysis of the apparently successful policies adopted in both South Korea and Singapore. They may well have had the effects that were intended, but our "hard" knowledge in this area is still limited. Second, there are some fundamental cultural and political differences between Asia and Canada, which could prove to be a stumbling block for the implementation of this type of policy in Canada. Canada values individual rights relative to collective rights, and Canadian governments are heavily constrained as a result. Moreover, the countries studied are small and have a unitary form of government. Because of the realities of Canadian federalism, "picking winners" without regard to regional implications is a "non-starter" for the federal government. Lipsey and Wills are well aware of these arguments, but they make the case that we should be more open-minded in looking at these policies as potential role models for more successful science and technology policies in Canada. Moreover, these policies are capable of being emulated at both the federal and provincial levels.

One country with which Canada has many characteristics in common is Australia. Australia has also been the most successful non-Asian country at exporting to East Asia. Three-fifths of Australian exports go to East Asia, with much of the growth in this area occurring in the past two decades. "What are they doing right that we aren't?" is a question that many North American observers of the region have asked. Moreover, much of the success in recent years is accounted for by exports to countries other than Japan.

In the last paper in the volume, Richard Pomfret reviews the Australian export experience and seeks to provide an explanation for this remarkable success story. He begins by noting that Australia's exports consist predominantly of primary products, imports manufactures, and services. It has a unique pattern of trade in that it runs a sizable surplus or deficit with almost every country – and is in balance with none. Australia imports manufactured products from the United States and has a bilateral trade deficit with that country, as with Germany and all other European countries. With Asian countries, the pattern is reversed and Australia has significant trade surpluses. He refers to this as a triangular trade pattern – importing from North America and Europe and exporting to Asia. Exports to Asia consist mainly of primary products, but Australia has also done quite well in tourism and education service exports, thanks to its proximity to the region.

In the early postwar era, Australia was fairly protectionist, with high tariffs that began to be dismantled seriously only during the 1980s. This resulted in a quite low trade/GDP ratio, given the small size of the economy: in 1984 Australia's trade ratio was 27 percent, while Canada's was 51 percent. Like Canada, however, Australia has discovered export promotion. Pomfret discusses the Team Australia approach to export marketing, expressing scepticism about its impact. His assessment is that, on balance, there has been a reduction in the importance of trade to the Australian economy. The direction of trade was influenced by the closing of the European market to Australian agricultural exports when the United Kingdom joined the Common Market. Technical change and resource discoveries during the 1970s and 1980s increased the available supplies of mineral resources. Pomfret argues that these relatively homogeneous products naturally found their way to the closest growing market – Asia. The Asian success story can be explained as a natural supply response to changing world market conditions.

Pomfret concludes that there is not much that Canada can learn from Australia about exporting to Asia. Australia's primary-product export boom to the region had little to do with policy. In my own discussions with Australians about these matters, however, my experience has been that this is not a majority view in policy circles. The more conventional view is that while much disappointment has been expressed at the lack of manufactured exports to Asia, the primary-product success story may have set the stage for the next step up the product cycle. It is clear that Australia has much greater experience at doing business in Asia than does Canada, and this experience is clearly worth something. If Australia can upgrade its human-capital base, then it may be well placed to export higher-value-added goods and services to East Asia. Canadians will be watching these developments with considerable interest.

CONCLUSION

THE POLICY IMPLICATIONS THAT FLOW from the research presented in this book are threefold.

First, economic growth in the Asia Pacific region will continue to create both opportunities and risks. Several papers in this book emphasize those opportunities. It is clear that there is a potential, as yet unrealized, for growth in export markets in higher-value-added services and resources. However, there are risks as well. The ever-growing share of world GDP accounted for by East Asia also implies that political or economic instability in the region could have important global consequences. Simply as a matter of balance, Canadians should be increasingly sensitive to developments in the region. This may mean diverting more of our bureaucratic and foreign-policy resources towards this region, at the expense of Europe.

Second, Canadian economic linkages are increasingly based in North America, in particular moving towards a deeper integration with the U.S. economy. While economic opportunities in Asia are important, their impact on many Canadians – for example, on those employed in the automobile industry – is likely to be through indirect linkages via the U.S. economy as North American business moves across the Pacific. Consequently, the management and deepening of the Canada-U.S. relationship remains as important a policy objective as ever.

Third, while Canada's trade and investment patterns remain skewed towards the United States, there are clear opportunities in the Asia Pacific region, particularly in services and resources. It is only prudent that Canada pursue these opportunities to the extent that is reasonably possible, keeping in mind, however, that there are problems as well, as a number of papers point out. First and foremost is the issue of marketing. If Canada is to succeed in Asia, Canadian business must establish a more visible presence there. Government policy directed towards reducing the barriers to establishing a market presence in Asia may be useful. The promotion of educational links with Canadian universities is another mechanism through which Canada could create economic linkages and at the same time leverage its human capital-based comparative advantage. Finally, recent immigration from the Asia Pacific region, particularly from Hong Kong and Taiwan, provides Canada with a natural pool of entrepreneurs who are familiar with the region and can facilitate the building of investment and export linkages.

There are other methods by which economic linkages with East Asia can be fostered. One is a reduction in existing Canadian trade barriers towards imports from Asia. We can only reasonably expect progress if Canada is seen as a "fair trader" in the region. Another approach is to promote liberalization within APEC, at least until something better comes along. As a number of the papers indicate, the world trading system is in flux as the pressures generated by globalization and regionalization – and by policy shifts within the United States and Europe – play themselves out. For the Asia Pacific countries, APEC is the only trade institution that matters. As a modest economic power on the Americas side of the Pacific, with long experience in dealing with the United States, Canada should let APEC institutional evolution follow its course, all the while resisting movement towards the building of a trade bloc but facilitating policy initiatives that have liberalizing effects. As emphasized in a number of the papers, the barriers to trade and investment are formidable in most Asia Pacific economies. While export growth has been strong, two-way trade and investment links are relatively weak. APEC can also be a vehicle through which the opening of domestic markets in the region can be promoted.

The essays in this volume cover a number of aspects of the Asia Pacific region's impact on the Canadian economy and on economic policy in Canada. It is clear that not all the evidence is contained in the disappointing trade statistics, although the data do pose both a puzzle and a challenge. Other factors – U.S. trade policy interventions, financial flows, immigration, geopolitical concerns over the environment, and changing business strategies in a global environment – also exert influence on trade patterns. There is little doubt that the Asia Pacific region will continue to be the major growth pole in the world economy for some time to come. Understanding the implications of that development is important both in assessing Canada's role in the world economy and in developing economic strategies aimed at generating future Canadian economic growth.

BIBLIOGRAPHY

Inglehart, R. "The Impact of Culture on Economic Development: Theory, Hypotheses and Some Empirical Tests." Ann Arbor: University of Michigan, 1994.

World Bank. *The East Asian Miracle: Economic Growth and Public Policy*. Washington: World Bank, 1993.

John F. Helliwell
Department of Economics
University of British Columbia[1]

2

Economic Growth and Social Capital in Asia

THIS PAPER WILL CONSIDER SOME OF THE SPECIAL FEATURES and implications of Asian economic growth experiences. After updating and extending the comparative analysis of Asian economic growth developed in Helliwell (1995), I shall introduce some new elements into the discussion to see if an analysis of social capital and institutions of the type emphasized by Putnam for the regions of Italy (Putnam, 1993; Helliwell and Putnam, 1995) and for the United States (Putnam, 1995, forthcoming) can help to explain growth differences among Asian economies or between South-East Asia and the rest of the world.

ASIAN GROWTH IN A GLOBAL PERSPECTIVE

AS SHOWN IN FIGURE 1, economic growth in China and in Asia as a whole has been much faster, on average, than in the rest of the developing world. Indeed, the developing countries as a group have grown faster than the industrial countries. Over the 1987-94 period, for example, real gross domestic product (GDP) growth in the developing countries averaged 2.75 percentage points per annum more than in the industrial countries, and this growth difference was projected to continue at 3 points through 1995 and 1996. Over the same period, population growth averaged about 0.5 percent in the more developed countries and 2 percent in the less developed countries. Thus both per capita and aggregate growth rates have been faster in the developing countries than in the industrial countries, thereby narrowing the proportionate income gap between rich and poor.

While Figure 1 shows fast economic growth in Asia, it is important to remember that per capita income growth in Asia started from a low base. As Figure 2 shows, although more than half of the world's population lives in Asia, it produces less than one quarter of the world's GDP, so that GDP per capita in Asia is still less than half as high as world per capita GDP. While this

FIGURE 1

ANNUAL GDP GROWTH IN DEVELOPING REGIONS, 1987-96

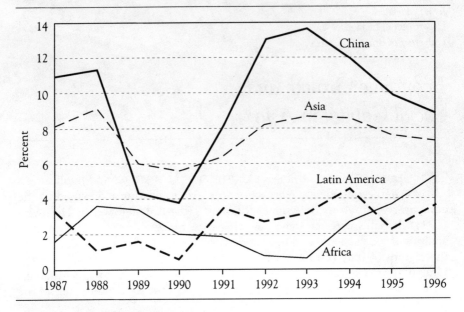

proportion is much higher than a decade ago, thanks to the high Asian growth rates in the intervening period, the fact remains that per capita incomes in the Organisation for Economic Co-operation and Development (OECD) are still three times the world average, or eight times as high as those in Asia. Incomes in the transition economies – the countries of the former Soviet Union and of Eastern Europe – were more than the world average at the beginning of the decade, but they have since dropped below the average in the wake of several years of shrinking real GDP. Per capita incomes in Latin America are just equal to the global average, while those in Africa have fallen behind those in Asia. Even these continental averages obscure wide chasms in both levels and rates of growth of per capita incomes.[2]

Over the past 35 years, there has been[3] a fairly well documented convergence of both rates of growth and levels of per capita incomes among the industrial countries. Countries starting off with lower incomes have grown faster than the initially richer countries.[4] Some observers have interpreted this evidence[5] as support for the idea that all countries have the same rates of efficiency growth, and that convergence is therefore taking place through higher levels of investment in the poorer countries. However, numerous studies that use explicit production functions to define levels and rates of growth of technical progress show that most of the convergence among the industrial countries takes place through faster measured technical progress in the initially poorer

FIGURE 2

REGIONAL SHARES OF WORLD EXPORTS, GDP AND POPULATION, 1994

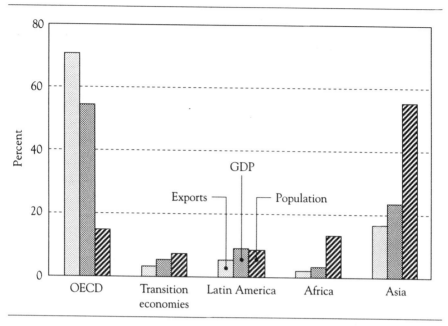

countries.⁶ Attempts to find convergence among global samples of countries have found no unconditional convergence of income levels, whether or not allowance is made for the possibility that statistical problems may lead some estimates to find convergence of income levels where none exists. When allowance is made for differences in education levels and investment rates, however, the global samples do show some evidence of conditional convergence. Although this finding seems fairly robust to any of the specific variables used to explain the factors other than convergence that influence growth rate differences, it is still sensitive to the statistical methods used and is certainly not equally applicable throughout the developing world.

Of particular relevance to students of Asian growth is the fact that the models so far used to show convergence among the industrial and global samples often show no evidence of convergence among the Asian economies. When the global samples are split among continents, they show evidence of conditional convergence in the industrial countries and in Latin America, slight signs of convergence in Africa, and divergent evidence in the sample of Asian economies.⁷ The Asian experience may be disconcerting for enthusiasts of convergence, but it is of no more comfort for those advocating models of endogenous growth based on national spillovers from domestic accumulation of human or physical capital,⁸ since the fastest-growing countries in East Asia included the very smallest and hence provided no signs of returns to scale. From more recent theoretical and empirical studies, it is becoming clear that where convergence is taking place, it is mainly through international transfers of knowledge and that these transfers can best be analyzed using theories of the production and distribution of knowledge that lie at the heart of the endogenous growth literature.[9]

A principal aim of this paper is to extend the analysis in Helliwell (1995) of the possible reasons for the different growth experiences within Asia and between Asia and the rest of the world. A substantial body of literature has emerged to explain why some Asian economies have grown faster than the rest and to search for lessons that may be transferable to other countries. The literature contains several types of argument:

- Growth experiences vary a great deal from decade to decade for each country, so that differences among countries, or groups of countries, are likely to be ephemeral and hence to depend more on "good luck" than on "good policy."[10]
- Once appropriate allowances are made for the large increases in labour force participation rates and high investment levels, the rates of total factor productivity increase in the "Four Tigers" (Hong Kong, Singapore, South Korea, and Taiwan) are not unusually high, especially when compared to those in the industrial countries during their convergence in the 1960s and 1970s.[11]
- The eight "high performing Asian economies" or HPAEs (the Four Tigers plus Japan, Malaysia, Thailand, and Indonesia) do display a remarkable mix of regionally concentrated and sustained high growth and low and declining levels of income inequality (World Bank, 1993). The World Bank study attributes that growth to high rates of accumulation of physical and human capital, good macroeconomic management, slower population growth,[12] and market-friendly government policies, perhaps aided in some countries by a number of "market-leading" interventions to promote exports and growth.[13]
- Among those who think that there is indeed an East Asian growth miracle, a variety of sometimes conflicting reasons are offered to explain growth in one or more of the Four Tigers or the eight HPAEs, including export promotion policies, industrial strategies, directed lending in financial markets, domestic competition policies, free trade and capital movements, economic freedoms, cultural factors, and a stable political and social order.

In a previous paper (Helliwell, 1995), I noted the need to explain why some of the Asian economies have taken off into sustained growth while others have not yet done so. Here, I propose to explore that question in greater institutional detail, focusing on issues of cultural differences and social organization that have received much attention but little quantitative analysis, mainly because of the difficulties in measuring how, and explaining why, differences in cultures and social institutions affect international differences in growth rates and the levels of well-being. But first I shall summarize the major empirical results of the previous paper and present more recent data on past and projected growth rates in the larger Asian economies.

COMPARATIVE GROWTH IN THE ASIAN ECONOMIES

THE EMPIRICAL RESULTS IN THIS PAPER CONTINUE where those in the previous paper left off, with differences in openness being used as a likely starting point in explaining growth differences among the Asian economies and in reconciling differences between the Asian growth experiences and those elsewhere in the world.[14] Figure 3 repeats the primary result from that earlier research, showing that dividing the Asian sample into three groups with differing degrees of openness at the beginning of the 1980s explains more than half of the variation in per capita growth rates among the Asian economies during that decade. The equation is as follows:

FIGURE 3

ASIAN GROWTH IN GDP PER CAPITA DURING THE 1980S

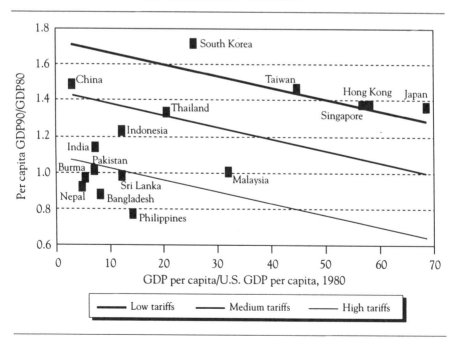

Per capita GDP90/per capita GDP80 = 1.78 − 4.42 tariff − 0.0065 GDP80
                                    (3.52)   (1.64)
                   R^2=0.555   SEE=0.185   N=16

where the tariff rate is the average proportionate import tariff in 1980 and GDP80 for each country is measured as a percentage of real per capita income in the United States.[15]

The lines in Figure 3 show the conditional forecasts for 1980s growth for countries in each of the three tariff categories. The low-tariff countries, with tariffs averaging 1.2 percent in 1980, include the Four Tigers and Japan; the medium-tariff countries, with 1980 import tariffs averaging 7.7 percent, are Indonesia, South Korea, and China; the high-tariff countries, with rates averaging 15.6 percent in 1980, are Bangladesh, India, Pakistan, Nepal, Sri Lanka, the Philippines, Thailand, Malaysia, and Myanmar (formerly Burma).[16]

The lines slope down from left to right, reflecting the conditional convergence effect implied by the estimated negative coefficient on per capita income at the beginning of the decade. The coefficient on the tariff variable suggests that a drop in tariff rates averaging 1 percentage point was associated with a per capita growth difference cumulating to 4.4 points over the decade. This is undoubtedly a substantial over-estimate of the consequences of tariff reductions viewed as an independent instrument of policy. It is more appropriate to view the size of the estimated coefficient on the tariff rate as the combined effect of a whole range of policies and events that encouraged the inflow and implementation of productive ideas from abroad. In general, the developing countries of Asia and elsewhere design and implement inward- or outward-looking strategies in packages, with varying mixes of tariffs, non-tariff barriers, exchange controls, foreign-investment controls, taxes, and regulations governing the flows of trainees and workers. Estimates based on the more comprehensive openness measure developed by Sachs and Warner (1995) also show large growth effects of openness, in both Asian and global samples. When the Sachs and Warner data for the extent of openness of the 16 Asian countries are used to repeat the 1980s growth regression reported above, the results are as follows:

Per capita GDP90/per capita GDP80 = 1.04 − 0.3286 open − 0.0001 GDP80
                                    (2.09)   (0.25)
                   R^2=0.348   SEE=0.225   N=16

where the openness variable takes the value 1.0 for countries classified as open throughout the 1980s, 0.0 for those that were still ranked as closed in 1990, and an intermediate value for the one country becoming open part-way through the decade.[17] The coefficient on the openness variable implies that an economy that was open rather than closed during the 1980s grew by a cumulative 33 percentage points more over the decade − an average of 2.8 points per year, very close to the 2.5 points per year estimated by Sachs and Warner for a global sample of 89 countries over the period 1970 through 1989. These estimates, if they are correct, tend to support those based on tariff classes, although they are somewhat smaller. The earlier estimate of 4.4-point cumulative growth for each percentage point drop in average tariffs, multiplied by the high-tariff countries' average tariff rate of 15.6 percent, would give a cumulative 10-year growth effect equivalent to 68 points. This is higher than the 33-point difference based on the Sachs and Warner variable, but it is based on moving to zero tariffs whereas the Sachs and Warner open economies generally still have positive tariffs. If the high-tariff countries were to lower their average tariffs from 15.6 percent to the average rate for the low- and medium-tariff countries – 4.9 percent – the implied increase in the accumulated 10-year growth would be 47 points – i.e., in the same range as the 38 points implied by the Sachs and Warner variable. The results using the Sachs and Warner variable do not show significant convergence effects. Sachs and Warner would argue that this is because the sample of 16 countries still includes too many closed economies, which have many fewer opportunities for the technological transfer on which convergence is primarily based. When they divide their global sample into two subsets – one open and one closed – they find both conditional and unconditional convergence to be evident over the 1970-89 period for the open economies, but not for the closed ones.

Turning to examine the past and current growth experiences of individual Asian economies as a prelude to a consideration of some of the other possible determinants of faster growth, Figures 4, 5 and 6 show past and projected

FIGURE 4

ANNUAL GDP GROWTH IN JAPAN AND THE FOUR TIGERS, 1987-98

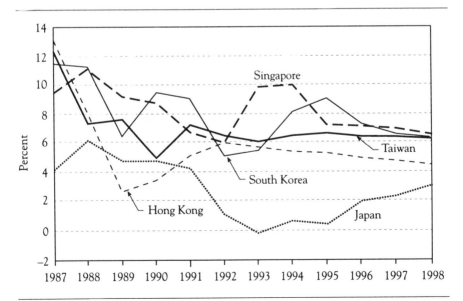

FIGURE 5

ANNUAL GDP GROWTH IN CHINA AND OTHER EAST ASIAN COUNTRIES, 1987-98

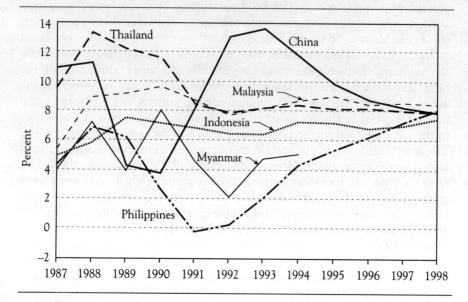

FIGURE 6

ANNUAL GDP GROWTH IN SOUTH ASIA, 1987-98

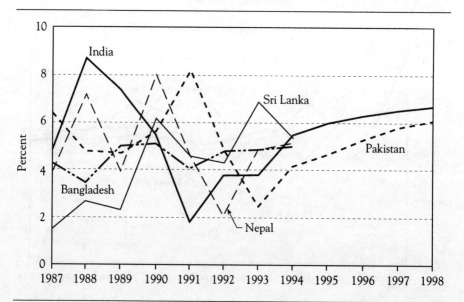

growth rates of total GDP for three groups of Asian economies over the period 1987-98.[18] Figure 4 shows the Four Tigers outperforming Japan for most of the 1990s – a result consistent with conditional convergence, since their average productivity levels are still below those in Japan, but also driven by the sharp drop in Japanese growth in the early 1990s. A broader range of growth experiences is shown in Figure 5, with China on average having the fastest rate of growth, and Myanmar the lowest, and with four ASEAN (Association of South-East Asian Nations) countries being in the middle. Figure 6 shows the narrower and generally lower range of growth rates in South Asia, for economies that were all in the high-tariff group during the 1980s.

To help put the three groups in a comparative context, Figure 7 shows average growth rates for China, the Four Tigers, the four ASEAN economies, and the major economies of South Asia.[19] Although all three groups had slower growth in 1991-94 than they did in 1987-90 (with the exception of China, where growth was slower in 1989 and 1990 in the aftermath of the events in Tienanmen Square), the reduction was in all cases proportionately less than corresponding reductions in the industrial countries and in the world as a whole. For example, GDP growth for the world economy fell from 3.6 percent in 1987-90 to 2.4 percent in 1991-94, while that in the industrial countries was almost halved, from 3.3 to 1.6 percent. By contrast, average growth in the Four Tigers fell from 8.5 percent in 1987-90 to 6.8 percent in 1991-94 – more

FIGURE 7

ANNUAL GDP GROWTH IN FOUR ASIAN REGIONS, 1987-98

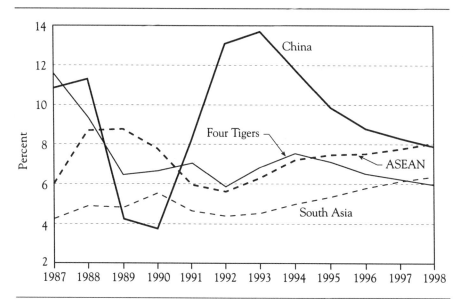

in percentage points but much less as a fraction of the earlier growth. Growth reductions were even smaller for the four ASEAN countries, from 7.8 to 6.3 percent. The same story applies to South Asia, where growth declined from 4.9 percent in 1987-90 only to just under 4.7 percent in the 1991-94 period. Clearly, there has been a convergence of growth rates among the Asian economies, and this is expected to continue over the rest of the 1990s. Over the 1995-98 forecast horizon, the Project LINK projected average growth rates of 6.4 percent for the Four Tigers, 7.7 percent for the four ASEAN countries, and 5.9 percent for India and Pakistan.

All else being equal, we might expect to see higher growth rates in South Asia than in ASEAN, and higher growth in ASEAN countries than among the Four Tigers, since existing income levels and comparative costs are much lower in South Asia than in ASEAN, and lower in ASEAN than among the Tigers. Of course, all else is not equal, and the Four Tigers, which were not far from South Asian income levels in 1960, clearly grew faster over the next 30 years. If being poor were all it took to ensure higher growth rates, then Africa would be growing fastest of all and there would remain the puzzle of how to explain why income differentials managed to arise in the first place. Much work on comparative growth has emphasized various sets of conditions and institutions[20] that might tend to favour rapid growth. In the 1960s, there was more faith in internally-oriented import-substitution policies, as exemplified in many of the high-tariff countries in our Asian sample. In the 1970s and 1980s, much greater hopes were pinned on outward-oriented strategies, sometimes focused on export-led growth, but more often on general increases in openness to world markets. This has been accelerated and accompanied by the several rounds of GATT (General Agreement on Tariffs and Trade) agreements and by donors' increasing insistence that aid be made conditional on the establishment of more open trade and more stable macroeconomic conditions. The growth experiences of the Asian economies over the 1980s tend to support the view that the more open economies have indeed grown faster, but neither the rationales for nor the robustness of these results have been fully established.

The idea that growth can be faster in more open economies is supported by certain versions of endogenous growth theory, whereby technical progress largely depends on the importation and use of technologies initially discovered and used elsewhere. There is uncertainty about whether these flows of productive ideas are linked to inward direct investment, exports, imports, high levels of domestic education, research and development at home or abroad, or some combination of these facilitating influences. A relatively new strand of empirical research has turned to consider whether there are measurable international differences in the quality of domestic institutions and in the structure of society more generally that might facilitate either the more efficient use of ideas from elsewhere or the more productive use of domestic resources, or both. To give some taste of preliminary results in this vein, I shall discuss briefly some data and results on the effects of institutional quality, and then deal with measures of social capital more broadly conceived.

How Important Are Institutions?

THERE IS A LARGE AND GROWING LITERATURE on the growth effects of democratic and other political institutions. I shall concentrate here on data collected and analyzed by Mauro (1995), which attempt to measure the quality of governmental institutions pertaining to foreign and domestic businesses. The data are from surveys of business opinions, chiefly among those who have operations in the countries in question. Mauro collected data for 68 countries, 13 of which are among the Asian economies under consideration in this paper. Mauro developed an index of bureaucratic efficiency that is the sum of three separate measures of institutional quality: the efficiency of the judicial system, the absence of bureaucratic red tape and the absence of corruption. It might have been more helpful if questions had been asked about the presence of good institutions as well as the absence of bad ones, but the data are what they are. The three component indices are fairly highly correlated in the global sample (0.78 or 0.79 between each of the three measures), so the use of Mauro's combined index is not likely to be different from the use of one or more of the components. He prefers the combined index to help smooth possible errors of measurement that may be specific to a particular measure.

When the Mauro index of bureaucratic efficiency in 1981-83 is added to the 1980s cross-sectional growth equation shown earlier, it has no effect, whether on its own or in combination with the tariff and/or initial-income variables. However, there is a strong positive correlation ($r = 0.86$) between the level of real GDP per capita and the measures of institutional quality, with richer countries having higher degrees of bureaucratic efficiency. Figure 8 shows the cross-sectional plot of Mauro's index and 1980 GDP per capita for the 13 Asian countries for which data are available. It would appear that countries acquire more efficient institutions as they get richer, but there is a somewhat puzzling lack of any influence from the institutional quality back to subsequent economic growth. Mauro obtains similar results with his global sample, but he does find an indirect role for his measure of bureaucratic efficiency in being associated with higher rates of investment in physical capital and education, both of which have been found to have positive effects on subsequent economic growth. If these indirect linkages were strong, however, we would expect to find them showing up in the reduced-form equation for growth in the Asian sample, since neither investment nor education variables are included in the equation, leaving the institutional variable free to take all the credit that might be due.

This result is reminiscent of my earlier attempts (Helliwell, 1994b) to disentangle the two-way linkages between democracy and economic growth,

Figure 8

Income and Government Efficiency
1980 GDP per Capita vs Mauro's Index

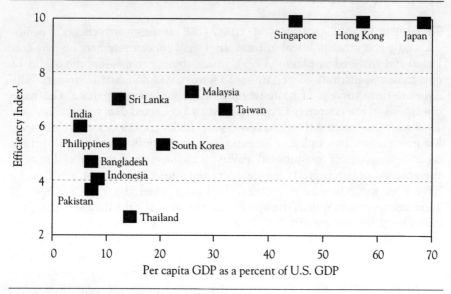

1   Maximum = 10.

with a strong effect found from the level of income to the presence of democratic institutions, but a weak negative or zero effect of democratic institutions back to subsequent economic growth. The latter result poses no great puzzle, as there are reasons to expect democracy to have mixed effects on subsequent economic growth. On the other hand, it is more difficult to imagine why the efficiency of governmental institutions could be bad for economic growth. Indeed, research measuring the quality of governmental institutions at the regional level in Italy (Putnam, 1993) has led to subsequent estimates showing that regional growth is higher in regions with better governmental institutions, with the eventual size of the income gap restrained by conditional convergence effects.[21] Thus it remains somewhat surprising that differences in institutional quality, especially using measures aimed at the services supporting business activities, do not appear to explain any of the growth differences among the Asian economies in the 1980s. Measurement errors are likely to be important, of course. However, the variables used are likely to be biased in a way that would make them positively rather than negatively correlated with subsequent growth, since they are based on the opinions of those planning future business decisions and are therefore likely to be influenced by other determinants of growth optimism or pessimism.

Social Capital and Cultural Differences

IT HAS OFTEN BEEN SUGGESTED THAT PART of the explanation for the development successes in Asia, and especially in South-East Asia, lies in a tighter and more robust set of social institutions than those found elsewhere in the modern world. An example of this view is found in Mahbubani (1995, 107): "The glue that holds Asian societies and families together has not been weakened by modernization." If this is true, it stands in contrast to what has been documented by Putnam (1995, forthcoming) for the United States, where key indicators of social glue – he emphasizes associational memberships and survey measures of social trust – have fallen by about 10 percent each decade over the past 30 years. He regards social trust and associational membership – trust and engagement – as two key facets of social capital. The U.S. social survey shows a strong positive correlation across individuals between associational memberships and the extent to which individuals believe that people can be trusted. Both variables are also strongly positively correlated with education, with a marginal effect that increases with the level of education. This correlation with education increases the significance of the large aggregate reductions in trust and engagement, since average educational levels in the United States have grown substantially during the post-1960 decades, just as social capital has been declining.

If social glue is either stronger or longer-lasting in Asia than elsewhere, that might provide another hypothesis to consider as part of the explanation for sustained fast economic growth in the region. Those devoted to the harder edges of the social sciences, economists among them, tend to resist the use of social and cultural variables as part of the explanation for economic differences. This may in part be due to insularity, but it no doubt also reflects a resistance to variables that are difficult to measure and explain and whose role in a causal structure may be only vaguely defined. However, there is increasing scope for more organized quantitative measurement and evaluation of social structures and their differences over time and among countries. Inglehart (1994) and others have developed a large-scale international survey that covers some of the key variables emphasized by Putnam in his analysis of social capital in the United States. For 35 countries in 1990-91, Inglehart has data (shown in Figure 14 of Putnam, 1995) on both trust and associational memberships. The correlation between the two measures is high across countries ($r = 0.65$), just as Putnam found across individuals in the United States.

Unfortunately for our present purposes, Inglehart's 35-country sample includes only three Asian economies (China, South Korea and Japan), and observations are not available for dates before 1990. Thus lack of country coverage and the absence of earlier observations foreclose the possibility of directly testing whether cross-country differences in the extent or durability of social capital help to explain growth differences among Asian economies or between them and the rest of the world. Although the United States was subject to significant reductions in social capital between 1960 and 1990, the Inglehart data still show U.S. social capital in 1990 to be high relative to that of most other countries – and about equal to that in Canada, except that the level of trust is slightly higher in Canada and the extent of associational memberships is slightly higher in the United States – and generally higher than in the three Asian economies. China has a higher degree of trust (0.60 compared to 0.50 for the United States, 0.52 for Canada, 0.42 for Japan, and 0.34 for South Korea), but all three Asian countries have lower associational memberships[22] than in North America. Aside from the high level of trust in China, only the Nordic countries and the Netherlands have higher levels of social capital, by either of the two measures, than do Canada and the United States.

It might be useful to attempt, for the 35-country sample, to see if some combination of the two measures helps to explain the residuals from more conventional cross-sectional growth equations. The risks of simultaneous bias will be considerable, as social capital, like the Mauro efficiency measure discussed earlier, is correlated with GDP per capita and has not been observed except at the end of the period for which per capita GDP data are available. Even with this bias working in its favour, however, associational membership has a negative rather than positive correlation with 1950-88 economic growth ($r = -0.22$ in Inglehart, 1994), so it is unlikely to become significantly positive in a more complete model of comparative growth. Inglehart finds, however, in partial support of Olson's (1982) view that long-established organizations tend to encourage wasteful rent-seeking, that the partial correlation between growth and associational membership is negative among the richer countries and positive among the poorer countries.

Inglehart (1994) also tried to use the World Values Survey data to supplement the social capital approach by asking if there are systematic differences in the values thought important to teach to children. His hypothesis is that growth prospects are likely to be brighter in societies where children are motivated to achieve. Three goals (thrift, determination and hard work) are thought to favour growth, and three others (obedience, religious faith and respect) to hold it back. The average of the first three minus the average of the second three is his index of achievement motivation. He finds a significant positive correlation across 43 countries between a simpler version of the index (thrift minus obedience) and economic growth over the 1950-88 period. Among the 43 countries, China and South Korea have the highest index values, with Japan ranking sixth, Canada 28th, the United States 30th, and India 41st. This index is less obviously subject to simultaneous feedback than are some of the measures of institutional structure. Both halves of the index reflect values that are culturally determined, but they are probably not influenced by economic and policy variables, although actual savings behaviour, in contrast to attitudes toward thrift, is influenced by taxation and other policies. It is not clear exactly how parental attitudes about what should be taught to children would influence the prospects for growth, but the general line of the argument is clear. The data, being available only for one survey date, provide no guidance about how these opinions and attitudes change over time.

We do not have enough data to test the likely growth effects of either social capital or cultural values among the Asian countries, but it might nonetheless be worthwhile to see if the hypothesized relations hold among the industrial countries for which more complete data are available. Table 1 shows the results of regressions explaining the cumulative log differences in aggregate efficiency measures in 17 OECD countries over the period 1962-89.[23] The independent variables include the 1962 logarithmic gap between each country's efficiency level and that in the United States, capturing convergence effects, and each of the measures described above for social capital and educational values. The primary measures are associational memberships, trust, thrift and obedience. The social capital measure is an equally weighted combination of trust and associations. None of the measures have the expected signs. In fact, the measures of social capital in 1990 are significantly negatively associated with productivity growth from 1962 to 1990. Does this mean that the factors that may have induced faster growth have also tended to reduce trust and associational memberships? At the same time, however, Putnam reports that measures of social capital tend to be more stable over time than among communities, so that one might still expect to find levels of high social capital positively correlated with high growth if, as was found in Putnam (1993) and in Helliwell and Putnam (1995) in the case of Italian regions, social capital is good for growth.

Perhaps the positive growth effects of social capital and thrift, the latter especially, flow through their effects on investment. If so, then the use of efficiency measures, which already remove the per capita growth effects of different investment rates – except to the extent that there are positive spillovers from investment to efficiency levels – would miss some or all of the positive growth effects. To check this possibility, tests were run to see if the international differences in the 1963-89 ratios of gross investment to GDP were positively influenced by any of the measures of social capital or values. They were not, nor were measures of the average growth of GDP per capita.

Thus the results in Table 1 are not consistent with those found in Putnam (1993) and Helliwell and Putnam (1995) for the regions of Italy, or with those reported by Inglehart (1994) for his 35-country sample. The next step in unravelling the puzzle will probably require social capital measures for a broader range of countries and regions, and with more variation over time. This may be possible for the states of the United States, but is not likely to be possible for the countries of Asia.

What can be concluded about the importance of social capital and cultural attitudes in explaining international differences in economic growth? It is simply too early to make any assessment, in my view, but I suspect that direct attempts to measure and explain various aspects of how society works will play an increasing role in comparative economics. In some cases, it may

TABLE 1

EFFECTS OF SOCIAL CAPITAL AND VALUES ON PRODUCTIVITY GROWTH IN OECD COUNTRIES

EQUATION	(I)	(II)	(III)	(IV)	(V)
Number of observations	17	17	17	17	17
Estimation method	OLS	OLS	OLS	OLS	OLS
Dependent variable	dlneff 1962-89	dlneff 1962-89	dlneff 1962-89	dlneff 1962-89	dlneff 1962-89
Constant	0.447 (3.6)	0.384 (3.2)	0.441 (3.5)	0.180 (1.5)	0.004 (0.0)
Coefficients: gap63	0.563 (6.0)	0.514 (6.6)	0.545 (6.9)	0.594 (6.3)	0.607 (6.9)
Trust	−0.59 (2.7)				
Associations		−0.13 (2.3)			
Social capital			−0.23 (2.7)		
Thrift				−0.08 (0.3)	
Obedience					0.45 (1.5)
$\bar{R}^2$	0.807	0.787	0.806	0.705	0.745
SEE	0.105	0.111	0.106	0.131	0.121

OLS = ordinary least squares; SEE = standard estimation error.
Note: Absolute values of t-statistics are in parentheses. The dependent variable is the logarithm of total the ratio of the 1989 to the 1962 efficiency level in each country, using the same data as in Helliwell (1994c), with Australia and New Zealand removed from the sample because they were not the World Values sample. Thrift, trust and obedience are all measured as proportions of respondents, while associations is the average number of associations to which respondents belong. All four variables are from the World Values Survey database from the University of Michigan.

even be possible to make inferences about social as well as technological ideas that can usefully be imported from other societies.

In the meantime, the most parsimonious interpretation of the growth experiences of the individual countries and groups of countries within Asia is that early openness seems to have encouraged early growth, raising the possibility that the increasing openness of the countries of South Asia will cement and accelerate their recent encouraging growth prospects. Exactly which channels, and which institutions, are most important to the successful transfer of productivity from the richer to the poorer countries remains, for now, an intriguing mystery. The role of international differences in social capital and of changes in social capital from one generation to another remains another key question for research. The question is likely to be asked and answered first for the industrial countries, where there are longer and more complete sets of available data. On these issues, the Asian experiences remain, for now, inscrutable.

ENDNOTES

1 Revised version of a paper delivered at the Industry Canada Conference on the Growing Importance of the Asia Pacific Region to the World Economy, Vancouver, December 1-2, 1995. I am grateful for helpful comments by conference participants, in particular Francis Colaço.

2 The data in Figures 1 and 2 are mainly drawn from the IMF's *World Economic Outlook* for May 1995. It should be noted that the shares of world GDP make use of purchasing power parities (PPPs) rather than market exchange rates to compare and combine real output in different countries. The United Nations continues to use market exchange rates for its international comparisons, which reduces the relative incomes of developing countries and regions, since real exchange rates are systematically lower in poorer countries, as documented in Summers and Heston (1991) and many other studies. The UN figures for 1994 suggest that only 12 percent of world GDP originates in Asia, while the IMF data show 23 percent. The IMF data are much more meaningful, since the PPPs are designed to permit comparison of real incomes, while the measures used by the UN are systematically biased in a way that underestimates the real GDP of poorer countries.

3 This paragraph, as well as the next two and some of the points of view discussed later on, are adapted from Helliwell (1995).

4 Studies documenting the convergence among the industrial countries include Abramovitz (1979, 1990), Maddison (1982), Baumol (1986), Baumol and Wolff (1988), Dowrick and Nguyen (1989), Dowrick and Gemmell (1991), and Helliwell and Chung (1991a).

5 Studies treating convergence, or convergence conditional on rates of investment in human and physical capital, as evidence supporting the

idea of a common international production function with the same rate of technical progress in all countries include Barro (1991) and Mankiw, Romer and Weil (1992). Mankiw, Romer and Weil, however, conclude that the evidence requires that even if countries all have the same rate of growth of productivity, the average levels of GDP per capita, for given levels of physical and human capital, may differ among them because of differences in natural resources and other factors.

6   Such tests are carried out and reported in Dowrick and Gemmell (1991), Helliwell and Chung (1991a), Coe and Helpman (1993), and Helliwell (1994c). These studies typically follow the example of Solow (1956, 1957) by inverting a neoclassical production function to obtain a measure of total factor productivity now often referred to as "Solow residuals." An alternative procedure is followed by Boskin and Lau (1992), who use the notion of a meta-production function equally applicable to all countries, with differences among countries and factors in the efficiency of factor use. They also find convergence in rates of technical progress among the industrial countries, but in their setup it comes from their estimates of declining returns to capital.

7   These results are reported in Helliwell and Chung (1995). A more detailed study of the Asian sample, reported in Helliwell (1994a), shows apparent evidence of strong influence from investment and several measures of openness, with a possible negative effect from democracy, but with still no signs of conditional convergence. In another recent study, Fukuda and Toya (1994) argue that export-led growth is the secret of the growth successes in Asia and report some evidence of conditional convergence once differences in export intensity are accounted for.

8   See, for example, Lucas (1988, 1990) and Romer (1986). Evidence showing no economies of scale among developing countries is reported in Helliwell and Chung (1995); their small aggregate size in industrial countries is reported by Backus, Kehoe and Kehoe (1991) and Helliwell and Chung (1991b).

9   See, for example, Grossman and Helpman (1991), Coe and Helpman (1993), Romer (1994) and Lucas (1993). For an empirical example applied to the Asian growth experience, see Ruffin (1993).

10 For this view, see Easterly, Kremer, Pritchett and Summers (1993), and Easterly (1994).

11 This view is developed for Hong Kong and Singapore in Young (1992), and for all Four Tigers in Young (1995), where the average annual growth of total factor productivity from 1966 to 1990 is estimated to be 2.3 percent for Hong Kong (to 1991), −0.3 percent for Singapore, 1.6 percent for South Korea and 1.9 percent for Taiwan.

12 For evidence that fertility declines precede rather than lag higher growth rates, see Brander (1992, 809-10).

13 Rodrik (1994) is critical of the World Bank report for not presenting adequate empirical evidence to support its judgments about the types of interventionist policies under review. Rodrik himself gives more causal weight than does the Bank report to the equality of income distribution that is characteristic of the HPAEs, and which the World Bank reports as an attractive feature of the HPAE growth record.
14 Supporting references include Ben-David (1993, forthcoming), Dollar (1992), Edwards (1993), Harrison and Revenga (1995), Helliwell (1994c), Lee (1993) and Sachs and Warner (1995).
15 This is the variable Y from version 5.6 of the Summers and Heston International Comparison Project data.
16 The tariff data were not available for Taiwan, China and Myanmar, so these countries were assigned to the low-, medium- and high-tariff categories, respectively, on the basis of other measures of their openness to trade and investment.
17 The countries classified as open in 1980 are Taiwan, Hong Kong, Japan, Singapore, Malaysia, Thailand, South Korea and Indonesia. Those classified as closed include China, India, Nepal, Pakistan, Bangladesh and Sri Lanka. The Philippines is given the value 0.2, reflecting its Sachs and Warner reclassification from closed to open in 1988 (0.2 = [90-88]/[90-80]).
18 The data for 1987 through 1993 are from the IMF's *World Economic Outlook* for May 1995, while the figures for 1994-98 are from the September 1995 *World Economic Outlook* of Project LINK.
19 The ASEAN countries include Indonesia, Malaysia, Philippines and Thailand. While Singapore is also part of ASEAN, it is excluded here because it has been included among the Four Tigers. The South Asian average is based on Bangladesh, India, Pakistan and Sri Lanka through 1994, and India and Pakistan for 1995-98. The averages are not weighted by country size.
20 See, for example, Abramovitz (1979).
21 Helliwell and Putnam (1995).
22 Associational memberships per capita in 1990 were 1.94 in the United States, 1.68 in Canada, 1.47 in South Korea, 0.83 in China and 0.49 in Japan.
23 Except for the social capital and values variables, the data are the same as those developed and used in Helliwell (1994c). The efficiency measure is drawn from the application of a constant-returns Cobb-Douglas production function based on capital and employment. Using a Divisia index for the growth of measured factor inputs, rather than a Cobb-Douglas function, gives identical results, as described in Helliwell (1994c).

BIBLIOGRAPHY

Abramovitz, M. "Rapid Growth Potential and its Realization: The Experience of Capitalist Economies in the Postwar Period," in *Economic Growth and Resources*. Edited by E. Malinvaud. London: Macmillan, 1979, 1-30.

_____. "The Catch-Up Factor in Postwar Economic Growth." *Economic Inquiry* (1990):1-18.

Alesina, A. and R. Perotti. "The Political Economy of Growth: A Critical Survey of the Recent Literature." *World Bank Economic Review*, 8 (1994):351-71.

Backus, D.K., P.J. Kehoe and T.J. Kehoe. "In Search of Scale Effects in Trade and Growth." Working Paper 451. Minneapolis: Federal Reserve Bank, 1991.

Barro, R.J. "Economic Growth in a Cross Section of Countries." *Quarterly Journal of Economics*, 106 (1991):407-44.

Baumol, W.J. "Productivity Growth, Convergence and Welfare: What the Long-Run Data Show." *American Economic Review*, 76 (1986):1072-85.

Baumol, W.J. and E.N. Wolff. "Productivity Growth, Convergence and Welfare: Reply." *American Economic Review*, 78 (1988):1155-59.

Ben-David, Dan. "Equalizing Exchange: Trade Liberalization and Income Convergence." *Quarterly Journal of Economics*, 108 (1993):653-80.

_____. "Income Disparity Among Countries and the Effect of Freer Trade," in *Economic Growth and the Structure of Long-Term Development*. Edited by L.L. Pasinetti and R.M. Solow. London: Macmillan (forthcoming).

Boskin, M.J. and L.J. Lau. "Capital, Technology and Economic Growth," in *Technology and the Wealth of Nations*. Edited by N. Rosenberg, R. Landau and D. Mowery. Stanford, Cal.: Stanford University Press, 1992.

Brander, J.A. "Comparative Economic Growth: Evidence and Interpretation." *Canadian Journal of Economics*, 25 (1992):792-818.

Coe, D.T. and E. Helpman. "International R&D Spillovers." NBER Working Paper 4444. Cambridge, Mass.: National Bureau of Economic Research, 1993.

Dollar, D. "Outward-Oriented Developing Countries Really Do Grow More Rapidly: Evidence From 95 LDCs, 1976-1985." *Economic Development and Cultural Change* (1992), 523-44.

Dowrick, S. and N. Gemmell. "Industrialisation, Catching Up and Economic Growth: A Comparative Study Across the World's Capitalist Economies." *Economic Journal*, 101 (1991):263-275.

Dowrick, S. and D.-T. Nguyen. "OECD Comparative Economic Growth 1950-85: Catch-Up and Convergence." *American Economic Review*, 79 (1989):1010-30.

Easterly, W. "Explaining Miracles: Growth Regressions Meet the Gang of Four," in *Lessons From East Asian Growth*. Edited by T. Ito and A.O. Kreuger. Vol. 4 of the NBER-EASE conference series. Chicago: University of Chicago Press, 1994.

Easterly, W., M. Kremer, L. Pritchett and L. Summers. "Good Policy or Good Luck? Country Growth Performance and Temporary Shocks." NBER Working Paper 4474. Cambridge, Mass.: National Bureau of Economic Research, 1993.

Edwards, S. "Openness, Trade Liberalization, and Growth in Developing Countries." *Journal of Economic Literature*, 31 (1993):1358-93.

Fields, G.S. "Changing Labor Market Conditions and Economic Development in Hong Kong, the Republic of Korea, Singapore, and Taiwan, China." *World Bank Economic Review*, 8 (1994):395-414.

Fukuda, S. and H. Toya. "Conditional Convergence in East Asian Countries: The Role of Exports in Economic Growth," in *Lessons From East Asian Growth*. Edited by T. Ito and A.O. Krueger. Vol. 4 of the NBER-EASE conference series. Chicago: University of Chicago Press, 1994.

Grossman, G.M. and E. Helpman. *Innovation and Growth in the Global Economy*. Cambridge, Mass.: MIT Press, 1991.

Harrison, A. and A. Revenga. "The Effects of Trade Policy Reform: What Do We Really Know?" NBER Working Paper 5225. Cambridge, Mass.: National Bureau of Economic Research, 1995.

Helliwell, John F. "International Growth Linkages: Evidence from Asia and the OECD," in *Macroeconomic Linkage: Savings, Exchange Rates and Capital Flows*. Edited by T. Ito and A.O. Krueger. Chicago: University of Chicago Press, 1994a, 7-28.

_____. "Empirical Linkages Between Democracy and Economic Growth." *British Journal of Political Science*, 24 (1994b):225-48.

_____. "Trade and Technical Progress," in *Economic Growth and the Structure of Long-Term Development*. Edited by L.L. Pasinetti and R.M. Solow. London: Macmillan, 1994c.

_____. "Asian Economic Growth," in *Pacific Trade and Investment: Options for the 90s*. Edited by F. Flatters. Kingston, Ont.: John Deutsch Institute, 1995.

Helliwell, John F. and Alan Chung. "Macroeconomic Convergence: International Transmission of Growth and Technical Progress," in *International Economic Transactions: Issues in Measurement and Empirical Research*. Edited by P. Hooper and J.D. Richardson. Chicago: University of Chicago Press, 1991a, 388-436.

_____. "Are Bigger Countries Better Off?" in *Economic Dimensions of Constitutional Change*. Edited by R. Boadway, T. Courchene and D. Purvis. Kingston, Ont.: John Deutsch Institute, 1991b, 346-67.

_____. "Tri-Polar Growth and Real Exchange Rates: How Much Can be Explained by Convergence?" in *A Quest for a More Stable World Economic System*. Edited by L.R. Klein. Boston: Kluwer, 1991c, 151-205.

_____. "Convergence and Growth Linkages Between North and South," in *Macroeconomic Linkages Between North and South*. Edited by D. Currie and D. Vines. Cambridge, Mass.: Cambridge University Press, 1995.

Helliwell, John F. and Robert D. Putnam. "Social Capital and Economic Growth in Italy." *Eastern Economic Journal*, 21, 3 (1995):295-307.

Inglehart, R. "The Impact of Culture on Economic Development: Theory, Hypotheses and Some Empirical Tests." Ann Arbor: University of Michigan, 1994.

International Monetary Fund. *World Economic Outlook*. Washington: IMF, May 1995.

Keefer, P. and S. Knack. "Why Don't Poor Countries Catch Up? A Cross-Country Test of an Institutional Explanation." IRIS Working Paper 60. College Park, Maryland: Center for Institutional Reform and the Informal Sector, 1993.

Lee, Jong-Wha. "International Trade, Distortions, and Long-Run Economic Growth." *IMF Staff Papers*, 40, 2 (1993):299-329.

Lucas, R.E. "On the Mechanics of Economic Development." *Journal of Monetary Economics*, 22 (1988):3-32.

_____. "Why Doesn't Capital Flow from Rich to Poor Countries?" *American Economic Review*, 90, 2 (1990):92-96.

Maddison, A. *Phases of Capitalist Development*. Oxford: Oxford University Press, 1982.

Mahbubani, K. "The Pacific Way." *Foreign Affairs* (January/February 1995):100-11.

Mankiw, G., D. Romer and D. Weil. "A Contribution to the Empirics of Economic Growth." *Quarterly Journal of Economics*, 107 (1992):407-37.

Mauro, P. "Corruption and Growth." *Quarterly Journal of Economics*, 110 (1995):681-712.
Olson, M. *The Rise and Decline of Nations*. New Haven: Yale University Press, 1982.
Park, Y.C. "Export-Led Growth and Economic Liberalisation in South Korea," in *The Transition to a Market Economy*. Edited by P. Marer and S. Zecchini. Vol. I, 150-2. Paris: OECD, 1991.
Putnam, Robert D. *Making Democracy Work: Civic Traditions in Modern Italy*. Princeton, N.J.: Princeton University Press, 1993.
_____. "Tuning In, Tuning Out: The Strange Disappearance of Social Capital in America." Ithiel de Sola Pool Lecture to the American Political Science Association, 1995.
_____. "Bowling Alone: Democracy in America at the End of the Twentieth Century" (forthcoming).
Rodrik, D. "King Kong Meets Godzilla: The World Bank and the East Asian Miracle." CEPR Discussion Paper 944. London: Centre for Economic Policy Research, 1994.
Romer, P.M. "Increasing Returns and Long-Run Growth." *Journal of Political Economy*, 94 (1986):1002-37.
Ruffin, R.J. "The Role of Foreign Investment in the Economic Growth of the Asian and Pacific Region." *Asian Development Review*, 11, 1 (1993):1-23.
Sachs, J.D. and A. Warner. "Economic Reform and the Process of Global Integration." *Brookings Papers on Economic Activity*, 1 (1995):1-118.
Solow, R.M. "A Contribution to the Theory of Economic Growth." *Quarterly Journal of Economics*, 70 (1956):65-94.
_____. "Technical Change and the Aggregate Production Function." *Review of Economics and Statistics*, 39 (1957):312-20.
Summers, R. and A. Heston. "The Penn World Table (Mark 5): An Expanded Set of International Comparisons, 1950-1988." *Quarterly Journal of Economics*, 106 (1991):327-68.
World Bank. *The East Asian Miracle: Economic Growth and Public Policy*. Washington: World Bank, 1993.
Young, Alwyn. "A Tale of Two Cities: Factor Accumulation and Technical Change in Hong Kong and Singapore." National Bureau of Economic Research, *Macroeconomics Annual*, 1992.
_____. "The Tyranny of Numbers: Confronting the Statistical Realities of the East Asian Growth Experience." *Quarterly Journal of Economics*, 110 (1995):641-80.

Comment

Francis X. Colaço
East Asia and the Pacific Region
The World Bank

PROFESSOR HELLIWELL'S PAPER is part of the growing literature, to which he has been a major contributor, on the "political economy of growth."[1] While the new endogenous-growth theories have focused on the role of such economic factors as human capital, investment, openness to trade, macroeconomic policies, physical infrastructure, and institutional development, the political-economy literature brings into play such non-economic factors as political decision-making structures, political stability, and initial social conditions (such as initial income distribution, the role of ethnic and other power groups, etc.).

The division between these two approaches is increasingly more apparent than real. And, particularly for a development practitioner, the importance of social and political conditions as a context for, and a determinant of, economic outcomes is becoming an increasingly important part of the policy language. The critical importance of the "ownership" of reform programs and the "participation" of civil society in the design and implementation of policies, programs, and projects is now increasingly realized and accepted.

Academic research, of which this paper is a part, is useful in illuminating some of the critical variables of importance to the growth and development process. At the same time, the literature illustrates the difficulties – of both methodology and measurement – in arriving at robust conclusions. Some factors (such as culture) may be invariant over very long periods of time, while others (redistributive policies, for example) may vary across countries and over time.

It is therefore not surprising that Professor Helliwell reaches the conclusion that "it is simply too early to make any assessment about the importance of social capital and cultural attitudes in explaining international differences in economic growth."

There are two approaches that further work could take: 1) a further careful specification of the underlying structural model that is being explored (including both economic and non-economic variables); 2) country analyses at both the macro and micro levels to provide a stronger understanding of the nature and role of non-economic factors in economic development. The literature indicates high correlation between some economic and non-economic variables, and the absence of clear specification of the underlying structural model can lead to difficulties in interpreting results. Also, some non-economic variables that are considered important are not susceptible to easy quantification.

The single-equation approach used in this paper is less than satisfying. To take one example, a regression that takes "openness" as the only explanatory variable downplays the importance of high savings and investment rates, education, and labour force participation as the dynamic explanatory factors for the rapid growth that has characterized the Asian experience from Japan to South Korea to the newly industrialized economies (NIE). One dominant characteristic of these economies has been higher savings and investment ratios and education levels (cognitive achievement levels) at lower levels of per capita income than for other comparable economies. The economic history literature attributes a considerable amount of the growth impetus from policy and institutional reforms (including "openness") to this strong base. Are these economic measures good proxies for the three cultural values (thrift, determination, and hard work) that Inglehart (1994) notes as favouring growth? Or is there a need to adopt hybrid indices of the kind posited?

Similarly, East Asian economies have been characterized, although this is not universally true, by greater distributive equality at comparable income levels. The evidence is that greater equality contributes to growth and that more inequality works against growth. This is a policy-induced result that may reflect either culture or pragmatism. In addition, there is a cadre of economic technocrats insulated from narrow political pressures, as well as institutions and mechanisms to share information and win the support of the business elites.

Project-level analysis clearly supports the view that an institutional capacity for managing projects is a critical determinant of the success and sustainability of projects, and hence of economic growth. It is therefore disappointing that Professor Helliwell finds a "somewhat puzzling lack of any influence from institutional quality back on subsequent economic growth." This, however, mirrors the inconclusiveness of results of some of the background work for the World Bank's study of the East Asian "miracle."[2] But the explanation may lie as much in the inadequacy of specification as in any other factor. It is difficult to understand modern Japanese economic development without understanding the role of its Ministry of International Trade and Investment (MITI), of policy deliberation councils, and of the *keiretsu*, or South Korean development without the *chaebol* and the close and cooperative relations between government and business. Similar institutional structures have existed in West European economies, but their specific Asian manifestation is sometimes attributed to culture.

The point, nevertheless, remains that institutional functioning is critical (as the experiences of Africa and East Asia indicate) to the development process, with complex two-way interactions. Further work on these interactions is justified. The kind of regression analysis undertaken does not yet seem to be yielding results, and that is disappointing – but probably not unexpected, given the complexity of the relationships. A research strategy could proceed on two fronts: detailed country-specific analysis, and cross-country analyses with more detailed structured comparisons.

In summary, even if cross-section analysis does not at present, for both methodological and measurement reasons, throw sufficient light on the inscrutability of the Asian experience, other evidence is helpful in illuminating that experience and providing lessons for other developing countries.

Before concluding, I would like to focus on prospects for future growth in the East Asian economies. The prospects are good for continuing rapid growth. But there are some structural factors that are emerging as critical constraints. First, there is the infrastructure constraint. The World Bank estimates that East Asian economies will have to invest between $120 and $150 billion a year, on average, over the next decade, if infrastructure is not to constitute a brake on economic growth. About half of these investments would have to be undertaken by China; about one-third of total investments would be energy-related, and about 40 percent would be for transport. The private sector has a critical role to play in providing investment, management, and technology.

A second critical constraint is human resources development. One aspect that needs attention is skills development aimed at developing a flexible labour force for more technology-intensive development. Labour migration among the East Asian economies has contributed to output expansion. It has also permitted some postponement of critical decisions about skills development, particularly in small- and medium-sized enterprises (SME). Another issue is social policy: with predictions of more than 1 billion additional urban residents over the next 20 to 25 years, East Asia countries will face budgetary pressures as well as questions about the mobilization and efficient allocation of savings. And here, financial-sector development, particularly development of capital markets, is becoming critical.

A final area is natural resource management. Both "green" and "brown" environmental impacts are already amply evident. Governments are beginning to address them. But the rapid changes taking place in these economies are making it increasingly urgent that attention be given to these issues, for both social and economic reasons.

These issues will have to be dealt with during periods when domestic political transitions are taking place in some economies. The successful management of both these transitions will determine outcomes for economic growth and development.

ENDNOTES

1   For a good survey of the literature on this subject, see Alberto Alesina and Roberto Perrotti, "The Political Economy of Growth: A Critical Survey of the Recent Literature," *The World Bank Economic Review*, Vol. 8, No. 3, pp. 351-71. Discussions with Michael Walton have been useful in clarifying the issues.
2   World Bank, *The East Asian Miracle: Economic Growth and Public Policy*, Washington: World Bank, 1993.

Walid Hejazi & Daniel Trefler
Department of Economics   The Irving B. Harris Graduate School
University of Toronto   University of Chicago
   Department of Economics
   University of Toronto

3

Canada and the Asia Pacific Region: Views from the Gravity, Monopolistic Competition, and Heckscher-Ohlin Models[1]

RECENT EAST ASIAN GROWTH HAS BEEN SPECTACULAR, whereas Canadian growth has not. Canadians express their anxiety about this in the yearly rite of purchasing the latest World Bank ranking of countries by per capita income. When this ritual began a decade ago, Canada was second only to the United States and the svelte European economies. In the latest World Bank report, however, Canada ranked an ominous 13th. Meanwhile, Singapore is nipping at our heels and Japan is ranked second. Why these countries have grown so fast is a matter of considerable debate. Some argue that it is a passing stage of development – East Asia's 15 minutes of "Warhol fame." Others are more pessimistic and make knowing predictions about China's putative dominance in the next century. As always, the truth probably lies in between.

While economists still have much to learn about the sources of long-term growth, one possibility is that East Asian growth has been export-led. This is frustrating because it seems to imply that East Asian exports to Canada have robbed Canada of its growth potential. The argument operates through various channels. One is that East Asia is running a trade surplus with Canada that is about 1 percent of Canada's gross domestic product (GDP). At first glance, this means that Canada has lost 1 percent of its jobs and investment to East Asia. This is unlikely, however, because aggregate unemployment rates in Canada have not moved in tandem with trade deficits. Moreover, the argument is essentially mercantilist: it ignores the much larger trade surplus that Canada routinely chalks up against the United States. A more likely channel through which East Asia has possibly impacted Canada is by competitively forcing a shift in Canadian industrial composition towards low-growth activities.

Two aspects of this argument will be assessed in this paper. The first, concerned with the factor content of trade, is whether East Asia is, in fact, stealing the good jobs and profitable industries away from Canada. The prevailing view in the U.S. context is that international trade is too small a part of the domestic economy to have had a major impact on factor markets (e.g., Krugman and Lawrence, 1994): the tail simply cannot wag the dog. That argument does not hold in Canada, where international trade does play a large role. Canadian imports as a share of GDP stand at almost 25 percent and grew rapidly over the 1970-92 period covered by this study. The dissenting U.S. view is that conventional analysis understates the effects of trade by ignoring the way it forces producers to adopt labour-saving and skill-biased technology (Wood, 1994). We examine this issue in the Canadian context, using an entirely new approach. We find that changing international trade patterns have indeed had large effects on factor markets.

The second issue is whether changes in industrial composition can be explained by standard comparative-advantage reasoning. East Asia has pursued an industrial policy focused on export-oriented industries – an approach that is particularly beneficial when there are spillovers. These may take the form of spillovers to downstream industries that result in industrial agglomerations, or they may take the dynamic form of building up the technology base needed to launch the next generation of technology. Of course, industrial targeting is not always successful, as shown by the debate on high-definition television (HDTV). Some experts argue that HDTV is an essential stepping stone to future technology, while others maintain that it is a dead-end technology. If industry experts cannot agree, how are policymakers to make informed decisions about industrial targeting? We sidestep this difficult question by raising a more basic issue: if East Asian industrial targeting has been successful, then, at a bare minimum, East Asia must have distorted trade patterns. That is, the region must have trade patterns that cannot be explained by theories of comparative advantage. We examine whether this is true by following Frankel et al. (1995) in using multivariate analysis to determine whether the components of East Asian trade that cannot be explained by comparative-advantage considerations are systematic. If they are, then one explanation, though not the only one, is that trade policies can be effective in promoting the development of desirable industries.

We part company with previous researchers in one major respect. Previous studies examined one model of trade – usually the gravity model, which focuses on the transaction-cost determinants of trade. One is left with the uncomfortable feeling that the results from such research could be driven by model misspecification and might thus reflect omitted sources of comparative advantage rather than industrial policy. To obviate this difficulty, we consider all three models of international trade that have received wide attention among economists – namely, the gravity model associated with transaction costs of trade; the increasing-returns-to-scale model associated with monopolistic competition; and the factor-endowments model associated with the Heckscher-Ohlin theorem. Our conclusion is that Canada trades more natural resource-based goods and East Asia trades more high-end manufactures than can be explained by any of these models. This is consistent with the observation that East Asian industrial targeting has effectively distorted trade patterns. However, we caution against the wider interpretation that East Asian industrial targeting was a success or that industrial targeting can be applied to the Canadian context.

Broad Facts About Canada's Trade with East Asia

FIGURE 1 PLOTS CANADA'S EXPORTS AND IMPORTS over the 1972-92 period. Since the data are in 1987 constant U.S. dollars, the rapid growth in both exports and imports poignantly captures the growing Canadian integration into the world economy. A feature of the plots is the dominant role played by the United States in Canada's trading relations. The next major trading partner is Europe, followed by East Asia. The latter has been broken down into Japan, the "Four Tigers" (Hong Kong, Singapore, South Korea and Taiwan) and the rest of East Asia (China, Indonesia, Malaysia, the Philippines and Thailand). Trade with Japan grew rapidly in the early 1970s, trade with the Four Tigers grew rapidly in the 1980s, and trade with the rest of East Asia grew rapidly in recent years. However, since the level of trade with East Asia was initially low, years of growth have left trade at relatively low levels. This is illustrated in Figure 2, where Canada's exports by region are plotted as shares. There have been some changes in these shares, notably a growing reliance on the U.S. market. Nevertheless, the general impression conveyed by the figure is that the shares have not moved dramatically over the past 20 years, although that is not the impression one garners from Canada's business press.

Our mercantilist instincts often direct us to the merchandise trade balance. Figure 3 plots Canada's bilateral trade balance with East Asia and Europe. Since the surpluses of the early 1980s, there has been a tendency for these trade balances to deteriorate. This is cause for concern when viewed as a Keynesian withdrawal from the Canadian economy, but of less concern when viewed as a response to capital inflows induced by low savings rates and fiscal deficits. Moreover, these regional merchandise trade deficits are dwarfed by Canada's trade surplus with the United States, which has averaged about $10 billion over the period.

Labour and Capital Market Implications: The Factor Content of Canada's Trade

WE TEND TO FOCUS ON TRADE FLOWS when what we really care about is their factor-market consequences. For example, the number of cars that

Figure 1

Canadian Exports and Imports, 1972-92

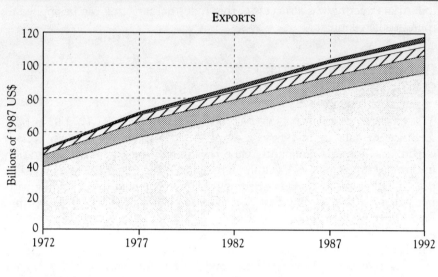

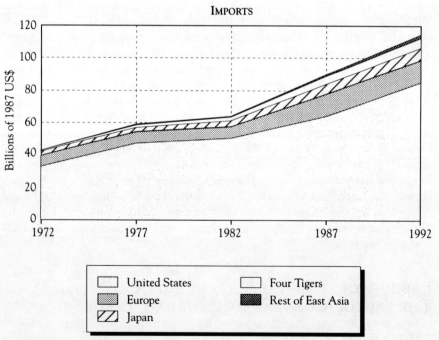

FIGURE 2

CANADA'S EXPORT SHARES, 1972-92

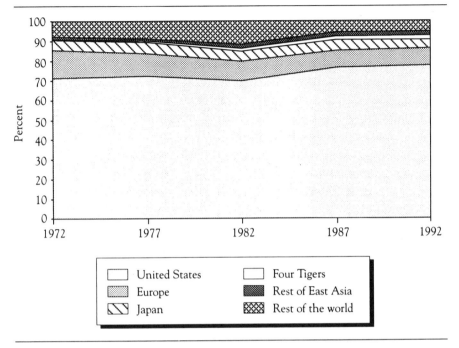

Canada imports from Japan is uninteresting in and of itself. What gives it significance is the impact it has on factor markets – the number of Canadian jobs displaced and the effect on wages; the supplanted investment and the effect on rates of return to capital – in other words, because of the factor content of trade. Factor-content calculations answer the question: How much skilled and unskilled labour is required to produce a country's exports and imports? The answer may be interpreted as revealing the impact of trade on the demand for skilled and unskilled labour.

Thus understanding the factor content of trade rather than trade in goods is not simply an academic exercise: it is central to the conduct of trade policy. Figure 4 provides factor-content calculations for three factors in 1972 and 1992: labour, capital and land. Note that we have taken the shortcut of evaluating the factor content of Canadian trade using the 1987 U.S. input-output table. The data are shown by region and are expressed as a percentage of Canada's factor endowment. For example, in 1992 Canada "exported" almost 1 percent of its labour force to the United States and "imported" almost 2 percent of its labour force from Europe and East Asia. (The terms

Figure 3

Canada's Merchandise Trade Balance, 1972-92

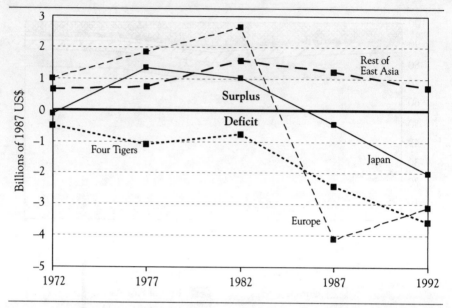

"exported" and "imported" are used in a factor-content sense.) These are not small numbers. They represent about 10,000 person-years of labour exported to the United States and 20,000 person-years imported from Europe and East Asia. As can be seen, Canada moved from being a net exporter to a net importer of labour between 1972 and 1992. It also halved its net exports of capital. Land exports rose largely due to increased agricultural exports to Japan. The figure also conveys regional details, but we do not find any dramatic features in this dimension.

Figure 5 breaks down these trends separately for imports and exports. It is striking that Canada exports and imports about 20 percent of its labour force. Notice that the rise in net imports of labour has coincided with increased gross exports of labour to each region, including the labour-abundant rest of East Asia.

Table 1 provides details of the factor content of trade by educational attainment, as defined by Barro and Lee (1993). Again, these factor-content-of-trade data are expressed as a percentage of the Canadian factor market in 1972 and 1992. For example, the labour content of Canada's exports to the world was equivalent to 21.9 percent of the Canadian labour force. This corresponds to three million person-years of labour. Much of this effect comes from workers with some secondary education.

FIGURE 4

FACTOR CONTENT OF CANADA'S NET EXPORTS, 1972 AND 1992

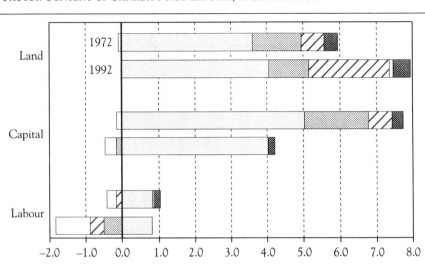

Input-output tables aggregate across subsectors in each industry in ways that can distort the factor content of trade. One example drawn from Feenstra and Hanson (1995) starts with the observation that each final good is produced from many different processes, some skilled-labour-intensive and some unskilled-labour-intensive. One expects Canada to import unskilled-intensive processes such as assembly and to export skill-intensive processes such as design (see Figure 6). Over time, more skill-intensive processes are done abroad, as illustrated in the figure.

Consider any good g and let U_{xgt} and U_{mgt} be the amounts of unskilled labour needed to produce exports and imports of good g in year t. The usual factor content of net exports calculation is $\sum_g (U_{xg92} - U_{mg92})$ and one concludes that trade has depressed unskilled wages if $\sum_g (U_{xg92} - U_{mg92}) - \sum_g (U_{xg72} - U_{mg72})$ is negative and large relative to the endowment of unskilled labour. These are the types of calculations that appeared in Table 1 and reappear in Table 2 under the column "domestic calculations." However, one would like to use the U_{xgt} and U_{mgt} that would have been used in the absence of increased East

FIGURE 5

FACTOR CONTENT OF CANADA'S IMPORTS AND EXPORTS, 1972 AND 1992

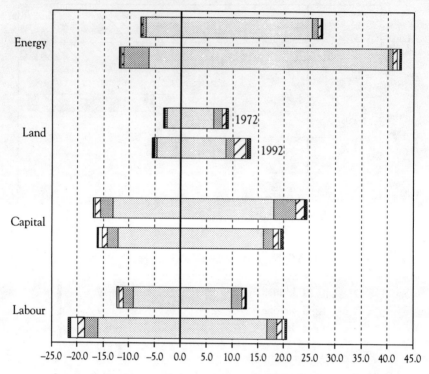

Imports (−) and exports (+) as a percentage of Canada's factor endowment

- United States
- Europe
- Japan
- Four Tigers
- Rest of East Asia

Asian exports to Canada. Call these U^*_{xgt} and U^*_{mgt}. In particular, following Wood (1994) one expects U_{mgt} to be very different from U^*_{mgt}.

We have used this observation to adjust the U.S. technology matrix to accommodate lesser-developed-country factor-content calculations. In particular, we have assumed that the amount of each type of labour used to produce a good in the United States depends on the educational percentile rather than on the absolute level of education. For example, a U.S. janitor may have a Grade 8 education but is doing a job that only requires a fifth percentile level of education. We assume that a fifth percentile worker in any country can do

CANADA AND THE ASIA PACIFIC REGION

TABLE 1
CANADA'S FACTOR CONTENT OF TRADE AS A PERCENTAGE OF THE DOMESTIC FACTOR MARKET, 1992[1]

TRADE REGION		BROAD ENDOWMENT CATEGORIES					EDUCATION				ADJUSTED EDUCATION			
		LABOUR	CAPITAL	LAND	ENERGY	NONE	PRIMARY	SECONDARY	COLLEGE	NONE	PRIMARY	SECONDARY	COLLEGE
								Exports					
4	Japan	1.0	1.0	2.3	0.8	0.7	0.5	1.8	0.5	12.8	2.4	0.5	0.1
2	Four Tigers[2]	0.5	0.5	0.3	0.5	0.3	0.2	0.9	0.3	5.5	1.2	0.3	0.1
2	Other East Asia[3]	0.5	0.4	0.6	0.4	0.4	0.2	0.8	0.3	5.5	1.0	0.3	0.1
9	Europe	1.9	1.9	1.5	0.9	1.1	0.8	3.4	1.2	21.6	4.7	1.1	0.3
78	United States	16.8	16.1	8.7	40.1	8.8	6.5	29.0	10.6	181.5	40.4	10.2	2.9
100	World	21.9	21.0	14.8	43.9	12.0	8.8	37.9	13.5	241.1	52.4	13.0	3.7
								Imports					
6	Japan	1.4	1.0	0.1	0.4	0.7	0.5	2.4	1.0	14.2	3.4	0.9	0.3
5	Four Tigers[2]	1.4	0.8	0.2	0.3	1.2	0.8	2.5	0.8	17.5	3.2	0.8	0.2
2	Other East Asia[3]	0.4	0.3	0.1	0.1	0.4	0.3	0.8	0.2	5.8	1.0	0.2	0.1
11	Europe	2.5	2.1	0.4	4.8	1.4	1.0	4.2	1.6	26.4	5.8	1.5	0.4
72	United States	16.0	12.0	4.6	6.3	8.6	6.0	27.4	10.3	169.6	38.5	9.9	2.8
100	World	22.5	17.0	5.8	14.7	12.9	9.0	38.7	14.3	243.7	53.9	13.7	3.9
								Net exports					
−45	Japan	−0.4	0.0	2.2	0.4	−0.0	0.0	−0.6	−0.4	−1.4	−1.0	−0.4	−0.1
−79	Four Tigers[2]	−1.0	−0.3	0.1	0.2	−0.9	−0.6	−1.6	−0.5	−12.0	−2.1	−0.5	−0.1
15	Other East Asia[3]	0.0	0.2	0.5	0.3	−0.1	−0.0	0.0	0.0	−0.3	0.1	0.0	0.0
−69	Europe	−0.5	−0.2	1.1	−3.9	−0.3	−0.2	−0.8	−0.4	−4.8	−1.1	−0.4	−0.1
234	United States	0.8	4.0	4.1	33.8	0.2	0.5	1.5	0.3	12.0	1.8	0.3	0.1
100	World	−0.7	4.0	9.0	29.2	−0.9	−0.1	−0.8	−0.8	−2.6	−1.5	−0.7	−0.2

1 All factor-content-of-trade data are expressed as a percentage of the Canadian factor market. For example, the labour content of Canada's exports to the world were 21.9 percent the size of the Canadian labour force in 1992.
2 Hong Kong, Singapore, South Korea, and Taiwan.
3 China, Indonesia, Malaysia, the Philippines, and Thailand.

Figure 6

Distortions in Factor-Content Calculations, 1972 and 1992

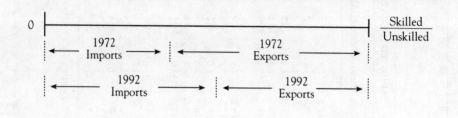

the job regardless of the absolute level of education. These are the numbers we treat as U^*_{xgt} and U^*_{mgt}. In Table 2, we report three different calculations:

$$\sum_g (U_{xg92} - U_{mg92}) - \sum_g (U_{xg72} - U_{mg72}) \quad \text{"domestic calculations,"} \quad (1)$$

$$\sum_g (U^*_{xg92} - U^*_{mg92}) - \sum_g (U^*_{xg72} - U^*_{mg72}) \quad \text{"foreign calculations," and} \quad (2)$$

$$\sum_g (U_{xg92} - U^*_{mg92}) - \sum_g (U_{xg72} - U^*_{mg72}) \quad \text{"hi-lo calculations."} \quad (3)$$

Equation (1) is the usual calculation. Equation (2) is the relevant calculation if we assume that even at high levels of education, all that matters is the percentile – e.g., Canadian and Thai scientists are equally well-trained and productive. Equation (3) drops this high-end assumption.

The results appear in Table 2. For Canadian trade with the world, such calculations alter the conclusions about the effect of trade on wages. Under the usual calculations, international trade competition has reduced the net demand for no-education workers by 1.3 percent or 3,000 of 233,029 jobs. Under the corrected method, international trade competition has reduced the net demand by 140.3 percent or 326,800 jobs. This is remarkably similar to Wood's (1994) calculation using a very different, more ad hoc procedure. The implication is that even a small amount of international trade can lead to very large domestic wage implications for less educated workers. None of this effect comes from trade with East Asia, and most of it comes from trade with the United States and, to a much lesser extent, from trade with Europe. Note that no-education workers form a small group representing less that 2 percent of the work force. As Table 2 shows, there is also a large effect for workers with only some primary education, who represent one-fifth of the labour force. For the largest groups – workers with more than primary education – international trade has increased demand. However, the size of the increase is small relative to the inaccuracies of the methodology. In contrast with the world results, the results for Canadian trade with East Asia are uniformly small.

TABLE 2
Canada's Adjusted Factor Content of Trade by Education

		DOMESTIC CALCULATIONS		FOREIGN CALCULATIONS		HI-LO CALCULATIONS	
EDUCATION	1992 ENDOWMENT	THOUSANDS OF WORKERS	% OF 1992 ENDOWMENT	THOUSANDS OF WORKERS	% OF 1992 ENDOWMENT	THOUSANDS OF WORKERS	% OF 1992 ENDOWMENT
			World				
None	233,029	−3,000	−1.3	−29,300	−12.6	−326,800	−140.3
Primary	2,956,720	−14,600	−0.5	−96,400	−3.3	−838,100	−28.3
Secondary	5,295,230	−118,000	−2.2	−50,300	−1.0	683,800	12.9
College	5,223,951	−53,400	−1.0	−15,500	−0.3	285,400	5.5
			East Asia				
None	233,029	−400	−0.2	−4,000	−1.7	−11,300	−4.8
Primary	2,956,720	−3,200	−0.1	−4,100	−0.1	−21,000	−0.7
Secondary	5,295,230	−8,700	−0.2	−1,000	0.0	17,100	0.3
College	5,223,951	−200	0.0	−100	0.0	6,900	0.1

These results are also consistent with the literature on income distribution. Kuhn (1995) shows that over the past 15 years, there has been an increase in the earnings differential between educated and less educated workers, as well as an increase in unemployment among the least skilled workers. In particular, during the 1980s unskilled earnings fell by 30 percent whereas skilled earnings remained unchanged.

FORMAL MODELLING

THE PROBLEM WITH DESCRIBING THE EFFECTS OF TRADE POLICY is the problem of modelling the counter-factual: What would trade have been in the absence of the current policy or in the presence of a policy under consideration? The answer depends on the underlying sources of comparative advantage. There are three popular approaches to controlling for comparative advantage – the gravity model, the monopolistic competition model, and the Heckscher-Ohlin model. We will consider all three. The need to consider them all is clear, as each makes different assumptions about the counter-factual world – i.e., about the sources of comparative advantage. Accordingly, we must convey how our conclusions depend on the modelling assumptions used. This is typically done in a limited fashion – e.g., a test for inclusion of an additional variable. We will consider it in the very broadest possible terms. The data set to be used consists of bilateral trade between 103 countries for 37 tradeables over the period 1970-92. (The countries and industries are listed in Tables A-1, A-2 and A-3.) Throughout, observations corresponding to unrecorded and zero bilateral trade are omitted.

TRANSACTION COSTS: THE GRAVITY MODEL

Let t index years, g index goods, i index the importing country, j index the exporting country, and r index regions. We will write that country i or j is a member of region r. Let M_{ijgt} denote bilateral imports. Using Greek letters to denote unobservables to be estimated, the gravity model divides determinants of M_{ijgt} into three types – transaction costs, a regional catchall, and unobservables:

$$ln(M_{ijgt}) = \alpha_g + ln(X_{ijt})\beta_g + \sum_r D_{ijr} \delta_{rg} + \sum_r D^*_{ijr} \delta^*_{rg} + \varepsilon_{ijgt}. \qquad (4)$$

X_{ijt} captures transaction costs broadly defined. It includes (letting gdp denote gross domestic product):

$gdp_{it} \times gdp_{jt}$ product of GDPs in countries i and j,
$gdp_{it}/pop_{it} \times gdp_{jt}/pop_{jt}$ product of per capita GDPs in countries i and j,

ppp_{ijt}	purchasing-power-parity index between countries i and j,
$distance_{ij}$	a measure of distance between countries i and j,
$neighbours_{ij}$	a dummy variable equal to unity for adjacent countries, and
$language_{ij}$	a dummy variable equal to unity if countries i and j share the same language.

The idea is that countries of similar size and per capita GDP have similar needs in terms of both intermediate input needs (Ethier, 1982) and consumption patterns. In addition, neighbouring countries, countries that are close together, countries with similar language and countries with small exchange-rate deviations from fundamentals will have small transaction costs of doing business and correspondingly large levels of bilateral trade.

The D_{ijr} and D^*_{ijr} are dummies whose coefficients capture systematic but unobserved differences between regions. Table 3 illustrates their definition. D_{ijr} is unity if both i and j are members of region r. Its coefficient is the average level of good g bilateral trade *within* the region unexplained by observed transaction costs. D^*_{ijr} is unity if only either i or j is a member of region r. Its coefficient is the average level of industry g bilateral trade *between* the region and the rest of the world that is unexplained by observed transaction costs. The definition of regions follows Frankel et al. (1995) and appears in the appendix.

The reader familiar with the literature will recognize that in this section we are trying to follow as closely as possible the work of Frankel et al. This allows for simple comparisons with previous work. We entertain only one important departure from their work, in that we recognize that the importance of transaction-cost motives for trade vary across goods. For example, transaction-cost motives are likely to be entirely unimportant for homogeneous goods with low transportation costs and to be important for goods with variable quality or high transportation costs.

TABLE 3

DEFINITION OF BETWEEN- AND WITHIN-REGION EFFECTS

		COUNTRY i	
		IN REGION r	OUTSIDE REGION r
Country j	In region r	$D_{ijr} = 1$, $D^*_{ijr} = 0$	$D_{ijr} = 0$, $D^*_{ijr} = 1$
	Outside region r	$D_{ijr} = 0$, $D^*_{ijr} = 1$	$D_{ijr} = 0$, $D^*_{ijr} = 0$

To focus ideas, Table 4 presents the results of a particular regression of the form in Equation (4). In this case, all industries are aggregated. All the transaction-cost variables have the expected sign. An analysis of variance (ANOVA) decomposition indicates that the transaction-cost variables explain between 22 and 58 percent of the sample variation. This is impressive for a sample with almost 150,000 observations. On the other hand, we have not scaled the regression. Preliminary work suggests that this has little effect on coefficient estimates but raises the sample fit considerably. (This is to be expected if there is heteroscedasticity.) In contrast, the regional dummies explain only between 4 and 30 percent of the sample variation. This is mirrored in the standardized beta coefficients reported in the table. A standardized

TABLE 4

GRAVITY MODEL, ALL INDUSTRIES, 1970-92

DEPENDENT VARIABLE: $\ln(M_{ijgt})$		ESTIMATED COEFFICIENTS	t-STATISTICS	BETA COEFFICIENTS
Transaction costs				
$\ln(gdp_{it} \times gdp_{jt})$		0.74	264.02	0.52
$\ln(gdp_{it}/pop_{it})(gdp_{jt}/pop_{jt})$		0.49	92.22	0.20
$\ln(ppp_{ijt})$		−0.23	−30.72	−0.05
$\ln(distance_{ij})$		−1.01	−119.55	−0.26
$neighbours_{ij}$		0.71	21.43	0.04
$language_{ij}$		0.69	43.13	0.08
intercept		−17.69	−123.18	0.00
Regional dummies				
EAEC	within	0.06	0.89	0.00
	between	0.12	4.93	0.02
NA	within	−0.81	−4.58	−0.01
	between	−0.26	−9.73	−0.02
EEC	within	1.42	31.51	0.06
	between	1.19	81.47	0.17
EFTA	within	1.62	20.56	0.03
	between	0.58	33.34	0.06
WH	within	0.50	13.12	0.02
	between	−0.09	−5.76	−0.01
APEC	within	2.70	48.06	0.14
	between	0.83	34.56	0.12
N	146,203			
Adjusted R^2	0.62			

Note: Estimates of Equation (4).

beta is the estimated coefficient from a regression in which each variable is scaled by its standard deviation. It is of interest because it is proportional to the reduction in mean squared error from adding the variable – i.e., it is proportional to the improvement in within-sample fit. The only large beta coefficients are those associated with the European Community (EC). Although the regional dummies as a group are statistically significant ($F = 1103.0$) and economically large, the ANOVA and standardized beta coefficient results suggest that regional idiosyncrasies are not important.

This said, we turn to the details of the regional dummies. With the exception of the East Asia Economic Caucus (EAEC) and North America (NA), the obvious pattern is that the within-region effects dominate the between-region effects. That is, unexplained trade within regional blocks exceeds unexplained trade between regional blocks. In addition, the dummies with large coefficients and standardized betas are all positive. That is, both within-region and between-region trade exceed unexplained trade outside regional blocks. Some of the between-region dummies – those for North America and the Western Hemisphere (WH) – are negative. This has been interpreted by some as evidence of trade diversion. As will be seen, however, all the regional effects are positive when disaggregated by good. One explanation of the results is that our regional dummies are correlated with omitted transaction-cost variables. Indeed, the very definition of region is endogenous, the transaction-cost controls are crude, and 38 percent of the sample variation remains unexplained. Another explanation is that the model is wrong and all the results are spurious. A third interpretation is that the regional dummies capture a feature of regional policy such as the promotion of economic integration through liberalized trade. In this interpretation, regionalism also confers important spillovers to other regions – i.e., the between-region effects are positive, not just the within-region effects – so that concerns about regional integration are somewhat mitigated.

All of this requires one to accept that transaction costs are uniformly important across industries. To investigate this, we disaggregated the regression by 37 industries – 28 in manufacturing and 9 in other sectors (see Table A-3). In all cases, the regional dummies are jointly significant. To get at the heart of the Canada-East Asia issue, we only report the regional dummies for North America (NA) and EAEC. (Full details on coefficients for regional dummies and transaction-cost variables appear in Tables A-5 and A-4, respectively.) In Table 5 the italicized numbers represent the difference between the two regional dummies. Industries are sorted by this difference, and only industries with large absolute values of the difference are reported. From the "high within-region trade/North America" section of the table, it is apparent that trade within North America is dominated by natural resources. In contrast, trade within East Asia is dominated by high-end manufactures such as electrical and electronics.[2] A similar picture emerges from between-region trade except that East Asia focuses on mid-range manufactures. To the extent that the regional

TABLE 5
Gravity Model, by Industry, 1970-92, Regional Dummies

	Within-region dummies δ_{rg}				Between-region dummies δ_{rg}^*		
	EAEC	NA	EAEC – NA		EAEC	NA	EAEC – NA
	High within-region trade				High between-region trade		
North America (NA)				**North America (NA)**			
Crude oil	3.61	80.26	−76.65	Crude oil	1.38	3.37	−1.99
Electricity	8.24	42.40	−34.18	Non-metal mining	0.71	2.22	−1.51
Non-metal mining	0.83	16.91	−16.08	Tobacco	0.79	2.00	−1.21
Fishing	8.24	24.02	−15.78	Forestry	0.60	1.51	−0.91
Liquors	1.53	12.26	−10.73	Liquors	0.66	1.53	−0.87
Forestry	1.85	10.15	−8.30	Electricity	3.13	3.97	−0.84
Vehicles	1.22	7.30	−6.08	Petroleum refineries	0.97	1.76	−0.79
Petroleum refineries	3.60	8.12	−4.52	Petroleum and coal products	0.43	1.06	−0.63
Apparel	2.74	5.91	−3.17	Crops	0.29	0.82	−0.53
Sawmills	0.46	3.17	−2.71	Food	0.34	0.62	−0.28
East Asia (EAEC)				**East Asia (EAEC)**			
Electrical and electronics	16.34	5.47	10.87	Electrical and electronics	6.46	1.55	4.91
Rubber	6.67	3.13	3.54	Misc. manufacturing	4.35	1.63	2.72
Instruments	5.95	2.84	3.11	Plastic products	3.65	1.18	2.47
Misc. manufacturing	7.55	4.65	2.90	Rubber	3.59	1.13	2.46
Pharmaceuticals, etc.	3.96	1.40	2.56	Metal products	2.48	0.96	1.52
Basic chemicals	3.58	1.36	2.22	Furniture and fixtures	3.26	1.74	1.52
Plastic products	3.64	1.60	2.04	Pottery and china	2.41	1.04	1.37
Iron and steel	1.99	1.10	0.89	Leather	1.87	0.52	1.35
Non-ferrous metals	1.70	1.08	0.62	Instruments	2.78	1.47	1.31
Metal products	2.36	1.75	0.61	Glass products	2.54	1.29	1.25

Note: Regional dummy coefficient estimates of Equation (4), by industry.

dummies capture regional differences in trade policy, this paints a dismaying picture of how East Asian policy has more successfully shifted production up the high-tech ladder.

In each of the three models that we estimate, trade is modelled as a function of comparative-advantage variables. The coefficients on the dummy variables capture the average levels of trade within and between regions that the comparative-advantage variables do not explain. It is in this sense that the coefficients on the dummy variables reflect the residual error associated with econometric misspecification. Our interpretation of the dummies may therefore be misplaced.

Increasing Returns to Scale: The Monopolistic Competition Model

NOT ALL TRADE IS TRANSACTION-COST BASED: much of international trade stems from economies of scale that lead to specialization. Empirically, this is most often examined in the context of the monopolistic competition model, which focuses on internal economies of scale at the level of the firm.[3] The model is typically associated with an aggregate demand for variety in manufactured goods. This demand for variety can come from variety-loving or heterogeneous "ideal-type" consumers (Helpman and Krugman, 1985). Alternatively, it can come from demand for specialized, differentiated intermediate products (Ethier, 1982). Because of the interaction between scale economies and the demand for variety, in equilibrium each firm in an industry produces a single differentiated product. Since all production occurs at a single location, aggregation across varieties yields import demands that depend on country-level outputs of all varieties of the good – i.e., on country j's output of good g, Q_{jgt}. Under certain restrictive assumptions, the model yields the following precise prediction:

$$M_{ijgt} = s_{it} Q_{jgt} \tag{5}$$

where $s_{it} = gdp_{it}/\sum_j gdp_{jt}$. That is, country i's imports of good g produced in country j (M_{ijgt}) is proportional to country j's production of good g (Q_{jgt}). The constant of proportionality (s_{it}) is country i's share of world GDP.

This strong hypothesis cannot be correct: when Equation (5) is summed across exporters and goods, one obtains the prediction that the ratio of imports-to-GDP is the same in all countries. This is far from accurate: countries such as Japan and the United States, for example, have much lower import-to-GDP ratios than Canada. A weaker version of the model is that a country's bilateral imports are proportional to the exporting country's production, but that the constant of proportionality is unrestricted across goods. Thus we estimate the following model:

$$\ln(M_{ijgt}) = \alpha_g + \ln(s_{it} Q_{jgt})\beta_g + \sum_r D_{ijr} \delta_{rg} + \sum_r D^*_{ijr} \delta^*_{rg} + \varepsilon_{ijgt} \qquad (6)$$

where the constant of proportionality is now $s_{it}\beta_g$, which varies across goods. In the strong version of the model, β_g is unity for all goods. Under the weaker hypothesis, it must be positive and is likely to be significantly less than unity, a reflection of well-documented Armington home bias (see, for example, Trefler, 1995). Unlike the other results in this paper, we restrict attention to manufacturing because of data availability for Q_{jgt}.

To fix ideas, Table 6 reports the results across industries for the period 1972-92. Only data for every fifth year are used, in part to reduce the sample size down to a manageable 353,905 observations. The model finds some support in the R^2 of 0.27 and in the fact that β_g lies between zero and unity. The regional dummies as a group are statistically significant ($F = 341.8$). An ANOVA decomposition reveals that $s_{it} Q_{jgt}$ explains 5 to 16 percent of the bilateral trade variance and the regional dummies explain 12 to 19 percent of the variance. There are a number of striking results. First, the regional dummies remain predominantly positive and the within-region dummies are larger than the between-region dummies. That is, the monopolistic competition model

TABLE 6

MONOPOLISTIC COMPETITION MODEL, ALL MANUFACTURING INDUSTRIES, 1970-92

DEPENDENT VARIABLE: ln (M_{ijgt})		ESTIMATED COEFFICIENTS	t-STATISTICS	BETA COEFFICIENTS
Monopolistic competition				
$\ln(s_{it} Q_{igt})$		0.51	215.49	0.33
intercept		3.23	101.56	0.00
Regional Dummies				
EAEC	within	0.62	12.88	0.03
	between	−0.05	−2.25	−0.01
NA	within	3.10	29.54	0.05
	between	1.03	45.08	0.10
EEC	within	4.69	165.48	0.26
	between	1.53	128.13	0.23
EFTA	within	3.82	71.35	0.11
	between	0.72	51.56	0.08
WH	within	0.66	23.01	0.04
	between	−0.48	−32.44	−0.06
APEC	within	2.40	62.30	0.17
	between	0.44	22.58	0.06
N	353,905			
Adjusted R^2	0.27			

Note: Estimates of Equation (6). Industry dummy coefficients are not reported.

under-predicts both between and within-region trade and also under-predicts within-region trade more than between-region trade. Surprisingly, there are only two differences from the gravity model results. First, only two of the between-region coefficients are now negative, and one of them (EAEC) is economically small. Second, the regional coefficients are much larger for the monopolistic competition model than for the gravity model.

Table 7 reports some of the Equation (6) coefficient estimates for the case where the regressions were estimated by industry. All the regression coefficients appear in Table A-6. A typical industry-level regression has about 10,000 observations. Within-region trade is higher in North America than in East Asia for manufactures that are tightly associated with natural resources – e.g., petroleum refineries, food and sawmills. This mirrors the results of the transaction-cost model. This similarity of results is partly disguised because Table 5 reports all industries while Table 7 reports only manufacturing industries. To emphasize the similarity, we have placed asterisks in Table 7 where the industry ranking, excluding non-manufacturing, is the same for both the gravity and monopolistic competition models. An interesting exception is machinery. Within-region trade is relatively higher in East Asia for items ranging from high-end plastic products and pharmaceuticals, etc. to lower-end textiles. Thus the pattern of within-region specialization is less stratified by technology intensity than for the gravity model. For between-region trade, North America tends to be more specialized in high-end manufactures than is East Asia, even though we have controlled for scale effects via the monopolistic competition model.

The pattern of between-region and within-region trade has been the focus of intense policy debate (Lorenz, 1992). The question is whether or not regional trading arrangements complement multilateral trading arrangements such as the World Trade Organization (WTO). If regional arrangements and the WTO are complementary, then when δ_{rg} is large, δ_{rg}^* should also be large. We approach this problem from the perspective of East Asia versus Canada by comparing the East Asian trading block with the North American trading block. Figure 7 plots $\delta_{EAEC,g} - \delta_{NA,g}$ against $\delta_{EAEC,g}^* - \delta_{NA,g}^*$. For both the transaction-costs and increasing-returns-to-scale models, when East Asian within-region trade is unusually large (relative to North American, $\delta_{EAEC,g} - \delta_{NA,g}$), East Asian trade with other regions is also unusually large (relative to North America, $\delta_{EAEC,g}^* - \delta_{NA,g}^*$). That is, the regional arrangements complement multilateral arrangements.

Factor Endowments:
The Heckscher-Ohlin Model

THE RESULTS FOR THE TWO PREVIOUS MODELS suggest that there is still considerable sample variation left unexplained. In particular, it may be that the natural

TABLE 7
MONOPOLISTIC COMPETITION MODEL, 1970-92, REGIONAL DUMMIES BY INDUSTRY

	WITHIN-REGION DUMMIES δ_{rg}			BETWEEN-REGION DUMMIES δ_g^*		
	EAEC	NA	EAEC − NA	EAEC	NA	EAEC − NA
	High within-region trade			High between-region trade		
North America (NA)						
* Liquors	−0.48	4.41	−4.89			
* Vehicles	0.22	4.66	−4.44			
* Petroleum refineries	0.37	4.26	−3.88			
* Petroleum products	−0.11	3.54	−3.65			
* Furniture and fixtures	1.34	4.79	−3.45			
Food	−1.09	2.10	−3.20			
* Apparel	0.78	3.93	−3.14			
* Sawmills	0.14	3.18	−3.04			
Machinery	0.79	3.56	−2.78			
Printing and publishing	−0.18	2.53	−2.71			
East Asia (EAEC)						
Plastic products	0.70	3.03	−2.33			
Textile	−0.08	2.13	−2.21			
Pottery and china	0.79	2.95	−2.16			
* Iron and steel	1.08	3.10	−2.02			
* Non-ferrous metals	1.09	3.08	−1.99			
Pharmaceuticals, etc.	1.16	3.08	−1.92			
* Instruments	0.37	2.27	−1.90			
* Basic chemicals	1.36	2.89	−1.52			
* Rubber	2.00	3.48	−1.48			
Tobacco	0.71	1.55	−0.84			
North America (NA)						
* Petroleum products				−1.32	1.41	−2.72
* Petroleum refineries				−0.84	1.41	−2.24
* Liquors				−1.08	1.03	−2.10
* Printing and publishing				−0.52	1.09	−1.60
* Vehicles				−0.09	1.42	−1.52
Machinery				−0.12	1.40	−1.51
* Tobacco				−0.40	1.10	−1.50
* Food				−1.15	0.26	−1.41
Pharmaceuticals, etc.				−0.08	1.28	−1.36
* Clay and cement products				−0.13	1.20	−1.33
East Asia (EAEC)						
* Plastic products				0.35	1.29	−0.94
Textile				−0.30	0.62	−0.92
* Glass products				0.66	1.46	−0.81
* Misc. manufacturing				0.57	1.32	−0.75
* Metal products				0.50	1.21	−0.71
* Leather				0.14	0.80	−0.66
Iron and steel				−0.04	0.52	−0.56
Footwear				0.13	0.61	−0.48
* Pottery and china				0.42	0.89	−0.48
* Rubber				0.80	0.96	−0.15

Note: Regional dummy coefficient estimates of Equation (6), by industry. An asterisk indicates that the industry's rank is similar to its gravity model rank (after excluding non-manufacturing industries from the list).

Figure 7

Regionalism versus Multilateralism

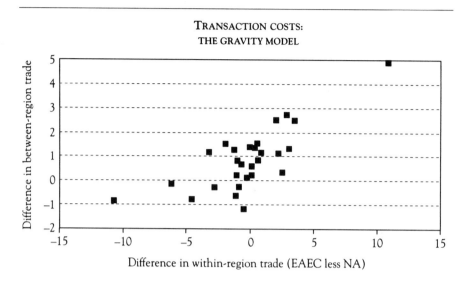

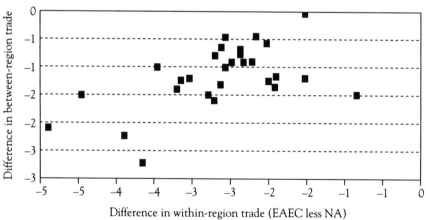

resources of a country as well as its capital and human capital resources are picked up in the regional dummies, especially the North American dummies. To purge the regional dummies of such effects, we turn to the Heckscher-Ohlin model. The theory states that a capital-abundant country exports capital-intensive goods. Let T_{igt} denote net exports by country i of good g in year t, and let E_{fit} be the endowment of factor f in country i in year t. The theory predicts a precise relationship between net exports and factor endowments:

$$T_{igt} = \sum_{f=1}^{F} E_{fit}\beta_{fg}. \tag{7}$$

Two elements stand out. First, we are now dealing with net exports, whereas previously we dealt with bilateral trade M_{ijgt}. Second, the β_{fg} do not depend on characteristics of country i. The importance of this is that one can fix the industry g and compute the cross-country regression of T_{igt} on $(E_{1it}, ..., E_{Fit})$ because the β_{fg} are stable across observations (countries), as required by regression analysis. That the β_{fg} are the same for all countries falls out of the Heckscher-Ohlin theorem. Under the usual assumptions of the model (equal numbers of goods and factors, internationally identical technology and factor prices, constant returns to scale, etc.), the β_{fg} depend solely on per unit factor input demands, and these factor input demands are the same in all countries (see Leamer, 1984 for details).

This discussion suggests the following regression:

$$T_{igt} = \alpha_g + \sum_{f=1}^{F} E_{fit}\beta_{fg} + \sum_r D_{ir}\delta_{rg} + \varepsilon_{igt} \tag{8}$$

where D_{ir} equals unity if country i is in region r, and zero otherwise. Notice that the region dummies do not distinguish between within-region and between-region effects because the dependent variable is net exports of country i, not bilateral trade.

To fix ideas about the estimation of Equation (8), consider the first row of Table 8, which reports the results of estimating Equation (8) for the livestock industry. With 103 countries and 23 years (1970-92) there are 2,369 observations. There are 11 factor endowments: the national capital stock is from Summers and Heston (1991), while data on levels of education are from Barro and Lee (1993). Land divides into three types, as do subsoil resource stocks. In addition, each regression contains six regional dummies and an intercept (see Table A-7). To aid comparison across factors measured in different units, we have standardized the coefficients. For example, a change of one standard deviation in a country's capital endowment leads to a decline of 0.23 standard deviations in livestock net exports. These standardized betas tend to be highly correlated with t-statistics.

Table 8 reports results for the nine non-manufacturing industries. The model performs reasonably well, as shown by the adjusted R^2. In addition, the coefficients in bold face are those identified a priori as sources of comparative

TABLE 8

HECKSCHER-OHLIN MODEL, NON-MANUFACTURING REGRESSION COEFFICIENTS, 1970-92

DEPENDENT VARIABLE: NET EXPORTS T_{igt}	CAPITAL	EDUCATION				LAND			SUBSOIL			R^2	F-TEST REGION
		NONE	PRIMARY	SECONDARY	COLLEGE	PASTURE	CROP	FORESTRY	COAL	OIL AND GAS	METALS		
Livestock	-0.23	-0.07	-0.15	0.05	0.05	**0.22**	0.18	-0.11	0.13	-0.09	0.09	0.24	26.1
Crops	-0.58	-0.40	-0.14	-0.16	**0.60**	0.13	**0.38**	0.05	0.22	-0.08	-0.05	0.41	20.5
Forestry	-0.27	-0.09	-0.04	-0.02	-0.09	-0.16	0.29	**0.19**	0.22	0.08	-0.02	0.13	26.2
Fishing	-0.48	-0.11	0.14	0.28	-0.12	-0.04	0.07	0.01	0.08	0.06	0.11	0.30	115.6
Coal	-0.31	-0.32	-0.16	-0.06	0.12	0.21	0.30	-0.14	**0.49**	0.00	0.09	0.47	15.0
Oil and gas	-0.19	0.04	0.01	0.10	-0.11	0.03	-0.02	0.03	0.01	**0.63**	0.00	0.44	4.6
Metal mining	-0.34	0.18	-0.12	-0.21	0.23	-0.06	-0.10	0.19	-0.03	0.02	**0.71**	0.55	2.3
Other mining	-0.26	-0.02	-0.05	-0.17	0.01	-0.05	0.15	0.03	0.41	-0.03	0.21	0.27	21.4
Electricity	-0.01	0.10	0.00	-0.10	-0.05	-0.09	0.00	0.08	-0.03	0.02	0.17	0.10	17.9

Note: Each row represents a separate ordinary-least-squares (OLS) estimate of Equation (8). Only the factor endowment coefficients are reported. The intercept and regional dummies appear in the appendix. Each regression has 2,369 observations (103 countries x 23 years). The "F-test region" column is the F-test for the joint significance of the six regional dummies. The 1 percent critical value is 2.1. Bold face indicates factors that are a priori a source of comparative advantage for the industry.

advantage. All of these are large and positive, as expected. Table 9 reports the results for the 28 manufacturing industries, sorted by the size of the capital coefficient. The results are again satisfying, as judged by the adjusted R^2 and by the sensible results for capital. For instance, capital is most important as a source of comparative advantage for machinery, electrical and electronics, pharmaceuticals, etc., instruments and vehicles. On the other hand, the education endowment coefficients provide a mixed picture of the performance of the model. While the "no education" stock is a source of comparative advantage only for low-end manufactures, as expected, the remaining education coefficients make little sense. One possibility is multicollinearity: the simple correlations among the education stocks and with capital are high.

The column "F-test regional" reports the F-statistic for the joint significance of the six regional dummies. In all cases, these are statistically significant at the 1 percent level. However, since factor endowments are likely most important for resource industries, it is reassuring that the regional dummies are smaller for such industries.

Table 10 reports the East Asia and North America regional dummies. As before, we have ranked industries by the size of the difference in these two regional dummies. The interpretation of the units is simple. For example, after controlling for factor endowments, North American countries export $5,186 million more in crude oil and gas than do East Asian countries. On the other hand, after controlling for factor endowments East Asian countries export $3,776 million more in machinery than do North American countries. As with the transaction costs and monopolistic competition models, East Asia tends to have unusually large net exports of some high-end industries as well as some low-end industries, while North America tends to have unusually large net exports of natural resource industries. This is similar to the results from the gravity and monopolistic competition models. Thus it does not appear that the conclusion depends on the choice of models.

CONCLUSIONS

IN THIS PAPER, WE HAVE ARGUED THAT any important effects of East Asia trade on Canada should appear in two ways. The first is through factor-market effects. We find that the labour and capital content of Canada's net exports each shrank by close to 5 percent over the 1972-92 period. This is just large enough to be cause for concern about the effects on Canadian wages and rates of return to capital. In addition, when adjusted to account for the fact that low-end competition from East Asia has forced Canadian producers to adopt labour-saving, skill-intensive technology, it appears that the unskilled-labour content of Canada's net imports has risen sharply. *Ceteris paribus*, this should have contributed to falling wages for less educated workers.

TABLE 9

HECKSCHER-OHLIN MODEL, MANUFACTURING REGRESSION COEFFICIENTS, 1970-92

DEPENDENT VARIABLE: NET EXPORTS T_{igt}	CAPITAL	EDUCATION				LAND			SUBSOIL			R^2	F-TEST REGION
		NONE	PRIMARY	SECONDARY	COLLEGE	PASTURE	CROP	FORESTRY	COAL	OIL AND GAS	METALS		
Food	-0.47	-0.19	0.16	0.17	0.06	0.10	0.09	0.11	0.02	-0.12	0.01	0.36	120.0
Non-ferrous metals	-0.36	-0.13	-0.02	0.21	-0.08	0.10	0.04	0.07	-0.01	-0.01	0.43	0.28	7.5
Tobacco	-0.28	-0.22	0.11	-0.16	0.31	-0.11	0.20	0.08	0.18	-0.17	-0.06	0.14	11.5
Apparel	-0.27	-0.12	0.05	0.54	-0.25	0.05	-0.01	-0.05	-0.09	-0.08	0.00	0.29	106.6
Sawmills	-0.21	-0.02	0.12	0.11	-0.14	-0.25	0.06	0.20	-0.04	-0.03	0.23	0.31	82.1
Leather	-0.11	0.05	0.05	0.16	-0.12	0.03	0.21	-0.09	-0.19	-0.05	-0.07	0.14	23.7
Petroleum refineries	-0.10	-0.01	-0.07	0.21	-0.21	0.01	0.00	0.01	0.01	0.34	0.03	0.19	26.4
Footwear	-0.08	-0.03	0.20	0.35	-0.22	0.06	0.00	0.01	-0.23	-0.07	0.03	0.12	13.6
Furniture and fixtures	-0.02	-0.09	0.16	0.11	-0.08	-0.06	-0.13	-0.04	-0.03	-0.19	0.00	0.09	7.7
Liquors	0.03	0.06	0.21	0.02	-0.14	0.01	-0.01	-0.02	-0.20	-0.06	0.12	0.12	33.1
Pulp and paper	0.05	0.19	-0.01	-0.19	0.10	-0.19	-0.07	-0.08	-0.13	-0.02	0.24	0.22	54.6
Textile	0.08	0.14	0.05	0.08	0.04	0.02	0.00	0.01	-0.16	-0.26	-0.13	0.20	19.9
Clay and cement products	0.11	-0.06	0.28	0.06	-0.06	-0.05	-0.17	-0.02	-0.07	-0.28	0.04	0.20	35.9
Misc. manufacturing	0.12	0.05	0.12	0.26	-0.28	0.03	-0.15	-0.01	-0.11	-0.22	-0.01	0.16	22.6
Rubber	0.16	0.03	0.11	0.14	-0.31	-0.07	-0.05	0.03	-0.11	-0.14	-0.04	0.38	165.8
Glass products	0.22	-0.09	0.21	0.02	-0.08	-0.01	-0.09	-0.01	0.01	-0.19	-0.20	0.23	29.2
Petroleum and coal products	0.28	0.11	-0.20	-0.31	0.05	-0.01	0.03	-0.04	0.24	-0.05	-0.12	0.15	41.5
Pottery and china	0.29	0.11	0.34	0.18	-0.22	0.01	-0.36	-0.03	-0.18	-0.22	-0.05	0.27	13.0
Plastic products	0.34	0.09	0.02	-0.03	-0.10	0.03	-0.14	-0.02	-0.11	-0.14	-0.12	0.17	30.4
Iron and steel	0.35	0.00	0.17	-0.04	-0.05	0.07	-0.31	-0.03	-0.16	-0.18	0.07	0.18	30.7
Metal products	0.37	0.07	0.19	-0.08	-0.02	-0.04	-0.26	-0.02	-0.13	-0.37	0.02	0.26	7.3
Printing and publishing	0.39	0.03	0.00	-0.32	0.09	-0.07	-0.02	-0.04	0.16	-0.11	-0.30	0.32	56.1
Basic chemicals	**0.43**	0.09	-0.18	-0.37	0.16	-0.06	-0.07	0.02	0.10	-0.02	-0.11	0.19	28.7
Vehicles	**0.46**	0.11	0.12	-0.21	0.09	-0.04	-0.22	0.02	-0.14	-0.22	-0.06	0.23	23.3
Instruments	**0.54**	0.14	-0.02	-0.20	0.07	0.04	-0.20	-0.02	-0.09	-0.24	-0.17	0.24	29.8
Pharmaceuticals, etc.	**0.61**	0.26	-0.10	-0.45	0.11	-0.02	-0.16	-0.02	0.08	-0.25	-0.19	0.29	25.4
Electrical and electronics	**0.63**	0.17	0.02	-0.09	-0.02	0.05	-0.30	-0.03	-0.26	-0.32	-0.10	0.33	28.2
Machinery	**0.74**	0.21	-0.05	-0.47	0.17	-0.03	-0.20	-0.03	-0.04	-0.31	-0.22	0.36	29.7

Note: Each row represents a separate ordinary-least-squares (OLS) estimate of Equation (8). Only the factor endowment coefficients are reported. The intercept and regional dummies appear in the appendix. Each regression has 2,369 observations (103 countries x 23 years). The "F-test region" column is the F-test for the joint significance of the six regional dummies. The 1 percent critical value is 2.1. Bold face indicates factors that are a priori a source of comparative advantage for the industry.

TABLE 10

HECKSCHER-OHLIN MODEL, 1970-92, REGIONAL DUMMIES BY INDUSTRY

	REGIONAL DUMMIES δ_r		
	EAEC	NA	EAEC – NA
	High regional net exports		
North America			
Crude oil	–1,062	4,124	–5,186
Pulp and paper	–521	2,483	–3,004
Crops	–451	1,036	–1,487
Sawmills	48	1,407	–1,359
Food	–2,385	–1,244	–1,142
Non-ferrous metals	234	1,118	–884
Basic chemicals	–324	407	–730
Iron and steel	–464	–190	–274
Livestock	–455	–189	–266
Other mineral mining	5	264	–260
East Asia			
Machinery	187	–3,589	3,776
Electrical and electronics	630	–2,690	3,320
Apparel	1,082	351	731
Instruments	0	–720	719
Rubber	474	–18	492
Misc. manufacturing	310	–84	393
Plastic products	201	–181	382
Metal products	–84	–421	337
Printing and publishing	102	–224	325
Textile	–424	–706	282

Notes: Each row represents a separate ordinary-least-squares (OLS) estimate of Equation (8). Only the factor endowment coefficients are reported. The intercept and regional dummies appear in the appendix. Each regression has 2,369 observations (103 countries x 23 years). The "F-test region" column is the F-test for the joint significance of the six regional dummies. The 1 percent critical value is 2.1. Italic face indicates factors that are a priori a source of comparative advantage for the industry..

The second effect is that East Asia trade may have altered Canada's industrial composition. We find that Canada has inexplicably large levels of trade in natural-resource-based products. This is inexplicable in the sense that we have controlled for Canada's sources of comparative advantage in those products: resource endowments, scale returns and transaction costs. On the other hand, East Asia has inexplicably large levels of trade in a few low-end manufactures as well as a few high-tech industries. This probably reflects the division of the region between the miraculous Tigers and labour-abundant countries such as China. Again, the results are inexplicable in the sense that we have controlled for the sources of comparative advantage associated with resource endowments, scale returns and the transaction costs of doing business outside a country's region. The implication is that East Asian industrial targeting has indeed distorted trade patterns.

These results make us wonder whether, like several East Asian economies, Canada does indeed have an industrial policy, albeit one that is buried in its tax structures and elsewhere. Our results suggest that such an implicit policy may exist and may have directed Canada towards inappropriate specialization in natural-resource-based industries. What may be called for is a new pattern of industrial incentives that would encourage Canadian entrepreneurs to migrate out of natural resources and into growth-oriented, high-wage technology industries. Both the factor-market and industrial-composition effects of East Asia trade with Canada are developments that will need to be examined in further detail. Industry Canada has wisely anticipated our conclusion by sponsoring this conference.

DATA APPENDIX

TRADE DATA ARE FROM THE STATISTICS CANADA WORLD TRADE DATABASE. Endowment data are from Barro and Lee (1993), the Food and Agriculture Organization of the United Nations and World Resources Institute (1994). Production data are from the UN's general industrial statistics database as reported on diskette and a variety of UN Yearbooks. GDP and PPP data are from Summers and Heston (1991). Exchange rates are from the IMF's *International Financial Statistics*. Distance, "neighbours," and language were kindly supplied by Werner Antweiler. The input-output matrix used for factor content calculations was constructed from the Bureau of Economic Analysis' input-output tables and the Bureau of Commerce's *Current Population Survey*, various years.

TABLE A-1
LIST OF COUNTRIES

THE AMERICAS	OCEANIA	EUROPE	ASIA	AFRICA
Developed economies	*Developed economies*	*Developed economies*	*Developed economies*	*Developed economies*
Canada	Australia	Austria	Israel	South Africa
United States of America	New Zealand	Belgium	Japan	
		Denmark		*Developing economies*
Developing economies	*Developing economies*	West Germany	*Other developing*	Algeria
Argentina	Fiji	Greece	Bangladesh	Burkina Faso
Bahamas	Papua, New Guinea	Finland	China	Burundi
Barbados		France	Hong Kong	Cameroon
Bolivia	**Middle East**	Iceland	India	Central African Republic
Brazil	Cyprus	Ireland	Indonesia	Egypt
Chile	Iran	Italy	Korea, Rep.	Ethiopia
Colombia	Iraq	Malta	Malaysia	Ghana
Costa Rica	Jordan	Netherlands	Myanmar	Ivory Coast
Dominican Republic	Kuwait	Norway	Nepal	Kenya
Ecuador	Qatar	Portugal	Pakistan	Madagascar
El Salvador	Saudi Arabia	Spain	Philippines	Malawi
Guatemala	Syria	Sweden	Singapore	Mauritius
Honduras	Turkey	Switzerland	Sri Lanka	Morocco
Jamaica	United Arab Emirates	United Kingdom	Taiwan	Mozambique
Mexico			Thailand	Niger
Panama		*Developing economies*		Nigeria
Paraguay		Bulgaria		Senegal
Peru		Czechoslovakia		Tanzania
Suriname		East Germany		Tunisia
Trinidad and Tobago		Hungary		Zambia
Uruguay		Poland		Zimbabwe
Venezuela		Romania		
		Former USSR		

TABLE A-2

REGIONAL GROUPINGS

Western Hemisphere (13)	European Community (11) EC	European Free Trade Area (6) EFTA
Canada	West Germany	Austria
United States	France	Finland
Argentina	Italy	Norway
Brazil	United Kingdom	Sweden
Chile	Belgium	Switzerland
Columbia	Denmark	Iceland
Ecuador	Netherlands	
Mexico	Greece	
Peru	Ireland	
Venezuela	Portugal	
Bolivia	Spain	
Paraguay		
Uruguay		
East Asia Economic Caucus (10) EAEC	**North American Free Trade Area (3) NAFTA**	**Asia Pacific Economic Cooperation (14) APEC**
Japan	Canada	EAEC
Indonesia	Mexico	Australia
Taiwan	United States	Canada
Hong Kong		New Zealand
South Korea		United States
Malaysia		
Philippines		
Singapore		
Thailand		
China		

Table A-3

List of Industries

NON-MANUFACTURING		MANUFACTURING	
ISIC	Industry	ISIC	Industry
.	Livestock	311	Food
.	Other agriculture	313	Liquors
.	Forestry	314	Tobacco
.	Fishing	321	Textile
210	Coal	322	Apparel
220	Crude oil	323	Leather
230	Metal mining	324	Footwear
290	Non-metal mining	331	Sawmills
411	Electricity	332	Furniture and fixtures
		341	Pulp and paper
		342	Printing and publishing
		351	Basic chemicals
		352	Pharmaceuticals, etc.
		353	Petroleum refineries
		354	Petroleum and coal products
		355	Rubber
		356	Plastic products
		361	Pottery and china
		362	Glass products
		369	Clay and cement products
		371	Iron and steel
		372	Non-ferrous metals
		381	Metal products
		382	Machinery
		383	Electrical and electronics
		384	Vehicles
		385	Instruments
		390	Misc. manufacturing

Note: ISIC = International Standard Industry Classification.

TABLE A-4

GRAVITY MODEL, 1970-92, MAIN REGRESSORS

	INTERCEPT	GDP	GDP PER CAPITA	PPP	DISTANCE	NEIGHBOUR	LANGUAGE
All industries	−17.69	0.74	0.49	−0.23	−1.01	2.04	1.98
Livestock	−9.98	0.46	0.29	0.07	−0.73	2.01	0.99
Other agriculture	−9.44	0.57	0.06	0.79	−0.77	2.03	1.31
Forestry	−7.33	0.40	−0.01	0.41	−0.44	2.56	1.24
Fishing	−12.36	0.24	0.61	1.59	−0.60	3.52	2.20
Coal	−10.88	0.63	−0.08	1.38	−0.49	1.45	0.40
Crude oil	−14.86	0.44	0.70	0.73	−0.73	1.64	1.13
Metal mining	−10.69	0.46	0.04	1.02	−0.22	2.25	0.82
Non-metal mining	−8.43	0.46	0.15	0.30	−0.76	1.91	1.17
Food	−9.13	0.48	0.21	0.52	−0.63	2.16	1.89
Liquors	−16.03	0.33	0.70	0.18	−0.54	2.78	3.05
Tobacco	−2.87	0.35	−0.09	0.70	−0.37	1.92	1.62
Textile	−12.30	0.57	0.29	0.25	−0.88	1.75	1.77
Apparel	−11.15	0.40	0.52	1.04	−0.96	2.21	2.21
Leather	−15.23	0.53	0.35	0.48	−0.70	1.91	2.40
Footwear	−9.73	0.30	0.42	0.84	−0.53	3.58	1.75
Sawmills	−6.92	0.39	0.31	0.63	−1.08	2.58	1.77
Furniture and fixtures	−21.67	0.54	0.78	−0.06	−1.07	2.98	2.37
Pulp and paper	−10.07	0.50	0.27	−1.00	−0.97	2.15	1.88
Printing and publishing	−18.74	0.55	0.52	−0.68	−0.90	2.30	9.34
Basic chemicals	−19.53	0.72	0.50	−1.02	−1.04	1.67	2.05
Pharmaceuticals, etc.	−16.31	0.57	0.44	−0.84	−0.76	2.40	3.22
Petroleum refineries	−9.85	0.41	0.55	0.49	−1.21	3.20	1.63
Petroleum and coal products	−4.89	0.39	0.26	−0.45	−1.28	2.81	1.98
Rubber	−12.22	0.49	0.31	−0.15	−0.83	2.74	1.68
Plastic products	−18.59	0.53	0.69	−0.62	−0.99	2.27	2.37
Pottery and china	−12.63	0.38	0.45	0.16	−0.69	2.59	2.03
Glass products	−16.52	0.56	0.51	−0.40	−0.98	2.57	2.14
Clay and cement products	−16.79	0.62	0.36	−0.17	−0.89	2.75	1.81
Iron and steel	−21.61	0.84	0.37	−1.08	−1.17	1.89	1.62
Non-ferrous metals	−19.63	0.67	0.42	−0.42	−0.91	2.61	2.71
Metal products	−17.61	0.61	0.54	−0.79	−0.99	2.39	2.86
Machinery	−21.57	0.69	0.59	−1.40	−0.90	2.35	2.43
Electrical and electronics	−22.37	0.62	0.79	−1.21	−1.01	2.49	2.37
Vehicles	−21.28	0.69	0.48	−1.06	−0.75	3.47	2.18
Instruments	−25.52	0.61	0.75	−0.91	−0.59	2.32	2.46
Misc. manufacturing	−19.58	0.50	0.67	−0.06	−0.62	2.72	2.53
Electricity	−8.55	0.29	0.40	0.20	−0.57	10.18	2.01

Note: Each row is an estimate of Equation (4). The transaction-cost coefficient estimates appear in this table. The regional dummy coefficient estimates appear in Table A-5.

Table A-5
Gravity Model, 1970-92, Regional Dummies

	EEC		EFTA		EAEC		WH		APEC		NA	
	Within	Between	Within	Between	Within	Between	Within	Between	Within	Between	Within	Between
All industries	4.15	3.27	5.03	1.78	1.07	1.13	1.65	0.91	14.85	2.30	0.45	0.77
Livestock	1.47	0.74	0.30	0.43	0.03	0.18	0.81	1.28	20.04	3.49	0.93	0.19
Other agriculture	6.93	1.55	0.57	0.58	0.24	0.29	1.61	1.39	26.43	3.13	1.76	0.82
Forestry	5.57	1.89	6.06	1.61	1.85	0.60	0.34	0.67	6.08	1.98	10.15	1.51
Fishing	13.51	1.68	1.81	0.56	8.24	3.09	0.61	0.90	6.95	1.65	24.02	2.08
Coal	0.30	0.41	0.03	0.48	0.00	0.06	0.41	0.90	40.18	7.07	0.16	0.28
Crude oil	0.11	0.14	0.01	0.08	3.61	1.38	0.35	0.67	0.13	0.17	80.26	3.37
Metal mining	6.28	1.10	4.10	2.32	0.39	0.53	3.30	4.37	10.16	2.29	1.74	0.28
Non-metal mining	3.64	1.40	1.48	0.86	0.83	0.71	0.73	0.76	3.35	1.39	16.91	2.22
Food	13.97	3.34	5.65	1.03	0.21	0.34	1.39	1.05	64.32	3.85	1.01	0.62
Liquors	27.45	3.85	1.30	0.77	1.53	0.66	0.69	0.80	4.71	1.15	12.26	1.53
Tobacco	3.70	1.13	2.83	0.75	1.37	0.79	0.83	1.44	1.30	1.28	1.85	2.00
Textile	6.99	1.92	3.71	0.91	0.66	1.11	0.80	0.87	18.93	3.13	0.50	0.53
Apparel	8.30	2.14	6.22	1.34	2.74	3.80	0.44	0.65	2.32	1.30	5.91	2.68
Leather	10.09	2.61	8.21	1.06	1.82	1.87	1.20	1.41	4.89	1.42	1.45	0.52
Footwear	4.20	1.51	4.54	0.99	1.93	2.26	0.63	0.71	1.40	1.20	2.88	1.43
Sawmills	7.75	3.13	16.26	1.94	0.46	0.99	0.82	0.81	33.43	3.35	3.17	1.25
Furniture and fixtures	5.65	2.25	10.21	1.43	3.48	3.26	0.33	0.61	3.58	1.26	5.34	1.74
Pulp and paper	9.38	1.92	23.04	4.83	0.47	0.91	2.79	1.72	19.59	2.02	1.46	0.68
Printing and publishing	13.15	3.40	11.51	1.62	1.48	1.21	2.14	1.15	5.49	1.72	1.31	1.00
Basic chemicals	4.91	2.09	3.40	1.29	3.58	2.26	1.48	0.93	2.39	0.80	1.36	1.18
Pharmaceuticals, etc.	9.92	3.10	6.78	1.72	3.96	1.55	1.19	0.87	2.89	1.00	1.40	1.23
Petroleum refineries	3.94	1.28	0.18	0.15	3.60	0.97	0.45	0.52	1.88	0.91	8.12	1.76
Petroleum and coal products	5.92	2.74	1.43	0.43	0.65	0.43	0.46	0.81	17.99	3.43	1.75	1.06

Table A-5 (cont'd)

	EEC		EFTA		EAEC		WH		APEC		NA	
	Within	Between	Within	Between	Within	Between	Within	Between	Within	Between	Within	Between
Rubber	11.75	2.67	8.67	0.93	6.67	3.59	1.06	0.94	3.35	1.45	3.13	1.13
Plastic products	7.81	2.32	7.70	1.29	3.64	3.65	1.21	0.83	4.22	1.26	1.60	1.18
Pottery and china	6.62	2.11	4.39	1.00	2.66	2.41	0.87	0.90	2.85	1.45	2.64	1.04
Glass products	6.01	2.04	4.22	0.87	4.89	2.54	0.82	0.81	1.95	0.83	6.07	1.29
Clay and cement products	4.93	1.93	3.34	1.31	2.39	1.22	0.69	0.66	4.09	1.26	2.61	1.11
Iron and steel	2.90	1.34	6.18	0.98	1.99	1.58	1.28	1.39	2.31	0.90	1.10	0.44
Non-ferrous metals	9.12	2.84	21.17	2.45	1.70	1.58	2.91	1.11	13.33	1.57	1.08	0.78
Metal products	5.73	2.29	9.85	1.48	2.36	2.48	0.95	0.79	4.91	1.40	1.75	0.96
Machinery	7.86	2.69	10.62	2.17	2.39	1.77	0.97	1.01	4.59	1.50	3.04	1.13
Electrical and electronics	7.51	2.45	10.29	1.63	16.34	6.46	0.71	0.80	2.35	1.05	5.47	1.55
Vehicles	9.39	2.67	9.39	1.32	1.22	1.30	0.94	0.93	4.60	1.41	7.30	1.45
Instruments	8.16	2.54	7.36	1.86	5.95	2.78	0.74	0.73	2.31	1.21	2.84	1.47
Misc. manufacturing	4.80	1.88	5.31	1.30	7.55	4.35	0.40	0.55	1.78	1.03	4.65	1.63
Electricity	0.31	0.29	1.70	1.26	8.24	3.13	0.19	1.15	0.05	0.22	42.42	3.97

Table A-6
Monopolistic Competition, 1970-92, All Coefficients

	Output	Intercept	EAEC Within	EAEC Between	NA Within	NA Between	EEC Within	EEC Between	EFTA Within	EFTA Between	WH Within	WH Between	APEC Within	APEC Between
Food	0.46	3.44	−1.09	−1.15	2.10	0.26	4.86	1.62	2.84	0.15	0.80	−0.14	4.67	1.32
Liquors	0.21	−1.96	−0.48	−1.08	4.41	1.03	6.26	2.24	2.76	0.84	0.78	−0.23	3.47	0.75
Tobacco	0.27	1.12	0.71	−0.40	1.55	1.10	2.59	0.38	1.59	−0.11	0.38	0.37	0.86	0.37
Textile	0.41	2.68	−0.08	−0.30	2.13	0.62	5.05	1.56	3.21	0.44	0.51	−0.46	3.75	1.16
Apparel	0.47	2.61	0.78	0.68	3.93	1.94	4.68	1.35	4.08	1.00	−0.32	−1.08	1.41	0.07
Leather	0.45	2.75	0.44	0.14	3.06	0.80	4.36	1.46	3.53	0.68	0.29	−0.39	2.44	0.38
Footwear	0.30	0.44	0.02	0.13	2.58	0.61	3.67	1.03	3.73	0.90	0.39	−0.45	1.61	0.56
Sawmills	0.49	2.17	0.14	−0.35	3.18	0.87	4.40	1.31	4.98	0.98	0.59	−0.68	3.18	0.69
Furniture and fixtures	0.28	−2.00	1.34	0.65	4.79	1.66	6.26	2.10	5.91	1.41	0.47	−0.59	2.84	0.52
Pulp and paper	0.45	2.54	0.06	−0.33	2.62	0.68	4.43	1.14	4.43	1.90	1.55	0.04	2.80	0.30
Printing and publishing	0.37	−0.07	−0.18	−0.52	2.53	1.09	5.39	1.93	3.78	0.36	1.81	−0.29	2.93	0.59
Basic chemicals	0.60	5.00	1.36	0.26	2.89	1.47	5.06	1.75	3.88	0.89	1.41	−0.46	2.22	0.08
Pharmaceuticals, etc.	0.27	0.66	1.16	−0.08	3.08	1.28	5.70	2.26	4.09	1.03	1.20	−0.23	2.66	0.45
Petroleum refineries	0.59	3.58	0.37	−0.84	4.26	1.41	4.48	0.66	1.58	−0.75	−0.64	−1.65	1.90	−0.18
Petroleum and coal products	0.10	−2.55	−0.11	−1.32	3.54	1.41	5.41	1.95	2.75	−0.30	−0.06	−1.00	3.49	0.86
Rubber	0.35	1.03	2.00	0.80	3.48	0.96	5.43	1.83	4.31	0.51	0.90	−0.45	1.92	0.38
Plastic products	0.41	1.37	0.70	0.35	3.03	1.29	5.75	2.00	5.22	1.20	1.29	−0.56	3.03	0.53
Pottery and china	0.26	−0.57	0.79	0.42	2.95	0.89	4.28	1.34	3.53	0.65	0.77	−0.29	2.10	0.57
Glass products	0.37	1.24	2.07	0.66	4.77	1.46	5.34	1.83	4.30	0.77	0.95	−0.43	1.53	−0.21

TABLE A-6 (CONT'D)

	Output	Intercept	EAEC Within	EAEC Between	NA Within	NA Between	EEC Within	EEC Between	EFTA Within	EFTA Between	WH Within	WH Between	APEC Within	APEC Between
Clay and cement products	0.38	1.17	1.19	−0.13	3.81	1.20	4.90	1.68	3.08	0.77	0.77	−0.56	2.39	0.40
Iron and steel	0.68	5.44	1.08	−0.04	3.10	0.52	4.50	1.36	4.02	0.76	1.19	0.03	1.47	−0.19
Non-ferrous metals	0.71	5.63	1.09	0.11	3.08	1.36	4.39	1.42	3.92	0.64	1.09	−0.76	2.57	0.08
Metal products	0.32	0.70	1.35	0.50	3.72	1.21	5.87	2.17	5.22	1.15	1.34	−0.38	3.09	0.63
Machinery	0.57	4.53	0.79	−0.12	3.56	1.40	5.25	2.02	4.81	1.30	1.30	−0.28	2.85	0.69
Electrical and electronics	0.57	3.36	1.73	0.64	4.21	1.58	6.01	2.33	5.69	1.46	1.01	−0.38	2.79	0.61
Vehicles	0.49	3.22	0.22	−0.09	4.66	1.42	5.55	1.99	4.68	1.07	1.05	−0.21	2.51	0.60
Instruments	0.64	5.31	0.37	−0.11	2.27	1.07	5.01	2.06	4.15	1.06	0.84	−0.39	2.67	0.88
Misc. manufacturing	0.34	1.11	1.08	0.57	3.45	1.32	5.17	1.89	4.60	1.12	0.48	−0.57	2.56	0.72

Note: Each row is an estimate of Equation (6).

TABLE A-7

HECKSCHER-OHLIN MODEL, 1970-92, REGIONAL DUMMIES

	EAEC	NA	EEC	EFTA	WH	APEC
			Non-Manufacturing			
Livestock	−455	−189	84	0	4	404
Crops	−451	1,036	−191	−31	30	205
Forestry	39	−71	23	13	−2	92
Fishing	−18	141	34	66	10	54
Coal	−116	−161	−123	−49	−28	132
Crude oil	−1,062	4,124	−800	182	101	194
Metal mining	23	98	14	−18	17	28
Other mining	5	264	5	−4	−1	−24
Electricity	−5	151	−10	29	0	11
			Manufacturing			
Liquors	−34	83	109	−21	−12	−9
Tobacco	−30	71	−16	−11	−9	−24
Textile	−424	−706	15	−50	69	222
Apparel	1,082	351	72	−86	7	−3
Leather	−51	−206	−33	−6	45	85
Footwear	99	−7	6	−13	−0	−1
Sawmills	48	1,407	−93	70	19	236
Furniture and fixtures	20	−17	27	−24	1	13
Pulp and paper	−521	2,483	−254	292	−12	368
Printing and publishing	102	−224	−6	−18	−2	−91
Basic chemicals	−324	407	−131	−22	−94	−288
Pharmaceuticals, etc.	−166	−332	14	−15	−13	−100
Petroleum refineries	1,110	869	15	−165	78	−139
Petroleum and coal products	116	−33	−37	−30	1	−29
Rubber	474	−18	−36	−32	−4	5
Plastic products	201	−181	−35	−46	−7	−74
Pottery and china	0	−46	−8	−8	−0	−9
Glass products	−12	−195	44	−18	−5	−60
Clay and cement products	−90	16	32	−9	−4	−16
Iron and steel	−464	−190	48	−14	−87	−227
Non-ferrous metals	234	1,118	−5	65	82	−203
Metal products	−84	−421	−72	−82	−8	−52
Machinery	187	−3,589	−863	−217	−138	−904
Electrical and electronics	630	−2,690	−647	−211	−124	−533
Vehicles	−550	−300	−1,185	−234	−103	−659
Instruments	−0	−720	−120	−11	−28	−210
Misc. manufacturing	310	−84	−41	−44	−6	−58

Note: Each row is an estimate of Equation (8). The remaining estimated coeffcients for the equation appear in text Tables 8 and 9.

ENDNOTES

1 We are indebted to Werner Antweiler, Jr., whose name really should appear alongside ours. We are grateful to Richard Harris for inviting us to prepare this paper, to Gilles Mcdougall for the thankless job of organizing the conference, and to Industry Canada for generous financial support.
2 A very different interpretation is that, for example, the East Asia the electrical and electronics trade is concentrated on the labour-intensive assembly portion of the industry (see Figure 6) and so is not high on the high-tech ladder.
3 The following discussion draws heavily on Harrigan (1993).

BIBLIOGRAPHY

Barro, Robert J. and Jong-Wha Lee. "International Comparisons of Educational Attainment." NBER Working Paper 4349. Cambridge, Mass.: National Bureau of Economic Research, 1993.
Ethier, Wilfred J. "National and International Returns to Scale in the Modern Theory of International Trade." *American Economic Review*, 72 (1982):389-405.
Feenstra, Robert C. and Gordon H. Hanson. "Foreign Investment, Outsourcing, and Relative Wages." NBER Working Paper 5121. Cambridge, Mass.: National Bureau of Economic Research, 1995.
Frankel, Jeffrey A., Shang-Jin Wei and Ernesto Stein. "APEC and Regional Trading Arrangements in the Pacific," in *Pacific Trade and Investment: Options for the 90s*. Edited by Wendy Dobson and Frank Flatters. Kingston, Ont.: John Deutsch Institute, 1995.
Harrigan, James. "OECD Imports and Trade Barriers in 1983." *Journal of International Economics*, 35 (1993):91-111.
Krugman, Paul and Robert Z. Lawrence. "Trade, Jobs, and Wages." *Scientific American*, 270 (1994):44-9.
Kuhn, Peter. "Labour Market Polarization: Canada in International Perspective." Paper presented at the Bell Papers Conference, John Deutsch Institute for the Study of Economic Policy, Queen's University, Kingston, Ont., 1995.
Leamer, Edward E. *Sources of International Comparative Advantage: Theory of Evidence*. Cambridge, Mass.: MIT Press, 1984.
Lorenz, Detlef. "Economic Geography and the Political Economy of Regionalization: The Example of Western Europe." *American Economic Review, Papers and Proceedings*, 82 (1992):84-7.
Summers, Robert and Alan Heston. "The Penn World Table (Mark 5): An Expanded Set of International Comparisons, 1950-1988." *Quarterly Journal of Economics*, 106 (1991):327-68.
Trefler, Daniel. "The Case of the Missing Trade and Other Mysteries." *American Economic Review*, 85 (1995): 1029-46.
Wood, Adrian. *North-South Trade, Employment and Inequality: Changing Fortunes in a Skill-Driven World*. Oxford: Clarendon Press, 1994.
World Resources Institute. *World Resources 1994-1995*. New York: Oxford University Press, 1994.

Comment

Marcus Noland
Institute for International Economics
Washington, D.C.

THIS WELL-WRITTEN PAPER contains a huge number of results generated from a variety of statistical models. For understandable reasons, the authors focus on the results rather than on the details of how the results were obtained. My comments will largely take the form of queries on this latter issue.

Before getting into this, however, it is worth commenting on the trade and wages motivation the authors use to justify the paper. I was surprised that they do not report any evidence on traded-goods *prices*: after all, price changes are the mechanism by which international trade alters factor remuneration (the Stolper-Samuelson effect) in conventional models of international trade. This is doubly surprising since this very point has proved quite controversial in research conducted on the U.S. economy. I would also question whether *bilateral* trade is really salient: in this context does anyone really care from which country the imports are coming from? Finally, I would simply note that in the "high/low" calculation, the median Canadian worker is a net exporter of labour services.

My questions about methodology fall into three groups: data, specification, and estimation. With respect to the data, the authors do not indicate whether the income data they use are in purchasing-power-adjusted terms or not (the World Bank data referred to in the introduction are not). This distinction is important because for the pure differentiated-products model, the pattern of trade is a function of world income shares. If the data are not adjusted for purchasing power, the results for areas with very high incomes (such as North America), as well as those with very low incomes, will be biased.

Also, the authors pool their data, which requires some adjustment for price changes over time. For example, the physical version of the Heckscher-Ohlin-Vanek model, which the authors employ, does not make sense if there have been changes in traded-goods prices over time.

Regarding specification of the regressions, the authors do not indicate how trade imbalances have been handled. Again, an adjustment needs to be made if aggregate trade is not balanced. This could be particularly important in the comparisons between North America, which has run large trade deficits, and East Asia, which has run surpluses.

In the gravity model, the regression specification should be linked more tightly to the underlying theory. The product of trade-partner countries incomes is not a measure of their similarity. Likewise, physical distance may be a poor proxy for transactions costs. And finally, it is unclear whether the dummy variables are supposed to be capturing policy or pure geographical effects (the latter could be important if, as is likely, distance is a poor proxy for transactions costs). If they capture policy, then the changing membership of the EU and its predecessor organization should be taken into account.

Applications of the gravity model on more limited data sets suggest that it is advisable to consider trade and investment as a simultaneous system. Unfortunately, the joint estimation of trade and investment is not feasible for the authors' large data set. Nevertheless, the parameters estimated by the authors are surely biased, and this bias could be significant for major investors such as the North American economies.

On the Heckscher-Ohlin-Vanek model, it would be desirable to consider possible measurement error or international factor productivity differences, either of which could change the results of the authors' analysis of residuals.

With regard to estimation, the data are pooled and the ordinary-least-squares (OLS) specification is biased. A variety of corrections are possible. These fixes will affect the residuals, however, and along with them the analysis of regional dummies.

With these caveats in mind, it is nonetheless striking that the authors obtain essentially consistent results regardless of the model approach they use – gravity, Heckscher-Ohlin-Vanek, or differentiated-product.

Finally, the authors attempt to use their results to comment on industrial targeting, though targeting per se is not discussed in any detail. In this regard, the authors appear to be working from some unexamined notions of the extent and character of industrial interventions around the world. For example, the U.S. government spends more on R&D than all Asian governments combined. The government of Japan has spent more subsidizing rice farming than semiconductor development. According to the World Bank, the East Asian developing countries as a whole had *less* price distortion than other developing countries. These facts do not seem to accord with the image of predatory Asians picking off Canadian industries.

Moreover, formal modeling of Asian (or at least Japanese) industrial policies appears to indicate that these interventions were probably not welfare improving (Noland, 1993). In fact, the authors' results can be stood on their head. Suppose countries around the world engage in a variety of interventions, subsidizing "emerging" industries and protecting the "senescent" from foreign competition. This means that governments would be shifting resources out of contemporaneously competitive sectors and into those which might be competitive in the future and those which may have been competitive in the past. In trade terms, one could think of this as resources being pulled out of net export sectors and into net import sectors.

Suppose a regression is run on cross-national data. The estimated regression will reflect this pattern of intervention around the world: countries will exhibit both lower exports and lower imports relative to what they would display if there had been no intervention.

Now add a non-interventionist country to the sample. This country will exhibit both greater exports and greater imports than predicted by the model, since the model is conditioned on the interventionist country norm. Hence the finding that some country or countries are outliers in cross-national regressions is not an indication of their trade interventions. Just the opposite: the outliers may be the free-traders! We cannot know for sure without more detailed investigation.

I would conclude by noting that in their discussion of the targeting issue, the authors continue the disturbing Canadian tendency to equate "North America" with "Canada." They are not the same, even if Canadian nationalists wish they were.

REFERENCE

Noland, Marcus. "The Impact of Industrial Policy on Japan's Trade Specialization." *Review of Economics and Statistics*, 75, 2 (1993):241-48.

Keith Head & John Ries
Faculty of Commerce and Business Administration
University of British Columbia

4

Rivalry for Japanese Investment in North America[1]

IN 1986, THE STATE OF KENTUCKY attracted a Toyota assembly plant by offering an incentive package valued at over US$300 million.[2] To the north, the Suzuki-GM plant in Ingersoll, Ontario received combined federal, provincial, and municipal assistance totalling C$100 million in that year. Also in this period, bidding wars among states and provinces resulted in large subsidies for Japanese automobile assembly plants across North America. These bidding wars underscored the active rivalry for manufacturing investment. In addition to bidding for specific large investments, states and provinces adopted general measures to encourage manufacturing to locate within their jurisdictions. Ultimately, in 1994 Canadian provinces agreed in principle to stop competing with each other for investments. Since U.S. states were not party to the agreement, Canadian provinces are still vulnerable to poaching for investment from south of the border.

The large subsidies to foreign manufacturing investments indicate that governments believe they bring significant economic benefits. Beyond the direct employment and output foreign investment creates, it may stimulate economic activity in supporting and related industries. Moreover, there is evidence that these firms pay higher wages and operate more efficiently than domestic establishments. Other benefits may come through technological spillovers to other manufacturers that lead to higher productivity throughout the economy.

This paper examines competition among North American states and provinces for Japanese investment. We evaluate the economic impact of foreign investment to identify the stakes of the competition, then assess the attractiveness of individual provinces and states for receiving new Japanese greenfield investment. We focus on whether government promotion increases the likelihood of receiving investment. The analysis will contribute to our understanding of the role played by government promotion of foreign investment and shed light on the consequences of continued promotion.

Our study introduces an important new dimension to the discussion of how foreign investment affects the welfare of recipient and non-recipient countries. When the promotional policies of national governments attract investors, they may do so at the expense of welfare in other countries. State or provincial governments may also lure investment from other locations within their borders. For example, if Nova Scotia secures investment that otherwise would have gone to Ontario, one Canadian province may gain at the expense of another. Provincial governments understand these beggar-thy-neighbour consequences of competition between provinces. However, even if the 1994 agreement to refrain from such activity is adhered to by the premiers, the rivalry with U.S. states for inward investment will continue until a continent-wide agreement is reached.

Foreign manufacturers have a large presence in Canada and the United States. The United States is the largest recipient of foreign investment, while foreign-controlled firms have historically produced roughly half of Canada's manufacturing output. In 1994, the stock of foreign direct investment in the United States and Canada stood at US$504.4 and C$148.0 billion, respectively.

Japanese direct investment came to the fore in the 1980s when that country became the leading source country of foreign investment. According to Japan's Ministry of Finance, Japanese direct investment abroad peaked in 1989 at US$67.5 billion, then levelled off to average US$41.9 billion annually in 1990-94.[3] (For a description of the major data sources used in this study, see the Appendix 1.)The bulk of Japanese direct investment, 43 percent in 1994, goes to North America. While Asia's share is increasing, in 1994 almost twice as much went to North America as Asia (US$17.8 versus US$9.7 billion). Japan is the principal Pacific Rim investor in North America. According to U.S. and Canadian statistics, Japanese investment comprises 88 percent of Pacific Rim investment in the United States (U.S. Department of Commerce, 1995) whereas the figure is 56 percent in Canada (Statistics Canada, 1994). Japanese direct investment is 20 percent of the United States' total foreign direct investment but only 4 percent of Canada's.

The United States has received disproportionately more Japanese investment than is warranted by its (10 to 1) economic size advantage. Host-country sources show that in 1994 the stock of Japanese direct investment in the United States was US$103.1 billion compared with C$5.8 billion in Canada. In that year the flow of Japanese investment into the United States outstripped that in Canada by US$17.8 billion to US$492 million. As these data include both non-manufacturing investment and acquisitions, they may mask crucial differences between the composition of investment in the two countries.[4]

Our statistical analysis focuses on Japanese greenfield establishments in North America. In this category of direct investment the Canadian performance is somewhat better, with 51 established in Canada in the 1980s versus 751 in the United States (a ratio of 15 to 1). We focus on greenfields for two reasons. First, we believe that local governments are most interested in new plants, and second, greenfields are more likely than acquisitions to be influenced by government promotion and other locational attributes.

The next section of this paper identifies the stakes of the competition by examining how foreign investment may enhance host-country welfare. We focus on wages and productivity and review recent empirical evidence that foreign-controlled establishments tend to be more productive and pay higher wages than domestically controlled firms. The third section discusses locational attributes that favour investment and compares Canadian provinces and U.S. states vis-à-vis these attributes. Data on actual Japanese investments enables us to assess the relative importance of location characteristics and the relative attractiveness of states and provinces for new investment. In the fourth section, we use our estimates to simulate the effects policy changes may have on the probability of particular locations receiving new investment. In the final section we summarize the policy implications of the study.

The Stakes of the Rivalry

GOVERNMENTS EXPEND CONSIDERABLE RESOURCES courting foreign manufacturing greenfields because of their perceived economic benefit. Their potential benefits, such as job creation, productivity increases, and high wages, as well as their potential costs, have been widely discussed in Canada, where a high degree of foreign control of manufacturing assets has led to close scrutiny of the economic and social consequences. In the United States, foreign control is less extensive. There the debate about the welfare effects of foreign investment has only recently emerged as a consequence of the rapid increase in foreign – principally Japanese – investment in the 1980s. In this section we focus on productivity and wages, two important determinants of the real income of the population of the host country. The social, political and economic impacts of foreign investment are dealt with in several other studies.[5]

It is important to note that rivalry for foreign investment is only part of a more general competition for manufacturing investment, regardless of the nationality of the investor. Indeed, many of the inducements used to attract foreign investors, such as low-interest loans, subsidized training and site preparation are also available to domestic investors. Some government programs, such as foreign trade zones (FTZ), are not valued highly by domestically owned plants, but we are not aware of any North American policy that restricts the benefits of such programs to foreign investors.[6] There are two principal reasons why we focus on foreign investment. First, foreign manufacturers may bring new management techniques that yield more efficient plants. In particular, the labour and production management techniques of Japanese companies are considered to be highly efficient. Second, foreign manufacturing investment is more likely to be "footloose." In other words, a foreign investor entering North America for the first time will not be as tied to a specific location as a domestic investor, who may be expanding an existing facility or prefer to locate new facilities near headquarters.

There are a number of ways in which foreign investment may increase real income in the host economy. It may increase the amount of labour and capital equipment used to produce goods, which directly increases national output. The extent to which direct investment can increase capital and labour inputs depends on slack in the macroeconomy. In an economy with full employment, foreign investment would most likely bid resources away from other sectors, leaving overall output unaffected. In an economy with slack, foreign direct investment (FDI) is likely to directly increase output as well as generate indirect gains by stimulating related and supporting industries. Van Loo (1977) estimates that a dollar of foreign investment in Canada yields $1.40 in total new investment, indicating that foreign capital complements domestic capital. This complementarity is corroborated by Borensztein, De Gregorio, and Lee (1995) who studied foreign investment in 69 developing countries and found that it has the effect of increasing investment more than one-to-one. Thus evidence supports the contention that, at least in some countries, direct investment does increase total capital inputs, thereby directly increasing economic output.

Direct investment increases real incomes by enhancing productivity. Even if capital and labour inputs are, on net, unaffected by injections of foreign capital, productivity gains would lead to additional national output, primarily through more consumer goods at lower prices and, possibly, higher wages for workers.

Two studies use establishment-level manufacturing data to assess the relationship among foreign ownership, value added per employee, and wages. Globerman, Ries and Vertinsky (1994) examine 21 four-digit SIC-level manufacturing industries where Japanese firms had investments in 1986. After controlling for industry, they found that foreign affiliates had higher value added per employee. There were no significant differences between Japanese, U.S., and European-owned establishments. The foreign productivity premium disappeared once controls for capital intensity and size were included. Wages per employee exhibited similar patterns. Doms and Jensen (1995) conduct a similar analysis using 1987 U.S. establishment data (with a much larger sample) and obtain somewhat stronger results. They find that employees of foreign-owned firms generate about 50 percent more value added and obtain 19 percent higher wages. They show that higher productivity and wages persist in foreign-owned plants, even after they are controlled for industry, size, capital intensity, plant age, and location (state), albeit to a lesser extent (12 percent for productivity and 4 percent for wages with the controls).[7]

Higher value added per worker in foreign-controlled establishments may be a source of welfare gain if this output rise is not "paid for" through use of additional capital. Since these two studies find differences after controlling for capital intensity, there is indeed evidence of gain. The fact that the productivity differences are largely attributable to size differences does not lessen the economic benefits, although it does lead to the question of why foreign firms better exploit economies of scale.

The finding of higher wages points towards real income gains to the domestic work force. However, it may reflect a higher skill mix among workers employed by foreign-controlled companies if, for example, their production techniques (such as flexible manufacturing) require advanced skills. In this case, foreign firms might pay higher wages to workers who would have received them anyway because of their superior skills. Foreign firms may also demand greater effort from their employees, perhaps because multinationals have a greater stake in maintaining a reputation for product quality. For example, Japanese firms' use of just-in-time manufacturing depends on high employee effort levels. Where it is not feasible to fully monitor workers, foreign firms may be required to pay wage premiums to workers to maintain a high level of performance.[8] However, workers' private costs of exerting more effort may offset most of the wage premium they receive from working at a foreign-owned plant. These considerations suggest that we still do not know enough about the mechanism behind the wage and productivity premiums to conclude that foreign-controlled plants generate net benefits for host-country workers.

Another way to interpret these results is that multinational companies perform better than purely domestic companies. Doms and Jensen demonstrate that in the United States, while foreign plants on average do better in terms of productivity and wages paid than domestic ones, they perform somewhat worse than the plants of U.S. multinationals. In general, these two studies are in favour of state and provincial government promotion of foreign investment. However, they also indicate that the source of the gain from foreign investment is not only foreign management techniques, but also the efficiencies associated with the large establishments of multinational enterprises.

We have considered the impact of foreign ownership on the productivity of the foreign-owned plant. An influx of efficient, foreign-owned plants may also lead to productivity gains for other firms in the economy, most likely through technological spillovers. The knowledge and skills of a foreign manufacturing plant may flow horizontally to other firms in the industry as well as vertically to suppliers and distributors. Blomstrom (1991) cites empirical evidence showing that productivity levels increased with foreign affiliates' share of the market in a number of countries, and Globerman (1979) establishes the same result for Canada. Wolf and Taylor (1991) study how the labour and production management techniques of Japanese automobile assemblers in North America are transferred to domestic assemblers and suppliers.[9] Aitken, Harrison and Lipsey (1995) find that the higher wages paid by foreign-owned firms in the United States push up wages in domestic firms, which may be a positive spillover.

If technology transfer depends on personal contact between firms, the extent that domestic suppliers benefit from foreign manufacturing plants is likely to hinge on the degree to which foreign-controlled companies source from domestic suppliers. Studies by Mersereau (1991) for Canada and Zeile (1995) for the United States reveal extremely high use of imported inputs by foreign-controlled firms. Imports were 32.4 percent of sales in 1986 in Canada among foreign affiliates as opposed to 7.9 percent among Canadian-controlled firms. The import propensities of foreign-controlled companies in the United States were significantly lower, at 11 percent, but higher than the 7 percent recorded for U.S.-controlled firms. Both studies show that Japanese import propensities were somewhat higher than average and that the Japanese, like other foreign investors, exhibit a strong tendency to source from home, especially from affiliates. Foreign firms' preference to source from home-country affiliates may also lead to foreign investment by affiliates who want to locate production near to assemblers. It also means that foreign investment does not necessarily bring more business and potential technology transfer to host-country suppliers.

Doms and Jensen's finding that workers in foreign-owned firms are 50 percent more productive but are paid only 20 percent more than workers in domestic plants suggests that foreign firms may earn higher profits. This would provide the opportunity for tax revenues, which might partly explain state and provincial governments' promotion of foreign investment. The gain to the domestic economy depends on the extent that profits are recorded in the accounts of the foreign affiliates rather than the parent firms' headquarters. Multinationals have long been criticized for using transfer pricing and other mechanisms to shift profits to low tax jurisdictions. Grubert, Goodspeed and Swenson (1993) investigate why the ratio of taxable income to assets in the United States in 1987 was only 0.58 for foreign-controlled companies compared to 2.14 for domestically controlled companies. After considering various plant characteristics such as the age of investment, they conclude that transfer pricing is being used to shift profits. This finding suggests that the desire to increase tax revenues is not a viable basis for promotional efforts.

THE RELATIVE ATTRACTIVENESS OF PROVINCES AND STATES TO JAPANESE INVESTORS

INVESTORS WILL LOCATE IN THE STATES AND PROVINCES that yield the highest profits. The profitablity of a location depends on the presence of certain economic attributes. We group the attributes desired by investors into four categories: labour market conditions, access to markets, government policies, and agglomeration. Following the framework developed by Head, Ries and Swenson (1994) in their examination of Japanese investment in the United States, we discuss how each attribute may affect the profitability of a new Japanese plant. We then show how Canadian provinces compare with U.S. states regarding these attributes, in order to indicate their competitiveness in attracting new foreign investment.

Labour Market Conditions

WAGE LEVELS DIRECTLY AFFECT the costs of production. At an exchange rate of 0.75, wages in Canada are low. Manufacturing census data in the two countries show that in 1992 the average manufacturing wage was US$32,000 in the 50 states and US$24,000 in the 10 provinces, giving Canadian provinces a decided wage advantage. For example, the average wage in Ontario was US$27,000 compared to US$37,000 in Michigan and US$33,000 in California, Ohio, and Illinois. The average manufacturing wage has several deficiencies as a summary measure of labour costs. First, workers' skills differ. One should also consider labour productivity, which may be reflected in manufacturing value added per production worker. Second, unionization may also influence the total cost of labour. Traditional work practices demanded by unions, such as strict job classifications, may raise the cost of personnel management for a foreign manufacturer. This may be particularly important to Japanese companies who prefer flexible job descriptions and frequent rotation of workers.

A comparison of U.S. states and Canadian provinces reveals that Canada has low productivity and high unionization. On average, Canadian value added per worker is 64 percent of that in the United States (at a purchasing power parity exchange rate of 0.84), more than offsetting Canada's wage advantage. Moreover, unionization is higher in Canada. Union membership relative to employment ranges from 26.2 percent (Alberta) to 54.4 percent (Newfoundland) in Canada, while the range is 5.2 percent (South Carolina) to 29.2 percent (New York) in the United States. The importance of these factors to Japanese investors, however, is debatable. Apart from joint ventures with domestic partners, none of the Japanese auto assembly plants in North America is unionized. Toyota's plant in Ontario won the 1991 J.D. Power and Associates award for highest quality in North America. Thus Japanese manufacturers may be able to avoid the low productivity and high rates of unionization characteristic of the regions in which they operate.

Canadian provinces also have high unemployment. A manufacturer may prefer to locate in an area with a high unemployment rate, since hiring may be easier when there is a large pool of unemployed workers. Moreover, high unemployment raises the cost of joblessness since it reduces the probability of finding new employment. This may encourage workers to exert more effort on the job. Conversely, investors will avoid high unemployment locations where there are generous unemployment benefits because they may reduce worker effort. Finally, a government in a high unemployment area is probably more eager to attract job-creating investments. Thus unemployment may be a proxy for unobserved government efforts to attract investors.

Market Access

AS PREVIOUSLY DISCUSSED, Japanese-owned plants import substantial amounts of intermediate goods from the parent firm and traditional suppliers based in

Japan. Also, manufacturers of some goods such as fish and wood products export to the Japanese market. Plant locations along the Pacific coast are attractive because of lower transportation costs to and from Japan.

It is also important to locate plants close to consumers in North America, since one would expect that proximity to final demand would reduce transport costs. Thus the total amount of personal income within a state or province will influence the investment decision. Demand from neighbouring areas should also be considered. Thus, while New Brunswick may not generate large demand, it may be a suitable place to locate a manufacturing plant because it is adjacent to Quebec.

In Figure 1, we show total personal income in the states and provinces. The darkest shaded locations are those with a total income of over US$114 billion. Within this group California, New York, Texas, and Florida are the top four. Ontario ranks 10th overall. These high incomes should be attractive to manu-

FIGURE 1

TOTAL PERSONAL INCOME IN U.S. STATES AND CANADIAN PROVINCES, 1993

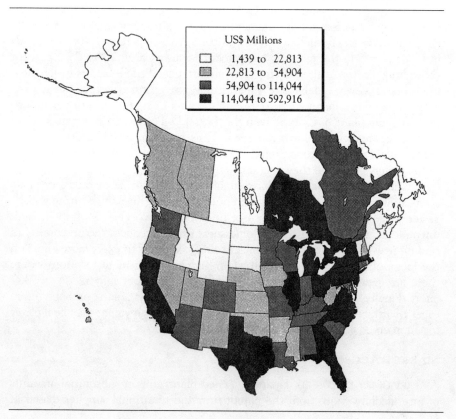

facturers. Manufacturers may also locate in a state such as Indiana because it is near places with large demand.

Government Policies

States and provinces stepped up efforts to attract foreign investment in the 1980s. By the end of the decade, 19 states had investment offices in Japan to provide information to potential investors. Similarly, by 1985 four Canadian provinces – Quebec, Ontario, British Columbia, and Alberta – had established investment offices in Japan, although Ontario later closed its office.

Almost all U.S. states also established FTZ.[10] These zones offer reduced duties to manufacturers through three mechanisms. First, they allow establishments to delay payment of tariff duties until goods are shipped from the FTZ to final market destinations. Second, operating in a foreign trade zone enables firms to avoid all tariffs on imported intermediate goods that are re-exported in final products. Third, costs are reduced as a result of reclassification when goods assembled within the zones are subject to a lower tariff than are the component parts imported into the zone. There are no FTZ in Canada, although various provinces have lobbied for their creation.[11]

Another important consideration for foreign investors is the corporate tax rate, which directly affects after-tax profits. Tax rates vary at the state, provincial, and federal levels in Canada and the United States (Figure 2).[12] Most Canadian provinces and a couple of U.S. states impose rates as high as 10-17 percent. However, these high provincial corporate tax rates are offset by relatively low federal tax rates in Canada. While the top U.S. marginal rate is 34 percent, Canada imposes a flat tax rate of 22 percent. Corporate taxes in the United States are slightly reduced by the rule that allows state taxes to be deducted from federal taxable income.[13]

Some U.S. states employ unitary taxation, a method of taxing firms based on a proportion of their worldwide profits rather than the accounting profits of the affiliate (see Figure 2). Foreign firms actively opposed the tax on the grounds that it exposes them to the possibility of positive tax payments even in states where they earned no direct profits. Unitary taxes also eliminate any tax saving benefits from transfer pricing strategies.

Finally, governments provide labour and capital subsidies to induce investors to locate new plants to their jurisdictions. The *Directory of Incentives for Business Investment* provides data on the general subsidies offered by states. The capital subsidies offered by states in our sample were specified as a percentage of investment value. Labour subsidies reduce labour costs in the first year of operation. Ten U.S. states offered them, and they were no more than 10 percent of the first-year wage bill and were usually much less.

While the federal and provincial governments in Canada devote considerable resources to economic development, the incentives they offer to new

FIGURE 2

CORPORATE TAX RATES IN U.S. STATES AND CANADIAN PROVINCES, 1994

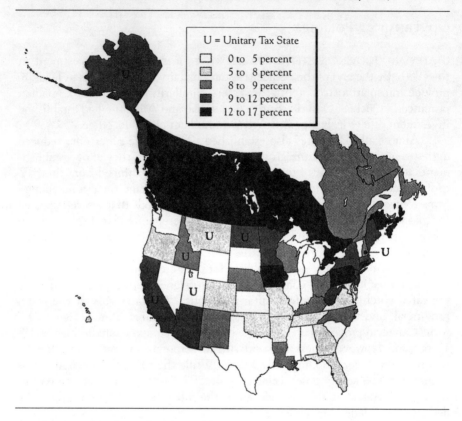

manufacturing establishments are difficult to quantify. In 1994, Industry, Science and Technology Canada (ISTC); the Atlantic Canada Opportunities Agency (ACOA); and Western Economic Diversification (WED) were important federal operations that subsidized economic development, with a particular focus on alleviating regional disparities. In general, these federal programs offered loan guarantees, interest-rate reductions, and cash grants, and all the assistance was largely repayable.[14] The Investment Development Program combined resources from Investment Canada, External Affairs, and ISTC in an effort to encourage foreign investment, primarily by disseminating favourable information about Canada's business environment.[15]

Provincial departments of industry and crown corporations actively solicit new investment, although their activities are highly secret (Tupper, 1982, p. 26; Savoie ,1992, p. 176). In 1971, Nova Scotia won a bidding war for Michelin Tire's first North American plant. Ontario won its Toyota assembly plant in competition with other provinces as well as U.S. states. In 1994 the Committee of Ministers on Internal Trade agreed to refrain from competing with each other for investment. However, interprovincial competition did not end there. In 1995, Newfoundland Premier Clyde Wells introduced an investment incentive package, which includes a 10-year tax holiday and a subsidy of C$2,000 per job while simultaneously criticizing Premier Frank McKenna for New Brunswick's investment promotion programs. Clearly, provinces and states compete for new investment, but it is difficult to measure the assistance they offer.

Agglomeration

Agglomeration economies are positive externalities that flow among firms that locate close to each other. They imply that manufacturers will be attracted to areas with concentrations of similar manufacturers to take advantage of these spillovers. Marshall (1920) identified interfirm knowledge flows, specialized labour, and diversified intermediate inputs as major sources of agglomeration economies. The first, and most frequent, source of agglomeration economies is technological externalities in the form of knowledge spillovers. Useful technical information seems to flow among entrepreneurs, designers, and engineers in a variety of industries. Physical proximity may enhance knowledge flows by making casual communication less costly. Arguably, most important knowledge inflows and outflows occur between the headquarters of corporations, where their most knowledge-intensive functions (research, marketing, etc.) take place. Locating overseas affiliates close to other manufacturers to tap technical knowledge spillovers "in the air" may therefore be of small importance to Japanese manufacturers. However, Japanese companies may want to locate near to other Japanese manufacturers to learn how to operate efficiently in a new foreign location. For example, they might want to learn how to meet local government regulations, adapt to the local climate and labour force, and find low-cost transportation options.

Specialized labour is the second source of agglomeration economies. Workers develop skills and choose places to live, which at least partly commits them to a specific industry and location. This makes them potentially vulnerable in two ways. First, they could be exploited by employers who may offer wages that are very low but just high enough that workers cannot gain by making the large sunk investments to change careers and/or location. Second, workers are at risk that their employer will suffer a business downturn and go bankrupt. Having multiple firms in the same industry in the same location mitigates both problems. Firms will make competing offers to obtain skilled workers, driving up the wage. And if an employer suffers a business setback, it is likely that another employer will be able to employ the displaced workers. Thus, by reducing monopsony power and pooling risk, agglomeration of firms in the same industry makes a location more attractive for specialized workers.[16]

The third source of agglomeration economies is the availability of diverse and efficient suppliers.[17] Proximity to suppliers is valued because it reduces freight costs, inventory overhead, and customs duties. Diversity in component supply is good because it increases the likelihood that each firm will be able to find the variety of input that exactly matches its needs. Agglomerations of input-users (assemblers) reduce total transportation costs and generate large enough levels of demand to warrant the production of highly specialized components. This will attract more input users, which will in turn stimulate the entry of new specialized input providers. The just-in-time inventory system employed by many Japanese manufacturers raises the cost of transporting parts over long distances because it requires flexible and punctual deliveries. Because this system places great importance on reliability and trust, it may also encourage specialization in the form of long-term relationships. These arguments suggest that input-based agglomeration economies will exert a particularly strong influence on Japanese manufacturers as they attempt to transplant their production systems to North America.

FIGURE 3

MANUFACTURING ESTABLISHMENTS IN THE UNITED STATES AND CANADA, 1992

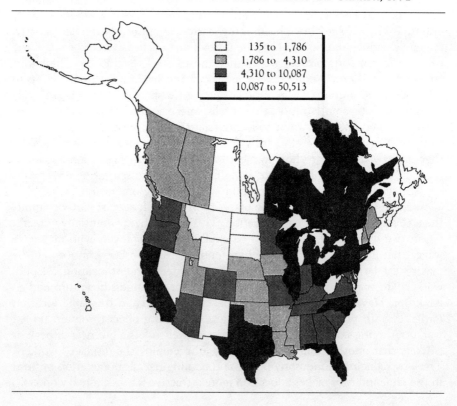

A Japanese investor may be attracted to concentrations of manufacturing establishments in general, as well as clusters of Japanese investment. In general, manufacturing is concentrated in the Great Lakes region of North America (Figure 3). The Japanese pattern deviates somewhat from the overall pattern (Figure 4). While Japanese and non-Japanese establishments are concentrated in Ontario, Michigan, Ohio, and Illinois, the Japanese have also shown a preference for Kentucky, Georgia, and Indiana and an aversion to the mid-Atlantic, New England, and Florida. Thus agglomeration economies do favor traditional manufacturing locations in the Midwest, but they also favor states stretching south from Kentucky to Georgia. Manufacturing areas in the Northeast may be unattractive to Japanese investment at least in part because of the lack of Japanese investment presence there. Furthermore, the Canadian Prairies and states in the western United States will also be unattractive due to lack of agglomeration. A location with a low level of agglomeration may be attractive, however, if it is near concentrations of manufacturing.

FIGURE 4

JAPANESE GREENFIELD INVESTMENT IN THE UNITED STATES AND CANADA, 1991

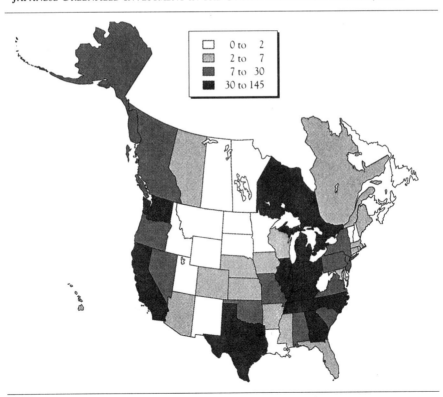

Weighing the Attributes: A Measure of Competitiveness

WE HAVE IDENTIFIED A NUMBER OF ECONOMIC ATTRIBUTES that might influence Japanese investment. In order to ascertain the overall competitiveness of individual locations for attracting investment, we have to weigh the importance of each attribute. Some may be vitally important and others immaterial. For instance, even though Canadian manufacturing is characterized by low productivity and high rates of unionization, Japanese firms in North America may be able to transfer their own high-productivity technology and establish nonunion plants. Thus prevailing unionization rates or productivity levels might be irrelevant for the prospective Japanese investor. And among the location attributes that do matter, it is likely that the investor will face important trade-offs. For example, states and provinces with high incomes and agglomeration also tend to have high wages. An ideal measure of competitiveness would aggregate the economic attributes in a way that reflects the degree to which each attribute matters to the prospective investor.

We use the revealed preference of Japanese investors for certain locations to infer the relative importance of the economic attributes. Head, Ries and Swenson (1994) use U.S. data on Japanese greenfield investment in the 1980s to relate the probability of choosing a state to its characteristics. They employ a conditional logit specification, which can be derived from the assumption that the prospective investor chooses the site that is expected to yield the greatest profits. The model assumes that all firms respond in a systematic way to variation in location characteristics. However, all investors do not choose the same location, for two reasons. First, the manifestations of the relevant attributes will be different according to the individual investors. For instance, the pattern of agglomeration in the sawmill industry differs markedly from that in the auto parts industry. Second, the model assumes that idiosyncratic factors – which are unobservable to the econometrician – influence each investor's location decision.[18] Provided that this hidden heterogeneity is distributed according to the double exponential, the probability that investor j chooses location s' will be

$$\text{Probability } (j \text{ choose } s') = \frac{\exp(\text{Profit}(s', j))}{\sum_{s \in S} \exp(\text{Profit}(s, j))}$$

where S is the set of potential location sites and "Profit" is a function of measured site characteristics.

Table 1 shows the results of the regression analysis conducted by Head, Ries and Swenson. The sample consists of the 760 new manufacturing plants opened by Japanese companies in the United States between 1980 and 1991.[19] On the whole, the explanatory variables have sensible and significant relationships

Table 1

Conditional Logit Results

Variable	Estimate	Standard Error
Labour market conditions		
Log of manufacturing wage	−2.111*	0.711
Log of manufacturing productivity	0.585	0.497
Unionization rate	−0.009	0.007
Unemployment rate	0.119*	0.033
Proximity to markets		
Pacific Rim dummy	0.793*	0.157
Log of state income	0.406*	0.099
Log of adjacent-state income	−0.001	0.084
State policies		
Japanese office dummy	−0.023	0.098
FTZ dummy	0.996*	0.336
Log (1 − corporate tax rate)	9.162*	1.659
Unitary tax dummy	−0.414*	0.154
Log (1 − labour subsidy)	−4.127*	1.502
Log (1 − capital subsidy)	1.214	1.375
Within-state effects		
Log of U.S. industry count	0.466*	0.056
Log of Japan industry count	0.876*	0.091
Log of *keiretsu* count	0.936*	0.129
Adjacent-state effects		
Log of U.S. industry count	0.297*	0.067
Log of Japan industry count	0.466*	0.091
Log of *keiretsu* count	0.386*	0.137
Log-likelihood	−2,273.58	
Number of choosers	760	
Number of choosers	42	

* Significance in a two-tail test at the 1 percent level.

with the location decisions. High wages reduce the probability of receiving investment. Low productivity and high unionization rates appear to deter investment, although neither variable is statistically significant. Unemployment enters positively and is very significant, suggesting that high unemployment levels reduce hiring costs or raise worker effort. As expected, the Japanese prefer locations on the Pacific coast and high-income states.[20]

Among the policy variables, favourable taxes and labour subsidies have a strong influence on the choice of location. In addition, the presence of FTZ significantly increases the probability of investment. As for agglomeration, new Japanese greenfields are attracted to areas with concentrations of firms in the same industry, and this effect is enhanced if those firms are Japanese-controlled. Moreover, the presence of *keiretsu* (Japanese corporate group) affiliates increases a location's attractiveness. Note that adjacent-state counts of firms also influence investment, although to a lesser degree than within-state counts. This result is particularly relevant for Ontario, given its proximity to highly industrialized areas such as Quebec, New York, and Michigan. The weighting implied by these estimates provides a means to measure the competitiveness of states and provinces for attracting future investment. We inserted current state and provincial data into the equation above to calculate the probability of receiving investment. These probabilities indicate the competitiveness of each state or province in attracting new Japanese investment. We note the following advantages of using the probability of receiving investment as a measure of competitiveness.

- It has a simple and unambiguous interpretation. It represents the predicted proportion of investments in an industry that would select a particular state or province. Thus it is possible to use the computed probabilities for both ordinal and cardinal comparisons.

- It is based on a non-arbitrary weighting of attributes. Rather than impose our own biases about the direction and magnitude of each of the effects that collectively reflect competitiveness, we let the actual pattern of decisions made by foreign investors speak for themselves. In contrast, the *World Competitiveness Report 1994* aggregates 381 criteria ranging from computing power per capita to population per nurse. While all 381 factors may be important in different contexts, any aggregate statistic based on them will lack a clear interpretation.[21]

Before presenting our results, we note two important caveats. First, our method assumes that the Japanese weigh each attribute equally whether the site is a state or a province. This assumption poses a problem when the data from the two countries are not comparable. In general, the U.S. and Canadian statistical agencies employ similar methods in collecting and reporting data. However, there could be pitfalls. For instance, suppose that Canada's higher unemployment rate stems from a more generous system of unemployment insurance. In that case, higher rates of unemployment might have little effect on the labour supply pool and be little inducement to high effort on the job. In addition, the provincial and state corporate tax codes, which we pare down to a single tax rate, are in fact very complex. It is conceivable that depreciation allowances or various tax credits entirely offset differences in statutory corporate tax rates.

A second caveat concerns the Canada-U.S. border. Japanese firms located in Canadian provinces may have less access to the larger U.S. market than firms located in the United States. In practice, Japanese firms in Canada do ship to the U.S. market. A survey by Japan's Ministry of International Trade and Industry reveals that Japanese firms in Canada ship on average 17 percent of their output across the border.[22] The *Globe and Mail* (1985) reports that 36 percent of Honda Canada's sales are exports (presumably, all to the United States). Nonetheless, transaction costs associated with the border will make this access imperfect. In addition, agglomeration effects may be impeded by the national border. Our regression results show that adjacent-state agglomeration is about half as attractive as within-state agglomeration. But are there really agglomeration effects between, say, Ontario and Michigan?

In principle, we could address the national border issue statistically by re-estimating the model with all 60 states and provinces. We could then determine from the data how national borders affect the probability of investment. Two considerations make this approach unfeasible. First, data for the dependent variable, Japanese manufacturing establishments, are not available from a single source, and using separate sources for the United States and Canada may lead to sample selection bias. Second, the agglomeration variables are based on four-digit SIC classifications that are different in the United States and Canada. Thus industry-level agglomeration measures across the two countries are not readily comparable and cannot even be created for many of the industries.

The predicted probabilities for each province as well as a complete listing of the top 20 states and provinces are reported in Table 2. The first set of results corresponds to a "representative" industry whose domestic and Japan-owned agglomeration counts are distributed according to the aggregate distribution of manufacturing establishments and Japanese greenfields. *Keiretsu* effects are omitted, since they are parent-firm specific. The finding that California is the most attractive location does not come as a surprise, given its population and manufacturing base. Note, however, that the runners-up do not include such populous states as Texas, New York, and Florida. This is partly because these states have inherent disadvantages. However, each state or province's current competitiveness also depends on the amount of Japanese investment it has already attracted. Thus history matters, in the sense that previous bad policy or just bad luck can exert a negative influence.

Only one Canadian province ranks in the top 20: Ontario. While according to the model it attracts less than half the investment of neighbouring Michigan, it nevertheless ranks higher than Texas. Outside the three largest provinces, the probability of attracting investment falls precipitously to less than 3 per 10,000. Summing the probabilities attached to each province yields an aggregate for Canada that is substantially below California's, despite the approximate equality of their GDPs.

Perhaps our representative industry is not really representative. In some industries, Canada is relatively strong and has a concentration of industrial activity. It is possible that the provinces have strengths that would allow them to be much more attractive to firms in certain industries. If this is the case,

TABLE 2

COMPETITIVENESS OF STATES AND PROVINCES: PROBABILITY OF ATTRACTING NEW INVESTMENT IN ALL MANUFACTURING AND THE MOTOR VEHICLE PARTS AND PULP AND PAPER INDUSTRIES

	ALL MANUFACTURING			MOTOR VEHICLE PARTS			PULP AND PAPER	
				(PERCENT)				
1	California	20.35	1	Indiana	24.83	1	Washington	10.03
2	Indiana	9.37	2	Michigan	19.18	2	Illinois	8.56
3	Illinois	9.03	3	Ohio	18.62	3	Indiana	7.55
4	Georgia	7.29	4	Illinois	8.51	4	California	6.40
5	Ohio	7.11	5	Kentucky	7.19	5	Georgia	5.55
6	Michigan	6.00	6	Tennessee	6.00	6	Michigan	4.81
7	Tennessee	4.95	7	Ontario	3.38	7	Florida	3.86
8	New York	3.73	8	California	2.73	8	Alabama	3.76
9	North Carolina	3.68	9	Missouri	1.83	9	New York	3.73
10	Kentucky	3.67	10	Georgia	1.45	10	Quebec	3.40
11	Oregon	3.39	11	North Carolina	0.99	11	Tennessee	2.94
12	Washington	3.25	12	Virginia	0.76	12	British Columbia	2.84
13	Ontario	2.42	13	New York	0.57	13	Ohio	2.70
14	Texas	2.17	14	Texas	0.56	14	Ontario	2.57
15	Virginia	1.71	15	Alabama	0.38	15	Texas	2.56
16	Pennsylvania	1.67	16	Mississippi	0.36	16	Missouri	2.51
17	South Carolina	1.37	17	South Carolina	0.31	17	North Carolina	2.31
18	Alabama	1.28	18	Pennsylvania	0.27	18	Oregon	2.18
19	Missouri	1.23	19	Arizona	0.24	19	Pennsylvania	1.72
20	New Jersey	1.14	20	Oregon	0.23	20	Kentucky	1.64
21	Quebec	0.93	21	Quebec	0.22	34	Newfoundland	0.53
28	British Columbia	0.24	41	Manitoba	0.01	39	Alberta	0.36
42	Alberta	0.02	42	British Columbia	0.01	48	Manitoba	0.12
47	Newfoundland	0.01	44	Newfoundland	0.01	50	New Brunswick	0.10
50	Manitoba	0.01	50	Alberta	0.00	56	Nova Scotia	0.06
55	New Brunswick	0.00	51	New Brunswick	0.00	57	Prince Edward Island	0.05
56	Saskatchewan	0.00	54	Nova Scotia	0.00	58	Saskatchewan	0.02
59	Nova Scotia	0.00	55	Prince Edward Island	0.00			
60	Prince Edward Island	0.00	60	Saskatchewan	0.00			

policy makers might do well to concentrate their efforts on those industries. In the second and third columns of Table 2 we explore the importance of industry differences using the motor vehicle parts and pulp and paper industries as examples. Auto-related goods are Canada's single largest export category, and the industry is heavily concentrated in Ontario. To calculate domestic auto-related agglomeration, we use the U.S.-Canada SIC concordance and match the eight four-digit Canadian motor vehicle parts industries to 11 corresponding U.S. industries. We base Japanese agglomeration counts on product descriptions from the Japan Economic Institute survey and the Ontario Japanese Investment Profile. Our results bear out our expectation that Ontario's competitiveness in this industry would be better, and in fact it would surpass California's.

For pulp and paper data we aggregated over the entire paper and allied products industry (USIC 26, CSIC 27). This combines pulp plants, which tend to be concentrated in the forested states and provinces, with paper products factories, whose locations more closely reflect the aggregate manufacturing distribution. As expected, Washington and British Columbia rank much higher in attractiveness. British Columbia and Quebec are on a par with California. In summary, the provinces appear to be much more competitive in their traditional industries than they are for the "average" manufacturing industry.

The Effects of Investment Promotion Policies

WE NOW TURN TO THE ISSUE of what policies might increase Canadian provinces' ability to attract investment. In so doing, we restrict our attention to the representative industry. This may be the relevant experiment if provinces are interested in diversifying away from their current industrial structures.

We present the forecast results of four policy initiatives in Table 3. The first column repeats the baseline outcomes from Table 2. The next column reports what the effects of Newfoundland's recent aggressive strategy, a grant of C$2,000 per job created and a tax holiday in return for a sufficient number of new jobs, would be on each province.

We find that Newfoundland's strategy would cause large increases in the probability of attracting investment for all the provinces. However, Newfoundland, the other Atlantic provinces and the Prairies would probably attract little extra investment since their odds of attracting investment remain less than 1 in 1,000. The same policies pursued by Ontario would have striking results. The province would become the second most attractive location for investment in North America. This dramatic difference in the efficacy of the policy arises partly because Ontario was already much more attractive than Newfoundland. It also reflects the fact that the elimination of Ontario's 13.5 percent tax matters more than the elimination of Newfoundland's 7.5 percent tax. One should keep in mind that while in Newfoundland the policy is unlikely to have an impact, it is also unlikely to be costly. On the other hand, Ontario would lose tax revenues from and pay subsidies to investors who preferred Ontario, even without the inducements. Furthermore, existing Ontario manufacturers would probably not tolerate a tax advantage that solely benefited new foreign investors. While our simulations demonstrate one way Ontario could attract substantially more investment, they do not consider real-world concerns that would lower the attractiveness and feasibility of that option.

TABLE 3

SIMULATIONS OF THE EFFECT OF FOUR POLICY INITIATIVES ON THE PROBABILITY OF ATTRACTING INVESTMENT, BY PROVINCE[1]

Province	Baseline[2]	Subsidies[3]	FTZ[4]	Depreciation[5]	Closed Border[6]
British Columbia	0.24 (28)	2.61 (13)	0.62 (23)	0.30 (26)	0.05 (37)
Alberta	0.02 (42)	0.19 (32)	0.06 (37)	0.03 (42)	0.02 (41)
Saskatchewan	0.00 (56)	0.02 (43)	0.00 (56)	0.00 (56)	0.00 (54)
Manitoba	0.01 (50)	0.06 (37)	0.01 (47)	0.01 (50)	0.00 (49)
Ontario	2.42 (13)	15.16 (2)	6.20 (6)	3.00 (13)	0.48 (23)
Quebec	0.93 (21)	3.56 (11)	2.39 (14)	1.15 (20)	0.54 (21)
New Brunswick	0.00 (55)	0.04 (40)	0.01 (49)	0.00 (54)	0.00 (53)
Nova Scotia	0.00 (59)	0.01 (50)	0.00 (57)	0.00 (58)	0.00 (59)
Prince Edward Island	0.00 (60)	0.00 (60)	0.00 (60)	0.00 (60)	0.00 (60)
Newfoundland	0.01 (47)	0.03 (41)	0.02 (43)	0.01 (47)	0.01 (44)

1 Numbers in parentheses are rankings among the 60 states and provinces.
2 "Baseline" refers to the current situations in the provinces.
3 Each province is allowed to offer Newfoundland's tax holiday and a C$2,000 per job subsidy.
4 In this simulation all provinces provide foreign trade zones.
5 A 10 percent depreciation is allowed, which lowers relative Canadian wages.
6 In this simulation two locations are considered adjacent only if they share a border and are in the same country.

The next policy we consider is the establishment of FTZ, which has been discussed in Canada, particularly with respect to British Columbia and Alberta. This time we run the experiment of FTZ being allowed in the whole country and assume that all provinces take advantage of the option. The introduction of FTZ, while not as impressive in its predicted effects as the tax holiday, still offers substantial potential benefits to Ontario and Quebec. The probabilities of attracting investment to Alberta and British Columbia increase, but remain at less than one investor in 100.

The last two experiments do not correspond to specific policies but instead consider hypothetical circumstances. First, suppose the dollar depreciated by 10 percent. As we write this chapter – the day after the "No" side carried a slight majority in the last Quebec referendum – such a possibility seems remote. However, given the elasticity of Japanese location choice to wages, it appears that even an exchange rate of US$0.65 per Canadian dollar would not have a major impact on the allocation of Japanese investment in North America. The flip side of this result is that if the Canadian dollar trends back toward purchasing power parity (roughly US$0.84), we would not expect a large decline in competitiveness.

Finally, consider the scenario in which Canada closes its border with the United States. We implement this idea by considering two locations as adjacent only if they share a border and lie in the same country. Of course it is possible that rather than being a counter-factual assumption this is actually the appropriate one for the current state of the North American economy. While most goods cross the international border duty free, province-state trade is subject to numerous transaction costs. Furthermore, if agglomeration arises from sharing a common pool of specialized labour or from casual exchanges of knowledge, the Canada-U.S. border may already present a sizeable obstacle. In any case, it is instructive to note that without the benefit of agglomeration in neighbouring U.S. states, the attractiveness of Ontario and British Columbia plummets. Since it is difficult to attribute our estimated agglomeration effects to a particular mechanism, we can only speculate as to what policies Canada might undertake to increase the agglomeration benefits derived from neighbouring states. If trade in inputs and temporary visits by skilled workers are important components of agglomeration, then the 1988 Free Trade Agreement with the United States may promote investment in Ontario, British Columbia, and to a lesser extent Quebec.

In conducting the above experiments, we considered only the direct effect of changing provincial attributes. In the long run, these static effects will be magnified by agglomeration. For instance, a tax holiday that succeeded in attracting an initial Japanese investment to Newfoundland would make it easier to attract subsequent investors. Once a core had been built, the province might find itself able to attract investment without resorting to subsidies.

The initial flood of Japanese investment in North America opened a window of opportunity in which states with relatively small domestic agglomeration counts, such as Kentucky, Indiana, and Georgia, built up critical masses of Japanese investment in specific industries. With Japanese agglomerations already established elsewhere, it will be difficult for the Atlantic provinces and the Prairies to catch up. Perhaps a more fruitful route would be to concentrate efforts on attracting investors from countries like South Korea, whose outward direct investment is in its infancy.

Conclusion

THE FIRST PART OF THIS PAPER reviews recent evidence showing that foreign-controlled establishments are more productive and pay higher wages than domestically controlled establishments. In addition, foreign investment may provide spillover benefits to domestic establishments in terms of higher productivity. Thus there may be a strong case for investment promotion policies that succeed in attracting foreign manufacturers.

Japanese direct investment continues to be of particular interest to governments in North America. While Japanese direct investment abroad has abated somewhat in recent years, from 1990 to 1994 about US$20 billion flowed into North America annually. Japanese manufacturers are known to bring efficient plants and innovative labour and production management techniques to the host country.

After reviewing the attractiveness of Canadian provinces as sites for new Japanese investment, we use earlier econometric results to rank the competitiveness of states and provinces. We then proceed to estimate the effects of investment promotion policies on the likelihood of drawing investment. We show that offering FTZ, (Newfoundland's current) job subsidies, and tax holidays substantially increases the probability of receiving investment. However, since the probabilities start out quite low for most provinces, dramatic increases in the number of investments may only be expected for Ontario.

The term "competitiveness" has become loaded with a variety of connotations. We define it solely in terms of ability to attract Japanese investment, although we think that our measure is probably a reasonable indicator of a site's attractiveness to most footloose manufacturers. It is not an indicator of overall welfare, even if welfare is defined narrowly to include only the material standard of living of a state or province's population. To see why this is the case, consider the wage. Lower wages attract investment but also tend to lower real incomes. Similarly, tax holidays and subsidies may involve significant opportunity costs.

This paper has described recent empirical evidence that suggests that there are positive benefits from inward foreign investment. Our analysis indicates that promotional policies can draw investors. We are not willing to conclude that governments *should* try to attract investment, because we have not quantified the costs of government policies. In addition, it seems likely that the main effect of state and provincial policies is to redistribute investment within North America.

Appendix 1

Table A-1

Canadian and U.S. Sources of Data Used in Examining Japanese Investment in North Amercia, by Variable

Variable	Description	U.S. source	Canadian source
Japanese greenfield investment	Manufacturing greenfields	Japan Economic Institute	Japanese Overseas Investments, 1992 (Toyo Keizai)
Domestic manufacturing establishment counts	Count of manufacturing establishments	Census of manufacturers	Census of manufacturers CANSIM #5378
Wage	Payroll divided by employees for manufacturing establishments	Census of manufacturers	Census of manufacturers CANSIM #5378
Productivity	Value added per production worker for manufacturing establishments	Census of manufacturers	Census of manufacturers CANSIM #5378
Unionization rate	Union membership divided by total employment	Compiled from *Current Population Survey*	CALURA Statistics Canada Cat. no. 71-202
Unemployment rate	All adults	*Statistical Abstracts of the United States*	CANSIM #2066, 2076
Personal income	Disposable income (total)	*Statistical Abstracts of the United States*	CANSIM #5089-5097, 6965
Tax rate	Highest marginal corporate tax rate	KPMG Tax Facts	KPMG Tax Facts
Labour and capital subsidy	Share of initial labour or capital costs	*Directory of Incentives for Business Investment and Development in the United States*	Not available

Notes on Sources

WE USED DATA FOR U.S. STATES to estimate coefficients for our regression model. To construct the probability that states and provinces receive investment, we collected recent comparable data for states and provinces.

Japanese Investments

NO SINGLE SOURCE lists Japanese greenfield investment in Canada and the United States. Japanese manufacturing establishments in the United States are recorded in the Japan Economic Institute 1990 Updated Survey. For Canada, we were unable to find a source of Japanese establishment-level data that allowed us to distinguish between acquisitions and greenfields. Therefore, we derived counts of Japanese investments in Canadian provinces from *Japanese Overseas Investments*, published by the Toyo Keizai Company. Thess data are listed for each Canadian affiliate of Japanese companies, although affiliates may not necessarily correspond to establishments. In particular, it is possible that a Japanese affiliate headquartered in Ontario may have manufacturing establishments in other provinces. The Japanese affiliate data are used to calculate Japanese agglomeration for each state and province. In the simulations, agglomeration levels correspond to the total number of Japanese greenfields.

State and Province Characteristics

THE REGRESSION MATCHES each Japanese greenfield investment in 1980-90 to a time series of data for U.S. states. However, not all of the data are recorded annually, and when data for particular years were unavailable, we matched each Japanese investment to the data that correspond most closely to the year the plant began operations. For example, investments from 1980 to 1984 were matched to 1982 Census of Manufacturing data while later investments were matched to the 1987 Census data. See Head, Ries and Swenson (1994) for details. In the case of the agglomeration variables, Japanese investments are matched to counts of establishments according to four-digit SIC designations. The following four variables are included in the regression but do not appear in the table: dummy variables indicating the presence of FTZ, unitary taxation, investment promotion offices in Japan, and counts of *keiretsu* affiliates. Data on FTZ are collected from the Foreign Trade Zones Board at the U.S. Department of Commerce. The Japan investment promotion office data are from the *Directory of Incentives for Business Investment and Development in the United States*, which is assembled by the National Association of State Development Agencies (NASDA). Data on the unitary tax were taken from Tannenwald (1984), the *Wall Street Journal* and the *New York Times*. Finally, we computed *keiretsu* counts based on affiliations found in *Kigyo Keiretsu Soran*, published by the Toyo Keizai Company.

To calculate competitiveness rankings and run simulations, we matched recent data for U.S. states and Canadian provinces. We gather 1992 manufacturing census data for both countries. The data on unionization rates are from 1991 for the United States and 1992 for Canada, while unemployment rate data are from 1993 and 1994, respectively. We used 1993 personal income data for both countries and 1994 tax data. Labour and capital subsidy data for U.S. states are from 1990.

ENDNOTES

1. The authors gratefully acknowledge the helpful comments of John Craig, Richard Harris and other conference participants.
2. This figure, as well as other estimates of financial packages offered to Japanese assemblers in the United States, can be found in Glickman and Woodward (1989), p. 231.
3. These data are based on notifications, not actual transactions.
4. Multi-billion dollar acquisitions such as Matsushita's purchase of MCA have had a significant impact on the investment figures.
5. For a good summary of the social, political, and economic issues associated with foreign investment in Canada, see Brander (1992) and Ries, Globerman and Vertinsky (1994). Graham and Krugman (1991) and Glickman and Woodward (1989) assess the impact of foreign investment in the United States.
6. The Chinese government, on the other hand, gave tax advantages to foreign-funded joint ventures, a policy which stimulated many indigenous investors to set up shell corporations in Hong Kong so as to qualify as foreign for tax purposes.
7. The result that foreign affiliates pay higher wages is not unique to studies of U.S. and Canadian industries. Aitken, Harrison and Lipsey (1995) find foreign firms in Mexico and Venezuela pay higher wages to both skilled and unskilled workers.
8. These wages are termed efficiency wages. An excellent summary of efficiency wages is found in Yellen (1984).
9. Both Blomstrom's and Wolf and Taylor's papers are part of an Industry Canada Research Series. See other papers in the volume for further analysis of foreign investment and technological spillovers.
10. By 1990 only three states – West Virginia, South Dakota and Idaho – did not have FTZ.

11 Since tariffs are determined at the national level, tariff exemptions must be approved by federal authorities. In the United States, the federal government ultimately approves each zone, although all applications are made with the support of state and local governments. Likewise, provinces must convince Ottawa before implementing plans to create FTZ.
12 Tax rates in the United States are graduated, and we refer to the top marginal rate.
13 A comparison of U.S and Canadian taxes is found in Shoven and Whalley (1992).
14 A good summary of regional economic programs can be found in Savoie (1992).
15 We found no evidence that the promotional activities of the program consciously favoured individual provinces, although particular industries are targeted. Details of the program can be found in Deigan (1991).
16 For earlier models of this mechanism, see Rotemberg and Saloner (1990) and Krugman (1991b).
17 Krugman (1991a) models industrial concentration through this type of agglomeration economy.
18 Our discussant, John Craig, a business lawyer whose clients include Japanese direct investors in Canada, provided us with a few examples of idiosyncratic factors influencing Japanese investment. For instance, Honda's decision to locate a plant in Canada where it sold a large number of cars not only provided low transport costs to consumers but also satisfied an implicit obligation Honda felt towards Canada for making it the company with the highest sales among Japanese producers.
19 See the Appendix for descriptions and sources of the data.
20 Adjacent income, however, has no measurable effect on the choice of location.
21 Furthermore, the behavior of investors can follow some surprising patterns. For instance, while the *World Competitiveness Report* considers a recession to be negative, Japanese investors tended to locate in states that, other things equal, had higher unemployment rates.
22 Kaigai Toshi Tokei Soran: Dai 3-Kai Kaigai Kigyo Katsudo Kihon Chosa, Tsusho Seisakukyoka Kokusai Kigyoka Henshu, 1989 (in Japanese).

Bibliography

Aitken, Brian, Ann Harrison and Robert E. Lipsey. "Wages and Foreign Ownership: A Comparative Study of Mexico, Venezuela, and the United States." NBER Working Paper 5102. Cambridge, Mass.: National Bureau of Economic Research, 1995.

Blomstrom, Magnus. "Host Country Benefits of Foreign Investment." In *Foreign Investment, Technology, and Economic Growth.* Edited by D. McFetridge. Industry Canada Research Series. Calgary: University of Calgary Press, 1991.

Borensztein, Eduardo, José De Gregorio and Jong-Wha Lee. "How Does Foreign Direct Investment Affect Economic Growth?" NBER Working Paper 5057. Cambridge, Mass.: National Bureau of Economic Research, 1995.

Brander, James. "Canadian Foreign Investment Policies: Issues and Prospects." In *Canadian Foreign Policy and International Economic Regimes*. Edited by A. Clair Cutler and Mark W. Zacher. Vancouver: University of British Columbia Press, 1992, pp. 113-29.

Deigan, Russell. *Investing in Canada: The Pursuit and Regulation of Foreign Investment*. Toronto: Thomson Professional Publishing, 1991.

Doms, Mark and J. Bradford Jensen. "A Comparison Between the Operating Characteristics of Domestic and Foreign Owned Manufacturing Establishments in the United States," paper prepared for the NBER Conference on Geography and Ownership as Bases for Economic Accounting, 1995.

European Management Forum. *World Competitiveness Report 1994*. Geneva: The EMF Foundation.

Glickman, Norman J. and Douglas P. Woodward. *The New Competitors: How Foreign Investors are Changing the U.S. Economy*. New York: Basic Books, 1989.

Globe and Mail (The), Report on Business Magazine (July 1995):89.

Globerman, Steven. "Foreign Direct Investment and Spillover Efficiency Benefits in Canadian Manufacturing Industries." *Canadian Journal of Economics*, 12, 1 (1979):143-56.

Globerman, Steven, John C. Ries and Ilan Vertinsky. "The Economic Performance of Foreign Affiliates in Canada." *Canadian Journal of Economics*, 27, 1 (1994):143-56.

Graham, Edward M. and Paul R. Krugman. "Foreign Direct Investment in the United States." Washington, DC: Institute for International Economics, 2nd edition, 1991.

Grubert, Harry, Timothy Goodspeed and Deborah Swenson "Explaining the Low Taxable Income of Foreign-Controlled Companies in the United States." In *Studies in International Taxation*. Edited by Alberto Giovannini, R. Glenn Hubbard and Joel Slemrod. Chicago: University of Chicago Press, 1993.

Head, Keith, John C. Ries and Deborah Swenson. "The Attraction of Foreign Manufacturing Investment: Taxes, Subsidies and Agglomeration Economies." NBER Working Paper 4878. Cambridge, Mass.: National Bureau of Economic Research, 1994.

Krugman, Paul R. "Increasing Returns and Economic Geography." *Journal of Political Economy*, 99 (1991a): 483-99.

_____. *Geography and Trade*. Cambridge, MA: The MIT Press, 1991b.

Marshall, Alfred. *Principles of Economics*. London: MacMillan, 8th edition, 1920.

Mersereau, Barry. "Longitudinal Analysis of Canadian Imports by Characteristics of Importing Firms, 1978-86." Research paper. Ottawa: Statistics Canada, 1991.

Ries, John C., Steven Globerman and Ilan Vertinsky. "Foreign Investment from the Asia-Pacific Rim: Implications for Canadian Investment Policy." Kingston: John Deutsch Institute for the Study of Economic Policy, 1994.

Rotemberg, J. and G. Saloner. "Competition and Human Capital Accumulation: A Theory of Interregional Specialization and Trade." NBER Working Paper 3228. Cambridge, Mass.: National Bureau of Economic Research,1990.

Savoie, Donald J. *Regional Economic Development: Canada's Search for Solutions*. Toronto: University of Toronto Press, 2nd edition, 1992.

Shoven, John B. and John Whalley (eds). *Canada-U.S. Tax Comparisons*. A National Bureau of Economic Research Project Report. Chicago: University of Chicago Press, 1992.

Statistics Canada. *Canada's International Investment Position*. Cat. No. 67-203, 1994.

Tannenwald, Robert. "The Pros and Cons of Worldwide Unitary Taxation." *The New England Economic Review*, (July/August 1984):17-28.

Tupper, Alan. *Public Money in the Private Sector.* Kingston: Institute of Intergovernmental Relations, Queen's University, 1982.

U.S. Department of Commerce. *Survey of Current Business*, August 1995.

Van Loo, Francis. "The Effects of Foreign Direct Investment on Investment in Canada." *The Review of Economic Studies* (1977):274-81.

Wolf, B.M. and G. Taylor. "Employee and Supplier Learning in the Canadian Automobile Industry: Implications for Competitiveness." In *Foreign Investment, Technology, and Economic Growth.* Edited by D. McFetridge. Industry Canada Research Series. Calgary: University of Calgary Press, 1991.

Yellen, Janet L. "Efficiency Wage Models of Unemployment." AEA *Papers and Proceedings,* 74, 2 (1984):200-05.

Zeile, William J. "Imported Inputs and the Domestic Content of Production by Foreign-Owned Manufacturing Affiliates in the United States." Paper prepared for the NBER Conference on Geography and Ownership as Bases for Economic Accounting, 1995.

Comment

John W. Craig
Barrister and Solicitor
Toronto

THERE IS COMPETITION BETWEEN American states and Canadian provinces for Japanese manufacturing investment, and it has had a variety of results. Head and Ries attempt to delineate the issues relevant to foreign direct investment, particularly Japanese foreign direct investment, and draw a number of conclusions as to what the provinces can or should do to increase or attract it. While the paper is a very practical attempt to determine the best investment promotion policies, its conclusions are qualified and subject to insufficient and/or incomparable data and problems presented by attempting to assess intangible factors using an econometric model.

Their focus is vastly different from mine, which is that of a business lawyer who has spent the last half of his practice concentrating to a significant degree on Japanese direct investment in Canada. This includes commercial and corporate work with and assistance in establishing Canadian operations for dozens of Japanese companies, over 40 visits to Japan, involvement with Japanese-Canadian business associations, and long-standing relationships with the senior executive officers of Japanese investors and referral sources. During my most recent trip to Japan (some six weeks before the conference in Vancouver), my discussions with clients and referral sources interestingly covered objectives and proposals involving Asia and the United States, but none involved

Canada. While some contacts sensed potential in Canadian automotive parts and silicone manufacturing, the only significant investments in Canada discussed were in the automobile manufacturing sector, and involved Toyota and Honda.

Assessment of the Paper

HEAD AND RIES FOCUS ON whether, and if so how, provinces and states can influence Japanese manufacturers' decisions on investments in manufacturing operations in North America. The authors clearly outline the conundrum for provincial governments. What are foreign corporations trying to achieve in their investment decisions? What policy and economic factors influence those decisions? What programs could be implemented that might affect those decisions? What infrastructural changes would make it easier to invest? There is no real answer, or no single answer, to this conundrum. An answer that is relevant in one year or one economic environment is likely no longer relevant in the next.

The authors begin with a set of premises with which I cannot disagree. There is an active rivalry among provinces and states for manufacturing investment; government promotion can and has made a difference in attracting foreign investment; and the United States has received disproportionately more Japanese direct investment than is warranted by its economic size advantage. While the authors question whether foreign investment enhances host-country welfare, I assume, for the purposes of this commentary, that the impact of foreign direct investment on employment, output, productivity, wages, use of resources, technological spillovers and the like is positive. Although the authors focus on productivity and wages, they seem to agree with this assumption.

When dealing with what the authors refer to as an *attractiveness* assessment, they focus on four specific categories – labour market conditions, access to markets, government policies, and agglomeration. It is difficult to argue with their conclusions that:

- the general low level of productivity and high level of unionization in Canada (compared with the United States) cannot be helpful;

- the high unemployment levels in Canada can be an attribute on a net basis;

- proximity to large consumer markets is a decided advantage (for example, southern Ontario's connections with the northeastern United States and the rust belt);

- having an office in Tokyo disseminating information about regional attributes is an asset (I question the authors' failure to note the impact of Bob Rae's closing of Ontario's long-standing Japanese office on the province's profile in the Japanese business community);

- foreign trade zones have been an advantage to many American states;

- corporate tax rates are relevant, and aggregate taxes in Canada must be competitive with those in contiguous jurisdictions;

- unitary tax regimes are confusing, potentially discriminatory, and often negative;

- labour and capital subsidies have successfully attracted investment; and

- agglomeration (attracting manufacturers to areas with concentrations of similar manufacturers) or clustering (Japanese investors, sometimes competing, locating near each other) are very positive factors for Japanese investors generally, and the resulting knowledge spillovers, sourcing of specialized labour, and availability of diverse input suppliers are all very attractive to concentrated manufacturing.

There is no doubt, however, that the authors' preliminary statement under this section (that "investors locate in places that yield the highest profits") is a *sine qua non* for Japanese investors.

When assessing the policy and program variables, the only possible conclusion is that there are only a few provinces in which investment can be contemplated, British Columbia and Alberta because of their resource base and Ontario and Quebec because of their infrastructure base and proximity to large markets. Head and Ries' measure of competitiveness results in a complicated economic formula from which they draw a number of conclusions. Since it is over 30 years since my last course in economics, I cannot comment on the appropriateness or otherwise of the formula. I should point out, however, that the conclusions as to the impact of policy variables seem to be those one might draw based on a liberal application of plain old common sense. Simply stated, the statistical data seems to support the obvious.

The authors then test the effect of various investment promotion policies on the likelihood of Canadian provinces attracting investment. The difficulty I have with their conclusions is that they are premised on a static situation (all other criteria remaining the same) and the impact on it of a change in one of the relevant criteria. Nevertheless, it is no doubt a valid attempt. It results in three rather interesting conclusions, namely:

- Most provinces should concentrate their efforts on attracting investors from countries such as South Korea "whose outward direct investment remains in its infancy."

- Ontario is the only contender in the North American sweepstakes, as "dramatic increases in the number of investments may only be expected for Ontario."

- The main effect of state and provincial investment attraction policies is simply to redistribute planned investment within the North American market.

These are hardly conclusions that will drive provinces to adopt particular programs or policies.

Anecdotal Comments

WHILE I CAN MENTION SOME INVESTMENTS made by Japanese companies for rather quixotic reasons, in my experience Japanese investors basically look at the whole business environment, including all of the criteria raised by Head and Ries. Japanese efforts are focused on finding a business environment that is likely to generate the best profit. However, their decisions are made within a context that belies statistical categorization.

Honda – Honda Manufacturing of Canada Ltd. was the first major automobile manufacturer of Japanese origin to establish a plant in Canada. It is not dealt with substantively in the paper, perhaps because it is too representative of the idiosyncratic nature of Japanese investment. Why and how did Honda choose Ontario for its investment? The primary reason seems to be that Honda believes it is in its best interest to invest in production facilities where it distributes. In other words, production should be located in a market already served by a distribution network selling large quantities of its products. Canada is, I believe, the only major market in the world where Honda outsells Toyota. Does this create an obligation in the corporate mind of Honda? Clearly so. The fact that the government was going to give Honda Junior Auto Pact treatment, that its labour market was satisfactory, that its tax and business environment compared with that of the United States was reasonable, that the required infrastructure was in place, that regulation was not excessive, and that the currency was relatively stable, all contributed to the decision. That a significant market was accessible from the Alliston plant at a reasonable cost no doubt also contributed to the decision. In other words,

Honda tested the environment and found it acceptable. The idiosyncratic nature of this particular Japanese company is probably evidenced best, however, by its refusal to look for government incentives. By not taking grants or low interest loans, Honda does not have the government in its back pocket and can deal with policy issues independently and at arm's length. This is an interesting, even provocative attitude in this incentive-seeking corporate world.

Toyota – As is well known, Toyota Manufacturing Canada Inc. recently announced a $600 million investment to increase the annual production capacity at its Cambridge plant from 80,000 to 200,000 vehicles. In an interesting interview in the September/October 1995 issue of the *Canada-Japan Business Review*, the new President of Toyota Canada Inc. (the Canadian distribution company for Toyota), Mr. Yoshio Nakatani, disclosed a significant amount of information about the decision to complete Toyota's first plant in Canada when he said:

> I arrived in February of 1985. Sales of Toyota vehicles in Canada in 1984 were 67,000 units and the mood among our dealers was somewhat low. They were uncomfortable with the continuing concerns about imported cars in Canada and Honda was already building an assembly plant in Alliston. The competition was fierce. Our dealers thought that Toyota's presence in Canada would be significantly enhanced if we had some commitment to manufacture here. We agreed and asked Toyota executives in Japan to consider investing in Canada for a plant which could supply the North American auto market.

Mr. Nakatani then went on to talk about site location, giving further insight into their decision-making process:

> Our major criteria for site selection were: availability and skills of the local labour force, size of the market in Canada and North America accessible within a certain geographic radius of the site, and the degree of cooperation the municipal government showed towards us. Other things like access to cultural and educational facilities were also considered.

While these are recollections about a decision made some years earlier, they reflect some of the factors that are germane to the Japanese investor and are difficult to show in an econometric model (protection of competitive position, enhancement of market position through commitment).

Michelin – While Michelin is not a Japanese company, the authors described the establishment of its first North American plant in 1971 in Nova Scotia as a "victory" for Nova Scotia. Where is Michelin now? The largest tire manufacturer in the world (with sales in excess of $12 billion) recently made a decision (described in a press release dated November 16, 1995) to consolidate its future North American expansion in South Carolina. The reasons given publicly for this decision are the favourable business environment, the excellent location and technical training, the willingness of the local government to act as a partner (read excellent tax and funding arrangements) and, perhaps most important, the existence of a "right to work" law, which went some way towards minimizing labour difficulties in the state. In an article in the *Report on Business* of the same date, Peter Cook contrasted the Ontario Federation of Labour's proposal to paralyze London, Ontario on December 11 in an effort to raise Ontario's consciousness of Mike Harris' agenda of sending Ontario "straight down the road to South Carolina" with Michelin's decision. He noted that South Carolina has been very successful in creating an environment with lower tax burdens and "the right climate" for investment, jobs and growth. While not particularly relevant to the Japanese decision-making process, it is germane to the criteria upon which major international investors, including the Japanese, base their investment decisions.

It is interesting to note that in the Seventh Annual Survey of Japanese companies in Canada issued in October 1995 by the Japan External Trade Relations Organization (JETRO), the three most important concerns of Japanese affiliated companies in Canada about their investment in this country were 1) the instability of the exchange rate (72.5 percent), 2) the inability to source quality employees (29.6 percent), and 3) intensified competition between Japanese affiliates (25.9 percent). The first two are concerns the federal and provincial governments may well want to address.

Conclusion

HEAD AND RIES MAKE IT CLEAR THAT there is no perfect combination of environment and incentives that will guarantee Japanese foreign direct investment in Canada. In my view, it is very important for the federal and provincial governments to develop good working relationships with potential Japanese investors and to build an economic, business and tax environment conducive to investment. It seems clear that relationships, cultural understanding and the local business environment are key elements, perhaps the key elements in this process.

Governments should assess the compatibility of the relevant policies with increased foreign direct investment. One particular area that should be examined is the continuing impasse over import relief for the transplanted automobile manufacturers in Canada (the recent public announcement about tariffs on auto parts represents a start only on this issue). The level of intransigence on the part of the federal government on this issue may foreclose future substantial investment in the automotive sector.

Perhaps it is sufficient to conclude by saying that certain unique attributes of Japanese business, including such factors as consensus decision making, inordinately active data and information consumption, obligation-driven responses and keen competitive instincts, give Japanese management a different focus and direction, one that the federal and provincial governments should understand clearly and address before launching incentive programs, changing policies or working to create a receptive economic and business environment.

Robert N. McRae
Department of Economics
University of Calgary

Canada's Natural Resource Exports to the Asia Pacific Region

THE ASIA PACIFIC REGION[1] consists of countries in North America, South America and Asia that border on the Pacific Ocean, as well as Australia and New Zealand. This paper examines Canada's exports of natural resource products to the Asian countries, including South Asian countries.[2] It focuses on both non-renewable resources such as energy and mineral products, and renewable ones such as forestry, agricultural and fish products.[3]

Economic activity in most of the Asian countries in the Asia Pacific region has been growing extremely rapidly. For instance, the real GDP in South Korea has been growing at a rate of 8 percent per annum for the last decade. The region includes countries that are at varying stages of development. With the exception of Japan and, just recently, Singapore, they are all classified as developing countries. In 1992 the real per capita GDP ranged from a low of $1,282 in India to a high of $15,105 (in 1985 US dollars) in industrialized Japan.

The economies of the developing Asian countries are growing at a faster rate than those of the industrialized countries of the OECD, and this trend is expected to continue. As these countries develop, their populations tend to shift from rural to urban areas and their economies become more industrialized. Such trends are usually accompanied by an increase in the per capita consumption of natural resources.

The paper is organized as follows. In the second section, I describe Canada's historical natural resource exports. By themselves, historical export patterns are not very useful for projecting resource export potential because of the consumption shift mentioned above. The relationship between economic development and resource use can be established by utilizing the country data in a cross-sectional analysis. Therefore, the third section contains the results of a cross-sectional analysis that relates resource consumption per capita to economic development. In the fourth section, I use the analysis from the previous two sections to forecast resource consumption as the Asian countries become more developed. This section also contains a brief discussion of the reliability of the technique. Finally, the last section is a summary of the main findings.

HISTORICAL NATURAL RESOURCE EXPORTS

DATA ON CANADA'S RESOURCE EXPORTS, in value terms, are taken from the Statistics Canada publication *Exports by Country*. These data were used because they can be aggregated to represent the natural resource industries (energy, minerals, agriculture, forestry, and fishery) and can be identified by destination country.[4] They are classified under the Harmonized Commodity Description and Coding System. The 97 categories (or Chapters, as they are called) are based on economic activity or component material.

The Chapters that were included in the aggregation are listed in Table 1. Data on Canada's exports in 1988-94 to the following 14 Asian countries were collected: Bangladesh, China, Hong Kong, India, Indonesia, Japan, South Korea, Malaysia, Pakistan, the Philippines, Singapore, Sri Lanka, Taiwan, and Thailand. Although these data are available on a quarterly basis, the annual values are used in this paper. The province of origin is also specified, so we were able to examine the regional importance of resource exports.

Natural resource products account for a large share of the value of Canada's overall exports. In 1994, they accounted for about 43 percent. Asia is an important export market for agricultural, fish, and forestry products. In 1994, the continent accounted for about 27 percent of the value of Canadian world exports of agricultural products, 34 percent of fish products, and 17 percent of forestry products. It also accounted for 8 percent of energy products and 10 percent of mineral products.

The real value of Canadian exports of agricultural, fish, forestry, energy, and mineral products from 1988 to 1994 to each of the 14 Asian countries individually, all the countries together, and all of the developing countries (all countries except Japan) are shown in Figures A-1 to A-16. The real export values were created by dividing the nominal values by the National Accounts deflator for exports.

Several observations can be made. In general, there is a substantial amount of variability, especially for agricultural, mineral, and forestry products. Except for Japan, and to a much lesser extent Hong Kong, the value of fish exports is very low. The only countries that import Canada's energy are Japan, South Korea, and Taiwan. Coal is the only energy fuel being exported to these countries. In aggregate terms, forestry products have the highest value among the resource exports, and they exhibit a strongly positive growth trend. The strength of forestry exports is due to the very large value for Japan (Figures A-15 and A-16). The values of agricultural and mineral products are very volatile, even in the aggregate. In general, the value of mineral exports has declined.

TABLE 1

CLASSIFICATION OF NATURAL RESOURCES ACCORDING TO COMMODITY CHAPTERS USED BY STATISTICS CANADA

RESOURCE		HARMONIZED CHAPTERS
Agriculture	01	Live animals
	02	Meat and edible meat offal
	04	Dairy products; birds' eggs; natural honey; edible products, nes
	05	Products of animal origin, nes or included
	06	Live tree and other plant; bulb, root; cut flowers; etc.
	07	Edible vegetables and certain roots and tubers
	08	Edible fruit and nuts; peel of citrus fruit or melons
	09	Coffee, tea, maté and spices
	10	Cereals
	11	Produce of mill industry; malt; starches; inulin; wheat gluten
	12	Oil seed, oleagi fruits; miscellaneous grain, seed, fruit; etc.
	13	Lac; gums, resins and other vegetable saps and extracts
	14	Vegetable plaiting materials; vegetable products nes
	15	Animal/vegetable fats and oils and their cleavage products
	24	Tobacco and manufactured tobacco substitutes
Fish	03	Fish and crustacean, mollusc and other aquatic invertebrate
Forestry	44	Wood and articles of wood; wood charcoal
	47	Pulp of wood/of other fibrous cellulosic material; waste, etc.
	48	Paper and paperboard; art of paper pulp, paper/paperboard
Energy	27	Mineral fuels, oils and product of their distillation; etc.
Minerals	25	Salt; sulphur; earth and stone; plastering materials; lime and cement
	26	Ores, slag and ash
	71	Natural/cultured pearls, precious stones and metals, coin; etc.
	72	Iron and steel
	73	Articles of iron and steel
	74	Copper and articles thereof
	75	Nickel and articles thereof
	76	Aluminum and articles thereof
	78	Lead and articles thereof
	79	Zinc and articles thereof
	80	Tin and articles thereof
	81	Other base metals; cermets; articles thereof

Source: Statistics Canada, *Exports by Country*, Cat. No. 65-003.

Cereals and oil seed are the most important agricultural products exported to the Asian countries. Presumably, the volatility in agricultural exports is a function of changes in both commodity prices and domestic supply. This appears to be the case for Bangladesh, China, India, Indonesia, Pakistan and the Philippines. On the other hand, agricultural exports to South Korea have grown spectacularly over the sample period, so that they are now slightly below those to China and about one-third of those to Japan. Most of the agricultural products are produced in western Canada: cereals and oil seed in Saskatchewan, Manitoba, and Alberta; meat and animal fat in Alberta and British Columbia; and fruit and oil seed in British Columbia. Agricultural products exported from central Canada are tobacco and some oil seed from Ontario and meat from Ontario and Quebec.

The value of fish exports is relatively low, except for those to Japan, which amounted to $750 million in 1994. Nevertheless, the value of fish exports to the next largest markets of Hong Kong, China, South Korea and Taiwan range from $10 to $50 million per year. About half the value of these exports originated in British Columbia and the other half in eastern Canada, primarily Nova Scotia, Newfoundland, and to a lesser extent, New Brunswick.

In 1994, almost $4 billion of forestry products were exported to Japan; over $1 billion to China, South Korea and Taiwan; and over $0.5 billion to Hong Kong, India, Indonesia, Malaysia, Singapore, and Thailand. Most originate in western Canada. The forestry exports to Japan consist mostly of wood, pulp, and some paper from British Columbia; some wood and pulp from Alberta; and some paper from New Brunswick. China, South Korea, and Taiwan import mostly pulp, with smaller values of wood and paper. These originate primarily in British Columbia, but some come from Alberta and Quebec. There is a similar product mix and province of origin for the other forestry importing countries.

Japan is by far the largest purchaser of Canada's energy exports, specifically coal from British Columbia and Alberta, which accounted for around $1.3 billion in 1994. South Korea is also a significant market, with expenditures of about $0.33 billion in 1994, and China also imports some coal from these provinces.

Japan is also the largest importer of Canadian mineral exports, at about $1 billion in 1994. Hong Kong, South Korea, and Taiwan together account for about $1 billion in mineral sales. China, India, Indonesia, the Philippines, Singapore, and Thailand account for about $0.33 billion. The mix of minerals and the province of origin varies. Most of the aluminum originates in British Columbia and Quebec; the ores and slag in British Columbia; precious stones in Ontario; nickel in Ontario and Manitoba; zinc in British Columbia; copper in British Columbia, Ontario, and Quebec; iron and steel in British Columbia, Ontario, and Quebec; and salt in British Columbia, Alberta, and Quebec. The value of mineral exports from western Canada is slightly greater than those from central Canada. The value of mineral exports to almost all the Asian countries declined significantly over the sample period.

Exports have been measured in value terms. It was not possible to get access to resource price data, so I could not calculate export volumes. I was able to obtain price indices for world exports of resource products from the United Nations (1995). These price indices can be used as a proxy for the

Canadian resource export prices. Unfortunately, they cannot be used to recreate an exact price index for agricultural, mineral, and forestry products because they are an aggregation of numerous products. The price indices for the aggregate resources as well as some of the major component products are shown in Figures B-1 to B-3.

With the exception of apples, the price indices of agricultural (food) and fish products were relatively constant from 1987 to 1993. Therefore, most of the large changes in the value of Canada's agricultural exports to the region are probably related to changes in volume. The price index for forestry products peaked in 1990 and then declined quite sharply. Hence the increases in the value of Canada's forestry exports are definitely related to significant increases in volume. The coal price index was fairly flat between 1987 and 1993, but the non-fuel mineral products price indices have fluctuated wildly, especially for nickel and zinc. The general trend has been for the mineral products price index to decline after 1989. Therefore, the decline in the value of Canada's mineral exports has been at least partially a result of the decline in mineral prices.

The value of Canada's resource exports is predominantly accounted for by relatively rich Japan, South Korea, Taiwan, and to a lesser extent, Hong Kong (but not Singapore); and by populous China, at least for agricultural products. In order to demonstrate the relationship between resource exports and economic activity, Figures C-1 to C-5 indicate the real value of the various resource exports per real GDP and real GDP per capita. Canadian export values were converted into real U.S. dollars using purchasing power parity (PPP) indices obtained from Penn World Tables, and real GDP values using PPP were obtained from the same source.[5]

One could hypothesize that as economies become more developed, as measured by real per capita GDP, they consume more natural resource products per GDP (or per capita). Two factors that might explain this are the rates of urbanization and industrialization, both of which increase with development. This effect is shown in Figures C-1 to C-5, but only partially, as they are export values rather than consumption values. Nevertheless, they generally support the hypothesis. The strongest support for a link between natural resource product consumption and economic development is in forestry and mineral products, and there is some support in energy and agricultural and fish products. There is less evidence that the use of agricultural and fish products are positively related to economic development, because food is a basic necessity.[6] The next section, which examines natural resource consumption in the Asian countries, includes strong evidence for the existence of the relationship for energy products. It is not evident in Figure C-4 because the only energy fuel exported to these countries is coal. There is a significant difference between Canada's exports and the importing countries' domestic energy consumption.

Graphically, in Figures C-2 to C-5, the poorest countries – Bangladesh, China, India, Indonesia, Pakistan, the Philippines, and Sri Lanka – are generally in the lower left-hand corner. Presumably, as the real GDP per capita increases in these countries, their use of natural resources per real GDP will also increase to levels observed in the Asian countries with the higher levels of real GDP per capita. It is impossible to predict exactly how large a shift in export growth there will be, since the upward trend is not uniform. But there is no doubt that on average an increase in real per capita GDP is associated with an increase in real exports per real GDP.

Rather than examine Canada's real *exports* per real GDP as a function of real per capita GDP, it is more appropriate to observe the importing country's domestic *consumption* per real GDP or per capita. It is domestic consumption that should be related to economic development; imports are a residual: domestic consumption net of domestic production and inventory changes. This is the focus of the next section.

THE RELATIONSHIP BETWEEN NATURAL RESOURCE CONSUMPTION AND ECONOMIC DEVELOPMENT

IN ORDER TO ASSESS THE PROSPECTS for natural resource trade with Asian countries, we would have to estimate future resource consumption in these countries. Since it was not possible to get access to detailed econometric models for each country, a simple cross-sectional analysis was employed to provide a crude estimate.

Cross-sectional analysis requires consistently defined consumption data over time for each resource for each country.[7] I was able to obtain implicit resource consumption data for selected years from the World Resources Institute (1992) and explicit consumption data for various energy fuels for all years from the Asian Development Bank (1992) and the International Energy Agency (1982, 1992). These data are in natural units rather than value units, so I measure resource consumption in natural units per capita, rather than per real GDP.

Data are available on aggregate energy consumption and energy fuel consumption, so it was possible to observe coal and oil consumption separately.[8] I was only able to obtain mineral consumption data for individual minerals, so the most important Canadian mineral exports – aluminum, copper, lead, nickel, zinc, iron ore, and crude steel – are analyzed separately. Forestry product consumption data consist of roundwood, which includes pulpwood but not paper. Cereals (wheat and wheat flour, rice, barley, maize, rye, and oats) was the only agricultural product for which I could obtain consistent consumption data. However, it is the most valuable of Canada's agricultural exports.

It is hypothesized that resource consumption per capita will increase with economic development.[9] This implies that there should be a positive trend in Figures D-1 to D-13, which depict a cross-section of country data for consumption per capita of various resource products and real GDP per capita. In order to provide a perspective to the analysis, resource data for Canada and the United States are also included. For all resources, except for energy, I was only able to obtain data on resource consumption for a few selected years, and the most recent data were for 1989 or 1990.[10] It is possible to test the hypothesis that economic development causes an increase in resource consumption per capita *within each country* by studying the changes in resource consumption over time. Since there are only two data points for each country,[11] the results are not conclusive.[12] The only resource for which it was not possible to conduct this test was fish, because there was only one data point.

There is a definite positive trend toward higher energy consumption per capita as real per capita GDP increases (Figure D-1). Each line represents the change within a country between 1980 and 1990. Except for Canada and the United States, each country increased its energy consumption per capita.[13] Figure D-2 depicts coal consumption, which is more relevant since coal is the only energy resource Canada exports to the region. The same observations can be made about coal consumption that were made about energy consumption. However, there are two differences. First, there was a slight decline in coal consumption per capita in Japan.[14] Second, China, and to a lesser extent India – both low-income countries – consume very high volumes of coal per capita. South Korea and Taiwan also consume high volumes of coal per capita. This is because in these countries it is an indigenous fuel, and it is used not only as a substitute for oil in most end-uses but also in steel production. Oil is the dominant energy fuel used in the Asian countries, and its consumption trends are similar to those for energy.

Figures D-4 to D-10 show mineral consumption per capita in a subset[15] of the Asian countries and, for comparative purposes, in Canada and the United States. Canada's main mineral exports to the Asian countries are examined separately, specifically: aluminum, copper, lead, nickel, zinc, iron ore, and crude steel. For all minerals there is a definite positive trend showing more mineral consumption per capita with higher real per capita GDP. Except for a few minerals in the case of some of the high-income countries (especially Canada and the United States), mineral consumption per capita increased between 1980 and 1990. Generally, the developed OECD countries experienced a decline in the intensity of mineral use, whereas the developing countries experienced an increase (see Tilton, 1990). This trend is especially noticeable in South Korea. Japan consumes more minerals per capita, with the exception of lead, than does the United States, which probably indicates that a higher proportion of economic activity in Japan is devoted to industrial production than to services.

There was a fairly flat rate of increase in consumption of forestry products per capita as real GDP per capita increased (Figure D-11). Canada is clearly an outlier, and the Asian developing countries are more likely to approach the consumption levels of the United States or Japan than those of Canada.

Within the Asian countries there does not appear to be much of a trend toward higher consumption of cereals per capita as real per capita GDP rises, but in Canada and the United States this trend is evident (Figure D-12). Since cereals are only a subset of agricultural consumption, it is not possible to base a conclusion regarding a shift in aggregate agriculture consumption with economic development on it. Nevertheless, one would not expect a very large shift, since most of these countries are sufficiently developed that the population consumes adequate amounts of a basic food group such as cereals.

There is a clear indication that fish consumption per capita increases with real per capita GDP (Figure D-13). It is interesting to note how the fish consumption habits of Asian countries differ from those of Canada and the United States. North American consumption of fish per capita is very low.

Using the 1990 data presented in Figures D-1 to D-13, I constructed a simple econometric model to test the hypothesis that there is a relationship between resource consumption per capita and real GDP per capita. The results are reported in Table 2. The 1990 data are used for the cross-section of countries listed in the figures. One would expect the marginal impact on resource consumption from an extra dollar of real GDP per capita to decrease as countries become more developed (see Tilton, 1990). This declining intensity of use effect can be observed for most of the resources. Consequently, a model that has the independent variable in logarithmic form is used for most of the equations.[16]

The relationship between resource consumption per capita and real GDP per capita is statistically significant (at the 95 percent level) for most of the resource products, with the exception of nickel and iron ore – for whose equations there were not many observations. For most of the equations, between 50 and 80 percent of the variation in resource consumption per capita is explained by real GDP per capita. It is no surprise to discover that other factors need to be included in this simple model. Finally, a Chow test was performed to determine whether the parameters of the relationship could be considered statistically stable if the 1980 rather than the 1990 cross-section data had been used. The results reported in Table 2 show that the relationship is stable in all cases except for oil.

In summary, this section has shown that natural resource consumption per capita is a function of real per capita GDP, and the relationship is different for various natural resource products.

THE POTENTIAL FOR THE EXPANSION OF NATURAL RESOURCE EXPORTS TO ASIAN COUNTRIES

IF WE KNEW THE RELATIONSHIP between resource consumption and economic development for the Asian countries, we could estimate the future potential

TABLE 2

THE RELATIONSHIP BETWEEN RESOURCE CONSUMPTION PER CAPITA AND
REAL GDP PER CAPITA, SELECTED ASIAN COUNTRIES,
THE UNITED STATES AND CANADA[1]

RESOURCE	CONSTANT	REAL GDP PER CAPITA	R^2	N	F^2	
Energy	−0.341 (−0.91)	$0.270*10^{-3}$ (6.61)	0.7574	16	F_{Chow} = 3.18	$F_{2,28}$ = 3.34
Coal	−0.579 (−0.90)	$0.873*10^{-1}$ (2.38)	0.3200	14	F_{Chow} = 0.07	$F_{2,24}$ = 3.40
Oil	−0.128 (−0.79)	$0.129*10^{-3}$ (7.19)	0.7990	15	F_{Chow} = 5.67	$F_{2,26}$ = 3.37
Aluminum	−0.0466 (−6.47)	$0.650*10^{-2}$ (7.96)	0.9268	7	F_{Chow} = 0.12	$F_{2,10}$ = 4.10
Copper	−0.0278 (−3.15)	$0.396*10^{-2}$ (3.88)	0.7149	8	F_{Chow} = 0.25	$F_{2,12}$ = 3.89
Lead	−0.0117 (−4.64)	$0.163*10^{-2}$ (5.67)	0.8426	8	F_{Chow} = 0.003	$F_{2,12}$ = 3.89
Nickel	−0.00192 (−1.50)	$0.273*10^{-3}$ (1.88)	0.5397	5	F_{Chow} = 0.03	$F_{2,6}$ = 5.14
Zinc	−0.0125 (−3.61)	$0.179*10^{-2}$ (4.54)	0.6959	11	F_{Chow} = 0.54	$F_{2,18}$ = 3.55
Iron ore	−0.859 (−0.63)	0.150 (0.98)	0.3232	4	F_{Chow} = 0.03	$F_{2,4}$ = 6.94
Crude steel	−1.439 (−2.63)	0.209 (3.27)	0.7809	5	F_{Chow} = 0.04	$F_{2,6}$ = 5.14
Forestry products	0.0734 (0.15)	$0.203*10^{-3}$ (3.31)	0.4995	13	F_{Chow} = 0.11	$F_{2,22}$ = 3.44
Cereal products	0.209 (4.71)	$0.271*10^{-4}$ (5.09)	0.7023	13	F_{Chow} = 1.76	$F_{2,22}$ = 3.44
Fish products[3]	−0.159 (−4.64)	$0.233*10^{-1}$ (5.46)	0.7304	13	n.a.	

Note: N is the number of observations.
n.a.: not applicable.
The number in parentheses is the t-value.
1 The independent variable in the equations for coal, all mineral products, and fish products is in logarithmic form, i.e., ln (real GDP per capita).
2 F_{Chow} is the calculated F-value for the Chow test; $F_{i,j}$ is the critical F-value.
3 The observations for Canada and the United States are dropped from the fish equation.

resource consumption in these countries. Canadian resource exporters could then use this information to gauge the size of future market opportunities.

The key variable in this analysis is real GDP per capita. The World Bank (1995, pp. 763-64) has classified the 14 Asian countries into three categories according to their income level: Bangladesh, China, India, Pakistan, and Sri Lanka are low income; Indonesia, South Korea, Malaysia, the Philippines, and Thailand are middle income; and Hong Kong, Japan, Singapore, and Taiwan are high income. Real GDP per capita from 1980 to 1992 in the three income classes is shown in Figures E-1 to E-3. It is clear that real economic growth has been spectacular in Hong Kong, Singapore, Taiwan, Thailand, and especially in South Korea. The average annual growth rate of real GDP per capita[17] in these countries in 1980-92 was 5.5 percent for Hong Kong, 4.8 percent for Singapore, 6.4 percent for Taiwan, 5.2 percent for Thailand, and 8.0 percent for South Korea (Table 3). Most of the other countries had growth rates of between 3 and 4 percent. The main exception is the Philippines, whose growth rate was negative 1 percent. For comparison, the growth rate in Canada over the period was 1.9 percent.

TABLE 3

POPULATION, POPULATION GROWTH AND REAL GDP PER CAPITA, SELECTED ASIAN COUNTRIES AND CANADA, BY INCOME GROUP, 1980-92[a]

COUNTRY	POPULATION 1992 (MILLIONS)	POPULATION GROWTH (1980-92)	REAL GDP/CAPITA 1992 (1985 US$)	REAL GDP/CAPITA GROWTH (1980-92)
Low income				
Bangladesh	114.0	0.0224	1,510	0.0287
China	1,162.0	0.0144	1,493	0.0371
India	884.0	0.0210	1,282	0.0338
Pakistan	119.0	0.0307	1,432	0.0241
Sri Lanka	17.4	0.0140	2,215	0.0231
Middle income				
Indonesia	184.0	0.0180	2,102	0.0349
South Korea	43.7	0.0109	7,832	0.0801
Malaysia	18.6	0.0253	5,746	0.0281
Philippines	64.3	0.0238	1,689	−0.0088
Thailand	58.0	0.0180	3,942	0.0524
High income				
Canada	27.4	0.0106	16,362	0.0185
Hong Kong	5.8	0.0118	16,471	0.0549
Japan	124.0	0.0052	15,105	0.0358
Singapore	2.8	0.0171	12,653	0.0477
Taiwan	20.9	0.0130	9,129	0.0641

a The growth rates were calculated as the slope of a regression equation that has the logarithm of dependent variable as a function of a time index for the years 1980 to 1992.
Source: Penn World Tables, PWT 5.6.

It is obvious that there is a large income gap between the low-, middle- and high-income countries. For instance, in 1992 the average real per capita GDP was about US$1,600 (1985 base) in the low-income countries, about US$4,300 in the middle-income countries, and about US$14,000 in the high-income countries (see Table 3).

I focus here on estimating resource consumption in China, India, and Indonesia once they attain the average income levels of the middle-income countries.[18] These are the three most populous countries in the sample (with a total of 2.23 billion people, they account for around 40 percent of humankind) and they have recently begun to encourage market reform. An average real GDP per capita like that of the middle-income countries would be slightly higher than that of Thailand in 1992 (see Figure E-2). The length of time required to attain it would depend upon the growth rate of real GDP per capita. I will consider two cases: the status quo case assumes a continuation of the historical growth rate in 1980 to 1992, and the high growth case assumes a real growth rate equal to South Korea's historical rate from 1980 to 1992 (namely, 8 percent).

To obtain resource consumption at the new level of real GDP per capita, the new value of per capita resource consumption is multiplied by the population. The new value for per capita resource consumption can be calculated from the coefficients reported in Table 2. The population level can be estimated using the historical growth rates (given in Table 3) and the number of years necessary to attain the new income level.

At a real GDP per capita of US$4,300, the estimated equations reported in Table 2 show the following per capita consumption levels: for energy, 0.82 metric ton of oil equivalent (TOE); for coal, 0.151 metric ton of oil equivalent; for oil, 0.427 metric ton of oil equivalent; for aluminum, 0.00778 metric ton; for copper, 0.00533 metric ton; for lead, 0.00194 metric ton; for nickel, 0.000364 metric ton; for zinc, 0.00248 metric ton; for iron ore, 0.396 metric ton; for crude steel, 0.310 metric ton; for forestry products, 0.946 cubic metre; for cereals, 0.326 metric ton; and for fish products, 0.0359 metric ton.

Under the status quo case, it would take China 29 years (until 2021) to reach the level of US$4,300 (1985-based) real per capita GDP, it would take India 36.4 years (until 2028), and it would take Indonesia 20.9 years (until 2013). The estimated population in China in 2021 would be 1,759 million; in India in 2028, 1,883.6 million; and in Indonesia in 2013, 267.1 million. Under the high growth case it would take China only 13.7 years (until 2006) to reach the level of US$4,300 (1985-based) real GDP per capita, it would take India 15.7 years (until 2008), and it would take Indonesia 9.3 years (until 2001). The estimated population in China in 2006 would be 1,413.4 million; in India in 2008, 1,225.1 million; and in Indonesia in 2001, 217.2 million.

The forecast of aggregate consumption in China, India and Indonesia of all of the resource products, given the status quo growth case, is shown in Table 4. If available, historical reference values for resource consumption are also included. The forecast values are quite large, especially for China[19] and India. However, the forecast values for China are for the year 2021 and for India the year 2028. To put the size of the forecast values in context, those for

TABLE 4

HISTORICAL[1] AND FORECAST[2] RESOURCE CONSUMPTION IN CHINA, INDIA AND INDONESIA: STATUS QUO GROWTH CASE

	CHINA		INDIA		INDONESIA	
RESOURCE (UNITS)	HISTORICAL	FORECAST	HISTORICAL	FORECAST	HISTORICAL	FORECAST
Energy (10^6 TOE)	536.5	1,442.4	128.9	1,544.6	40.6	219.0
Coal (10^6 TOE)	347.2	265.6	51.7	284.4	1.4	40.3
Oil (10^6 TOE)	95.7	751.1	57.0	804.3	30.1	114.1
Aluminum (10^6 tons)	0.7	13.7	0.4	14.7		2.1
Copper (10^6 tons)	0.5	9.4	0.1	10.0		1.4
Lead (10^6 tons)	0.3	3.4	0.08	3.7		0.5
Nickel (10^6 tons)	0.03	0.6	0.01	0.7		0.1
Zinc (10^6 tons)	0.5	4.4	0.1	4.7		0.7
Iron ore (10^6 tons)	184.0	696.6		745.9		105.8
Crude steel (10^6 tons)	69.5	545.3	20.0	583.9		82.8
Forestry (10^6 m^3)	289.2	1,664.0	265.3	1,781.9	172.5	252.7
Cereals (10^6 tons)	386.3	573.4	194.7	614.1	52.3	87.1
Fish (10^6 tons)	9.5	63.1	2.6	67.6	2.4	9.6

Note: The historical data for consumption of mineral products in Indonesia were not available.
1 For energy and mineral products, the data are from 1990; for forestry products, the 1987-89 average; for cereals, the 1988-90 average; and for fishery products, the 1986-88 average.
2 The forecast year is 2021 for China, 2028 for India, and 2013 for Indonesia.

China can be compared to those of the United States in 1990. The forecast consumption of energy and oil in China is similar to that of the United States in 1990; the coal consumption forecast is more than 4 times the 1990 value; the forecasts for most of the mineral products are 3 to 4 times the 1990 value, except iron ore, which is 10 times and steel which is about 5 times; the forecasts for forestry and cereal products are about 3 times the 1990 values; and forecast fish products consumption is almost 12 times the 1990 value for the United States (although only about 7 times the value for Japan). Although the forecast for India is for seven years later (2028), the forecast values for India's resource consumption are greater than China's because of the relatively high population growth in India. The forecast values for Indonesia can be compared to Japan's 1990 values. Most of the forecast values are between 60 and 100 percent of Japan's 1990 values, except forestry products, which are 3 times greater, and cereals, which are 2 times greater. Put differently, the forecast for total resource consumption in the three countries is roughly equivalent to 40 percent of the energy consumed *in the world* in 1990, between 140 and 180 percent of minerals, over 100 percent of forestry products, about 70 percent of cereals, and over 200 percent of fish products.

Obviously, there will be a huge impact on resource consumption if these predictions are upheld. I have only concentrated on three countries and have made the rather modest assumption that the economies will expand to the same level of economic development as the current average real GDP per capita of the middle-income countries. Even that level would leave these countries at only about 50 percent of the 1992 levels of real GDP per capita in Taiwan and South Korea, and about 30 percent of the level in Japan. The status quo assumption for the growth of real per capita GDP may be somewhat low. Under the high growth assumption, the forecasts in Table 4 would be somewhat smaller,[20] but they would occur much sooner (between the years 2001 and 2008).

Another hypothetical experiment was performed that assumed the population and real per capita GDP in every country grew for 10 years at their historical rates. The results are reported in Table 5 for the low-, middle- and high-income countries. Except for energy and coal in the low-income countries (see Endnote 19), the forecast shows much higher consumption of resource products. Obviously, the forecast would be higher if I used the high-growth case (8 percent) for real GDP per capita. In fact, resource consumption would be about 50 percent higher with the high-growth case.

How confident can we be in the results? Clearly, this methodology is rather simplistic, and should not be considered a substitute for detailed country- and industry-specific econometric models. However, there is an advantage in using cross-sectional data in that the predicted rate of resource consumption is within the range of actual resource consumption in other countries. Nevertheless, the economies could in reality be very different. In particular, one should be sceptical of applying the prediction methodology for transport and

TABLE 5

Historical[1] and Forecast[2] Resource Consumption in Asian Countries, by Income Group, Status Quo Growth for 10 Years[3]

RESOURCE (UNITS)	LOW-INCOME COUNTRIES		MIDDLE-INCOME COUNTRIES		HIGH-INCOME COUNTRIES	
	HISTORICAL	FORECAST	HISTORICAL	FORECAST	HISTORICAL	FORECAST
Energy (10^6 toe)	691.8	535.6	158.6	456.1	345.4	878.0
Coal (10^6 toe)	401.9	229.7	22.6	62.9	44.0	47.6
Oil (10^6 toe)	164.3	352.0	110.3	233.4	200.5	425.2
Aluminum (10^6 tons)		7.5		3.1		3.0
Copper (10^6 tons)		6.2		2:2		1.9
Lead (10^6 tons)		1.8		0.8		0.7
Nickel (10^6 tons)		0.4		0.1		0.1
Zinc (10^6 tons)		3.0		1.0		0.9
Iron ore (10^6 tons)		768.6		168.2		104.0
Crude steel (10^6 tons)		404.5		127.1		105.2
Forestry (10^6 m³)	616.4	1,310.3	289.2	489.0		714.3
Cereals (10^6 tons)	636.4	723.1	108.1	153.6		128.1
Fish (10^6 tons)	13.4	49.0	8.1	14.8		12.0

1 For energy and mineral products, the data are from 1990; for forestry products, the 1987-89 average; for cereals, the 1988-90 average; and for fishery products, the 1986-88 average.
2 Forecast is for 10 years of growth at historical rates for population and real GDP per capita.
3 The historical consumption levels for mineral products are not included because the data are not available for most of the countries. The same is true for forestry, cereals, and fish for all of the high-income countries except Japan.

heavy industry usage (energy, oil, and some minerals) to Hong Kong and Singapore, because of their small land base.

The dramatic nature of the predictions made in this paper are reminiscent of those by Meadows et al. (1972). Obviously, there will be endogenous economic responses (which have not been modelled) that will dampen the magnitude of the forecast. If resource use were to increase by the amounts predicted in this paper, one would expect resource prices to increase and interfactor substitution to choke off some of the demand. It is also very probable that future technological developments will be resource-saving, which will also reduce expected demand. Per capita resource consumption will also be affected by government policies, which influence investment rates and the growth and composition of the manufacturing sector. There is room to dispute the extent of the growth in natural resource consumption by Asian countries, but there is no doubt that it will increase substantially.

Finally, we should ask the question: Will Canadian exporters be able to benefit from the expanded resource market in Asia? The answer depends upon the Canadian resource base and whether Canadian industry can be competitive, as well as the resource production capabilities of the Asian countries. Resource exports to Asia are expected to increase (see Labson et al., 1992, and Yamaguchi, 1995 for an analysis of energy and minerals exports). Canada has been successful in the past in exporting coal, mineral, forestry and agricultural products to the region. However, from 1988 to 1994 Canadian exports of agricultural, fish, mineral, and energy products failed to increase as fast as real GDP in the Asian economies, and only forestry products grew faster. Canadian producers probably could expand their production of resource-based products, but that analysis is beyond the scope of this paper. Canada's existing resource supply can be compared to other world producers by examining the data in World Resources Institute (1992).

Summary

THE ECONOMIES OF ASIA ARE EXPECTED TO GROW at a faster rate than those of the developed countries and therefore to consume more natural resource products. Since Canada is a major exporter of resources, it is important to examine Asia's potential resource requirements. In this paper I examined the historical value of exports of agricultural, fishery, forestry, mineral and energy products to 14 Asian countries. Resource exports are predominantly from western Canada. Trends in per unit prices of the resources were examined in order to differentiate changes in value from changes in quantity. The relationship between resource consumption per capita and real GDP per capita was captured using cross-sectional analysis. For each resource product, this relationship, along with various assumptions about the expected growth in real GDP per capita and in the population, was used to estimate potential resource consumption for each country. I placed special emphasis on the most populous of the low-income countries, namely, China, India, and Indonesia.

How can Canadian resource producers benefit from the likely growth of resource consumption in Asia? If Canadian suppliers can maintain or increase their market share, they will be able to expand the volume of resource exports to Asia. Therefore, it is crucial that they be capable of increasing production at market prices. Canadian producers and investors could also benefit by exporting resource production technology and services to Asian countries. For example, Canadian firms should be able to provide knowledge and capital to construct pulp mills, oil and natural gas pipelines, oil refineries, and electric power plants using hydro or natural gas cogeneration technology. They should also be able to assist in the exploration for and production of oil, natural gas, and minerals. Some of the Asian countries, especially China, have under-explored their resource potential in energy and minerals. Despite the potential, the level of Canadian foreign direct investment in Asian economies is very low (Flatters and Harris, 1995).

APPENDIX A

FIGURE A-1

CANADA'S RESOURCE EXPORTS TO BANGLADESH, 1988-94

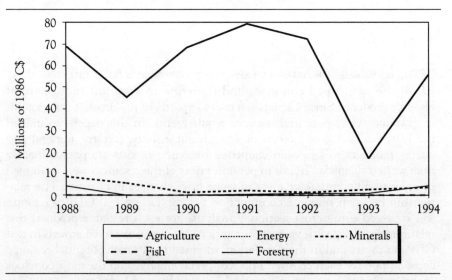

Source: Statistics Canada and Department of Finance (1995).

CANADA'S NATURAL RESOURCE EXPORTS

FIGURE A-2

CANADA'S RESOURCE EXPORTS TO THE PEOPLE'S REPUBLIC OF CHINA, 1988-94

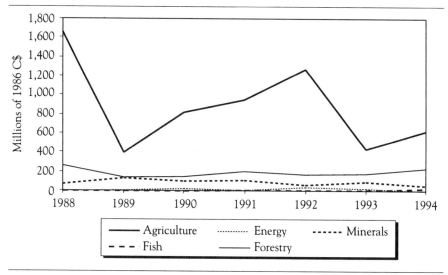

Source: Statistics Canada and Department of Finance (1995).

FIGURE A-3

CANADA'S RESOURCE EXPORTS TO HONG KONG, 1988-94

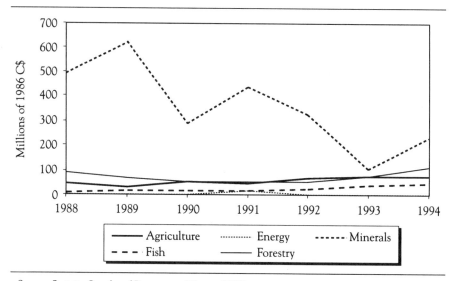

Source: Statistics Canada and Department of Finance (1995).

Figure A-4

Canada's Resource Exports to India, 1988-94

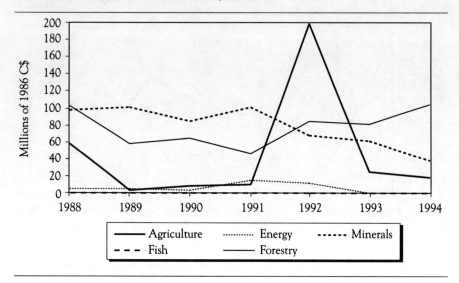

Source: Statistics Canada and Department of Finance (1995).

Figure A-5

Canada's Resource Exports to Indonesia, 1988-94

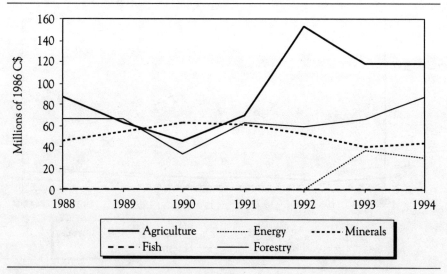

Source: Statistics Canada and Department of Finance (1995).

CANADA'S NATURAL RESOURCE EXPORTS

FIGURE A-6

CANADA'S RESOURCE EXPORTS TO JAPAN, 1988-94

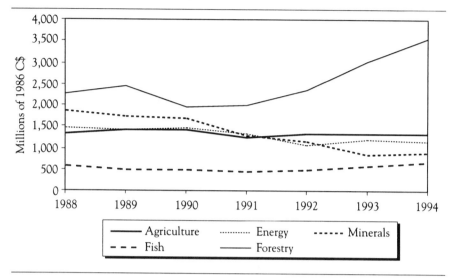

Source: Statistics Canada and Department of Finance (1995).

FIGURE A-7

CANADA'S RESOURCE EXPORTS TO SOUTH KOREA, 1988-94

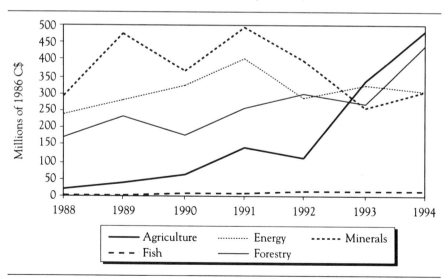

Source: Statistics Canada and Department of Finance (1995).

FIGURE A-8

CANADA'S RESOURCE EXPORTS TO MALAYSIA, 1988-94

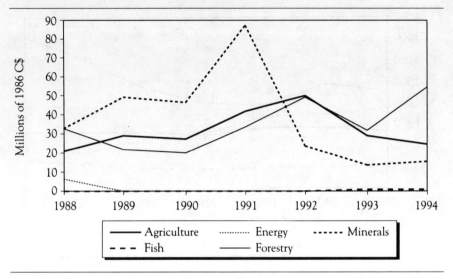

Source: Statistics Canada and Department of Finance (1995).

FIGURE A-9

CANADA'S RESOURCE EXPORTS TO PAKISTAN, 1988-94

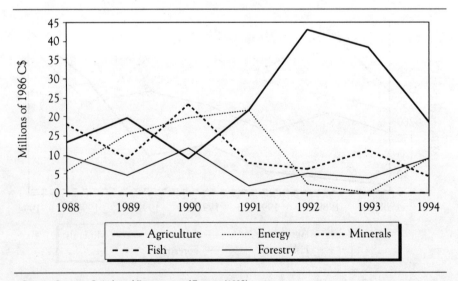

Source: Statistics Canada and Department of Finance (1995).

CANADA'S NATURAL RESOURCE EXPORTS

FIGURE A-10

CANADA'S RESOURCE EXPORTS TO THE PHILIPPINES, 1988-94

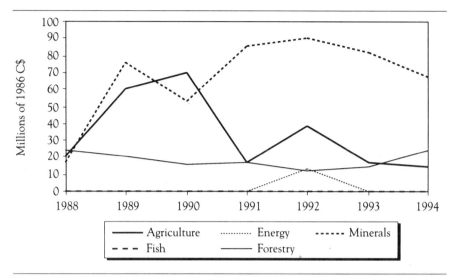

Source: Statistics Canada and Department of Finance (1995).

FIGURE A-11

CANADA'S RESOURCE EXPORTS TO SINGAPORE, 1988-94

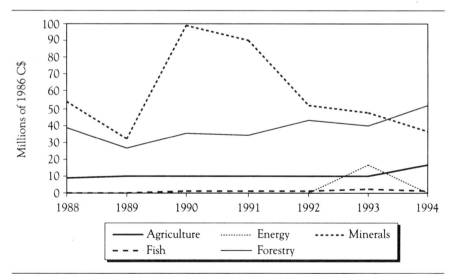

Source: Statistics Canada and Department of Finance (1995).

Figure A-12

Canada's Resource Exports to Sri Lanka, 1988-94

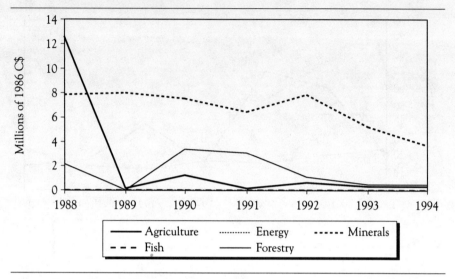

Source: Statistics Canada and Department of Finance (1995).

Figure A-13

Canada's Resource Exports to Taiwan, 1988-94

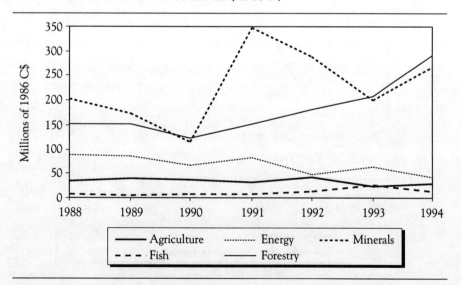

Source: Statistics Canada and Department of Finance (1995).

CANADA'S NATURAL RESOURCE EXPORTS

FIGURE A-14

CANADA'S RESOURCE EXPORTS TO THAILAND, 1988-94

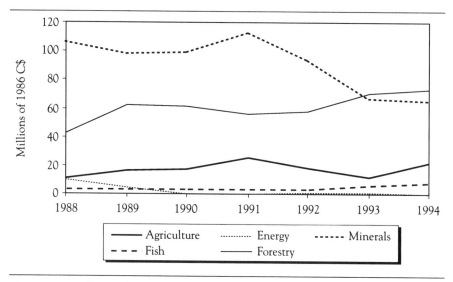

Source: Statistics Canada and Department of Finance (1995).

FIGURE A-15

CANADA'S RESOURCE EXPORTS TO ALL 14 ASIAN COUNTRIES, 1988-94

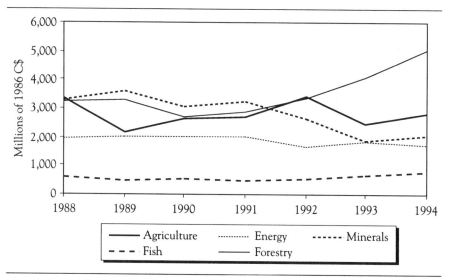

Source: Statistics Canada and Department of Finance (1995).

FIGURE A-16

CANADA'S RESOURCE EXPORTS TO ASIAN DEVELOPING COUNTRIES,[1] 1988-94

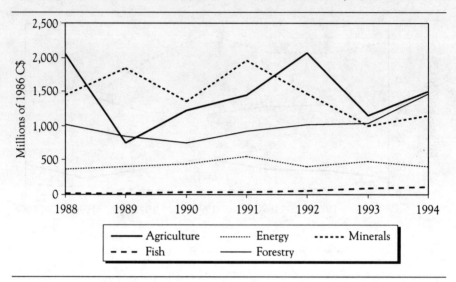

1 All countries except Japan.
Source: Statistics Canada and Department of Finance (1995).

CANADA'S NATURAL RESOURCE EXPORTS

Appendix B

Figure B-1

World Export Price Index for Agricultural and Fish Products, 1987-93

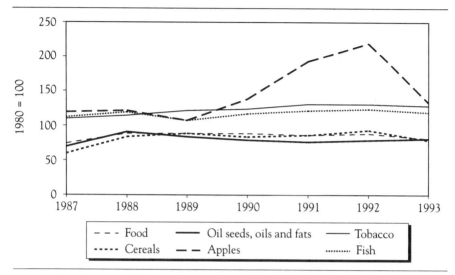

Source: United Nations (1995), pp. 5118-119.

Figure B-2

World Export Price Index for Forestry Products, 1987-93

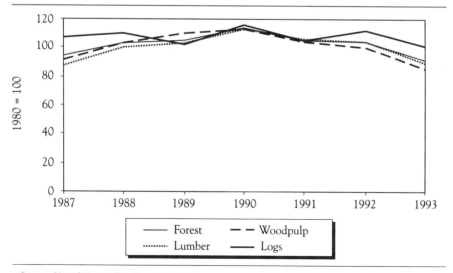

Source: United Nations (1995), pp. 5118-119.

145

Figure B-3

World Export Price Index for Energy and Mineral Products, 1987-93

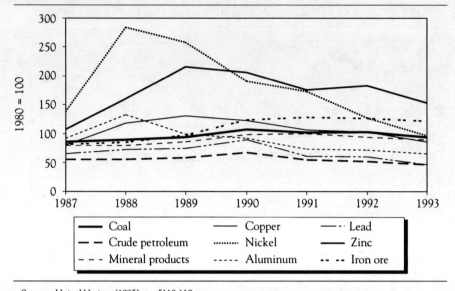

Source: United Nations (1995), pp. 5118-119.

APPENDIX C

FIGURE C-1

CANADA'S AGRICULTURAL EXPORTS TO SELECTED ASIAN COUNTRIES, 1992

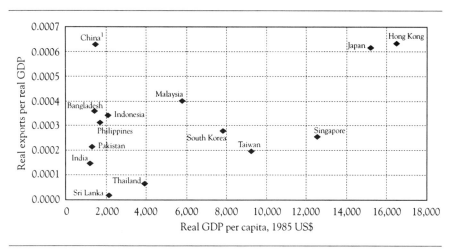

1 People's Republic of China.
Source: Penn World Tables, PWT 5.6, and Statistics Canada.

FIGURE C-2

CANADA'S FORESTRY EXPORTS TO SELECTED ASIAN COUNTRIES, 1992

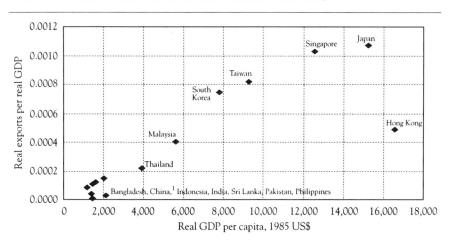

1 People's Republic of China.
Source: Penn World Tables, PWT 5.6, and Statistics Canada.

Figure C-3

Canada's Mineral Exports to Selected Asian Countries, 1992

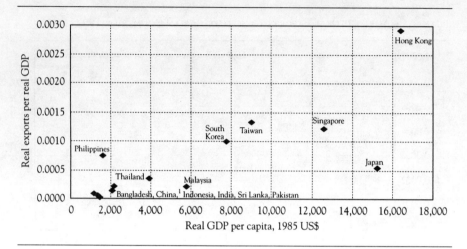

1 People's Republic of China.
Source: Penn World Tables, PWT 5.6, and Statistics Canada.

Figure C-4

Canada's Energy Exports to Selected Asian Countries, 1992

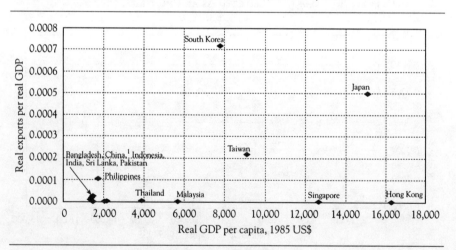

1 People's Republic of China.
Source: Penn World Tables, PWT 5.6, and Statistics Canada.

FIGURE C-5

CANADA'S FISH EXPORTS TO SELECTED ASIAN COUNTRIES, 1992

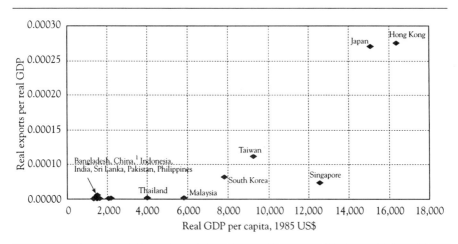

1   People's Republic of China.
Source: Penn World Tables, PWT 5.6, and Statistics Canada.

Appendix D

Figure D-1

Energy Consumption per Capita, Selected Asian Countries, the United States and Canada, 1980-90

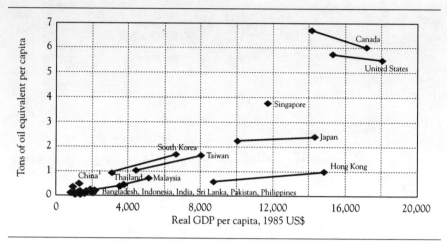

1 People's Republic of China.
Source: World Resource Institute (1992), Asian Development Bank (1992), International Energy Agency (1982, 1992), and Penn World Tables, PWT 5.6.

Figure D-2

Coal Consumption per Capita, Selected Asian Countries, the United States and Canada, 1980-90

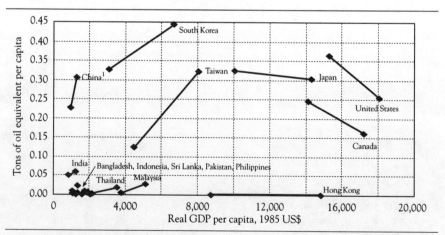

1 People's Republic of China.
Source: World Resource Institute (1992), Asian Development Bank (1992), International Energy Agency (1982, 1992), and Penn World Tables, PWT 5.6.

CANADA'S NATURAL RESOURCE EXPORTS

FIGURE D-3

OIL CONSUMPTION PER CAPITA, SELECTED ASIAN COUNTRIES,
THE UNITED STATES AND CANADA, 1980-90

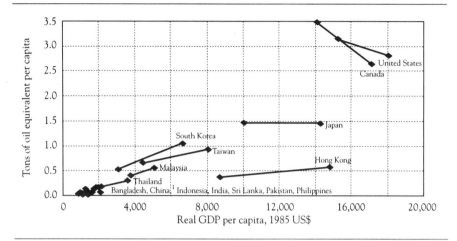

1   People's Republic of China.
Source: World Resource Institute (1992), Asian Development Bank (1992), International Energy Agency (1982, 1992), and Penn World Tables, PWT 5.6.

FIGURE D-4

ALUMINUM CONSUMPTION PER CAPITA, SELECTED ASIAN COUNTRIES,
THE UNITED STATES AND CANADA, 1980-90

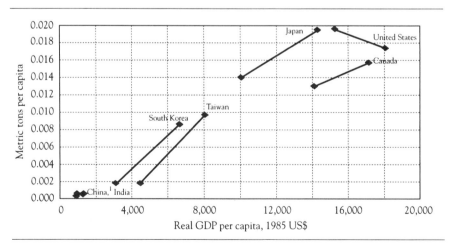

1   People's Republic of China.
Source: World Resource Institute (1992), Asian Development Bank (1992), International Energy Agency (1982, 1992), and Penn World Tables, PWT 5.6.

Figure D-5

Copper Consumption per Capita, Selected Asian Countries, the United States and Canada, 1980-90

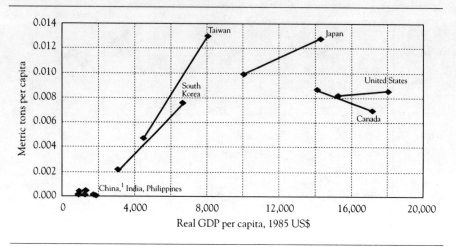

1 People's Republic of China.
Source: World Resource Institute (1992), Asian Development Bank (1992), International Energy Agency (1982, 1992), and Penn World Tables, PWT 5.6.

Figure D-6

Lead Consumption per Capita, Selected Asian Countries, the United States and Canada, 1980-90

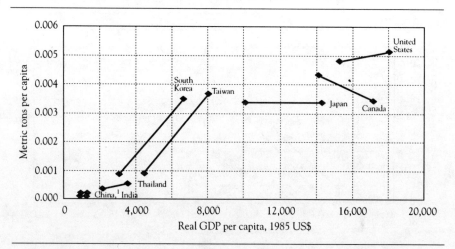

1 People's Republic of China.
Source: World Resource Institute (1992), Asian Development Bank (1992), International Energy Agency (1982, 1992), and Penn World Tables, PWT 5.6.

CANADA'S NATURAL RESOURCE EXPORTS

FIGURE D-7

NICKEL CONSUMPTION PER CAPITA, SELECTED ASIAN COUNTRIES, THE UNITED STATES AND CANADA, 1980-90

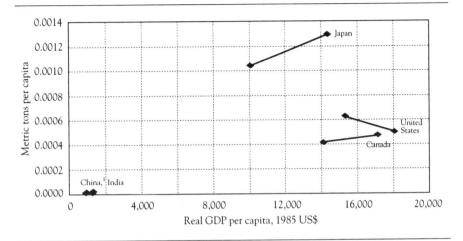

1   People's Republic of China.
Source: World Resource Institute (1992), Asian Development Bank (1992), International Energy Agency (1982, 1992), and Penn World Tables, PWT 5.6.

FIGURE D-8

ZINC CONSUMPTION PER CAPITA, SELECTED ASIAN COUNTRIES, THE UNITED STATES AND CANADA, 1980-90

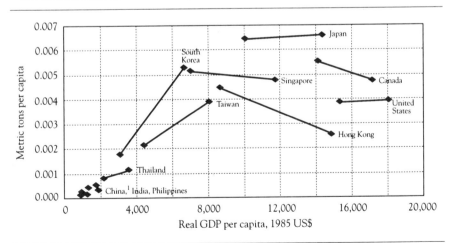

1   People's Republic of China.
Source: World Resource Institute (1992), Asian Development Bank (1992), International Energy Agency (1982, 1992), and Penn World Tables, PWT 5.6.

153

Figure D-9

Iron Ore Consumption per Capita, Selected Asian Countries, the United States, 1980-90

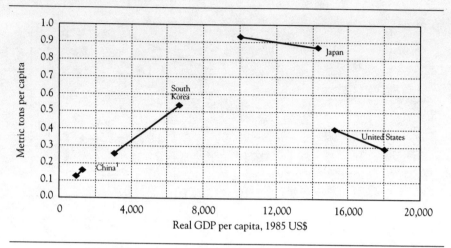

1. People's Republic of China.

Source: World Resource Institute (1992), Asian Development Bank (1992), International Energy Agency (1982, 1992), and Penn World Tables, PWT 5.6.

Figure D-10

Crude Steel Consumption per Capita, Selected Asian Countries, the United States, 1980-90

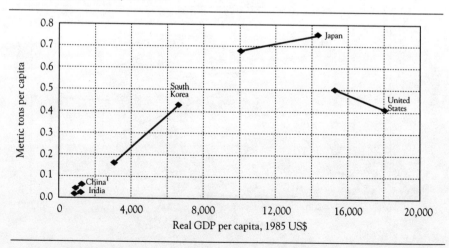

1. People's Republic of China.

Source: World Resource Institute (1992), Asian Development Bank (1992), International Energy Agency (1982, 1992), and Penn World Tables, PWT 5.6.

CANADA'S NATURAL RESOURCE EXPORTS

FIGURE D-11

AVERAGE FOREST PRODUCTS CONSUMPTION PER CAPITA, SELECTED
ASIAN COUNTRIES, THE UNITED STATES AND CANADA, 1977-79 TO 1987-89

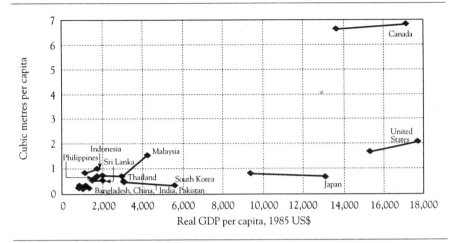

1   People's Republic of China.
Source: World Resource Institute (1992), Asian Development Bank (1992), International Energy Agency
         (1982, 1992), and Penn World Tables, PWT 5.6.

FIGURE D-12

AVERAGE AGRICULTURAL (CEREAL) CONSUMPTION PER CAPITA, SELECTED
ASIAN COUNTRIES, THE UNITED STATES AND CANADA, 1978-80 TO 1988-90

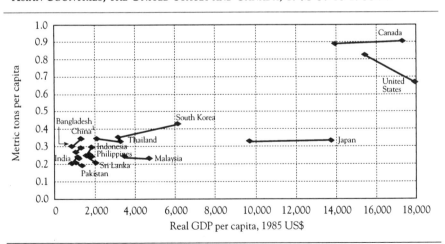

1   People's Republic of China.
Source: World Resource Institute (1992), Asian Development Bank (1992), International Energy Agency
         (1982, 1992), and Penn World Tables, PWT 5.6.

FIGURE D-13

AVERAGE FISH CONSUMPTION PER CAPITA, SELECTED ASIAN COUNTRIES, THE UNITED STATES AND CANADA, 1988

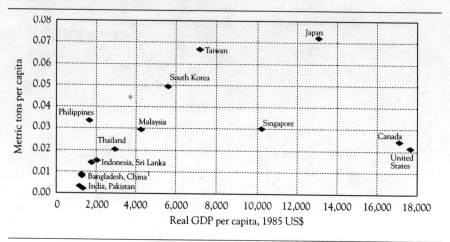

1 People's Republic of China.

Source: World Resource Institute (1992), Asian Development Bank (1992), International Energy Agency (1982, 1992), and Penn World Tables, PWT 5.6.

Appendix E

Figure E-1

REAL GDP PER CAPITA IN SELECTED LOW-INCOME ASIAN COUNTRIES, 1980-92

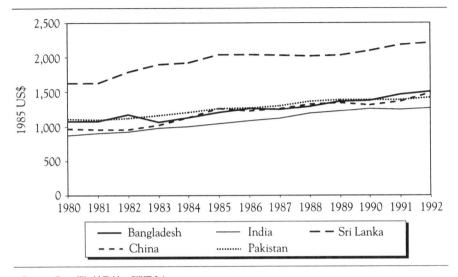

Source: Penn World Tables, PWT 5.6.

Figure E-2

Real GDP per Capita in Selected Middle-Income Asian Countries, 1980-92

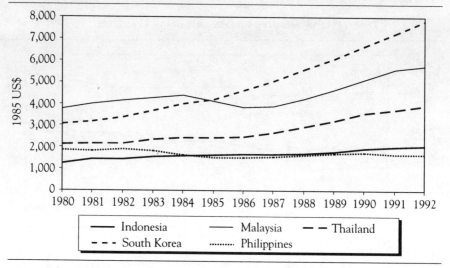

Source: Penn World Tables, PWT 5.6.

Figure E-3

Real GDP per Capita in Selected High-Income Asian Countries and Canada, 1980-92

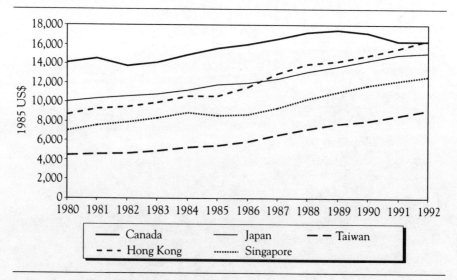

Source: Penn World Tables, PWT 5.6.

Endnotes

1. The major Asia Pacific countries are: in North America – Canada, Mexico and the United States; in South America – Chile, Columbia, Ecuador and Peru; in Asia – China, Hong Kong, Indonesia, Japan, South Korea, Malaysia, Philippines, Singapore, Taiwan, and Thailand.
2. The South Asian countries include Bangladesh, India, Pakistan and Sri Lanka. They are incorporated into the analysis because they are strategically linked to the other Asian economies.
3. These are aggregate classifications. For instance, energy includes various types of coal, oil products, and natural gas; minerals include aluminum, copper, lead, zinc, nickel, and iron ore; forestry products include lumber, sawmill products, and pulp and paper; and agricultural products include wheat, oil seeds, oils and fats, and many other products. Fish products are separated from other food products because their production uses a different technology.
4. The aggregation was performed using macros within Microsoft Excel.
5. Version 5.6 of the Penn World Tables was obtained on line from NBER Harvard, at nber.harvard.edu.
6. Perhaps it would have been more instructive to examine agricultural exports per capita rather than per real GDP.
7. The United Nations (1993) mostly collects resource production data. It does not collect resource trade data, therefore it is not possible to calculate consumption as domestic disappearance.
8. I concentrate on coal because it is an important fuel for several Asian countries, and it is the only fuel that Canada exports to the region. Oil consumption is also examined because it is the dominant energy fuel used in most of the countries.
9. Real GDP per capita is used to measure economic development. The relationship between resource consumption per capita and economic development for newly industrializing countries is described in Labson et al. (1992) and Tilton (1990). The ratio of resource consumption per capita – the intensity of use – is a function of the resource composition of products and the product composition of economic output.
10. For agricultural, forestry, and fishery products the consumption data points were a three-year average.
11. The Asian Development Bank has examined the relationship between energy consumption per capita and real per capita GDP, which was not adjusted for purchasing power parity. For each Asian developing country the results indicate an unambiguous increase in energy consumption per capita as economic development increases. They also indicate a strong relationship between urbanization and energy consumption per capita (1992, pp.7-8).

12. Graphically, in Figures D-1 to D-13 a number of countries are clustered in the lower left-hand corner, so it is not possible to visualize the upward shift in resource consumption per capita with development. However, it does occur in all cases.
13. There are numerous studies of energy consumption per GDP (energy intensity). McRae (1994) compares changes in energy intensity by end-use sector and by energy fuel in Canada with some Asian developing countries. Although that study uses energy consumption per GDP, while this one uses energy consumption per capita, the results should be similar. Since 1973, energy intensity has declined by about 2 percent per year in developed countries, whereas it has increased in most developing Asian countries.
14. Coal consumption per capita declined in Japan because liquefied natural gas was being substituted for coal in electric power generation and steel production declined.
15. The World Resources Institute (1992) only lists mineral consumption for the top 10 countries. The only Asian countries for which they report data are Japan, China, South Korea, and occasionally India. I supplemented the mineral consumption data with those reported in Bang (1992).
16. I experimented with alternative models, one of which had the independent variable in a linear form and the other in a second-order polynomial form. The polynomial model provided a better statistical fit, but the predicted decline in resource intensity for the high-income countries was so severe that the forecast levels for consumption were sometimes negative.
17. The average annual growth of real GDP is the sum of the average annual growth of real GDP per capita and the growth of the population.
18. Research by authors such as Helliwell (1995) and Mueller (1994) gives ample reasons why such growth rates are possible.
19. The forecast consumption values for cereals and coal in China are probably too low because the actual consumption rates for 1990 are well above those predicted by the equations. This effect is most noticeable with coal, for which the forecast value is not as large as the historical value in 1990! It is conceivable that coal's share of energy consumption will decrease as China becomes more developed, but I think the equation overstates this effect. The IEA (1994) provides a forecast of energy consumption for China in the year 2010. When I used the equivalent growth rate for real GDP per capita (6.45 percent) and the same population as the IEA, the forecast for energy was about the same, coal was about 36 percent and oil was over 200 percent of the IEA forecast. The over-estimate for oil consumption in China is also consistent with a forecast by Wang (1995). He predicts an average annual growth rate for oil products of about 4 percent until the year 2005, whereas my prediction is equivalent to an average annual growth rate of about 8 percent.

20 The forecast values are smaller because the population has not been growing for as many years. Specifically, the population forecast values for China and Indonesia are 20 percent lower, and for India they are 35 percent lower.
21 The only exception is the Philippines, whose growth of real per capita GDP was arbitrarily assumed to be 0 rather than the historical rate of –1 percent.

Bibliography

Asian Development Bank. *Energy Indicators of Developing Member Countries of ADB.* Manila, July 1992.
Bang, K.-Y. "The Structure of Demand and Supply for Major Metals in PECC/MEF Economies," in *New Limits to Growth: Conference Proceedings of the Fifth Minerals and Energy Forum.* Edited by C. Findlay and D. Parsons. Sydney, 1992, pp. 195-218.
Canada. Department of Finance. *Economic Reference Tables,* August 1995.
Flatters, F. and R.G. Harris. "Trade and Investment: Patterns and Policy Issues in the Asia-Pacific Rim," in *Pacific Trade and Investment: Options for the 90s.* Edited by W. Dobson and F. Flatters. Kingston: John Deutsch Institute for the Study of Economic Policy, 1995, pp. 111-40.
Helliwell, J. F. "Asian Economic Growth," in *Pacific Trade and Investment: Options for the 90s.* Edited by W. Dobson and F. Flatters. Kingston: John Deutsch Institute for the Study of Economic Policy, 1995, pp. 17-47.
International Energy Agency. *Energy Policies and Programmes of IEA Countries, 1982 Review.* Paris: OECD/IEA,1982.
_____. *Energy Policies of IEA Countries, 1992 Review.* Paris: OECD/IEA,1992.
_____. *World Energy Outlook 1994 Edition.* Paris: OECD/IEA, 1994.
Labson, B.S., B. Jones, P. Gooday and M.J. Neck. "Minerals and Energy to the 21st Century," in *New Limits to Growth: Conference Proceedings of the Fifth Minerals and Energy Forum.* Edited by C. Findlay and D. Parsons. Sydney, 1992, pp. 83-107.
McRae, R.N. "Energy Intensity: Canada and some Developing Asian Countries," in *Open Regionalism and Sustainable Development: Conference Proceedings of the Sixth Minerals and Energy Forum.* Edited by P. Crowley, Beijing, March 1994.
Meadows, D.H., D.L. Meadows, J. Randers and W.W. Behrens. *The Limits to Growth: A Report for the Club of Rome's Project on the Predicament of Mankind.* New York: Universe Books, 1972.
Mueller, R.E. "Determinants of Economic Growth in Developing Countries: Evidence and Canadian Policy Implications." Policy Staff Paper No. 94/08. Ottawa: Foreign Affairs and International Trade, 1994.
Republic of China. *Taiwan Statistical Data Book, 1994.* Council for Economic Planning and Development, 1994.
Statistics Canada. *Exports by Country,* Catalogue No. 65-003.
Tilton, J.E. *World Metal Demand: Trends and Prospects,* Resources for the Future, Washington, 1990.

United Nations. *Statistical Yearbook for Asia and the Pacific.* 1993.
_____. *1993 International Trade Statistics Yearbook: Volume I.* 1995.
Wang, H.H. "Petroleum Product Demand in China." Presented to 14th CERI International Oil and Gas Markets Conference, Calgary, September 1995.
World Bank. *World Tables 1995.* Baltimore: The Johns Hopkins University Press, 1995.
World Resources Institute. *World Resources 1992-93.* Oxford University Press, 1992.
Yamaguchi, N.D. "Dynamics of the Asia-Pacific Oil Market." Presented to 14th CERI International Oil and Gas Markets Conference, Calgary, September 1995.

Comment

Tim Hazledine
The University of Auckland
New Zealand

THE MAIN EMOTION INDUCED IN ME by reading this excellent paper was apprehension, even fear. In Table 2, McRae shows income elasticities of demand for a dozen or so resource commodities that are all positive. There is not an inferior good among them – not even dirty old coal. When these positive income elasticities are coupled with his rather conservative income growth predictions – just some of the large lower-income countries moving up into the medium-income group – we are faced, as he documents, with a quite stupendous demand for resources, if nothing else changes. The logic, of course, is simple: match the huge differences in current per capita resource use between poor and richer countries with the huge populations of India, China, and Indonesia, predict that these populations will continue to become significantly better off over time, and there simply must be a substantial increase in the demand for resources.

So why does this induce apprehension? There are at least two reasons. The first, which I will not dwell on here, stems from the ecological and environmental consequences of major increases in the extraction and transformation of primary resources. The second is more economic and possibly political. Where will the resources come from? McRae does not examine the supply side of the issue. He assumes that increased demand for the outputs of Canadian resource industries is good news for Canada. And so it is, up to a point. In another paper in this volume, John Helliwell writes of the obligation of those who have, say, an abundance of forests, to share their trees with less well-wooded countries, at an "appropriate" price. But what determines this price? Also in this volume, Hejazi and Trefler write of transaction costs being "entirely unimportant" for homogeneous goods with low transportation costs, which include most resources. In the sense that well-organized spot markets exist for such commodities, this is close to being correct. But many or even most large resource transactions are not effected through spot markets, but rather through bilateral bargaining between buyer and seller, often between governments or their agencies.

Under such circumstances, transaction costs can be very large indeed. We are dealing with "rent goods," which are worth more to consumers than they cost to produce, and the efforts to capture such rents can be rather costly. It is not, I fear, far-fetched to envisage the fight for resources breaking out of the more-or-less polite bounds of economic negotiations into the political sphere and beyond that into warfare. Many countries, including some of those in McRae's dataset, have resorted to military force in their quest for land and resources. The Western economies themselves did so recently in their Gulf War adventure.

What might save us from such a fate, as the relentlessly rising demand from Asia for natural resources confronts what must eventually be an inelastic supply? Of course, the *pax americana* may hold, as it has, more or less, for 50 years now, though we may doubt whether the foundations of this peace – which is based very much on military power and the imposition of a particular economic ideology – are secure.

The economists can offer their own panacea: elasticity. As McRae's diagrams well demonstrate, recent history has shown no signs of a sustained explosion or even augmentation of resource prices: the elasticity or substitutability in the system has operated on both the supply and demand sides to accommodate increases in demand for resources by finding new sources of supply as well as substitutes in demand.

As an optimist, I would be happy to carry on in the belief that elasticities of substitution will continue to save the day. But the explosion in incomes among the vast populations of Asia may present a challenge to the price system rather greater than anything that has occurred before. At the very least, it might be reasonable to plan for something of a regime shift in the terms of trade between resource-rich and resource-poor countries. In Canada, as in other net resource exporters such as New Zealand, the dominant marketing paradigm remains fixed in anxiety about selling and being competitive, with the presumption that the buying side of the market holds the whip hand. Perhaps this has been the case in the past. But how can it continue to be so? In the coming world of vastly expanded Asian incomes, it will be the buyers of resources (including resource-intensive activities such as tourism) that need to be competitive. Perhaps those Pacific Rim countries like Canada and New Zealand that are lucky enough to possess relatively abundant, but still finite and fragile, endowments of land should soon consider *taxing* the sale and export of these endowments, in place of their present crude mercantilist "move the product" attitude, which generates absurdities such as taxpayers' money in New Zealand being spent directly to subsidize the airfares of Japanese tourists visiting the country!

Two technical points should be noted. First, McRae's Figures D-1 and D-3 suggest that his cross-sectional regressions will over-estimate the income elasticity of demand for energy, given that the actual energy and oil consumption paths in 1980-90 of individual countries are much flatter than the line estimated by joining the points across countries. The resource-rich North American economies have much higher energy/GDP ratios than do the newly wealthy Asian economies. Second, if we really wanted to forecast demand for Canadian resources, we would have to model not just the behaviour of Canada's customers (as McRae does), but also the behaviour of other resource suppliers.

Lawrence L. Schembri
Department of Economics
Carleton University

6

Canadian Exports of Business and Education Services to the Asia Pacific Region[1]

INTERNATIONAL TRADE ALLOWS CANADIANS to benefit from the remarkable economic performance of countries in the Asia Pacific region (APR).[2] The gains from trade accrue not only from the increased supply of imports from this region that are either less expensive or of better quality, but from the increased demand for Canadian exports of goods and services. Greater demand translates into higher returns on Canadian factors used in the production of these goods and services as a result of static and dynamic gains to specialization.

The purpose of this paper is to examine the potential for growth of Canadian exports of business and education services to the APR.[3] Apart from the fact that the APR represents a large, growing and relatively untapped market for Canadian exports in general, this issue is of interest for several reasons.[4] Canada is seen as having a comparative advantage in certain types of business services and education services relative to the APR. In 1993, Canada ran its largest surplus in consulting and other professional services: $430 million, or $1,451 million in exports and $1,021 million in imports (Table 1). This surplus occurred despite Canada having an overall deficit of over $4 billion in business services (with exports of around $11 billion and imports of around $15 billion). Although Canada usually runs deficits on most categories of business services with the United States and the European Union, it runs large surpluses with the remaining countries, including the APR, in the categories of consulting and professional services ($540 million), research and development ($159 million), and commissions ($150 million).

Education services, which comprise the tuition fees paid by foreign students in Canadian institutions at all levels of education, as well as their local living expenses, are included in the "other" subcategory of "other services" in Table 1. Although disaggregated data for this category are not publicly available, it is reasonable to believe that the education of roughly 75,000 foreign students represents a significant proportion of the exports in this category and is also primarily responsible for the consistently positive balance.[5]

TABLE 1
Canadian International Trade in Services With the United States, the European Union and Other Countries,[1] by Service Category, 1993 (C$ Millions)

	EXPORTS				IMPORTS				BALANCE			
	United States	European Union	Other	Total	United States	European Union	Other	Total	United States	European Union	Other	Total
Travel				8,804				16,681				-7,877
Freight and shipping				6,343				6,431				-89
Business services												
Consulting and other professional	450	56	944	1,451	455	163	404	1,021	-5	-107	540	430
Transportation related	345	412	304	1,061	328	412	290	1,030	17	0	14	31
Management and administrative	427	114	79	619	1,788	129	52	1,969	-1,361	-15	27	-1,350
Research and development	945	87	180	1,212	759	177	21	957	186	-90	159	255
Commissions	654	148	329	1,130	515	77	179	771	139	71	150	359
Royalties, patents and trademarks	144	56	79	279	1,491	189	190	1,869	-1,347	-133	-111	-1,590
Films and broadcasting	176	13	5	194	366	47	30	443	-190	-34	-25	-249
Advertising and promotional	62	26	29	117	196	10	19	224	-134	16	10	-107
Insurance	797	215	332	1,343	795	410	398	1,604	2	-195	-66	-261
Other financial	396	258	183	837	572	549	138	1,259	-176	-291	45	-22
Computer services	426	107	55	588	726	14	4	744	-300	93	51	-156
Equipment rentals	183	36	6	224	300	24	5	328	-117	12	1	-104
Franchises and similar rights	1	6	15	22	123	1	12	136	-122	5	3	-114
Communications	302	70	345	717	110	129	404	643	192	-59	-59	74
Refining and processing	48	12	24	84	0	0	0	0	48	12	24	84
Tooling and other automotive charges	746	77	41	864	1,640	–	–	1,640	-894	77	41	-776
Other	177	24	100	302	220	38	188	447	-43	-14	-88	-145
Total	6,279	1,717	3,050	11,046	10,384	2,368	2,335	15,087	-4,105	-651	715	-4,041
Government services				759				1,489				-730
Other services												
Trade unions				147				129				18
Other				1,020				789				231
Total				1,167				918				241
Total, all services				28,118				40,606				-12,487

1 The regional disaggregation is only available for business services. "Other countries" includes the APR.
Source: Statistics Canada, *Canada's International Transactions in Services, 1993-94*, cat. no. 67-203.

Recent trends indicate that the potential for expanding international trade in services, especially business services, is very large. From 1985 to 1993, the global trade in business services grew more than 50 percent faster than the trade in merchandise (14.8 versus 8.8 percent per year), totalling US$350 billion or approximately 10 percent of merchandise trade in 1993. Over the same period, Canadian exports of business services increased almost 60 percent faster than merchandise exports (9.7 versus 6.1 percent), totalling C$11 billion or approximately 7.6 percent of merchandise trade exports in 1993.[6]

Several explanations for the rapid growth in trade in services have been put forward. Generally speaking, as economies develop and per capita incomes rise, the domestic service sector expands because of specialization in the production of intermediate services and increased demand for consumption services due to a positive income elasticity.[7] Based on the stylized fact that most industrialized countries have reached the point where approximately 60 percent of domestic GDP and employment is generated by the service sector, the volume of the international trade in services seems relatively small. In 1993 it was approximately US$1 trillion, whereas merchandise trade was roughly US$4 trillion.[8] Clearly, there are sizable impediments to international trade in most domestically provided services. These impediments are being reduced; communication, transportation, and information-processing costs have fallen steadily in real terms over the last two decades and significant government-imposed non-tariff barriers are slowly being reduced by regional and multilateral trade agreements. These trends bode well for Canadian exports of services to the APR, especially as the economies in this region continue to grow and per capita incomes rise.

The recently concluded Uruguay Round of trade negotiations and the resulting General Agreement on Trade in Services (GATS) will greatly facilitate international trade in services over the long term. Although critics of the agreement have argued that the extent of liberalization achieved is limited because most of the commitments are standstill provisions, the GATS is a remarkable breakthrough because it encompasses the complex nature of international transactions in services, thus establishing an essential framework for future negotiations. In particular, the GATS explicitly recognizes four modes for the international delivery of services:

- *cross-border*: those that do not require the physical movement of either the provider or consumer, e.g., data services and insurance;
- *consumption abroad*: those where the consumer goes to the provider, e.g., education and medical services;
- *presence of natural persons*: those which require the temporary movement of personnel, e.g., consulting; and
- *commercial presence*: those which require the provider to establish a local presence through foreign direct investment or representative offices, e.g., banking and retailing.

It is noteworthy that only the first three modes constitute trade as it is normally defined in the balance of payments accounts. Hence a distinction is often made between international trade in services, on the one hand, and international transactions in services, which also include local sales of services by a foreign affiliate, on the other.[9]

Strong arguments can be made that countries in the APR, chiefly those with the least-developed service sectors, require foreign-supplied services to maintain relatively high rates of economic growth. For example, Bhagwati (1987) argues that business services are primarily intermediate inputs in the production of goods and services, so limiting access to foreign-produced services will increase costs and reduce international competitiveness. Other observers (for example, UNCTAD and the World Bank, 1994) argue that increased business service exports can enhance productivity growth because service inputs, such as finance and communication, facilitate the creation and absorption of new technologies. In addition, because the provision of services abroad often requires the movement of skilled personnel or foreign direct investment (FDI), the transactions will increase the rate of technological diffusion, thereby raising productivity in the service-importing country. Thus the less-advanced countries in the APR should be looking to Canada and other OECD countries for these services.

In this paper, I analyze the prospects for Canadian exports of business and education services to the APR from several perspectives and using several different data sets. Although this multipronged approach produces a variety of insights, the research was limited by the quantity and quality of the data available on international trade in services, which precludes a detailed empirical/econometric analysis based on a single set of data.[10]

In the next section, data on service imports by APR countries are examined. An analysis of recent data on Canadian business service exports disaggregated by country and type of service follows in the subsequent section. This data set was prepared specially for this paper by Statistics Canada. Unfortunately, data on business service exports to China are not available, so I move on to an examination of some qualitative evidence on Canadian business exports to China. This is followed by an analysis of recent data on Canadian postsecondary education services to students from the APR. The implications of the GATS for Canada-APR trade in services are considered in the subsequent section. Finally, some concluding remarks are presented in the last section.

APR Service Imports: An Overview

ONE WAY TO ASSESS THE POTENTIAL OF CANADIAN EXPORTS of business and education services to the APR is to examine the region's total demand for imports of these services. Service imports in 1984-93 by APR countries except Taiwan (for which no data is available) are displayed in Table 2. Global totals

TABLE 2
IMPORTS OF SERVICES BY APR COUNTRIES,[1] 1984-93 (US$ BILLIONS)

COUNTRY		1984	1985	1986	1987	1988	1989	1990	1991	1992	1993	CAGR[2](%)
Japan	Total services	32,800	32,740	35,450	48,420	63,530	75,010	81,970	85,040	89,730	92,770	12.25
	Other private services (OPS)[3]	11,940	12,440	13,920	18,150	21,350	26,040	28,780	30,430	31,830	33,190	12.03
	OPS as % of total	36.4	38.0	39.3	37.5	33.6	34.7	35.1	35.8	35.5	35.8	
Singapore	Total services	4,024	3,976	4,165	5,015	6,227	7,306	9,297	9,863	10,231	11,635	12.52
	Other private services	1,883	1,913	1,898	2,284	2,653	3,012	3,678	3,803	4,362	4,610	10.46
	OPS as % of total	46.8	45.6	45.6	45.5	42.6	41.2	39.6	38.6	42.6	39.6	
Hong Kong	Total services			5,679	6,743	8,044	9,245	11,241	12,956	14,691		17.00
	Other private services			1,056	1,262	1,479	1,616	1,783	2,031	2,395		14.50
	OPS as % of total			18.6	18.7	18.4	17.5	5.9	15.7	16.3		
South Korea	Total services	4,192	4,094	4,598	5,397	6,829	9,354	11,151	13,324	14,695	16,643	16.56
	Other private services	1,657	1,489	1,599	1,850	2,300	2,972	3,601	4,259	5,014	5,899	15.15
	OPS as % of total	39.5	36.4	34.8	34.3	33.7	31.8	32.3	32.0	34.1	35.4	
Malaysia	Total services	4,418	4,048	3,714	3,770	4,453	5,039	5,751	6,878	7,640	8,244	7.18
	Other private services	1,386	1,248	981	731	925	1,094	1,412	1,847	2,170	2,252	5.54
	OPS as % of total	31.4	30.8	26.4	19.4	20.8	21.7	24.6	26.9	28.4	27.3	
Thailand	Total services	1,860	1,769	1,804	2,342	3,481	4,377	6,139	7,834	9,350		22.37
	Other private services	200	193	211	222	333	386	646	1,174	1,919		32.66
	OPS as % of total	10.8	10.9	11.7	9.5	9.6	8.8	10.5	15.0	20.5		
Philippines	Total services	1,168	850	824	1,124	1,281	1,527	1,723	1,748	2,253	3,030	11.17
	Other private services	652	413	344	444	460	513	556	581	923	1,427	0.09
	OPS as % of total	55.8	48.6	41.7	39.5	35.9	33.6	32.3	33.2	41.0	47.1	

TABLE 2 (CONT'D)

Country		1984	1985	1986	1987	1988	1989	1990	1991	1992	1993	CAGR[2] (%)
Indonesia	Total services	4,239	5,135	4,256	4,440	4,606	5,439	6,056	6,564	8,100	8,939	8.64
	Other private services (OPS)[3]	1,579	2,703	1,991	2,191	2,061	2,311	2,033	1,948	2,850	3,039	7.55
	OPS as % of total	37.2	52.6	46.8	49.3	44.7	42.5	33.6	29.7	35.2	34.0	
China	Total services	2,378	2,524	2,276	2,485	2,603	3,910	4,352	4,121	9,414		18.77
	Other private services	563	364	115	164	204	205	304	704	2,076		17.72
	OPS as % of total	23.7	14.0	5.1	6.6	7.8	5.2	7.0	17.1	22.1		
APR	Total services	55,079	55,136	57,087	72,993	93,010	111,962	126,439	13,572	151,413		11.03
	Other private services	19,860	20,653	21,059	26,036	30,286	36,533	41,010	44,746	51,144		10.91
	OPS as % of total	36.1	37.5	36.9	35.7	32.6	32.6	32.4	33.1	33.8		
APR excluding Japan	Total services	22,279	22,396	21,637	24,573	29,480	36,952	44,469	50,332	61,683		9.03
	Other private services	7,920	8,213	7,139	7,886	8,936	10,493	12,230	14,316	19,314		9.02
	OPS as % of total	35.5	36.7	33.0	32.1	30.3	28.4	27.5	28.4	31.3		
World	Total services	421,362	426,036	477,164	575,603	648,421	712,494	858,203	900,660	1,004,243	983,264	9.87
	Other private services	111,684	118,608	135,708	176,276	193,893	220,762	284,262	295,955	349,217	353,251	13.60
	OPS as % of total	26.5	27.8	28.4	30.6	29.9	31.0	33.1	32.9	34.8	35.9	

1 Excluding Taiwan, for which there are no data.
2 Compound annual growth rate for 1984-93, except for Hong Kong: 1986-92; Thailand and China: 1984-92; and APR and APR excluding Japan: 1984-92.
3 "Other private services" is primarily business and education services.

Source: International Monetary Fund, *Balance of Payments Statistics Yearbook*, 1994; and Hong Kong Census and Statistics Department, *Annual Digest of Statistics*, 1994.

are included for comparison. Except for Hong Kong, all the data are taken from the IMF Balance of Payments Database for the period from 1984 to 1992 (or 1993). (1993 data are not available for China and Thailand because they are slow to report their figures to the IMF.) The categories shown in the table are "total services" and "other private services," which are primarily business and education services. For Hong Kong, the definition of other private services is narrower than the IMF definition, so the volumes for Hong Kong seem relatively small.

Japan is by far the largest importer of other private services (over 60 percent of the total in 1992), which is not unexpected given its economic size and per capita income level. South Korea, Singapore, and Hong Kong are also major importers, together accounting for roughly 20 percent. A loose comparison of import volumes across countries in the APR indicates that aggregate and per capita income levels explain much of the variation. Indeed, a regression of imports of other private services on GDP and GDP per capita obtains positive and statistically significant estimated coefficients for each variable.[11] Thus, as the economies in this region grow and per capita incomes rise, the potential market for Canadian exports of business and education services will undoubtedly expand.

The average annual growth rates of other countries' imports of private services vary substantially. Thailand, China, South Korea, and Hong Kong's imports grew faster than the world average of 13.6 percent per year over the sample period. Japan and Singapore's growth rates appear relatively steady at just below the world level. Malaysia, the Philippines, and Indonesia's average growth rates calculated over the whole period 1984-93 are well below the world rate, but this comparison may be misleading as an indication of future trends. Their growth rates over the period 1990-93 are much higher: 16.8 percent in Malaysia, 36.9 percent in the Philippines, and 14.3 percent in Indonesia (the world growth rate over this period was 7.5 percent). This evidence indicates that in recent years the APR, especially when one excludes Japan, is the fastest growing market for other private services in the world. Moreover, growth rates of other private services imports are greater than of services as a whole, so the share of other private services in total services is generally rising. This observation supports the hypothesis that business services must play an important role in facilitating rapid economic expansion in these economies.

These data indicate that the APR represents an excellent potential market for Canadian exports of business and education services. The APR's demand for imports of other private services is likely to continue to grow faster than the world demand for the foreseeable future.

Canadian Exports of Business Services

Disaggregated data on Canadian exports of business services to the APR are limited to a small, recent sample. Nevertheless, it is still possible to glean some useful insights about patterns and trends from this Statistics Canada data, on which Tables 3-7 are based.[12] Data on Canadian business exports to China are not publicly available, for reasons of confidentiality and data quality, but because the Chinese economy will almost certainly dominate the region, it is too important to omit. Therefore I consider the potential for Canadian business services at the end of this section by examining some qualitative evidence as well as a relevant U.S. study.

Overall, this region, including China, probably represents almost 10 percent of total Canadian exports of business services. Not surprisingly, Japan is Canada's most important market; in 1993 it accounted for over one-third of exports of business services to this region (Table 3). South Korea and Hong Kong also took sizable shares. In comparison, Canada's business service exports to Singapore and Taiwan seem small, particularly in view of the size of their economies and their per-capita income levels.[13]

Canadian business service exports to the APR are expanding rapidly; over the period 1990-93, they increased by almost 50 percent from $539 to $811 mil-

TABLE 3

CANADIAN EXPORTS OF BUSINESS SERVICES TO THE APR,[1] 1990-93 (C$ MILLIONS)

COUNTRY	1990	1991	1992	1993	GROWTH RATE[2] (%)
Japan	297	258	260	284	2.1
Singapore	15	13	37	43	42.1
Hong Kong	105	92	123	140	10.1
Taiwan	12	15	26	23	24.2
South Korea	64	110	138	158	35.2
Malaysia	20	25	39	39	24.9
Thailand	20	48	38	35	20.5
Philippines	10	12	27	47	67.5
Indonesia	26	15	36	42	17.3
Total	539	588	724	811	14.6
Total excluding Japan	272	330	464	527	24.7
Total Canadian business service exports	8,322	8,930	9,814	11,046	9.9

1 Countries are listed in descending order by GDP per capita.
2 Compound annual growth rate for 1990-93.
Source: Statistics Canada, Balance of Payments Division, unpublished data.

lion. This is a growth rate of 14.6 percent per year, which is significantly greater than the 10.4 percent growth rate for total Canadian exports. When Japan is excluded, the growth rate in this region is almost 25 percent. Thus, while Japan may currently be the largest importer of Canadian business services in the APR, its potential for further growth seems somewhat limited, unless there is a significant reduction in trade barriers. The rest of the region is the fastest-growing segment of the Canadian export market. Apart from Japan and Hong Kong, demand in the APR countries has increased by almost 20 percent per year. In the Philippines, Singapore, and South Korea it has increased by over 35 percent per year. Although these growth rates are impressive, the absolute volume of exports is still relatively small. Nonetheless, further expansion of demand for Canadian business services seems very likely.

To analyze these data further, business service exports are disaggregated for Japan, the Association of South-East Asian Nations – ASEAN (Indonesia, Malaysia, Philippines, Singapore and Thailand); and Hong Kong, South Korea, and Taiwan as a group in Tables 4-6, respectively. It is important to note that "other business" is the remaining business services that cannot be disaggregated for reasons of confidentiality or data quality. This residual item also includes the business services category "other business services," which is a catch-all category that generally includes a statistical adjustment for undercoverage in the surveys employed to collect the data.

Over the past few years, Canada's business services exported to Japan have been concentrated mainly in the financial areas, with insurance and other financial services (chiefly banking) and commissions (primarily from sales of merchandise rather than financial assets) accounting for almost one-third in 1993 (Table 4). This is not unexpected, because there are several Canadian banks and insurance companies that are large enough and have the international reputation to gain access to Japanese markets.

The growth of Canada's exports of business services to Japan has been relatively slow: 2.1 versus 10.4 percent for its total business service exports over 1990-93. The most notable increases in terms of both absolute amounts and rates of growth have been in insurance and other financial services, royalties, management and administrative services (essentially head office functions), and computer services.

In contrast to Japan, the exports to the ASEAN countries were lower in absolute terms, but the growth rate was more rapid – 31.1 percent over 1990-93 (Table 5). Most of the increase was in consulting and other professional services and other business services. The annual growth rates of these two items were 34.3 and 65.6 percent, respectively (the comparable figures for total Canadian exports of these services were 32.0 and 6.2 percent, respectively). This pattern is also evident in the third grouping of Hong Kong, South Korea, and Taiwan. The growth rate of total business service exports was 21.1 percent, and consulting and other professional and other business services grew by 69.7 and 19 percent, respectively (Table 6).

TABLE 4

CANADIAN EXPORTS OF BUSINESS SERVICES[1] TO JAPAN, 1990-93 (C$ MILLIONS)

SERVICE	1990	1991	1992	1993	GROWTH RATE[2] (%)	GROWTH RATE FOR TOTAL[3] (%)
Consulting and other professional	7 (1.1)	8 (0.8)	8 (0.6)	8 (0.6)	4.6	32.0
Management and administration	2 (0.3)	2 (0.3)	4 (0.6)	8 (1.3)	58.7	−2.0
Research and development	4 (0.4)	9 (0.9)	6 (0.6)	7 (0.6)	20.5	11.0
Royalties	2 (1.3)	7 (3.6)	7 (2.8)	9 (3.2)	65.1	23.0
Films and broadcasting	1 (0.5)	1 (0.6)	1 (0.4)	1 (0.5)	0	−1.0
Advertising and sales promotion	16 (14.5)	15 (12.9)	14 (11.1)	19 (16.2)	5.9	2.0
Insurance and other financial	28 (1.8)	23 (1.4)	35 (1.8)	40 (1.8)	12.6	11.0
Commissions	51 (5.3)	53 (5.6)	53 (5.2)	52 (4.6)	0.7	5.0
Computer	3 (1.6)	4 (1.1)	4 (0.7)	7 (1.2)	32.6	47.0
Communications	16 (2.7)	21 (3.3)	16 (2.9)	18 (2.5)	4.0	6.0
Other business	137 (3.4)	115 (2.8)	110 (2.7)	115 (2.4)	−5.7	6.0
Total	267 (3.3)	258 (2.9)	258 (2.6)	284 (2.6)	2.1	10.0

1 Exports of this service as a proportion of total Canadian exports in this category are in parentheses.
2 Compound annual growth rate for 1990-93.
3 Compound annual growth rate of Canada's total exports of these services for 1990-93.
Source: Statistics Canada, Balance of Payments Division, unpublished data.

To a large extent the increase in Canadian exports of consulting and other professional services to these two groups of countries consisted of new contracts to Canadian engineering and telecommunications firms, many stemming from large infrastructure projects such as the sale of nuclear reactors to South Korea by Atomic Energy of Canada Limited (AECL). The reason for the rapid growth of exports of other business services are more difficult to discern because their composition is unknown. However, a substantial proportion

TABLE 5

CANADIAN EXPORTS OF BUSINESS SERVICES[1] TO THE ASEAN COUNTRIES,[2] 1990-93 (C$ MILLIONS)

SERVICE	1990	1991	1992	1993	GROWTH RATE[3] (%)	GROWTH RATE FOR TOTAL[4] (%)
Consulting and other professional	45 (7.1)	68 (6.6)	88 (6.9)	109 (7.5)	34.3	32.0
Management and administration	6 (0.9)	6 (0.9)	6 (0.9)	5 (0.8)	−5.6	−1.6
Royalties	0	0	2 (0.8)	4 (1.4)	n.a.	23.0
Insurance and other financial	10 (0.6)	5 (0.3)	6 (0.3)	8 (0.4)	−7.2	11.1
Commissions	16 (17.0)	14 (1.5)	15 (1.5)	19 (1.7)	5.9	5.4
Computer	1 (0.5)	1 (0.3)	2 (0.4)	1 (0.2)	0.0	46.5
Other business	13 (0.3)	18 (0.4)	58 (1.4)	59 (1.2)	65.6	6.2
Total	91 (1.1)	112 (1.3)	177 (1.8)	205 (1.9)	31.1	10.4

1 Exports of this service as a proportion of total Canadian exports in this category are in parentheses.
2 Indonesia, Malaysia, Philippines, Singapore, and Thailand.
3 Compound annual growth rate for 1990-93.
4 Compound annual growth rate of Canada's total exports of the services for 1990-93.
Source: Statistics Canada, Balance of Payments Division, unpublished data.

of the increases consist of research and development.[14] These large increases are consistent with Canada's comparative advantage in these services relative to the ASEAN countries, Hong Kong, South Korea, and Taiwan.

Canadian business service exports to the APR in 1990-93 are disaggregated in Table 7 on the basis of whether the sale was between affiliates or not. The most likely affiliation, if any, is between a Canadian parent firm and a foreign subsidiary. Affiliated service exports, which are primarily management and administration services, are often generated by firms whose primary line of business is manufacturing goods rather than providing services. The major banks and insurance companies are exceptions to this general rule.

Canada's non-affiliated sales are much less significant in Japan than they are in the ASEAN countries and Hong Kong, South Korea, and Taiwan. When we look at non-affiliated exports to the APR countries excluding Japan, they are clearly the fastest growing in terms of both absolute increases and

TABLE 6

CANADIAN EXPORTS OF BUSINESS SERVICES[1] TO HONG KONG, SOUTH KOREA AND TAIWAN, 1990-93 (C$ MILLIONS)

SERVICE	1990	1991	1992	1993	GROWTH RATE[2] (%)	GROWTH RATE FOR TOTAL[3] (%)
Consulting and other professional	18 (2.9)	57 (5.5)	85 (6.7)	88 (6.0)	69.7	32.0
Management and administration	3 (0.5)	0	0	2 (0.3)	-12.6	-1.6
Royalties	3 (2.0)	2 (1.0)	1 (0.4)	3 (1.05)	7.0	10.7
Insurance and other financial	8 (0.5)	7 (0.4)	9 (0.5)	6 (0.3)	-9.0	23.3
Commissions	33 (3.4)	36 (3.8)	32 (3.1)	42 (3.7)	4.3	11.2
Computer	2 (1.0)	1 (0.3)	2 (0.4)	3 (0.5)	14.5	5.4
Communications	34 (5.7)	32 (5.1)	29 (5.3)	43 (6.0)	8.0	46.5
Other business	79 (2.0)	80 (2.0)	130 (3.2)	133 (3.0)	19.0	6.2
Total	**180 (2.2)**	**215 (2.4)**	**288 (3.0)**	**320 (2.9)**	**21.1**	**10.4**

1 Exports of this service as a proportion of total Canadian exports in this category are in parentheses.
2 Compound annual growth rate for 1990-93.
3 Compound annual growth rate of Canada's total exports of the services for 1990-93.
Source: Statistics Canada, Balance of Payments Division, unpublished data.

rates of growth. They are growing at almost double the rate as Canada's total non-affiliated exports, at 23.0 versus 11.8 percent over 1990-93. These observations indicate that most of the increase in service exports to the APR is not being generated by Canadian firms that have established a commercial presence in the region, but by new contracts to Canadian engineering and consulting firms, which operate on a project basis.

There are several possible explanations for the relative importance of affiliated exports to Japan. First, because Japan is the largest and most mature economy in the region, Canadian multinationals in manufacturing and services (primarily banking and insurance) have a greater incentive based on expected return and risk to undertake the fixed costs of establishing a permanent commercial presence there. Second, in certain services there is a positive

Table 7

Canadian Exports of Business Services to APR Country Groups, by Affiliated or Non-Affiliated Sale, 1990-93 (C$ Millions)

Country Group	1990			1991			1992			1993			Growth Rate[1] (%)		
	Affiliated	Non-Affiliated	Total	Affiliated	Non-Affiliated	Total	Affiliated	Non-Affiliated	Total	Affiliated	Non-Affiliated	Total	Affiliated	Non-Affiliated	Total
Japan	102	165	267	103	155	258	90	170	260	96	188	284	-2.0	4.4	2.1
ASEAN	25	66	91	30	83	113	43	134	177	43	163	206	19.8	35.2	31.3
Hong Kong, South Korea, and Taiwan	12	169	181	12	205	217	44	243	287	47	321	321	57.6	17.5	21.0
Total	139	400	539	145	443	588	177	547	724	186	625	811	10.2	16.0	14.6
Total excluding Japan	37	235	272	42	288	330	87	377	464	90	437	527	34.5	23.0	24.7
Total Canadian business services exports	3,342	4,980	8,322	3,460	5,470	8,930	3,726	6,087	9,814	4,093	6,954	11,046	7.0	11.8	9.9

1 Average compound growth rate for 1990-93.
Source: Statistics Canada, Balance of Payments Division, unpublished data.

relationship between the volumes of trade in merchandise and services (e.g., commissions). Thus, because Japan is an important market for Canadian merchandise, and international transactions often occur between related entities, it is not surprising that exports of services to affiliated firms are also relatively significant.

Exports of Business Services to China

ALTHOUGH CHINA COULD WELL BECOME the dominant economy in the APR within the next decade, historical data on Canada's exports of business services to China are not available. Nonetheless, some evidence can be gathered from other sources. China's total imports of other private services has increased dramatically in recent years (see Table 2). In general, it is likely that the trends and patterns of Canadian business exports to China are similar to those of the ASEAN countries (excluding Singapore), because China's economy is at a comparable stage of development.

Figures on selected Canadian firms' sales of business services to China are given in Table 8. Although this evidence is not representative, several observations can be made. First, most of the sales can be attributed to a few large companies such as SNC-Lavalin and Northern Telecom. This indicates that the fixed costs associated with making a sale in China (gathering information, marketing, travel, legal costs, etc.) may be large and difficult to finance from external sources by small- to medium-sized firms, especially given the uncertainty and the information asymmetries surrounding contracts in China and the relatively small amounts of collateral that service-oriented firms typically possess.

Second, these sales are concentrated in technology-intensive service sectors such as engineering, telecommunications and financial services, which reflects Canada's comparative advantage in this field and also China's need for efficient intermediate services to further its economic development.[15]

Third, several of these export sales are related to infrastructure spending (e.g., telecommunications) and large individual projects such as the Three Gorges Dam, which reflects the fact that China and the less advanced countries in the APR lag well behind Canada and the other more developed countries in the APR in infrastructure development (Figures 1-4).

Finally, many foreign firms are entering into joint ventures with local firms to take advantage of their marketing expertise and gain immediate market access. One form of joint venture is the build-operate-transfer (BOT) process, in which foreign firms that have access to technology and capital build a plant, operate it for a temporary period, and then turn it over to their local partners.[16]

It seems that large firms and projects are likely to dominate sales of business services to China, at least in the short term. However, over the longer

TABLE 8
Examples of Canadian Business Exports to China, Various Years

Company	Amount	Type of service	Year	Contracting party/ destination
Monenco-Agra Inc. (Oakville)		Energy and construction expertise Computerized project management for Three Gorges Dam Project		China
Royal Bank, Bank of Nova Scotia, Bank of Montreal		Establish branches Technology transfer	Since 1981	China
Noranda Copper Smelting and Refining	Several million	Technology transfer Assistance to start up plant operations Sale and transfer of smelting reactor technology	1992	China National Nonferrous Metals Import and Export Corporation
SNC-Lavalin Group Inc./ BC Hydro/Hydro-Québec/ Acres International (engineering firm)	$14 million	Feasibility study, Three Gorges Dam Project Partially funded by CIDA	1992	Chinese government
Nortel Inc.	$130 million	Design/make computerized switches Buy stake in electronic circuit factory Open new electronic chip design and manufacturing facility Open university R&D lab	1992-97	China
Manufacturer's Life (Manulife)	$500 million	Life insurance	1993	China
Reid Crouther and Partners (with five other engineering consultants)		Design and construction of wastewater treatment plant	1993	Jinan, China
Stanley International Group Inc. (Stanley Technical Group Inc., Edmonton)		Six environment-related projects	1993	China

TABLE 8 (CONT'D)

Company	Amount	Type of service	Year	Contracting party/destination
Goodman and Goodman (law firm)		Opening of office in Beijing	1994	China
Canadian Airlines International		Ticket office in Shanghai	1994	China
Pacific Entertainment Group		Two skating rinks and operating responsibility	1994	Shanghai/Chongquing
Chai-Na-Ta Ginseng Products	$27 million	Produce/distribute pharmaceutical products	1994	China
Environmental Technologies Inc. (Calgary)	$40 million	Develop plant to convert sewage into methane	1994	China
Bennett & Wright (Toronto)	$140 million	Design and build 51-storey residential/commercial building	1994	Shanghai
Power Pacific (subsidiary of Power Corp.)	$60 million	Develop high-tech industrial park	1994	Shanghai
Dominion Bridge Inc.	$715 million	Design, engineer, and construct: – transit system – hydro station – expressway	1994	Chengdu City, China
MMC International (Toronto)	US$1 billion	Design, oversee construction of Shenyang City Centre, Shenyang	1995	China

Source: *The Globe and Mail*, 1992-95.

Figure 1

INDICATOR OF INFRASTRUCTURE DEVELOPMENT:
ELECTRIC POWER PRODUCTION

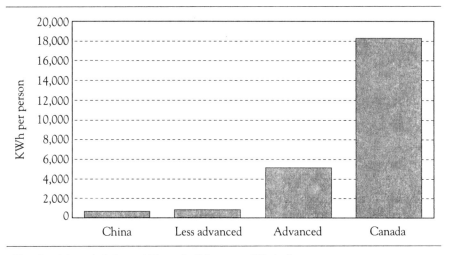

Note: Less Advanced: Indonesia, Malaysia, the Philippines, and Thailand.
Advanced: Hong Kong, Singapore, and South Korea

Figure 2

INDICATOR OF INFRASTRUCTURE DEVELOPMENT:
TELEPHONE MAINLINES

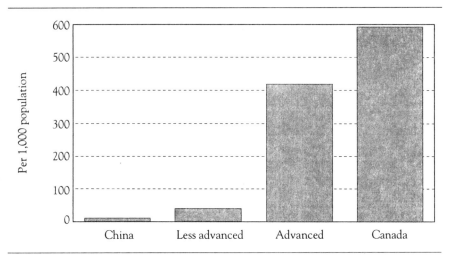

Note: Less Advanced: Indonesia, Malaysia, the Philippines, and Thailand.
Advanced: Hong Kong, Singapore, and South Korea

Figure 3

INDICATOR OF INFRASTRUCTURE DEVELOPMENT: POPULATION WITH ACCESS TO SAFE WATER

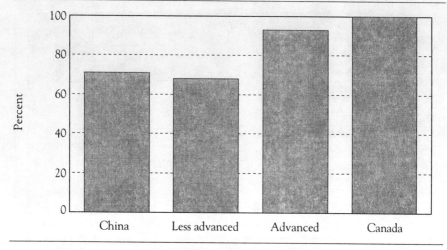

Note: Less Advanced: Indonesia, Malaysia, the Philippines, and Thailand.
Advanced: Hong Kong, Singapore, and South Korea

Figure 4

INDICATOR OF INFRASTRUCTURE DEVELOPMENT: TOTAL ROADS

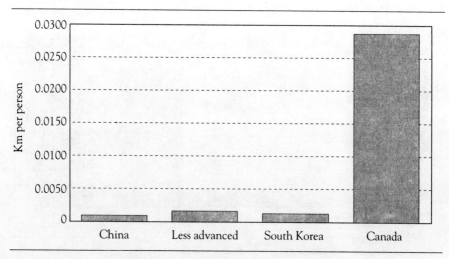

Note: Less Advanced: Indonesia, Malaysia, the Philippines, and Thailand.

term, as China's commercial practices become better known and the country develops economically, more small- and medium-sized Canadian firms may become involved.

In a study similar to this one, Cleveland (1994) performs a detailed analysis of U.S. exports of business services to China, Hong Kong, and Taiwan, a region he calls the Chinese economic area (CEA). This aggregation is meaningful because many of the exports destined for Hong Kong and Taiwan often end up in China. U.S. exports of business services to this region tripled over the five years between 1987 and 1992 to total almost US$2.1 billion, almost 3 percent of total U.S. exports of business services. The composition of U.S. exports of business services to these three countries reflects the industrial orientation of their economies. China, which has a large heavy industrial base, predominantly imports U.S. construction, engineering, and mining services. Hong Kong, which has a fairly diversified economy, imports a lot of U.S. commerce-related services such as telecommunications, legal, and management services. Taiwan, which has a technology-based manufacturing economy, imports significant amounts of U.S. computer, research and development, and telecommunications services. Cleveland concludes that because the CEA is forecast to grow by 5 to 8 percent per year until 2000, it will become a substantial market for U.S. business service exports. Moreover, China still has a long way to go to liberalize trade in services, and if trade barriers are reduced (depending in part on when China is accepted into the WTO and accedes to the GATS), U.S. exports of business services could grow even faster.

Most of Cleveland's observations would apply to Canada's exports of business services to the CEA, especially to China. Although Canada has a smaller number of large firms engaged in exporting business services, its comparative advantage in business services relative to the CEA and China is comparable to that of the United States.

CANADIAN EXPORTS OF EDUCATION SERVICES

AS THE ECONOMIES IN THE APR CONTINUE TO GROW and develop, real wage rates will rise and manufacturing will shift into the production of higher quality, technology-intensive goods and services. Concomitantly, the demand for skilled labour will expand, and it is unlikely that the supply of postsecondary education will meet the demand. Establishing colleges and universities that provide a high quality of education, particularly in the sciences and engineering, is costly and may take several years if not decades. Even in the more-developed economies in the region such as Hong Kong, there is a large excess demand for postsecondary education, which is accommodated by students going abroad to study. Given the importance of English as the language of international commerce, many of these students prefer to be educated in English-speaking countries, primarily Australia, Canada, the United States, and the United Kingdom.

Canada has fared reasonably well in attracting these students over the last two decades. Over the period 1975-92, total college/trade and undergraduate enrollment of foreign students from the APR has increased by almost 7 percent per year to a total of almost 23,000 students in 1992-93 (out of a total foreign student enrollment at these two levels of approximately 75,000 [Table 9]).[17] College/trade enrollment has increased at a faster rate than undergraduate enrollment, which may be due to several factors including relatively higher tuition fees at the undergraduate level. Although there is evidence that the enrollment of students from the APR has declined recently, Canadian exports of education services to the APR are still significant, because students from these countries probably spend over $200 million in Canada, almost 25 percent of total business service exports to the APR ($200 million compared with $811 million).[18]

In the 1990s, total enrollment of APR students in college/trade programs in Canada has been slightly greater than that in undergraduate programs, and the disaggregations by country of origin indicate different enrollment patterns. At both levels, students from Hong Kong are very significant; they account for approximately 60 percent of total APR undergraduate enrollment and 20 percent of all college/trade enrollment. The political uncertainty surrounding the future of Hong Kong has undoubtedly kept the number of its students seeking higher education in Canada relatively high, but it is interesting to note that enrollment at both levels fell substantially in 1992-93 from peak levels in the late 1980s and early 1990s. The reasons for this are not obvious. They could include higher fees, the expansion of postsecondary education in Hong Kong, and more aggressive recruiting from other English-speaking countries. While students from Hong Kong dominate undergraduate enrollment from the APR, students from the Philippines and Japan, surprisingly, account for roughly half of the students at the college/trade level. The relatively high enrollment from the Philippines might be explained by the large number of women working in child care and taking courses part time to upgrade their skills. There is some evidence that many of the Japanese students are enrolled in English language training. At the undergraduate level, students from Malaysia and Singapore account for roughly one-quarter of enrollment from the APR.[19]

In forecasting future trends in Canada's exports of education services to the APR, several other issues should be considered. First, the number of students from a small group of APR countries – Japan, South Korea, Taiwan, and, Thailand at both the college/trade and undergraduate education levels, and China at the undergraduate level – has increased monotonically over the sample period. The average growth rates in enrollments of students from Japan, South Korea, and Taiwan far exceed the APR average over the whole period and over the last few years. The fact that these three countries all have relatively high and growing per capita income levels may indicate that foreign education is a luxury consumer good. Alternatively, the increasing numbers of students may indicate a demand for specialized programs not widely available in their

TABLE 9

ENROLLMENT OF APR STUDENTS IN CANADIAN EDUCATIONAL INSTITUTIONS, BY TYPE OF PROGRAM AND STUDENTS' COUNTRY, FOR ACADEMIC YEARS 1975, 1980, 1985 AND 1990-92[1]

	1975	1980	1985	1990	1991	1992	ANNUAL GROWTH RATE[2] (%)
College/trade programs							
Japan	80	219	252	1,835	2,085	2,728	21.7
Singapore					145	126	−13.0
Hong Kong	1,818	779	1,309	2,827	2,313	2,496	1.8
Taiwan	33	34	68	885	1,004	1,415	23.2
South Korea					507	680	34.0
Malaysia	285	289	305	310	271	391	1.8
Thailand					76	84	10.5
Philippines	20	45	855	7,404	4,523	3,590	33.4
Indonesia					174	119	−33.5
China	85	66	157	1,436	867	798	13.2
Total	2,321	1,432	2,946	14,697	11,965	12,427	9.8
Undergraduate programs							
Japan	48	98	74	290	479	517	14.1
Singapore					1,118	1,057	−5.5
Hong Kong	3,761	5,173	7,661	5,921	6,451	6,187	2.8
Taiwan	61	49	67	192	242	305	9.4
South Korea					99	117	18.0
Malaysia	388	3,117	2,192	1,405	1,465	1,369	7.3
Thailand					25	32	28.0
Philippines	23	52	43	69	58	64	5.8
Indonesia					234	216	−7.6
China	199	139	225	383	418	448	4.6
Total	4,480	8,628	10,262	8,260	10,589	10,312	4.7
College/trade and undergraduate programs							
Japan	128	317	326	2,125	2,564	3,245	19.7
Singapore					1,263	1,183	−6.3
Hong Kong	5,579	5,952	8,970	8,748	8,764	8,683	2.5
Taiwan	94	83	135	1,077	1,246	1,720	17.5
South Korea					606	797	31.5
Malaysia	673	3,406	2,497	1,715	1,736	1,760	5.4
Thailand					101	116	14.9
Philippines	43	97	898	7,473	4,581	3,654	28.0
Indonesia					408	335	−17.9
China	284	205	382	1,819	1,285	1,246	8.5
Total	6,801	10,060	13,208	22,957	22,554	22,739	6.9

1 Disaggregations for Singapore, South Korea, Thailand, and Indonesia are only available for 1991-92.
2 Average compound annual growth rates are calculated over the available sample period.
Source: Statistics Canada, *International Student Participation in Canada*, cat. no. 81-261.

home countries such as aerospace engineering, environmental studies, and teaching English as a second language.

Second, although the demand for more specialized postsecondary education will undoubtedly increase in the APR, it is unclear to what extent this demand will be satisfied by students going abroad. Clearly, the supply of higher education in the APR will increase. Although this is occurring throughout the region, Hong Kong, Singapore, and Malaysia seem to have undertaken the most vigorous expansion programs. Canadian institutions and academics are playing and will probably continue to play a role in this expansion.[20] To the extent Canadians are compensated for their participation, these payments would represent an export of education services.

Third, the competition for students from the APR has increased in recent years. Postsecondary education in Canada is still attractive for several reasons, including the quality of education, relatively low tuition levels, and good living conditions. However, Cameron (1993) suggests that Canada is lagging behind Australia and the United Kingdom, especially in marketing postsecondary education to prospective students from the APR. In response to this greater competition the Canadian government, in conjunction with the Asia Pacific Foundation, has established Canadian Education Centres in major cities in the APR to coordinate marketing activities and raise the profile of Canadian education.

Although the existing data are limited, they indicate that Canada has a strong comparative advantage in certain postsecondary education programs relative to countries in the APR, and that the region is a significant potential market for Canadian education services.

TRADE BARRIERS AND
THE GENERAL AGREEMENT ON TRADE IN SERVICES

EXPORTS OF CANADIAN BUSINESS SERVICES to the APR are affected by the significant non-tariff barriers to international transactions in services.[21] The recognition that these barriers pose a substantial threat to the growing global trade in services was the impetus to include services for the first time in the Uruguay Round of multilateral trade negotiations. The resulting General Agreement on Trade in Services (GATS) came into force April 15, 1994, after almost eight years of difficult negotiations. This section provides a brief overview of the existing barriers to international transactions in services and the implications of GATS, with a particular focus on Canadian business exports to the APR.

Most of the impediments to international transactions in services stem from two distinguishing characteristics of services: the provision of services is typically more heavily regulated than the production of goods, and transactions in services often require the simultaneous presence of the provider and the consumer (i.e., "contact" as opposed to "long-distance" services), which requires that either the provider (i.e., one or more of the key factor inputs of capital, labour, and technology) or the consumer relocate. Domestic regulations often restrict who can provide services to those individuals and firms deemed qualified: for example, lawyers, engineers, accountants, doctors, banks, insurance companies, and telephone companies. In practice, these regulations not only restrict market access, but they also may discriminate against service providers from other countries. Some countries impede international transactions in services by restricting the cross-border flows of factors of production. For example, rules controlling foreign investment may prevent ownership in certain industry sectors or limit the repatriation of profits, and immigration rules may restrict the movement of skilled personnel.

To summarize the impact of these different types of regulatory impediments, it is useful to categorize them in one of two ways: 1) as restricting the access of foreign service providers to the domestic market – the issue of market access, or 2) as discriminating against foreign providers once they have entered the domestic market – the issue of national treatment (UNCTAD and World Bank, 1994). Although conceptually the difference between market access and national treatment appears clear, in practice, it can be obscure, especially in the case of cross-border services. The rule of thumb for such services is that quantitative restrictions on providers of foreign services in the domestic market, such as quotas or market-share limits, are viewed as market-access restrictions, whereas policy instruments that distort the relative prices of domestic and foreign-provided services, such as subsidies or price controls, violate national treatment. For the other three modes of transacting services internationally – consumption abroad, commercial presence, and the presence of natural persons – the distinction between market access and national treatment is more obvious. (Further examples are provided in Table 10.)

The most comprehensive source for impediments to international transactions in services is the GATS itself. This may seem ironic, because the agreement is supposed to liberalize international transactions in services, but in fact the bulk of the commitments made by the signatories are standstill provisions. Although it is disappointing that more commitments to reduce impediments were not attained, the agreement is still a remarkable achievement because it effectively encompasses many of the intricacies associated with international transactions in services. Not only does it recognize the variety of services that can be transacted internationally (155 in total) and the different modes of delivery, but it also acknowledges the complexity of the non-tariff barriers created by heavy government regulation by focusing on the crucial issues of market access and national treatment. To deal with this complexity, the GATS includes a schedule for each signatory country (87 in total) listing the service industries and the commitments the country is willing to make regarding restricting international transactions in that service. These commitments concern limitations on market access and national treatment by mode

TABLE 10

EXAMPLES OF IMPEDIMENTS TO INTERNATIONAL SERVICE TRANSACTIONS

MODE OF SUPPLY	MARKET ACCESS RESTRICTIONS	VIOLATION OF NATIONAL TREATMENT
Cross-border supply	Quotas on foreign films Bilateral air service agreements Local content rules on software	Port taxes Telecommunication price controls Subsidies for the film industry
Consumption abroad	Exit visas Travel documentation Entry restrictions	Limits on foreign currency Recognition of educational standards Taxes on travel
Commercial presence	Total exclusion of foreign banks Nationality requirements for lawyers Foreign equity limits for media	Limits on legal services provided "Buy-local" state procurement policy Technology transfer requirements
Presence of natural persons	Immigration controls Professional licensing requirements	Restrictions on living conditions Limits on overseas remittances Restrictions on workplace benefits

Source: Pacific Economic Cooperation Council (1995).

of delivery. Thus, for each industry for which a commitment is made, there is a four (delivery modes) by two (market access, national treatment) table. Entries in this table are either "none" (i.e., no limitation), "unbound" (i.e., no commitment is made), or "other" (i.e., a specification of a particular limitation, but the country is otherwise bound to the GATS). If a particular industry is not listed in the schedule, it is considered unbound from any GATS commitments.

Although researchers have had some success in assessing the impact of non-tariff barriers in merchandise trade by calculating, for example, tariff equivalents, the idiosyncratic nature of services coupled with the data limitations have precluded similar assessments for international trade in services. However, Hoekman (1995) has developed a frequency-based assessment derived from the GATS schedules. He assigns a value of 1.0 for a "none" response, a value of 0.0 for an "unbound" response, and a value of 0.5 for an "other" response. Summing these values over the eight cells in an industry table, for example, would give a score out of a possible eight; a higher score implying fewer restrictions and a lower score implying more. By summing scores across industries, a score for an individual country out of 1,240 (4 x 2 x 155) can be obtained. However, rather than taking a simple sum across industries, thus giving each industry equal weight, weights based on some measure of industry size and importance could be employed.

A 1995 report by the Pacific Economic Cooperation Council (PECC) employs a variant of Hoekman's method to assess the extent of impediments to international transactions in services among Asia Pacific Economic Cooperation

(APEC) organisation, which includes the countries in the APR plus the United States, Canada, Australia, New Zealand, Chile, and Mexico. Although this study provides little information on individual countries – the focus is APEC as a whole – it does find that the frequency measure of trade impediments is inversely related to GDP per capita. That is, countries with the lowest GDP per capita have the highest level of restrictions on market access and national treatment. Hence, because the APR sample of countries consists primarily of countries with lower GDPs per capita, they will have more restrictions, on average. The study's basic finding is that the extent of trade impediments among APEC members is relatively high, approximately 60 to 70 on a scale of 100. Therefore, there is still much scope for further negotiations to remove these barriers. Limitations on market access are found to be more prevalent than those on national treatment. The consumption abroad and cross-border delivery modes are the least restricted, while the presence of natural persons is the most restricted. This result is of particular concern for Canada, because consulting and other professional services often require the temporary movement of highly-skilled personnel. However, negotiations under the GATS on the movement of natural persons are currently underway.

Examining the level of impediments across service industries, the PECC study finds that services that are normally provided or regulated by governments are the most restrictive – for example, postal services, health and social services, basic telecommunications (i.e., telephone services), education, and cultural services (e.g., entertainment and audio-visual services). Although this conclusion is not surprising, the result for education services should be considered more carefully. In the GATS, education services consist of five subsectors: primary, secondary, higher, adult, and other. Given that the governments of APEC countries are generally heavily involved in the provision of primary and secondary education, the restrictions imposed by home governments on post-secondary students studying in Canada are probably small. Most other types of business services (e.g., computer services, value-added telecommunications, professional services, and insurance) are less restricted. Again, this finding is perhaps not surprising, given that these services are intermediate to the production process so that restricting them would raise the cost of domestic production.

Because the GATS provides a comprehensive multilateral framework for governing international transactions in services, it presents an unprecedented opportunity for reducing the barriers Canada faces in exporting services to the APR. Although detailed descriptions of the contents of the GATS and general assessments of its impact are available in other sources, it is worth highlighting several important features here.[22] The GATS was negotiated on a separate track from the GATT and will be administered independently by the Council on Trade in Services, which operates within the newly-established World Trade Organization (WTO). However, disputes over trade in services will be governed by the provisions of the WTO's dispute settlement mechanism, which under certain circumstances will permit countries to impose sanctions on merchandise imports if they win a dispute involving services.

The GATS consists of three main components: 1) a set of general obligations that apply to all services, 2) country schedules of specific commitments on market access and national treatment, and 3) sectoral annexes that provide details on obligations for specific sectors. The general obligations are for the most part similar to those in the GATT; for example, most favoured nation (MFN) treatment, transparency, increasing participation of developing countries, economic integration (i.e., regional agreements are allowed to the extent they do not raise trade barriers), the free flow of payments subject to balance of payments constraints, emergency safeguards, government procurement (i.e., government services are exempt), national treatment, market access, dispute settlement, and progressive liberalization. Apart from the innovation of incorporating country schedules to register commitments on market access and national treatment, the most noteworthy issue raised by these general obligations is that exemptions from MFN status are permitted on a temporary basis and subject to review. This exemption was included to prevent firms in countries that made few commitments on market access/national treatment to free ride on the commitments made by other countries. These exemptions were invoked for the sensitive areas of financial services, basic telecommunications, and marine transport, because negotiations on these services were not completed and the developed countries wanted to withhold their MFN commitments as leverage until a satisfactory conclusion to the negotiations is reached. Also, the obligation of progressive liberalization entails commitments to begin negotiations in the following areas: emergency safeguards, government procurement, financial services, basic telecommunications, marine transport, movement of natural persons, and subsidies.[23]

The GATS should facilitate Canada's exports of business and education services to the APR because it liberalizes trade in the services in which Canada appears to have a comparative advantage. In particular, commitments on reducing limitations to market access and national treatment in the areas of professional services (e.g., accounting, auditing, tax services, management consulting, engineering, and computer services) and advanced telecommunications (e.g., data processing, electronic mail, and electronic data transmission) were obtained. More benefits are likely to accrue in the future as outstanding negotiations are completed – especially in the areas of financial services, government procurement, basic telecommunications, and the movement of natural persons – and when Taiwan and China are admitted to the WTO and accede to the GATS.

Conclusion

THE COUNTRIES OF THE ASIA PACIFIC REGION, especially China, represent an extraordinary opportunity for Canadian firms engaged in exporting business services, especially engineering, consulting, financial, telecommunications, and R&D services; and for Canadian postsecondary institutions. The market for business and education services in the APR is large and likely to grow much larger as the economies of the region continue to expand rapidly and as trade impediments are reduced by improvements in communications and transportation technology and by trade liberalization.

The question that needs to be addressed is what are the appropriate roles for governments in Canada in promoting exports of business and education services to the APR? The most important and the least controversial is for the Canadian government to push for further trade liberalization on all fronts: bilateral, regional, and multilateral. Many of the countries in the APR have relatively high levels of non-tariff barriers to trade in services, and the Canadian government should encourage their unilateral reduction or removal. APEC countries have already agreed to eliminate all tariffs and other barriers to merchandise and services trade by 2020; Canada should hold them to this commitment and strive to accelerate the liberalization. The GATS has opened the door for further multilateral reductions in trade barriers; Canada should encourage the admission of China and Taiwan to the WTO and play a leadership role in ongoing and future negotiations.

Postsecondary education services are provided largely by institutions chartered and primarily supported by provincial governments. Although the federal government, which provides indirect support through block grants, student loans, and research funding, has recently taken useful steps to market Canadian education to the APR, provincial governments and the institutions themselves need to do more to attract foreign students. Given the economic significance of both the education services provided to foreign students and their local expenditures, there is little evidence to suggest that tuition fees and other costs are set optimally. Tuition fees charged to foreign students differ markedly across provinces, sometimes by a factor of five for similar programs of study. In an industry where costs are likely to be declining, it appears that some provinces are setting tuition fees at marginal cost while others are pricing at average cost. Indeed, since the elasticity of demand for education by foreign students is probably greater than one and rents undoubtedly are earned by domestic residents who provide other goods and services to foreign students, pricing at average cost may not be optimal. To facilitate the development of this market, further study is required. In particular, benefit-cost analyses of foreign students as well as comparative studies of APR student enrollment trends and tuition fee levels in Australia, the United States, and other English-speaking countries would be useful to shed light on Canada's competitive position and determine appropriate tuition fees.

Another important role for the federal government is to increase the quantity and improve the quality of the available data on trade in business and education services. Although improvements have been made in recent years, more needs to be done so that the data can be employed in analytical studies to assist in policy making.

Other appropriate roles for governments are harder to define. Market imperfections may exist that create barriers to the export of services to the APR (e.g., capital market imperfections that limit access by small, collateral-poor service-providing firms and economies of scale in marketing and promotion). In these cases, some form of government intervention may be justified: for example, the establishment of trade offices and education centres in the APR and recent prime-ministerial tours of China and India. However, there is much anecdotal evidence to suggest that the federal government has subsidized directly (e.g., by low-cost loans) or indirectly (e.g., via AECL), a significant portion of recent exports of consulting and other professional services to the APR, mainly to the least developed countries in this region – China, Indonesia, and Malaysia. Although a full analysis of the normative issues surrounding these types of subsidies is beyond the scope of this paper, it is worth recognising that some form of government support may be necessary if market imperfections exist that are preventing Canadian firms from exploiting markets in the APR. Unfortunately, the current system of awarding trade subsidies by different government departments and agencies is ad hoc and uncoordinated. Thorough and consistent analyses of possible market imperfections should be performed to determine the benefits and costs of subsidizing exports of services.

ENDNOTES

1 The author would like to thank Gilles Mcdougall and Someshwar Rao of Industry Canada and Hugh Henderson and Rick Murat of Statistics Canada for their assistance with the data; Frank Flatters, Richard Harris, and George Sciadas for their helpful comments; and Allan Seychuk for excellent research assistance.
2 For the purposes of this paper, the Asia Pacific region (APR) is defined to include the following countries: Japan, Singapore, Hong Kong, Taiwan, South Korea, Malaysia, Thailand, the Philippines, Indonesia, and China (in descending order of GDP per capita). These countries were selected on the basis of their economic importance and the availability of data. Japan has a relatively advanced economy as compared to the others in the APR, so it also serves as a basis of comparison.
3 Statistics Canada (1995) defines five broad categories of internationally traded services: travel, freight and shipping, business services, government services, and other services. (These definitions are roughly comparable to those of other data-collecting agencies; for example, the IMF categories are travel, freight and shipping, other transport, other private services, and other official services.) "Business services" includes 17 services,

mostly intermediate business services, although some final consumer services like legal services are included as well. Harris and Easton (1995) consider Canadian exports of travel (in particular, tourism) services to the APR.

4   Rao (1992) finds that levels of Canadian merchandise exports to the APR are less than expected given this region's rapid economic expansion.

5   If the average foreign student were to spend $10,000 per year on tuition and living expenses, for example, then 75,000 students (Statistics Canada, 1994) would account for exports of education services of roughly $750 million. This figure would explain almost 80 percent of the $953 million exports in the "other" subcategory of "other services."

6   Global trade is measured by import data taken from the annual IMF Balance of Payments Report; the Canadian export data are from Statistics Canada.

7   For a useful review of the explanations given for the growth of the service sector, see Grubel and Walker (1989).

8   The relatively low levels of reported trade in services can also be explained in part by problems in accurately measuring and classifying it. The problem is that the data must be collected primarily from business surveys, rather than from customs documents, as is done in the case of merchandise, because of the intangible nature of services. Hence limited or varying survey coverage will create biases; for example, the omission of small- to medium-sized firms, which are relatively significant in international trade in services, creates a systematic downward bias. Also, as coverage has expanded, a biased upward trend in the reported data may have been created. St-Hilaire (1988) argues that the trade in banking services is not properly measured because banks are compensated for providing intermediation services by the interest rate spread between deposits and loans, but this return is normally recorded as investment income in the balance of payments accounts. Data misclassification may also generate biases. For example, Statistics Canada records trade in maintenance and repair services as a component of merchandise trade. This type of misclassification would also occur when the traded products are actually commodity bundles of merchandise and services, for example, specialized equipment that requires training to operate. McRae (1992), Ascher and Whichard (1991), St-Hilaire (1988), Sciadas (1992), and Schembri (1992) provide surveys of the measurement and classification issues involving international trade in services.

9   Although sales of foreign services affiliates are not defined and measured as trade in services, any provision of head office services by the parent firm (e.g., advertising, accounting, and R&D) is considered trade. This is also true for foreign affiliates that manufacture goods.

10 Annual series for Canadian trade in services only began in 1983, and the data published by Statistics Canada tend to be highly aggregated.

Disaggregated data by country/region or type of service are typically not available. Data on trade in services are also known to be of poorer quality than comparable data on trade in merchandise because of the measurement and classification problems discussed in endnote 8.

11 The cross-sectional regression of other private services on GDP and GDP per capita for the nine APR countries in Table 2 for 1992 generates estimated coefficients for these variables that are positive and have t-statistic values of 11.5 and 2.5, respectively. The regression itself has an adjusted R^2 of 0.98, indicating that these two variables explain virtually all of the variation in imports of other private services across countries.

12 The data in Tables 3 to 7 are unpublished and were provided by the Balance of Payments Division, Statistics Canada.

13 Total U.S. exports of business services to the APR (excluding Japan and China) in 1993 were just over US$6 billion (approximately C$8 billion) as compared to Canadian exports of C$527 million (in Table 3), a ratio of roughly 16 to 1. Exports to these countries represented 8.8 percent of total U.S. and 4.8 percent of total Canadian exports of business services. Therefore, by both measures, the extent of Canada's penetration of business service exports in this region is significantly less than that of the United States. Note that the U.S. data on business service exports were taken from the *Survey of Current Business*, September 1995 and were altered to make them comparable with the Canadian data by removing education services and adding royalties.

14 This information was provided by an official of Statistics Canada.

15 Sutton (1995) provides an excellent overview of opportunities for Canadian financial institutions in China.

16 In his comments on this paper, Professor Flatters notes that a sizable portion of Canada's business services exports to China and other APR countries are likely to be related to the exploitation of natural resources. This insight, though not obvious from the data on business exports, is worth emphasizing because Canada has developed a comparative advantage in the technology associated with exploiting natural resources (for example, remote-sensing exploration and environmentally sound methods of manufacturing wood pulp). This technology is transferred to the APR either through joint ventures like BOT, hiring Canadian consultants, or licensing.

17 Data on the number of foreign students are published by Statistics Canada for five levels of education: primary, secondary, college/trade, undergraduate, and graduate. Unfortunately, data on tuition fees and other expenditures are not published by level of education. Some data, however, are available on undergraduate tuition fees. For the purposes of this paper, primary and secondary education are omitted because it is unclear whether students at those levels are enrolled in public or private schools. Graduate education is also omitted because graduate students often receive some form of support from their host institution.

18   The average undergraduate foreign student tuition fee in 1992-93 was approximately $5,700. This is a weighted average of tuition fees across provinces and programs. See Statistics Canada (1994) for more details.
19   It is important to note that the data published by Statistics Canada are based primarily on foreign student visas issued by the Canadian immigration authorities. Although they accurately measure foreign students enrolled full time in colleges and universities, they do not adequately capture the number attending for-profit institutions operated in the private sector, such as specialized language or technical schools. Anecdotal evidence suggests that these schools have been much more aggressive and as a result more successful in attracting students from the APR than traditional postsecondary institutions in Canada.
20   In his comments on this paper, Professor Flatters notes that Canadian and other foreign educational institutions are developing links with institutions in the APR to assist them in establishing and expanding local programs in postsecondary education.
21   Government-imposed barriers to APR students studying in Canada at the postsecondary level are relatively minor compared to the restrictions on importing business services from Canada. The most common restrictions are exit visa requirements and capital controls limiting access to foreign currency. However, the majority of countries in the APR do not restrict the movement of students abroad.
22   See, for example, Schott and Buurman (1994).
23   An interim agreement on financial services was reached on July 28, 1995. Unfortunately, the United States withdrew its offer in June 1995 because it felt insufficient progress was being made. The agreement is therefore only interim; it will expire in December 1997. Although Korea, the Philippines, Indonesia and Malaysia reduced some barriers to trade in financial services, the overall impact of the agreement on Canada is limited.

BIBLIOGRAPHY

Ascher, Bernard and Obie G. Whichard. "Developing a Data System for International Sales of Services: Progress, Problems and Prospects," in *International Economic Transactions: Issues in Measurement and Empirical Research*. Edited by Peter Hooper and J. David Richardson. Chicago: University of Chicago Press, 1991.

Bhagwati, Jagdish N. "Trade in Services and the Multilateral Trade Negotiations." *The World Bank Economic Review*, 1, 4 (1987):549-69.

Cameron, Catherine. *International Education: The Asia Pacific Region and Canada*. Ottawa: Department of Foreign Affairs and International Trade, 1993.

Cleveland, Douglas B. "U.S. Service Exports to China, Hong Kong and Taiwan: Emerging Growth Markets." *The Service Economy*, 8, 1 (1994):1-8.

Grubel, Herbert G. and Michael Walker. *Service Industry Growth*. Vancouver: The Fraser Institute, 1989.

Harris, Richard G. and Stephen T. Easton. "Canada and the Asia Pacific Region: Tourism and Related Services." Paper presented at "The Growing Importance of the Asia Pacific Region in the World Economy: Implications for Canada." Industry Canada Conference, Vancouver, 1995.

Hoekman, Bernard. "Tentative First Steps: An Assessment of the Uruguay Round Agreement on Services." Paper presented at "The Uruguay Round and the Developing Economies." World Bank Conference, Washington, January 26-27, 1995.

McRae, James J. "An Exploratory Analysis of Canada's International Transactions in Service Commodities." Working Paper No. 27. Ottawa: Economic Council of Canada, 1992.

Pacific Economic Cooperation Council. *A Survey of Impediments to Trade and Investment in the APEC Region*. Singapore: Pacific Economic Cooperation Council, 1995.

Rao, Someshwar. "The Asia Pacific Rim: Opportunities and Challenges to Canada." Working Paper No. 37. Ottawa: Economic Council of Canada, 1992.

St-Hilaire, Françoise. "Measuring Trade in Services," in *Trade in Services: Case Studies and Empirical Issues*. Edited by David Conklin. Ottawa: Institute for Research in Public Policy, 1988.

Schembri, Lawrence L. "International Trade in Services: An Assessment from a Quantitative Perspective." Unpublished manuscript. Department of Economics, Carleton University, Ottawa, 1992.

Schott, Jeffrey J. and Johanna W. Buurman. *The Uruguay Round: An Assessment*. Washington: Institute for International Economics, 1994.

Sciadas, George. "Volume Estimates of International Trade in Business Services." *National Income and Expenditure Accounts*, 39, 3 (1992):1-19. (Statistics Canada cat. no. 13-001).

Statistics Canada. *International Student Participation in Canadian Education, 1992*. (Cat. no. 81-261.) Ottawa: Statistics Canada, 1994.

──────. *Canada's International Transactions in Services, 1993-94*. (Cat. no. 67-203.) Ottawa: Statistics Canada, 1995.

Sutton, Brent. *Out From Behind The Great Wall: Emerging Opportunities for Canadian Financial Institutions in China*. Report 146-95. Ottawa: The Conference Board of Canada, 1995.

UNCTAD and the World Bank. *Liberalizing International Transactions in Services: A Handbook*. Geneva: United Nations, 1994.

Comment

Frank Flatters
Department of Economics
Queen's University

I FIND LITTLE TO QUARREL WITH IN THIS PAPER, and so I will take this opportunity to provide some additional observations on some of the themes of this session and of the conference. My comments will have a clear bias in favour of

South-East Asia, which is the part of this region about which I have some knowledge.

Growth and Potential Export Opportunities in the Asia Pacific Region

THE ASIA PACIFIC REGION (APR) is growing very rapidly, and its economies are becoming increasingly integrated with each other and with the rest of the world. This integration, and especially the international division and dispersal of production, has made it increasingly difficult to define "services." Bloom and Noor (1995) find that there has been a great increase in labour market integration in the APR recently, and in comparing the contributions to it of trade, international investment, and labour mobility, find that the former makes the largest contribution. It is certainly a complex process and one to which it is difficult to attribute one cause. Nevertheless, the policy environments in the APR economies, which are becoming increasingly open and market-oriented, play a central role.

The region's rapid growth has often been referred to as being "export-led." This is not inappropriate. According to estimates, up to 25 percent of the recent growth in Thailand and Indonesia has been due to the growth of manufactured exports (Brummitt and Flatters,1992; Flatters and Nader, 1995). However, despite their importance, manufactured exports have not been the fastest growing sectors of these economies. A number of service industries have grown much faster, especially construction and financial services. But demand for services is still outstripping supply in many of the APR countries, with the result that a number of South-East Asian countries are concerned about chronic deficits in their service accounts.

A major reason for growth in the APR has been liberalization, which has created an environment in which the economies can adapt to and take advantage of their own assets and the rapid changes occurring in the global economy. But the liberalization process is far from complete. In an attempt to describe the current phase of trade and industrial policies in Indonesia, Malaysia, and Thailand, Flatters and Harris (1994; 1995) coined the phrase "export-oriented protectionism." This is meant to describe regimes in which exporters are shielded, through a variety of special measures, from many of the costs of what the World Bank describes as "moderate import-substitution regimes." The success of this second wave of South-East Asian "tigers" will depend in large part on their ability to move away from export-oriented protectionism to become truly outward oriented.

Rapid growth, together with increased liberalization, certainly creates many opportunities for Canadian investors and providers of goods and services. Nevertheless, the liberalization of the service sector, especially with respect to foreign participation, has tended to lag behind.[1] In light of the crucial role these sectors play in their economies, failure to open them up could be quite costly. On the other hand, many of these economies are already becoming major investors in regional and even global service industries. I am thinking here of investors from Thailand and Malaysia, for instance – not to mention those from Japan, China and the four tigers – trooping to Mandalay, Rangoon, Ho Chi Minh City, Kunming and Shanghai to invest in telecommunications, tourism, banking and other service industries. One implication of this is that Canada's competition in this region will come in the future not only from Western developed economies but also from the more recently developed countries of the APR itself. For example, Malaysia, which has been one of the world's largest importers of educational services, has launched a series of policy changes in which it intends to become a net *exporter* of education in the near future.

ARE WE MAKING ENOUGH OF THE OPPORTUNITIES?

THE GENERAL MESSAGE OF SCHEMBRI'S and other papers presented in this volume is that Canada is not getting its fair or appropriate share of the action in this, the most rapidly growing part of the world economy. This claim really comprises three separate questions: 1) Is it true that we are participating less than should be expected in this part of the world economy? My general sense is that the answer is probably yes.[2] 2) If so, why is this the case? I have no general answers to this question, but suspect that our preoccupation in recent years with Canada-U.S. free trade and subsequently with its extension to NAFTA has created a business climate that has an unnatural and perhaps unhealthy focus on our southern neighbor.[3] 3) If we are concerned about our lack of participation in the APR, what should we do about it? Most of the papers in this volume, including Schembri's, deal with the first of these questions, and pay little attention to the other two.

To answer these questions properly, we need more and different data from that provided by Schembri. We need to know much more about the workings of the service markets in the relevant countries and especially about their domestic and international policy environments. It is incorrect to look at the APR as a single market. The policy regimes of the different countries vary considerably and in some cases are also changing very rapidly. What follow are a few examples from the markets for particular services.

Financial services – The markets for financial services are being liberalized and are growing at almost breakneck speed. This is especially true in South-East Asia. While Schembri's data show Canadian exports of financial services to the region to be concentrated in Japan, this could change very rapidly with the accelerating growth of financial service markets in South-East Asia. A clear distinction must be made, however, between liberalization for domestic suppliers and opening up the markets for foreign participants. The latter is occurring much more slowly, and the lack of agreement in the recent GATS talks is an example. It appears that liberalization might proceed more rapidly among members of the ASEAN group than with other outsiders. This is a factor that could work to the disadvantage of Canadian suppliers in the ASEAN.

Postsecondary education – Problems with both the quality and quantity of postsecondary education in many APR countries, together with the very rapidly growing demand for skilled labour, provides great potential for exports of educational services to the region. Several countries, notably Malaysia, are already undertaking major reforms of their systems of higher education. The Malaysian reforms include eventually privatizing universities and relaxing many financial and regulatory constraints, with the intention of making Malaysia a major *exporter* of higher education. In this new environment, future opportunities for exports to Malaysia will be largely through partnerships with Malaysians to provide postsecondary education in Malaysia. By looking at the enrolment of foreign students in Canada, Schembri misses this phenomenon entirely. Postsecondary education is already one of the most highly advertised products in the Malaysian media. Most of the programs offered are provided by "twinning" with foreign institutions. Despite Canada's comparative advantage in this area – its long history of high-quality postsecondary education in English and its commonwealth connections, etc. – Canadian institutions and partnerships are conspicuous by their absence in Malaysia. The same is true in Thailand.

Why are Canadians absent? It might be simply that we are not very entrepreneurial. Or there might be serious information gaps preventing Canadians from being informed of the opportunities. If this is true, then perhaps it is an area for policy improvement.[4] Another explanation almost certainly lies in the adverse incentives created by the system of governance and funding of Canadian universities. The almost exclusive reliance on provincial funding, and funding formulae that – in Ontario at least – effectively tax additional earnings from higher education exports at almost 100 percent, significantly deter initiatives to export Canadian postsecondary education.

In light of various educational and financial externalities associated with greater presence in international education, a strong case could be made for *subsidizing* rather than taxing Canadian exports of higher education. Policy adjustments such as major decreases in federal and provincial funding, accompanied by some relaxation of adverse regulatory and financial controls, could improve the situation.

Professional and consulting services – According to Schembri, Canada has been exporting these services with some success. He does, however, hint that government subsidization is one reason for this success. It would be very interesting to find out more about the amounts and types of subsidies involved and to see some discussion of their social costs and benefits. What are the benefits to Canada of subsidizing large engineering firms to undercut foreign competitors in the provision of engineering services for urban transport in Bangkok or hydroelectric development in China? Are there any externalities in Canada that would justify such subsidies? Are we actually lowering the costs of these infrastructure projects in poor countries, thus providing aid, or are we simply permitting Canadian firms to compete with lower-cost foreign firms, in which case no aid is being provided?

Services related to resource exploitation – Canada has been a major exporter of primary products to the world, including, increasingly, to the APR. The paper in the present volume by McRae ("Canada's Natural Resource Exports to the Asia Pacific Region") outlines some possible scenarios in which demand for these products in the APR, and hence Canadian exports, will increase. What McRae and Schembri neglect to mention is the fact that a number of the APR countries are also major primary product producers. According to my very casual personal observation, Canadian firms and individuals appear to have been rather successful in exporting exploitation services for resources like oil and gas, coal, forest products and minerals to this part of the world. If the demand for these products continues to grow, there could be additional opportunities for exports of these services. More information about exports in this segment of the service sector would be interesting, since it links traditional commodity exports with "new" service export possibilities.

Conclusion

SCHEMBRI'S PAPER IS AN OVERVIEW, based on macroeconomic data, of the patterns of Canadian service exports to the APR. At this stage, the policy questions raised in it far outnumber those that have been seriously considered, let alone answered. While this is certainly good news for researchers interested in this area, it is less helpful to those who are charged with formulating economic policies.

Endnotes

1   The recent refusal by the United States to sign the agreement on financial services negotiated as a follow-up to the GATS accord in the Uruguay Round Agreement is a sign of the distance between developed and developing countries on this issue.
2   Plans to carry out case studies of Canadian exporters and investors in the APR in 1992 by the Economic Council of Canada were cancelled because the researchers were unable to identify a sufficient number of Canadian entrepreneurs engaged in this kind of business.

3   McCallum (1995) reports on the importance of the U.S.-Canada border as a barrier to trade. He finds that, after controlling for other determinants of trade patterns, Canadians are more than 20 times more likely to trade with other Canadians than with Americans. Recent evidence (see Flatters and Hazledine, 1995) seems to suggest that Canada-U.S. trade has expanded far more than would have been expected or predicted from simple elasticity calculations. It is quite possible that we are witnessing a "regime shift" in which the psychological importance of the U.S. boundary has diminished, and those of the Pacific ocean have increased, at least comparatively.

4   The cooperative arrangements among Canadian universities and other institutes of higher education that are now being actively pursued are an alternative to direct government policy involvement.

REFERENCES

Bloom, David and Waseem Noor. "Is an Integrated Regional Labor Market Emerging in East and Southeast Asia?" NBER Paper 5174, National Bureau of Economic Research, 1995.

Brummitt, William E., and Frank Flatters. "Exports, Structural Change, and Thailand's Rapid Growth." Background report for Thailand Development Research Institute Conference, "Thailand's Economic Structure: Towards Balanced Development?" Bangkok: TDRI, 1992.

Flatters, Frank and Richard G. Harris. "Export Orientation, Trade Liberalization and Regionalism." Presented at Queen's University/TDRI conference, "AFTA and Beyond." Bangkok, April 1994.

_____. "Trade and Investment: Patterns and Policy Issues in the Asia Pacific Rim," In *Pacific Trade and Investment: Options for the 90s*. Edited by Wendy Dobson and Frank Flatters. International and Development Studies Program, John Deutsch Institute for the Study of Economic Policy. Kingston, Ont.: Queen's University, 1995.

Flatters, Frank and Tim Hazledine. "Industrial Restructuring and Liberalization: Some Lessons From International Experience." Presented at "National Outlook Conference." Kuala Lumpur: Malaysian Institute of Economic Research, December 1995.

Flatters, Frank and Joanna Nader. "Indonesia's Recent Export-Led Growth: A Simple Quantitative Assessment." Unpublished manuscript, Queen's University, 1995.

McCallum, John. "National Borders Matter: Canada-U.S. Regional Trade Patterns." *American Economic Review* (June 1995).

John F. Chant
Department of Economics
Simon Fraser University

7

Financial Sector Opportunities in the Asia Pacific Region: The Case for Canadian Banks[1]

WHAT IS THE IMPACT OF ECONOMIC DEVELOPMENTS in the Asia Pacific region on the demand for the financial services of the Canadian banks? This is an interesting question for two reasons. First, the prospect for strong growth in the Asia Pacific region makes it an appealing market in general. This region has already experienced remarkable economic growth that has led to rapid increases in trade flows with many other countries, including Canada.[2] Only recently, two large countries, China and Vietnam, have begun to share in this development by moving from centralized planning towards a market economy. Many expect these changes will stimulate further growth in the region and add to its potential for external trade. The second reason is more specific to the financial sector. Most countries in the Asia Pacific region have taken significant steps to reform their financial system in one way or another. While these reforms differ in their direction, degree, and especially their starting point, by creating new opportunities they have spurred the demand for financial services. This paper explores whether Canadian banks are well placed to take advantage of these opportunities.

Other segments within the financial systems in Asia Pacific countries can also be expected to provide opportunities for Canadian institutions – for brokers, underwriters, mutual funds, and insurance companies, in particular – as governments liberalize these segments and as the demands for financial services expand with income growth.[3] While many of the same forces will be shaping developments in these other sectors, each market is sufficiently distinct to deserve separate treatment. These other sectors fall beyond the scope of the present study.

The study is also limited with respect to its perspective on banking. Many banking activities are well documented from information provided for corporate reporting and regulatory purposes. Much of that information is found in the published balance sheets provided by banks, but other parts of banking activity are less well documented. Many new activities fall into the off-balance sheet category and, as a result, are less visible. Accordingly, this study is somewhat unbalanced because it focuses on those banking activities for which information can readily be obtained. That emphasis does not deny the potential opportunities that less visible activities can present for Canadian banks in the Asia Pacific region, but the study cannot assess that potential.

The demand for the services of Canadian banks in Asia Pacific countries depends on a number of factors:

- the overall level of demand for banking services in those economies;
- the regulatory environment that governs the ability of the banking system to meet the demand for banking services;
- the treatment of foreign banks in host countries; and
- the competitive strengths of the Canadian banks.

The Demand for Banking Services

THE DIVERSE NATURE OF BANKING SERVICES MEANS THAT there is no single measure of the demand for these services. The most commonly used indicator is a broad measure of the stock of money, defined to include both "narrow money" (currency and demand deposits) and "quasi-money" (savings and time deposits). This measure captures only the intermediation activities of the banking system and does not indicate the scale of banks' off-balance sheet activities, which are becoming increasingly important.

A significant determinant of the overall size of an economy's monetary sector is its level of aggregate income. Any expansion in the size of the economy will be paralleled by an expansion of the monetary sector, with an accompanying increase in the demand for banking services. Studies of the relationship between broad money and gross domestic product (GDP) indicate that the growth of the financial system may proceed more rapidly than the growth of income. The World Bank (1993), for example, found that "financial systems of higher-income countries are usually deeper (as measured by the ratio of liquid liabilities to GNP) than those in poorer ones." These issues pertaining to an economy's demand for financial services are examined more fully later on.

The Regulatory Environment

THE REGULATORY SYSTEM IN ANY COUNTRY will determine the degree to which the demand for financial services will be realized. Many regulations commonly found in developing countries retard the development of their financial systems. Regulations such as interest-rate ceilings on deposits lessen the attractiveness of monetary assets for the general public. Similarly, compulsory credit programs reduce the profitability of banks and weaken their ability to raise funds from the public. The Asia Pacific countries have all moved to liberalize their approaches to regulation to encourage further development of their banking systems, but they began the process from very different initial points (see below).

The Treatment of Foreign Banks

A SIGNIFICANT FEATURE OF BANKING REGULATION throughout the world has been the differential treatment of domestic and foreign banks. The liberalization of regulations governing domestic institutions will not increase the demand for services of foreign banks unless they gain access to domestic markets and can carry out banking activities on terms that are comparable with those enjoyed by domestic banks. In this respect, while the Asia Pacific countries have all made progress with respect to domestic financial reform, they have been slower to open their banking markets to foreign banks.

The Competitive Strengths of the Canadian Banks

AT BEST, THE FACTORS DISCUSSED ABOVE – the overall demand for financial services, the state of domestic regulation, and the policy towards foreign banks – make up an environment favourable to the supply of services from Canadian banks: they can only create market opportunities. Whether Canadian banks respond to these opportunities depends on their strengths relative to that of host-country banks and other international banks, and on the corporate strategies followed by individual institutions.

Trade in Financial Services

The Concept of Financial Services

THE CONCEPT OF TRADE IN BANKING SERVICES is not well understood because the supply of financial services by banks is frequently confused with the provision of finance itself. Finance consists of the flows of funds through which lending, borrowing, and investment take place. Through the process of finance, the suppliers of funds place command over resources at the disposal of the demanders of funds. Finance by itself does not directly produce current income. Rather, it transfers command over resources from ultimate lenders to ultimate borrowers. Financial services, in contrast, consist of productive activities that support the flows of funds between the ultimate suppliers and ultimate users of the funds. The contribution of the banks to the Canadian economy consists of the value placed on the services by the banks' customers. As with other goods and services, it also makes up the incomes earned by Canadians in producing these services – the wages, the rent, and the profit that make up the value-added of the banking sector.

This distinction between finance and financial services can be illustrated more clearly through an example. A $6 million loan represents a flow of finance from a bank to a borrower. The financial services provided in making that loan depend on the resources required to carry out the transaction. If the loan is part of a larger syndicated loan to a borrower with a strong credit rating, where each lender participates under standardized terms arranged by the lead bank, a participating bank may not need to devote many resources to assessing and administering the loan. On the other hand, if the loan is made to a new firm in an emerging industry in an economy without a clear commercial code or a developed legal infrastructure, the bank must direct many resources to assessing the borrower's prospects, negotiating the terms of the loan, and setting conditions to assure its repayment.

In the first case, the participants provide relatively few banking services. The borrower could probably raise the funds directly from the ultimate lenders without much additional expense and would be unwilling to pay a large amount for the bank's services. In the second case, lenders unfamiliar with the country, the specific industry, or the borrower's prospects would have to invest considerable resources before undertaking the loan. The bank provides a service by enabling the lender to make funds available to the borrower without having to bear such costs – a service that could not be duplicated by the lender without bearing considerable expense. Even though the two loans are of equal value, the financial services involved in the transactions differ substantially.

The incomes from supplying finance and supplying financial services also differ. Suppliers of finance earn future income from the returns resulting from the flow of funds. Suppliers of financial services, in contrast, earn income from the fees and service charges that they receive from the users and suppliers of the funds. This distinction may not always be clear, especially when suppliers of financial services participate in indirect finance. Intermediaries usually charge interest to the fund users at one rate and pay interest to the fund suppliers at another, lower rate. While the financial institution may appear to supply its own funds and to receive interest as a payment for the use of these funds, as an intermediary it actually channels funds from the ultimate supplier to the user. Its implied fee for this service often consists of the difference between the interest it receives from borrowers and the interest it pays to lenders.[4]

Some perspective on the difference between financial services and funding can be gained from examining different measures of the Canadian financial system in 1993. A common measure of the importance of deposit institutions to the economy compares their total assets to GDP. In 1993, banks, trust companies, and credit unions held assets of $1,084 billion. This measure exceeded Canada's GDP of $721 billion for the year by over 50 percent. From this perspective, this sector appears very large relative to the total economy. In contrast, this sector's contribution to GDP is much smaller, consisting of the difference between interest received and interest paid, together with service charges and other income. Over 1993, this contribution was $13 billion. This measure indicates that deposit institutions accounted for only 1.8 percent of GDP that year.

The Demand for Financial Services

Measuring the demand for financial services is complicated because it does not manifest itself directly in the same way as the demand for haircuts or transportation. The demand for financial services is derived from the demand for different types of financial products. A bank deposit itself is not a financial service. It is the means through which customers gain financial services. If the deposit is a checking deposit, it gives the customer access to payments services. The customer also gains the opportunity to earn a return on a collection of assets held indirectly through the intermediary services offered by the bank.

The starting point for an estimation of the demand for financial services would be the estimation of separate demands for different types of financial products. These would include the demands for consumer and corporate loans and different types of deposits, together with the demand for many other types of services, such as foreign-exchange services, portfolio management, and custodial services that do not show up on banks' balance sheets. Each of these demands must then be translated into a demand for financial services. This operation requires the determination of the value added by financial institutions in performing these services. As with other products, this value added represents the value of services provided by the financial institutions and also the sum of incomes earned by producing these financial services.

The limited availability of statistics practically rules out the calculation of any such measure of the demand for financial services. Instead, the usual benchmark has been some broad measure of money, such as the sum of narrow money and quasi-money, which indicates the size of banks' balance sheets. By measuring the funds raised and the funds loaned and invested by banks, this measure indicates the scale of intermediary services supplied by the banks.[5]

International Banking Activities

Lewis and Davis (1987) have identified the multifold activities of international banks:

- taking deposits and making loans in domestic currency to foreign governments, enterprises, and individuals;
- taking deposits and lending in foreign currencies to domestic and foreign entities;
- managing and acting as agents for syndicated loans, and designing special financing requirements for international trade and projects;
- conducting foreign-exchange transactions, and dealing in gold and precious metals, and international money transfers;
- providing documentary letters of credit, standby letters of credit, multiple credit lines, bank acceptances, and Euronote issuance facilities;
- trading in currency futures and options, financial futures and options, and interest-rate and asset swaps; writing interest caps; and
- underwriting and placement of Eurobond issues; distribution of Eurodollar commercial paper; assisting cross-border mergers, acquisitions, and sales; and financial advisory and investment services.

These activities differ in many different ways. The first two items represent the traditional intermediary activities of commercial banking, where banks serve as a principal, collecting deposits from one group and lending to another. In many of the remaining items, the banks participate as agent for their customers, managing their syndicated loans or underwriting or transacting on their behalf in markets for currencies or derivatives. Banks may also take positions as traders in currencies, securities, or financial derivatives, or in so-called off-balance sheet activities.

Typically, banks earn income on balance sheet activities through the spreads between the interest rates they charge and the interest rates they pay. The income from off-balance sheet activities usually comes from fees that banks charge for their services.

THE ORGANIZATION OF INTERNATIONAL BANKING

BANKING SERVICES CAN TAKE PLACE IN DIFFERENT MARKETS with different types of customers. Banks deal with other banks in the interbank market. Here, the size of individual transactions is large. Competition among banks and the typically standardized types of transaction lead to low spreads. Interbank activities consist of both balance sheet activities, where banks raise or place funds in the interbank market, and off-balance sheet activities, where banks trade foreign currencies, derivatives, or other financial instruments with other banks. Banks deal with governments and large corporations in the wholesale market. While the traditional intermediary business remains important in the wholesale market, increasingly banks are providing off-balance sheet services to meet the changing needs of their wholesale customers. The development of continuing relationships between banks and their customers remains an essential part of wholesale banking. Banks also deal with small- and medium-sized businesses and households in retail-banking relationships. Despite the recent broadening of the range of services offered to retail customers, this part of banking remains more concentrated in the intermediary business than are wholesale and interbank activities.

Banks meet the needs of their international customers in a variety of ways. Interbank activities most often take place in international banking centres where both parties to the transaction have some form of presence. The centres include the banking markets of countries with well-developed banking systems, especially those which permit foreign banks to operate with few restrictions. Much interbank activity also takes place in international financial centres, such as Singapore and the Bahamas, that are separate from national banking systems. Cross-border banking, where banks located in one country deal with nationals of other countries, may be suitable for many types of wholesale business, especially off-balance sheet activities. Other types of wholesale business and most retail business require the supplier to establish a presence in the customer's country.

Banking activity in other countries can be organized in many different ways – through representative offices, affiliates, branches, or subsidiaries; as members in consortiums; and as partners in joint ventures (Table 1). Pecchioli (1983) describes the functions of each (pp. 58-9):

> Representative offices serve as points of contact for providing market information and establishing business connections, as well as acting as an intermediary between the parent bank and customers without, as a rule, serving as booking locations. "Affiliates" are local institutions incorporated under host-country law in which the banks hold a minority interest. Subsidiaries are local institutions incorporated under host country law in which the bank has a direct or controlling interest. Foreign branches are a legal extension of the parent bank in a foreign country and are not separately-constituted locally chartered corporations. Hence branches are an integral part of the parent bank, without a separate identity, and, as such, are subject to home country control and regulation.

Consortiums are groups of banks organized specifically to finance large-scale projects. Joint ventures are usually organized to conduct particular activities of common interest to several banks.

Different organizational forms will be appropriate for different types of banking business. Retail business requires some form of establishment such as a branch, affiliate, or subsidiary authorized to carry on banking business in the host country. Much wholesale business can be conducted through representative offices that can serve as agents to attract business for other units of the bank.

The choice of form will also be governed by host-country regulations. The use of representative offices will be more difficult when the host country enforces foreign-exchange controls. Barriers to entry may prevent the use of

TABLE 1

RELATIVE ADVANTAGES OF DIFFERENT ORGANIZATIONAL FORMS

	REPRESENTATIVE OFFICE	SUBSIDIARY	BRANCH	CONSORTIUM AND JOINT VENTURE
Required investment	Modest	Moderate/ substantial	Moderate/ substantial	Moderate
Control over operation	Direct	Substantial/ direct	Direct	Minor/ moderate
Referral business	Favourable	Favourable	Favourable	Minor/ moderate
New business	Favourable	Favourable	Favourable	Minor/ moderate
Flexibility of operation	Relatively inflexible	Favourable	Favourable	Some flexibility
Staffing	Some staff required	*	*	Modest

* Difficult to characterize, since it depends on specific circumstances, the type of subsidiary or branch, and the characteristics of the business to be undertaken.
Source: Pecchioli (1983), p. 61.

subsidiaries, or other regulations may restrict their business. Finally, some countries, such as Canada, may restrict the form of organization for prudential reasons.

The relationship between international banks and local banking markets may also differ. Onshore banking markets are fully subject to domestic regulations and serve the needs of local residents. Offshore markets, in contrast, are subject to less regulation than domestic institutions and tend to serve the wholesale banking needs of multinational businesses. Offshore markets include both markets that are beyond the scope of national regulation (such as the Eurocurrency and Eurobond market), and international financial centres, where international banking activities are subject to less regulation than is banking business with domestic customers.

DEMAND FOR FINANCIAL SERVICES IN THE ASIA PACIFIC REGION

THE OVERALL DEMAND FOR FINANCIAL SERVICES in the Asia Pacific region will be one of the determinants of the opportunities for Canadian banks in the region. The demand for financial services in an economy, like other demands, depends on their price, the level of income as a representation of the demanders' resources, and other variables characterized as tastes and technology. As discussed earlier, the demand for financial services has usually been represented by some broad measure of the money supply as a proxy for intermediary services. Rarely have cross-country studies of the demand for financial services included measures of the price of these services. More attention has been placed on the relationship between income and the demand for financial services as represented by broad money.

The relationship between broad money and income has been estimated for a cross-section of 97 countries for 1990 to be

$$M/Y = 0.356 + 0.0000154 YCAP \qquad (1)$$
$$(t = 13.06) \quad (t = 6.26)$$
$$R^2 \text{ (adjusted)} = 0.28$$

where M/Y measures the ratio of broad money to national income and YCAP is per capita income measured in U.S. dollars. The measure of broad money used for the study consists of the sum of narrow money (currency and demand deposits) and quasi-money (saving and time deposits).[6]

Predicted M/Y ratios based on equation (1) are shown in Table 2 for different levels of per capita income. As might be expected, broad money appears to grow together with income, thus contributing to an increase in the demand for financial services. This growth can be expected to take place whether the growth in income is caused by higher per capita income or by population growth. The coefficient for YCAP also suggests further growth in the demand for financial services from financial deepening. It shows that the M/Y ratio varies directly with the level of per capita income: for every $100 increase in per capita income, there is a corresponding 1.5 percent increase in the M/Y

TABLE 2

RELATIONSHIP BETWEEN PER CAPITA INCOME AND M/Y

PER CAPITA INCOME (DOLLARS)	PREDICTED M/Y RATIO	PREDICTED BROAD MONEY
500	0.364	182
1,000	0.371	371
2,000	0.387	774
5,000	0.433	2,165
10,000	0.510	5,100
15,000	0.587	8,805
20,000	0.664	13,280

Source: Based on data from IMF, *International Financial Statistics*.

ratio. Thus a doubling of per capita income from $500 to $1,000 can be expected to increase holdings of broad money by 108 percent, whereas a doubling of per capita income from $10,000 to $20,000 can be expected to increase broad money by 160 percent.

The estimates can also be used to give some indication of further growth in the demand for financial services that might result from the removal of restrictive regulation such as controls of deposit and loan rates. Limits on these rates, for example, reduce the ability of banks to attract deposits from the public. Countries that maintain restrictive regulation can be expected to have lower M/Y ratios for any level of income than countries with liberalized financial systems. If this is so, financial reform can raise a country's M/Y by an amount additional to the effects of income growth.

Estimates of the effects of financial reform on broad-money holding can be obtained by using the results from equation (1). The data used for the estimates encompass countries with differing systems of financial regulation. Ideally, estimates of the effects of restrictive regulation could be made by comparing the M/Y ratios of different countries with reference to their regulatory approach. Unfortunately, an assessment of the regulatory system of each country is beyond the scope of this study.

Accordingly, rough estimates have been made by comparing the M/Y ratio of each country with the benchmark M/Y ratio that might be expected with liberalized financial regulation. For present purposes, the benchmark has been assumed to be the value of M/Y that is one standard error above the predicted value for each level of the dependent variable.[7] The estimates show that Japan and Malaysia both have higher M/Y ratios than are represented by the benchmark. In contrast, the M/Y ratio in 1990 would be predicted by the benchmark to be higher by 77 percent in South Korea, 66 percent in the Philippines, and 31 percent in Indonesia under a liberalized financial regime.[8]

In summary, three separate sources of possible increased demand for financial services in Asia Pacific countries can be identified. First, income growth by itself will increase the demand for financial services just through the maintenance of stable M/Y ratios. Second, growth in per capita income can be expected to stimulate financial deepening, which will result in higher M/Y ratios. Finally, the movement from financial repression to liberalized regulation can be expected to raise M/Y ratios in some countries. The importance of each of the factors will differ from country to country. For example, rapid income growth is expected to stimulate expansion of the financial sectors of China and Vietnam in the near future. Further growth in the M/Y ratio, however, can be expected only if substantial financial reform proceeds. The effects of financial reform may be more imminent for economies such as Indonesia, Malaysia, South Korea, and the Philippines, where reform is already underway. Japan, Hong Kong, and Singapore, on the other hand, with their higher levels of income and more fully developed financial systems, will experience increased demand, but not at as dramatic a rate.

Financial Reform in the Asia Pacific Region

WHAT FINANCIAL REFORMS HAVE BEEN UNDERTAKEN by Asia Pacific countries through the 1980s and 1990s? The features of regulation that are most likely to affect the potential business of foreign banks include interest-rate controls, the level of direct credit allocation by the central bank, and the degree of government involvement through ownership.

Financial reform is not a uniform process. As Caprio (1994, p. 2) observes, the terms deregulation, liberalization, innovation, privatization, and internationalization have been used to characterize different components of financial deregulation. The measures taken by a country will depend on its starting point. A country (such as China or Vietnam) that starts with a single state-owned bank intimately involved in economic planning will take different steps than a former colony (such as Malaysia or Singapore) that has inherited a system of private banks owned and controlled in the former colonial power.

The elements of a financial reform differ substantially across countries and may include any of the following:

- elimination or reduction of portfolio requirements that force financial institutions to hold minimum amounts of specified assets (e.g., minimum ratios of government debt);
- scaling back or elimination of directed credit programs, which require financial institutions to give credit for particular purposes (export credit) or to specific groups (small businesses) on favourable terms;
- removal of restrictions on both the structure and level of interest rates that are paid to depositors by financial institutions;
- division of state-owned monopolistic banks (or "monobanks") into separate banks;
- privatization of state-owned banks;
- commercialization of state-owned banks;
- removal of restrictions on the structure of interest rates charged to borrowers by banks and financial institutions;
- removal of restrictions on products offered by banks and other financial institutions;
- opening of entry to new domestic banks;
- opening up the cross-border supply of banking services;
- elimination or reduction of restrictions on the entry of foreign banks; and
- removal of constraints on the operations of foreign banks.

The banking systems of Asia Pacific countries had achieved different levels of development by the beginning of the 1980s. Only in Hong Kong and the offshore market of Singapore were they relatively free from government control. Even though privately owned, the major Japanese banks, the Philippine banks, and banks in the domestic Singapore market were subject to controls on the interest rates that they could pay and charge, and their lending was governed by credit ceilings. State-owned banks in Indonesia, South Korea, Malaysia, and Taiwan faced controls extending beyond interest rates and credit ceilings to credit allocation among groups and sectors. Finally, the banking systems of China and Vietnam each consisted of a monobank that supplied working capital to state enterprises according to the central plan.

Details of specific measures taken in the different countries are summarized in Appendix 2, which highlights the differences in financial reform throughout the Asia Pacific region. Each country's approach has been shaped by its starting point. Little reform was necessary in Hong Kong and the offshore Singapore markets, where banking operated under liberal regulation. In Japan, interest-rate controls on deposits and loans were removed gradually together with credit ceilings, so that all interest rates had become market-determined by 1995.

More comprehensive measures have been taken in Indonesia, South Korea, Malaysia, and Taiwan. Each of these countries entered the 1980s with large dominant state-owned banks that were subject to interest-rate controls and were required in differing degrees to allocate their credit to meet government priorities. Although their reforms have taken different directions, all governments have allowed a greater scope for market forces to determine interest rates. Virtually all interest-rate controls have been removed. Now, only Malaysia maintains any official control over the setting of interest rates. Restrictions on new entry and branching have also been reduced in all cases. The scope of directed credit programs has been reduced, but not eliminated. While only South Korea has succeeded in privatizing state banks, its government influence still remains strong. Perhaps the greatest decrease in direct state influence has occurred in Indonesia, where the dominance of state banks has been diminished by the rapid growth of private banks.

Finally, China and Vietnam have moved from monobanks to a system that assigns central- and commercial-banking responsibilities to different institutions. Neither, however, has moved significantly away from the centralized allocation of credit or removed state control over interest rates on deposits and loans.

Table 3 classifies Asia Pacific banking systems according to the present status of their financial regulation. Hong Kong, the offshore market of Singapore, and Japan share certain fundamental characteristics as *predominantly market systems*. While some differences in regulation exist between these economies, banks are predominantly under private ownership and remain outside the government's direct influence. The government neither sets interest rates nor determines the allocation of credit among sectors or among groups. In the *mixed market system*, the government either owns or strongly influences the actions of the major commercial banks. Deposit and loan rates are set primarily by market forces. The banks operate on commercial principles in much

TABLE 3

REGULATION OF DOMESTIC BANKING ACTIVITY
IN THE ASIA PACIFIC REGION, 1995

COUNTRY	INTEREST RATE CONTROLS DEPOSIT	LOANS	NEW ENTRY (DOMESTIC)	BRANCHING (DOMESTIC)	DIRECTED LENDING	GOVERNMENT CONTROL
Market benchmark	No	No	Yes	Yes	No	No
Predominantly market						
Hong Kong	No	No	Yes	Yes	No	No
Japan	No	No	Yes	Yes	No	No
Singapore offshore	No	No	Yes	Yes	No	No
Mixed market						
Indonesia	No	No	Yes	Yes	Yes	Yes
South Korea	No	No	Yes	Yes	Yes	Yes
Malaysia	No	Yes	Yes	Yes	Yes	Yes
Philippines	No	No	Yes	Yes	Yes	Yes
Singapore domestic	No	No	No	Yes	No	Yes
Taiwan	No	No	Yes	Yes	Yes	Yes
Centrally controlled						
China	Yes	Yes	?	?	Yes	Yes
Vietnam	Yes	Yes	Yes	Yes	Yes	Yes

Source: Country profiles in Appendix 2.

of their business, but are expected to meet targets with respect to priority credits. Finally, banks in *state-directed systems* still operate in a system where interest rates are set by the central bank or the government and the majority of credit is allocated in accordance with a central plan.

REFORM AND THE TREATMENT OF FOREIGN BANKS

A CRITICAL ASPECT OF FINANCIAL REFORM FOR THE CANADIAN BANKS is the host country's approach to the entry of foreign banks. Host-country regulations can affect the operations of foreign banks in a variety of ways. For example, regulations can determine the conditions under which foreign banks are allowed to enter a country's banking markets. These conditions may specify the organizational form that the entry can take, together with other restrictions, such as the need to form a joint venture with a partner from the host country. Entry alone does not assure that foreign banks can compete on an equal footing with domestic banks. Foreign banks may also be granted different powers or be subject to differential regulatory requirements than domestic banks.

Entry of Foreign-owned Banks

TABLE 4 SHOWS THAT ENTRY CONDITIONS for foreign banks differ substantially among Asia Pacific banking markets. Only Hong Kong, Japan, and South Korea offer unrestricted entry to foreign banks.[9] In addition, Singapore permits entry into its offshore banking centre. Three countries – China, the Philippines, and Indonesia – permit restricted entry. China allows entry only into 13 cities and special economic zones. Indonesia requires foreign banks to enter as joint ventures with Indonesian entities that must hold at least a 15 percent interest. The Philippines offers opportunities to invest in existing local banks or in a limited number of new branch opportunities. In contrast, Malaysia, Singapore, and Taiwan do not currently allow foreign banks to enter their domestic markets.

The conditions governing the entry of foreign banks shown in Table 4 can be compared with the classification of banking systems in Table 3. The

TABLE 4

Conditions on Foreign-Bank Entry

Country	New entry	Form	Number of foreign banks
China	Restricted	Branches and subsidiaries	98 banking offices
Hong Kong	Permitted	Branches	
Indonesia	Restricted	Joint ventures only	29 joint ventures
Japan	Permitted	Branches only	90 banks
South Korea	Permitted[1]	Branches only (in practice)	52 banks
Malaysia	Not permitted	Branches (prior to restriction)	14 banks
Singapore			
Offshore market	Permitted	Branches and subsidiaries	119 banks
Domestic	Not permitted		36 banks
Taiwan			
Offshore	Permitted	Branches	
Domestic	Permitted	De facto branches	37 banks

1 This classification may be *de jure* rather than *de facto* for reasons discussed in the text.
Source: U.S. Department of the Treasury (1994).

market-based systems have been more willing to permit unrestricted entry and grant comparable powers to foreign and domestic banks. In contrast, the centrally directed banking systems have allowed only restricted entry of foreign banks. The authorities in mixed systems have adopted restriction of differing degrees towards foreign banks. Some (such as Malaysia, Singapore, and Taiwan) have prohibited entry into domestic banking activities. Others (Indonesia, for example) have limited the form in which entry can occur. Such restrictions are not a matter of indifference to foreign banks. They may limit the scope of foreign-bank activity because some forms of entry may be unsuitable for particular types of banking activity.

THE REGULATION OF FOREIGN-BANK ACTIVITIES

TABLE 5 SHOWS THE POWERS GRANTED TO FOREIGN BANKS in different Asia Pacific countries. At one extreme, Japan gives foreign banks *de facto* national treatment, allowing them the same powers as domestic banks. In other countries, regulations directed towards foreign banks include restrictions on location of business, discriminatory capital requirements, and limits on types of funding and other types of business. As discussed earlier, rules on form of organization, limits to types of funding, and exclusion of parent's capital all hamper foreign banks in expanding their activities.

Table 5 reinforces the conclusion that the nature of domestic banking influences policies towards foreign-bank entry. The activities of foreign banks are least restricted in economies where private banks operate in a market environment. More significant restrictions on activities can be found in the economies with dominant state-owned banks, such as South Korea, Malaysia, Singapore, Taiwan, and Indonesia. Finally, formerly planned economies (such as China and Vietnam) are least open to the activities of foreign banks.

THE POWERS OF FOREIGN BANKS: A FORWARD LOOK

ALL ASIA PACIFIC COUNTRIES HAVE INITIATED A PROCESS of financial reform. Much progress has been achieved in some areas of domestic reform: interest rates have been deregulated; banks and other financial institutions have gained expanded powers; and new domestic banks have been allowed to enter the industry. Less progress has been made towards reducing the dominant position of government-owned banks and the allocation of credit by government.

A similar approach towards foreign banks can be found among countries at the same stages of financial-sector development. Countries with monobanks, directed lending, and controlled interest rates offer the least scope for foreign banking. In those countries, a larger, though still limited, role for foreign banks may be allowed as the banking system develops to a mixed market system. The

TABLE 5
MAJOR RESTRICTIONS ON FOREIGN-BANK POWERS
Japan *De facto* national treatment **Hong Kong** Limited to one branch in addition to main office No subsidiaries **Indonesia** Must be joint venture (minimum 15 percent domestic ownership by qualified bank) Higher capital requirement Branches only in seven largest cities Limits to foreign personnel Capital requirements exclude parent's capital **South Korea** *De facto* prohibition of subsidiaries Limits to foreign exchange swaps Capital requirements exclude parent's capital Branches must be capitalized separately Ceilings on issue of certificates of deposit Maturity limits on certificates of deposit Permission to offer new products **Taiwan** *De facto* prohibition of subsidiaries Prohibition on debentures Limits to foreign exchange positions Operational ratios tied to remitted capital **Malaysia** Required to be subsidiaries Maximum of 40 percent of bank credit to foreign owned companies Prohibition on new branches Limits to foreign personnel **Singapore** (domestic market) Limits in retail business *De facto* prohibition on foreign branch opening Lack of access to provident fund accounts
Source: U.S. Department of the Treasury (1994).

predominantly market systems typically make few distinctions between foreign and domestic banks.

The slow progress achieved with respect to the reduction of barriers to foreign banks' activities raises an important question. Can the liberalization of rules on foreign banking be expected to follow soon as a further step in financial reform? Or is the relaxation of constraints on foreign-bank operations distinct and separate from the current thrust of reform?

Several indications suggest that restrictive policies towards foreign banks will persist despite the progress of domestic reform. First, governments appear to be highly concerned with the solvency problems of state-owned banks. These banks typically hold many poor-quality loans, often made in response to past government priorities and directives. In many cases, the authorities hope that the transformation of state-owned banks into commercial banks will enable them to strengthen their portfolios and overcome their bad-loan problems. The presence of competition from foreign banks would make it more difficult for state-owned banks to earn the profits needed to restore their balance sheets.

Second, governments may also continue to hold priorities with respect to the allocation of credit. Despite the financial reforms already introduced, governments may still wish to use domestic banks as instruments to direct credit towards such priority uses. This becomes more difficult when state banks must face foreign competitors for deposits and for profitable lending opportunities.

Finally, governments may have long-run concerns about the evolution of the banking system. Banking appears to be an area where governments usually value a strong domestic presence. The operations of foreign banks may be seen as interfering with the eventual development of strong domestic institutions.

Any or all of these explanations may account for the continued restrictions on foreign banking that are found in mixed market systems. None of these explanations provide much optimism that these countries will lift the restrictions on foreign banks in the near future. Foreign banks operating in the Asia Pacific region will probably continue to face regulatory constraints that will prevent them from competing fully with domestic institutions in meeting the demand for banking services.

The opening of financial markets may not be determined by national considerations alone. Increasingly, the trade policies of individual countries are being influenced by multinational trade negotiations. While trade agreements under the General Agreement on Tariffs and Trade (GATT) have traditionally excluded services, the recent Uruguay Round of negotiations covered the financial sector for the first time. Agreement in this area proved to be very difficult and was eventually delayed beyond the conclusion of the rest of the negotiations. A tentative agreement was reached in 1995 that "is based on the traditional principles of Most Favoured Nation and national treatment, and entailed a large number of countries offering binding commitments having to do with specific areas of financial services" (White, 1996, p. 17). Countries have until early 1998 to determine the final terms of their participation under the agreement.

The same concerns that have slowed progress towards the unilateral opening of national markets appear to have led governments to be cautious in their undertakings under the multilateral framework. The United States has signalled its dissatisfaction with that outcome (U.S. Department of the

Treasury, 1994, p. 98): "About 15 additional commercially important GATT member countries, including Japan, Korea, other South and Southeast Asian countries, and major Latin American countries continue to have major barriers to market access or national treatment and, in some cases, their commitments do not cover important financial sector activity."

In reaction, the United States has taken an exception from the General Agreement on Trade in Services (GATS) obligation to grant most-favoured nation status to all participants. This exemption allows the United States to differentiate among foreign financial institutions on the basis of their respective home countries' commitment to open their domestic financial markets. GATS signatory countries will make their final commitments in early 1998. It remains to be seen whether the actions taken by the United States will cause them to accelerate the opening of their financial markets.

THE COMPETITIVE ADVANTAGES OF INTERNATIONAL BANKS

OTHER THINGS EQUAL, the prospects for Canadian banks in the Asia Pacific region depend on the advantages that they can offer relative to domestic and other foreign financial institutions. Aliber (1984) has identified two questions that the theory of international banking has sought to answer. Why do banks from one country enter the banking markets of other countries? What explains the over- or under-representation of different countries' banks in banking markets throughout the world?

The present study emphasizes the opportunities for Canadian banks in the Asia Pacific. Are Canadian banks in a favourable position to respond to the expected changes that will take place in the region's financial markets? Theories of international banking may provide guidance by indicating sources of advantage for Canadian banks in foreign markets.

Explanations of international banking typically stress the same factors used to explain foreign direct investment in non-financial industries (Caves, 1977; Grubel, 1977). Any firm investing in a foreign market starts at a disadvantage, relative to domestic firms, because of its limited knowledge of that market. The knowledge that is required goes beyond the narrowly economic to the cultural, legal, and social environment. To make a foreign investment, a firm must hold some specific advantage relative to its foreign competitors. In addition, direct entry must be a more effective way for the firm to exploit its advantage than exporting its products or signing licensing agreements with domestic producers in the foreign country.

Current theories of banking recognize the importance of information and knowledge for intermediation. Banks need two types of knowledge for successful lending and investment. First, they need general knowledge about economic conditions. Second, and more important, they need specific knowledge about the borrower – the type of product sold, the borrower's management skills, past credit record, and the quality and marketability of the borrower's security. A lack of this local knowledge harms foreign banks relative to local banks in dealing with domestic borrowers. To operate in foreign markets, international banks must possess some inherent advantage that enables them to overcome their lack of familiarity with local conditions. Among the theories that have been proposed to explain international banking are the "follow-the-leader" and "banking-advantage" models.

THE "FOLLOW-THE-LEADER" MODEL

CAVES (1977) PRESENTS A "FOLLOW-THE-LEADER" THEORY of international banking, which suggests that the foreign activities of a country's banks follow the foreign direct investment of that country's enterprises. He argues that the "goodwill relations" that banks have established with commercial and industrial customers at home provide them with an advantage that overcomes their lack of local knowledge in foreign markets. Banks enter foreign markets to continue servicing the banking needs of their customers. International banking, according to this theory, becomes an extension of the domestic banking industry into other countries.

The follow-the-leader theory offers specific predictions about patterns of international banking. It suggests that the entry of banks into foreign markets follows the entry into those market of their industrial and commercial customers in the home country. It also identifies which banks will take part in international banking activities – namely, banks from countries that are major suppliers of foreign direct investment and, specifically, banks that service the domestic banking needs of enterprises that invest abroad.

This theory also predicts where banks will expand their international activities: their foreign operations should correspond to the pattern of direct investment abroad by domestic enterprises. It also predicts that international banking will be concentrated in wholesale activities that meet the needs of multinational enterprises. Conversely, foreign banks will not be inclined to set up branch networks to service retail markets in foreign countries.

The follow-the-leader model explains only the pattern, not the scale or form, that banking activity may take in specific countries. A contrary view of the advantages to banks from following the leader has been expressed recently by Rutenberg and Neave (1996). They characterize the life cycle of a firm's international involvement (pp. 12-3):

> When a Canadian firm is merely exporting, a Canadian bank can service its needs for letters of credit ... The familiar bank provides psychological assurance to the client Canadian firm.

> Before the Canadian firm considers a joint venture, or a permanent facility abroad, it needs a local national bank.... The right national bank may be different from the Canadian bank's correspondent.
>
> Later, if the Canadian firm grows a network of many foreign subsidiaries, it faces the need to mobilize its global cash flow.... The firm may turn over its global treasury function to a multinational bank (e.g. Citibank, HKSB, Natwest).
>
> ... In summary, Canadian firms will learn most rapidly if they can wean themselves of Canadian banks.

If Rutenberg and Neave are correct, the follow-the-leader approach will not explain patterns of international activities by banks.

The costs of banking operations must also be considered. At low levels of activity, banking has been found to be subject to diseconomies of scale.[10] Thus foreign banking on a small scale would have high unit costs. Those costs would be higher still if foreign regulators imposed minimum capital requirements on new entrants. Foreign direct investment in a host country must reach a critical level to create enough demand for financial services to make the entry of home-country banks profitable. These conditions are more likely to apply when the home country is large and concentrates its foreign direct investment in a few countries.

The follow-the-leader approach suggests that Canadian banking activity will follow Canadian foreign direct investment. Table 6 shows that Canadian foreign direct investment in the Asia Pacific region was almost $13 billion in 1994 and accounted for just 10 percent of total Canadian foreign direct investment. Hong Kong, Japan, and Singapore each received $2 billion or more in direct investment and together accounted for 85 percent of Canada's direct foreign investment in the region. None of the other countries in the

TABLE 6

CANADIAN FOREIGN DIRECT INVESTMENT IN THE ASIA PACIFIC REGION, 1994

COUNTRY	FOREIGN DIRECT INVESTMENT ($ MILLIONS)	SHARE OF TOTAL (PERCENT)
Singapore	2,184	1.7
Indonesia	739	0.6
Hong Kong	1,979	1.6
Japan	3,050	2.4
Taiwan	60	
Malaysia	354	0.3
South Korea	166	0.1
Total foreign direct investment	125,247	100.0

Source: Statistics Canada, *Canada's International Investment Position*, cat. no. 67-202.

region received more than $1 billion, or 1 percent of the total. Given the moderate size of direct foreign investment, the follow-the-leader approach predicts that Canadian banks will be at a disadvantage relative to foreign banks from larger countries, especially when operating in small foreign economies. This scale of direct investment by Canadian enterprises may be too small in many Asia Pacific economies to support bank branches or subsidiaries. Even in those countries with a disproportionate share of Canadian investment, the banks may operate through low-cost alternatives such as representative offices. Subsidiaries and branches will be common only where there is a substantial presence.

The "Banking-Advantage" Model

Grubel (1977) offers a "banking advantage" theory of international banking as an alternative to the follow-the-leader approach. His theory focuses on the characteristics of the banks rather than of their customers. While he agrees that banks must possess some advantage in order to enter a foreign market, he identifies this advantage as being based in the inherent efficiency of a bank's operation. He suggests that the advantages of international banks arise from superior management skills and banking technology developed for their domestic market. They can transfer these advantages to foreign markets at very low marginal costs.

The predictions about international banking derived from the banking-advantage theory differ from those of the follow-the-leader approach. Banks would enter those markets where the return from their advantages would be the highest. Thus their choice of market need not be tied to those of home-country multinationals. The services that they offer in other countries would be those where they possess an advantage and could be retail, wholesale, or financial-centre services.

According to the banking-advantage approach, the ability of Canadian banks to enter foreign markets is not closely linked to the scale of the Canadian economy. Rather, features of both the Canadian and foreign banking environments are crucial. Canadian banks have an advantage in those areas where our regulatory framework and market structure foster efficiency in banking. A foreign country's regulation and banking structure also affect the potential entry of Canadian banks. Entry will be more difficult where the foreign-banking environment fosters the same efficiencies held by the Canadian banks.

A strength of the Canadian banks has been their extensive branch networks. They appear to hold an advantage over U.S. banks in this regard because the development of branch networks has been limited by regulation in the United States. Some states prohibit branch banking, while others severely restrict it. National banking laws have further prevented the establishment of branch networks across states. An awareness of this advantage has shaped the strategy of at least one Canadian bank, which plans to exploit the opportunities created by the North American Free Trade Agreement (NAFTA) to turn itself into a "North American" bank by building a branch network in the United States.[11]

Like many aspects of banking, the effective management of a branch network depends not only on management skills but also on a knowledge of local conditions. Existing branch networks in Canada have been built with personnel who possess certain traditions and levels of skills. Workers in a country with different cultural, institutional, and economic traditions will have an entirely different set of skills. Similarly, customers also vary from country to country. Techniques for attracting and serving depositors in a high-income Western economy may not work well elsewhere. Loan approvals and credit rating require different skills, according to the quality of accounting information and the legal framework governing foreclosure and recourse.

These difficulties associated with the transfer of banking expertise formed in one culture to another can be illustrated by the case of Japan. The banking sector in Japan has been open to foreign entry for some time, but foreign banks have not moved in to supply banking services to the local population. As the *National Treatment Study* observes (U.S. Department of the Treasury, 1994, p. 333), "foreign banks remain only marginal players in the Japanese banking market. This is partly a reflection of Japan's regulatory environment and partly the consequence of exclusionary business practices. Strong relationships between members of related business groups, often involving cross-shareholding arrangements make it extremely difficult for nonmembers (i.e., foreign banks) to effectively compete in the Japanese banking market."

In light of these problems, it appears that one of the main competitive advantages enjoyed by Canadian banks may not be easily transferable to many of the economies of the Asia Pacific region.

Canadian banks are also based in an economy that is particularly open to foreign trade. Canadian enterprises will seek finance for export credit. Moreover, extensive foreign trade and investment flows will create a demand for foreign-exchange services. As a result, Canadian banks would be expected to have developed expertise in exchange operations and the financing of international trade. Much of the economic development of the Asia Pacific economies is concentrated on foreign markets. To the extent that competitive advantages are derived from national experience, Canadian banks then would be expected to have established a capacity for meeting the specialized banking demands arising from these activities.

COMPETITIVE ADVANTAGE: SUMMARY

THE TWO COMPETING THEORIES OF INTERNATIONAL BANKING discussed here are very general and do not apply to any economy in particular. Both approaches stress that home-country conditions determine which country's banks will be suppliers of financial services. For Canadian banks, the small size of the economy, the importance of banking activities related to international trade and its finance, and the extensive branch network developed by major Canadian financial institutions will be important. The former matters for the follow-the-leader approach: the relatively small direct investment by Canadians in the region limits their scope for serving as bankers for Canadian businesses in the region. Branch networks and experience in activities related to foreign trade matter for the banking-advantage approach.

THE INTERNATIONAL BUSINESS OF THE CANADIAN BANKS

SOME PERSPECTIVE ON THE PROSPECTS OF CANADIAN BANKS in the Asia Pacific region can be gained from examining their existing international business, with special attention to that region.[12] Several questions can be addressed. What has been the role of Canadian banks in international banking? How significant has been their participation in the Asia Pacific region? What types of business do Canadian banks perform in the region?

OVERALL INTERNATIONAL BUSINESS

A BANK'S COMMITMENT TO INTERNATIONAL BANKING will affect its ability to develop business in the Asia Pacific region. If it is already active in international banking, it will have the experienced personnel and established management techniques required for expanding into new markets. When these are available, the bank may be expected to be well placed to take advantage of emerging opportunities in the Asia Pacific region.

The development of the international business of Canadian banks is compared with that of other countries' banks in Table 7 for the period 1975 to 1995. The table shows that the foreign business of world banks has grown substantially, increasing 13 times over the period. The overall foreign assets and liabilities of Canadian banks have also grown steadily over all the periods shown. Still, this growth has failed to match that of any reference group (all countries, industrialized countries, and the United States) from 1975 through 1990, and it has continued to fall behind that of all groups other than the United States from 1990 to 1995. Over the entire period, the effects of the slower growth have been striking. Canadian banks have seen their share of foreign assets of all international banks decline to less than one-third and their share of foreign liabilities to less than one-half of their values in 1975.

TABLE 7

BANKS' FOREIGN ASSETS AND LIABILITIES (US$ BILLIONS)

	ASSETS, BY COUNTRY OF LENDING BANK				
	1975	1980	1985	1990	1995
All countries	557	1,836	2,984	6,788	7,890
Industrial countries	446	1,374	2,228	5,120	5,860
United States	55	177	417	578	628
Canada	14	35	44	52	52
Canada as a percentage of:					
All countries	2.5	1.9	1.5	0.8	0.7
Industrial countries	3.1	2.5	2.0	1.0	0.9
United States	25.5	19.8	10.6	9.0	8.3
	FOREIGN LIABILITIES, BY RESIDENCE OF BORROWING BANK				
	1975	1980	1985	1990	1995
All countries	596	1,901	3,843	7,149	8,071
Industrial countries	158	1,362	2,205	5,459	5,970
United States	63	151	381	733	944
Canada	14	43	65	78	81
Canada as a percentage of:					
All countries	2.3	2.3	1.7	1.1	1.0
Industrial countries	8.9	3.2	2.9	1.4	1.3
United States	22.2	28.5	17.1	10.6	8.3

Source: International Monetary Fund, *International Financial Statistics*, Tables 7yd and 7xd.

ASIA PACIFIC BUSINESS OF CANADIAN BANKS

A NUMBER OF PERSPECTIVES CAN BE OFFERED on the Asia Pacific business of Canadian banks. How committed to the region are they, relative to other international banks? Has their commitment been changing as opportunities in the area have become apparent? What types of business do they undertake? Does their involvement in those markets correspond to their expected strengths?

The banks' prospects in the Asia Pacific region will be influenced by both their expertise and their management's commitment to the region. The needed expertise goes beyond knowledge of the banking techniques and practices that will be demanded in the region. Many types of banking require the establishment of continuing relationships with customers to assist banks in assessing credit worthiness and in supervising and monitoring the repayment of loans. Such relationships are especially important in economies that lack an adequate legal infrastructure and a well-developed commercial code. They will be more difficult to establish where cultural differences exist between the bankers and their customers. Canadian banks will be better placed to meet Asia Pacific opportunities when they are already committed to international banking and to developing expertise in the region.

Comparison with All Banks

Tables 8 and 9 show the share of the Asia Pacific markets in the cross-border business of Canadian and other international banks.[13] The first panel in each table shows that other international banks hold almost one-quarter of their cross-border claims and raise almost one-fifth of their cross-border funds in the region. The tables also show that their business is heavily concentrated in Japan, Hong Kong, and Singapore. The second panel shows that the interbank business is distributed among the three centres, while the third panel shows that only Japan accounts for a significant share of the non-bank business.

The comparisons also show that Canadian banks carry out a substantially smaller proportion of their cross-border business in the region than other international banks. The share of the Asia Pacific region in their total cross-border claims is less than one-third, and the share of their total liabilities coming from the region is less than half that of the other banks.

The composition of the Asia Pacific business of the Canadian banks also differs from that of other banks. Canadian banks carry out less of their regional business with Japan and more with the non-financial centres. The data also show that Canadian banks have raised more of their funds from non-bank sources than have other banks. As in the rest of their international business, Canadian banks have relied more heavily on non-bank sources in raising funds from the region.

Composition of Business

Canadian banks carry on international banking activities throughout the world. As already seen, their banking business in the Asia Pacific region differs from that of other international banks. Do these differences in turn reflect differences in Canadian banks' approach to international business? Or do they reflect differences in their approach towards the Asia Pacific region relative to their other international business? Tables 10 and 11 compare the features of Canadian banks' Asia Pacific business with those of their worldwide business. These tables also indicate the share of the business that is carried out with other banks and the share of the business that is non-local.[14]

TABLE 8

CLAIMS HELD BY INTERNATIONAL BANKS

RESIDENCE OF BORROWER	ALL BANKS		CANADIAN BANKS	
	($ BILLIONS)	(PERCENT)	($ BILLIONS)	(PERCENT)
Total claims by residence of borrower				
All countries	8,444	100.0	215	100.0
Asia Pacific	2,011	23.8	16.5	7.7
China	47	0.6	0.8	0.4
Hong Kong	551	6.5	4.4	2.0
Indonesia	41	0.5	0.0	0.0
Japan	981	11.6	5.7	2.7
South Korea	57	0.7	1.3	0.6
Malaysia	13	0.2	0.4	0.2
Philippines	7	0.1	0.6	0.3
Singapore	314	3.7	2.5	1.2
Taiwan	n.a.		1.0	0.5
Vietnam	0.2	0.0	0.0	0.0
Cross-border interbank claims by residence of borrowing bank				
All countries	6,134	100.0	66	100.0
Asia Pacific	1,637	26.7	8.5	12.9
China	31	0.5	0.3	0.5
Hong Kong	532	8.7	1.4	2.1
Indonesia	10	0.2		0.0
Japan	703	11.5	3.7	5.6
South Korea	42	0.7	0.5	0.8
Malaysia	6	0.1	0.1	0.2
Philippines	3	0.0	0.3	0.5
Singapore	310	5.1	2.2	3.3
Taiwan	n.a.		0.2	0.3
Vietnam	0	0.0	0.0	0.0
Claims on non-banks by residence of borrower				
All countries	2,310	100.0	149	100.0
Asia Pacific	374	16.2	8	5.4
China	16	0.7	0.5	0.3
Hong Kong	19	0.8	3	2.0
Indonesia	31	1.3		0.0
Japan	278	12.0	2.0	1.3
South Korea	15	0.6	0.8	0.5
Malaysia	7	0.3	0.3	0.2
Philippines	4	0.2	0.3	0.2
Singapore	4	0.2	0.3	0.2
Taiwan	n.a.	0.0	0.8	0.5
Vietnam	0.2	0.0	0.0	0.0

n.a. = not available.
Source: International Monetary Fund, *International Financial Statistics*, Tables 7xd and 8yad. Bank of Canada, *Bank of Canada Review*, Table C10.

TABLE 9

CLAIMS ON INTERNATIONAL BANKS

RESIDENCE OF LENDER	ALL BANKS ($ BILLIONS)	(PERCENT)	CANADIAN BANKS ($ BILLIONS)	(PERCENT)
Total claims on banks by residence of lender				
All countries	8,038	100.0	238	100.0
Asia Pacific	1,533	19.1	20.7	8.7
China	2	0.0	0.8	0.3
Hong Kong	394	4.9	11.3	4.7
Indonesia	20	0.2	0.0	0.0
Japan	786	9.8	2.9	1.2
South Korea	42	0.5	0.4	0.2
Malaysia	2	0.0	1.3	0.5
Philippines	8	0.1	0.4	0.2
Singapore	279	3.5	1.6	0.7
Taiwan	0	0.0	2	0.8
Vietnam	0.1	0.0	0.0	0.0
Cross-border interbank liabilities by residence of lending bank				
All countries	6,051	100.0	120	100.0
Asia Pacific	1,457	24.1	10.7	8.9
China		0.0		0.0
Hong Kong	357	5.9	4.7	3.9
Indonesia	17	0.3		0.0
Japan	764	12.6	1.5	1.3
South Korea	40	0.7	0.3	0.3
Malaysia		0.0	1.0	0.8
Philippines	7	0.1	0.3	0.3
Singapore	272	4.5	1.1	0.9
Taiwan		0.0	1.8	1.5
Vietnam				
Claims on non-banks by residence of lender				
All countries	1,987	100.0	118	100.0
Asia Pacific	76	3.8	10.0	8.5
China	2	0.1	0.8	0.7
Hong Kong	37	1.9	6.6	5.6
Indonesia	3	0.2		0.0
Japan	22	1.1	1.4	1.2
South Korea	2	0.1	0.1	0.1
Malaysia	2	0.1	0.3	0.3
Philippines	1	0.1	0.1	0.1
Singapore	7	0.4	0.5	0.4
Taiwan		0.0	0.2	0.2
Vietnam	0.1	0.0	0.0	0.0

Source: International Monetary Fund, *International Financial Statistics*, Tables 7yd and 8yad. Bank of Canada, *Bank of Canada Review*, Table C10.

TABLE 10

TOTAL CLAIMS BOOKED WORLDWIDE VIS-À-VIS NON-RESIDENTS BY CANADIAN BANKS

	TOTAL CLAIMS		ON BANKS	SHARE OF ASSETS ON NON-RESIDENTS	ON BANKS	ON NON-RESIDENTS
	($ Millions)	(Percent)	($ Millions)		(Percent)	
China	664	0.3	173	651	26	98
Japan	5,684	2.6	3,683	3,886	65	68
South Korea	1,241	0.6	489	938	39	76
Malaysia	416	0.2	137	153	33	37
Philippines	417	0.2	15	417	4	100
Taiwan	1,120	0.5	342	923	31	82
Thailand	578	0.3	167	573	29	99
Off-shore banking centres						
Hong Kong	4,427	2.1	1,391	2,916	31	66
Singapore	2,498	1.2	2,156	2,389	86	96
Total (East Asia and banking centres)	17,045	7.9	8,553	12,846	50	75
Worldwide	215,522	100.0	65,817	109,972	31	51

Source: *Bank of Canada Review*, Table C10.

Claims held. Table 10 compares the banks' Asia Pacific claims with their worldwide business. These claims differ substantially from the banks' other international business in a number of respects. Claims on banks are much more important in the Asia Pacific region, accounting for 50 percent of regional claims, relative to just 31 percent worldwide. These claims are especially important for both Japan (65 percent of loans) and Singapore (86 percent of loans). A similar difference can be found with respect to non-local claims. They account for 75 percent of all claims in the region – a much higher proportion than their 51 percent share in worldwide business. Only in Malaysia, where the Bank of Nova Scotia operates in the domestic banking market, is the share of non-local claims lower than for worldwide business. Japan and Singapore also have lower shares of claims on non-locals than other Asia Pacific countries. In both cases, their shares of non-local business still exceed the share for overall international business by a substantial margin.

Funding. Table 11 shows the sources of funds for each of the Asia Pacific markets. With the exception of Hong Kong, Canadian banks rely heavily on non-local sources of funds in all markets other than Japan. Overall, non-local funds account for 86 percent of all funds raised in the Asia Pacific region,

TABLE 11

TOTAL LIABILITIES BOOKED WORLDWIDE VIS-À-VIS NON-RESIDENTS BY CANADIAN BANKS

	TOTAL LIABILITIES		TO BANKS	SHARE OF LIABILITIES TO NON-RESIDENTS	TO BANKS	TO NON-RESIDENTS
	($ Millions)	(Percent)	($ Millions)		(Percent)	
China	751	0.3	702	751	93	100
Japan	2,892	1.2	1,547	1,564	53	54
South Korea	418	0.2	332	343	79	82
Malaysia	1,273	0.5	990	1,029	78	81
Philippines	395	0.2	314	395	79	100
Taiwan	2,401	1.0	1,897	2,209	79	92
Thailand	481	0.2	451	481	94	100
Off-shore banking centres						
Hong Kong	11,261	4.7	4,737	10,006	42	89
Singapore	1,591	0.7	1,125	1,588	71	100
Total (East Asia and banking centres)	21,463	9.0	12,095	18,366	56	86
Worldwide	237,579	100.0	120,415	174,728	51	74

Source: *Bank of Canada Review*, Table C10.

compared with 74 percent for the banks' worldwide business. Though this aspect is less pronounced, Canadian banks also rely on other banks for the majority of their financing in most Asia Pacific markets. Only in Hong Kong and Japan did the share of funds raised from other banks fall below 70 percent. With the heavy weighting of these two countries, funding from banks in the Asia Pacific region accounted for 63 percent of total funding, compared with 50 percent for the banks' worldwide business.

Trends. Table 12 shows that the funds raised by Canadian banks in the Asia Pacific region have increased 10 times, and their claims on the region six times, between 1974 and 1994. Over the whole period, the region's share of both claims and funding has increased. In each case, there was growth in the share of business through the early 1980s. The region's share of claims reached a peak in 1983 and has fluctuated ever since. The share of funding continued to increase to its peak in 1990 and has declined since.

There has also been a substantial shift in the net claims on the region. Claims exceeded funds raised in the region in every year through 1985, with the maximum difference reaching $2.7 billion in 1980. From 1986 onward, the relationship has reversed: funds raised exceeded claims on the region in

TABLE 12

ASIA PACIFIC BANKING BUSINESS OF CANADIAN BANKS, 1978-94 ($ MILLIONS)

	WORDWIDE			ASIA PACIFIC			ASIA PACIFIC AS SHARE OF TOTAL	
	ASSETS	LIABILITIES	NET ASSETS	ASSETS	LIABILITIES	NET ASSETS	ASSETS	LIABILITIES
							(Percent)	
1978	53,077	57,000	−3,923	2,827	2,121	706	5.3	3.7
1979	65,194	73,107	−7,913	4,245	2,694	1,551	6.5	3.7
1980	88,669	98,167	−9,498	6,867	4,134	2,733	7.7	4.2
1981	112,298	132,592	−20,294	9,167	6,984	2,183	8.2	5.3
1982	118,727	131,620	−12,893	9,183	8,234	949	7.7	6.3
1983	122,615	135,733	−13,118	10,490	10,475	15	8.6	7.7
1984	146,623	157,624	−11,001	12,387	10,364	2,023	8.4	6.6
1985	159,498	172,683	−13,185	12,898	12,496	402	8.1	7.2
1986	168,255	175,801	−7,546	13,512	15,956	−2,444	8.0	9.1
1987	159,042	171,189	−12,147	12,889	16,964	−4,075	8.1	9.9
1988	136,955	154,318	−17,363	10,320	14,934	−4,614	7.5	9.7
1989	138,541	157,387	−18,846	9,923	15,429	−5,506	7.2	9.8
1990	154,861	183,798	−28,937	13,170	20,216	−7,046	8.5	11.0
1991	161,959	193,591	−31,632	13,223	19,256	−6,033	8.2	9.9
1992	175,347	198,295	−22,948	14,627	19,553	−4,926	8.3	9.9
1993	193,538	218,993	−25,455	15,041	20,665	−5,624	7.8	9.4
1994	218,805	244,189	−25,384	17,281	21,521	−4,240	7.9	8.8

Source: *Bank of Canada Review*, Table C10.

every year. Net funds raised reached a maximum of $7 billion in 1990 and have since declined to $4.2 billion in 1994.

Tables 13 and 14 show the country-by-country patterns that lie behind the movement of the aggregate shares. Much of the early growth in share of claims on the region came from Japan. Up to 1987, claims on other Asia Pacific countries actually declined as a share of the banks' total international claims. This pattern was reversed from 1990 onward, as Japan's share of claims declined. During the same period, the shares of claims on China, Taiwan, and Hong Kong all grew, though not enough to offset the decrease in Japan's share. Interestingly, the shares of South Korea, Malaysia, the Philippines, and Singapore all decreased over the period.

Similar changes have taken place in the pattern of funding. Japan's share reached a peak in 1987 and has subsequently declined. Only the shares of Hong Kong and Malaysia have provided any significant offset. Over the entire period, the shares of Singapore and the Philippines have decreased, while those of China, Japan, Malaysia, Hong Kong, and Taiwan have all increased.

The net balance also differs by country. China, Japan, South Korea, the Philippines, Thailand, and Singapore have generally been net users of funds

TABLE 13
Claims Held by Canadian Banks by Country of Residence, 1978-94

	All Countries	China	Japan	South Korea	Malaysia	Philippines	Taiwan	Thailand	Hong Kong	Singapore	East Asia and the Pacific
						($ Millions)					
1978	53,077	4	323	324	173	508	113	63	599	720	2,827
1979	65,194	217	339	667	179	629	173	165	768	1,108	4,245
1980	88,669	264	672	964	215	1,124	294	232	1,660	1,442	6,867
1981	112,298	201	1,042	1,457	380	1,253	420	269	2,351	1,794	9,167
1982	118,727	24	1,648	1,727	523	1,232	452	309	1,660	1,608	9,183
1983	122,615	24	2,755	1,819	602	1,238	367	266	2,035	1,384	10,490
1984	146,623	105	3,113	2,006	627	1,167	267	289	2,729	2,084	12,387
1985	159,498	271	3,364	2,068	521	1,211	280	270	2,740	2,173	12,898
1986	168,255	130	4,545	1,577	484	1,210	244	168	2,977	2,177	13,512
1987	159,042	155	6,368	853	358	883	422	76	2,096	1,678	12,889
1988	136,955	100	5,551	551	286	461	295	65	1,729	1,282	10,320
1989	138,541	149	5,216	488	269	292	352	72	2,074	1,011	9,923
1990	154,861	123	7,125	631	355	320	416	138	2,626	1,436	13,170
1991	161,959	90	6,350	955	262	323	593	230	2,952	1,468	13,223
1992	175,347	126	6,129	1,121	415	349	924	290	3,213	2,060	14,627
1993	193,538	338	5,789	1,116	504	352	996	297	4,004	1,645	15,041
1994	218,805	573	5,745	1,355	328	451	1,475	534	4,412	2,408	17,281

TABLE 13 (CONT'D)

	ALL COUNTRIES	CHINA	JAPAN	SOUTH KOREA	MALAYSIA	PHILIPPINES	TAIWAN	THAILAND	HONG KONG	SINGAPORE	EAST ASIA AND THE PACIFIC
						(Percent)					
1978		0.0	0.6	0.6	0.3	1.0	0.2	0.1	1.1	1.4	5.3
1979		0.3	0.5	1.0	0.3	1.0	0.3	0.3	1.2	1.7	6.5
1980		0.3	0.8	1.1	0.2	1.3	0.3	0.3	1.9	1.6	7.7
1981		0.2	0.9	1.3	0.3	1.1	0.4	0.2	2.1	1.6	8.2
1982		0.0	1.4	1.5	0.4	1.0	0.4	0.3	1.4	1.4	7.7
1983		0.0	2.2	1.5	0.5	1.0	0.3	0.2	1.7	1.1	8.6
1984		0.1	2.1	1.4	0.4	0.8	0.2	0.2	1.9	1.4	8.4
1985		0.2	2.1	1.3	0.3	0.8	0.2	0.2	1.7	1.4	8.1
1986		0.1	2.7	0.9	0.3	0.7	0.1	0.1	1.8	1.3	8.0
1987		0.1	4.0	0.5	0.2	0.6	0.3	0.0	1.3	1.1	8.1
1988		0.1	4.1	0.4	0.2	0.3	0.2	0.0	1.3	0.9	7.5
1989		0.1	3.8	0.4	0.2	0.2	0.3	0.1	1.5	0.7	7.2
1990		0.1	4.6	0.4	0.2	0.2	0.3	0.1	1.7	0.9	8.5
1991		0.1	3.9	0.6	0.2	0.2	0.3	0.1	1.8	0.9	8.2
1992		0.1	3.5	0.6	0.2	0.2	0.4	0.2	1.8	1.2	8.3
1993		0.2	3.0	0.6	0.3	0.2	0.5	0.2	2.1	0.8	7.8
1994		0.3	2.6	0.6	0.1	0.2	0.7	0.2	2.0	1.1	7.9

Source: Statistics Canada, CANSIM.

TABLE 14
Claims Held on Canadian Banks by Country of Residence, 1978-94

	All Countries	China	Japan	South Korea	Malaysia	Philippines	Taiwan	Thailand	Hong Kong	Singapore	Asia Pacific Region
						($ Millions)					
1978	57,000	33	82	20	83	134	90	5	1,014	660	2,121
1979	73,107	52	126	24	130	212	118	21	1,124	887	2,694
1980	98,167	110	320	75	170	197	141	81	1,732	1,308	4,134
1981	132,592	357	626	101	130	149	244	144	2,943	2,290	6,984
1982	131,620	974	751	52	314	153	317	172	3,464	2,037	8,234
1983	135,733	1,259	2,051	68	252	121	443	150	4,239	1,892	10,475
1984	157,624	1,380	1,697	77	262	107	1,004	103	4,451	1,283	10,364
1985	172,683	600	2,901	96	312	158	1,176	142	5,290	1,821	12,496
1986	175,801	555	5,140	92	344	147	2,777	186	5,517	1,198	15,956
1987	171,189	458	6,798	115	166	100	2,677	79	5,589	982	16,964
1988	154,318	488	5,290	112	185	110	2,115	103	5,447	1,084	14,934
1989	157,387	255	4,458	179	320	141	1,932	111	6,596	1,437	15,429
1990	183,798	615	6,321	222	974	195	2,078	514	7,887	1,410	20,216
1991	193,591	1,082	4,678	364	953	248	1,902	772	7,937	1,320	19,256
1992	198,295	1,213	2,870	297	1,061	282	1,387	594	9,874	1,975	19,553
1993	218,993	1,013	2,650	315	1,471	153	1,635	334	11,223	1,871	20,665
1994	244,189	978	3,656	338	1,042	271	2,773	41	10,773	1,649	21,521

TABLE 14 (CONT'D)

	ALL COUNTRIES	CHINA	JAPAN	SOUTH KOREA	MALAYSIA	PHILIPPINES	TAIWAN	THAILAND	HONG KONG	SINGAPORE	ASIA PACIFIC REGION
						(Percent)					
1978		0.1	0.1	0.0	0.1	0.2	0.2	0.0	1.8	1.2	3.7
1979		0.1	0.2	0.0	0.2	0.3	0.2	0.0	1.5	1.2	3.7
1980		0.1	0.3	0.1	0.2	0.2	0.1	0.1	1.8	1.3	4.2
1981		0.3	0.5	0.1	0.1	0.1	0.2	0.1	2.2	1.7	5.3
1982		0.7	0.6	0.0	0.2	0.1	0.2	0.1	2.6	1.5	6.3
1983		0.9	1.5	0.1	0.2	0.1	0.3	0.1	3.1	1.4	7.7
1984		0.9	1.1	0.0	0.2	0.1	0.6	0.1	2.8	0.8	6.6
1985		0.3	1.7	0.1	0.2	0.1	0.7	0.1	3.1	1.1	7.2
1986		0.3	2.9	0.1	0.2	0.1	1.6	0.1	3.1	0.7	9.1
1987		0.3	4.0	0.1	0.1	0.1	1.6	0.0	3.3	0.6	9.9
1988		0.3	3.4	0.1	0.1	0.1	1.4	0.1	3.5	0.7	9.7
1989		0.2	2.8	0.1	0.2	0.1	1.2	0.1	4.2	0.9	9.8
1990		0.3	3.4	0.1	0.5	0.1	1.1	0.3	4.3	0.8	11.0
1991		0.6	2.4	0.2	0.5	0.1	1.0	0.4	4.1	0.7	9.9
1992		0.6	1.4	0.1	0.5	0.1	0.7	0.3	5.0	1.0	9.9
1993		0.5	1.2	0.1	0.7	0.1	0.7	0.2	5.1	0.9	9.4
1994		0.4	1.5	0.1	0.4	0.1	1.1	0.0	4.4	0.7	8.8

Source: Statistics Canada, CANSIM.

while Malaysia, Taiwan, and Hong Kong have been net suppliers of funds. At the end of 1994, Hong Kong ($6.3 billion), Taiwan ($1.3 billion), and Malaysia ($700 million) were all net suppliers. Japan ($2.1 billion), South Korea ($1 billion), and Singapore ($800 million) were the largest net users of funds.

Differences Among Banks

The analysis so far has concentrated on the aggregate activity of Canadian banks in the Asia Pacific region. Canadian banks may have pursued very different strategies towards the region according to their competitive strengths and their perceptions of the opportunities it offers. The use of aggregates may mask the activities of individual banks that may be pursuing opportunities more aggressively than Canadian banks overall. The information available can be used to indicate possible differences among Canadian banks in their approach to the region.

Table 15 shows that all of the major banks have branches in Hong Kong, Seoul, Singapore, and Tokyo; at least three have representative offices in Beijing. Subsidiaries, in contrast, operate in Singapore and Hong Kong. Among the subsidiaries in these markets, Bank of Nova Scotia Asia has the highest level of assets ($333 million) and CIBC Asia has the highest book value ($140 million). The *National Treatment Study* (U.S. Department of the Treasury, 1994) indicates that the Bank of Nova Scotia's operation in Malaysia has assets of $390 million and accounts for approximately 0.6 percent of total banking assets in that country.

Canadian banks, as shown in Table 16, appear to have followed different strategies with respect to business in the Asia Pacific region. The Bank of Nova Scotia, traditionally active in international business, holds over 6 percent of its total assets in the Asia Pacific region. At the other extreme, the Toronto-Dominion Bank holds only 1.6 percent of its earning assets in the region. Comparable data are not available for the Bank of Montreal, but it holds only 2.0 percent of its total international assets outside developed economies.

Contribution to GDP

The Canadian banks' business in the Asia Pacific region has been analyzed so far only in terms of assets and liabilities – a perspective that gives no indication of the contribution of this business to Canada's GDP. Some limited indication of the importance of these activities can be gained from data available in the category "other financial services" in balance-of-payments statistics.[15]

Overall, Canada earned $837 million from trade in other financial services in 1993, a contribution of just 0.1 percent of GDP. Trade with the Asia

Table 15

Presence of Canadian Banks in the Asia Pacific Region

Canadian Imperial Bank of Commerce	
Subsidiaries	CIBC Asia Limited; book value = $140 million
	CIBC(Hong Kong) Ltd.
Branch offices	Hong Kong, Tokyo, Singapore, Taipei
Representative office	Beijing
Affiliate	CEF Limited – Hong Kong and Singapore
Royal Bank of Canada	
Subsidiary	Royal Bank of Canada (Asia); assets = US$160 million (1989)
Branch offices	Singapore, Hong Kong, Seoul, Tokyo, Shanghai
Representative offices	Beijing, Shenzhen, Guangzhou
Bank of Nova Scotia	
Subsidiaries	BNS International (Hong Kong) Ltd.; book value = $6.5 million
	Boracay; book value = $2.3 million
	Bank of Nova Scotia Asia Limited; assets = US$333 million (1990)
	Scotia McLeod (Hong Kong)
	Scotia McLeod (Tokyo)
	Bank of Nova Scotia, Malaysia; assets = US$390 million (1993)
Offices	Tokyo, Osaka, Seoul, Kuala Lumpur, Manilla, Singapore, Taipei, Beijing, Guangzhou
Bank of Montreal	
Subsidiary	Bank of Montreal Asia Limited; book value = $31 million; assets = US$286 million (1991)
Branches	Hong Kong, Tokyo, Seoul, Singapore, Guangzhou
Representative office	Beijing
National Bank of Canada	
Subsidiaries	Natcan Finance (Asia) Ltd, Hong Kong; book value = $6.7 million
	National Bank of Canada (Asia) Ltd, Singapore; book value = $104 million
Branches	Hong Kong, Seoul, Singapore, Tokyo
Representative offices	Taipei, Beijing
Toronto-Dominion Bank	
Subsidiaries	Tordom Nominees(HK) Ltd.
	Toronto-Dominion (South-East Asia) Limited, Singapore
Offices	Taipei, Singapore, Tokyo, Hong Kong

Source: *The Bankers' Almanac*, 1993; banks' annual reports for 1994 (except CIBC, 1993); U.S. Department of the Treasury (1994); and Sutton (1995).

TABLE 16

CANADIAN BANKS' SHARES OF ASSETS IN THE ASIA PACIFIC REGION

	PERCENT
Canadian Imperial Bank of Commerce	
Japan	1.2
Hong Kong	1.2
Other Asian and Pacific countries	0.6
Bank of Nova Scotia	
Earning assets	
Japan	3.7
Other Asia	2.5
Loans and acceptances	
Asia	3.4
National Bank of Canada	
Earning assets	
Japan	0.2
Other Asia and Pacific	1.7
Toronto-Dominion Bank	
Earning assets	
Japan	0.8
Other Asia and Australasian	0.8
Bank of Montreal	
Earning assets ("other countries")	
Designated lesser-developed countries	0.4
Other	1.6
Royal Bank of Canada	
Earning assets	
Asia Pacific region	2.7

Source: Annual reports, 1994 (except CIBC, 1993).

Pacific region, however, cannot be identified by itself. It is included in "other countries," a residual that remains after the removal of receipts from the United States and the European Union. During 1993, this category accounted for receipts of $138 million. These data set a ceiling for the contribution of exports of banking services to the Asia Pacific region of less than 0.025 percent of GDP. Given this dimension, it would take a substantial shift in the scale of Canadian banking activities in the region to create any appreciable impact on Canadian growth.

The Prospects for Canadian Banks in the Asia Pacific Region

THE PROSPECTS FOR GROWTH IN THE ASIA PACIFIC REGION make it one of the most attractive markets for enterprises seeking to expand their business. Yet the opportunities will differ from industry to industry and from country to country because of specific circumstances. This is particularly true of the banking industry. The state of banking among the Asia Pacific countries differs from the sophisticated centres of Hong Kong, Japan, and Singapore to rudimentary conditions in China and Vietnam. As in most other countries, banking is among the most regulated sectors in the Asia Pacific region, but differences in the regulatory approaches are significant. In some countries, banks remain under the direct control of the central government, whereas in others they operate in a competitive market environment.

Parts of the Asia Pacific market will present substantial opportunities for banking. Rapid growth in the newly emerging economies will expand their demand for banking services – for both existing services and new, sophisticated services. These demands are being further enhanced by the general liberalization that is taking place in financial regulation. Formerly, regulation inhibited the development of banking by enforcing low terms for depositors and maintaining conditions that rendered banks unable to respond to market opportunities.

To what extent will foreign banks, and Canadian banks in particular, be able to take advantage of these opportunities? Financial liberalization commonly opens up opportunities for domestic banks well before similar opportunities are made available to foreign banks. Hong Kong, Japan, and Singapore have passed beyond this stage to become international banking centres. These are now competitive banking markets where few opportunities remain for new entrants. Banking markets are opening up most rapidly in mixed market systems such as Indonesia, Malaysia, South Korea, and Taiwan. But in these countries, regulations still limit the activities of foreign banks. If this pattern continues, the opportunities for foreign banks will not keep pace with the growth of the banking demand. While recent trade negotiations, with their emphasis on financial services, put pressure on these countries to open their banking markets, their commitments under the GATS may permit a gradual process of adjustment.

Are Canadian banks well placed to meet market opportunities when they do develop? Analysis of the Canadian banks' international activities and comparisons with other international banks suggest that Canadian banks do not attach a high priority to the Asian Pacific market. This may be an appropriate strategic response: few opportunities remain in the highly competitive banking centres; the mixed banking systems remain protected against foreign banks. Local experience and expertise will be important as banking opens up in these economies. Canadian banks are not well placed at present to respond to new opportunities in the region. They may perceive their best opportunities as being elsewhere, where their competitive strengths are greater.

APPENDIX 1

Sources of Data

THIS STUDY RELIES ON A VARIETY OF STATISTICAL SOURCES for representing the banking activities of Canadian and other international banks in the Asia Pacific region. The data from these different sources are not always comparable. This Appendix describes the data sources and indicates differences in their coverage.

The International Monetary Fund's *International Financial Statistics* presents data on cross-border banking business "to form a more comprehensive picture of international banking activity, by residence of debtor or creditor" (p. xvi). Cross-border banking refers to banking activity between a bank sited in one country and a bank customer sited in another. These data have been the basis for the comparisons of trends in the international business of Canadian and other international banks.

The data on the nature of Canadian banking abroad come from Table C10 of the *Bank of Canada Review*, which presents data on the "foreign currency assets and liabilities (excluding bullion, foreign currency note and coin holdings, and foreign currency subordinated debt) on the books of the chartered banks, domestic and foreign branches, agencies and subsidiaries (excluding the investment dealer subsidiaries of the banks)." These figures are much more comprehensive than the IMF cross-border data in that they also include the banking business conducted by Canadian banking affiliates within a country. These data are also presented in more detail and for a longer time span in the C18000 and C19000 series of CANSIM.

In only one instance were these different sets of data compared. Tables 8 and 9 showed the distribution of claims and liabilities over the Asia Pacific region for all banks (from the IMF data) and for Canadian banks (from the Bank of Canada data). While this comparison may be affected by the differences in the data, the two sets of data were quite consistent where they described common phenomena such as trends in the development of the Canadian banks' international business.

Appendix 2

Financial Reform in the Asia Pacific Region[16]

Hong Kong[17]

AT THE BEGINNING OF THE 1980S, the banking system of Hong Kong was the most liberalized in the Asia Pacific region, having already abolished foreign-exchange controls in 1972. Today, Hong Kong freely allows entry by both foreign and domestic banks. The government generally applies the same rules and regulations regarding the supply of banking services to residents and non-residents alike. Hong Kong has developed into the major international banking centre in the region.

The government of Hong Kong does not have any ownership interest in any of the major banks. One of the three note-issuing banks, however, is owned by the government of the People's Republic of China. The present monetary and financial regulatory structure is supposed to remain autonomous under guarantees made by the government of China when Hong Kong becomes a Special Administrative Region of that country in 1997.

While interest rates are not controlled by regulators, interest rates on time deposits up to HK$500,000 have been governed by rules set by the Hong Kong Association of Bankers, an organization to which all major banks must belong. Recently, the Hong Kong Consumer Council and the Legislative Assembly have called on the government to abolish interest-rate ceilings. The Hong Kong Association of Banks announced a schedule to eliminate all controlled interest rates by the end of 1995.

Singapore[18]

SINCE 1973 THE SINGAPORE AUTHORITIES have actively encouraged the development of an international financial centre by granting offshore banking licenses to establish Asian Currency Units (ACU) quite freely. They keep the business of the ACU separate from domestic banking by limiting their ability to do business with residents through ceilings and other restrictions. Otherwise, both the setting of interest rates and the allocation of credit in the offshore market are determined by market forces. The freedom from control has allowed the offshore market to develop into a regional financial centre that is second only to Hong Kong.

The domestic market in Singapore has been treated differently than the offshore market in many ways, especially with respect to entry. While there are few restrictions on new entrants into the international financial centre, entry into the domestic market has been frozen since 1973 because of the fear of overbanking. Existing foreign banks are also restricted in their ability to expand in the domestic markets. Currently, they are unable to open new branches and also face limits on some services. Only one foreign bank has been allowed to offer special low-interest accounts that qualify for the Central Provident Fund. Domestic banks, in contrast, must meet both reserve and liquidity requirements and are required to keep unprofitable branches in business. Like the offshore market, the domestic banking market has been free of interest-rate controls and direct credit allocation.

Unlike in Hong Kong, the Singapore government participates in banking through ownership and other means. It is the principal shareholder of one "Big Four" bank and of another medium-sized bank. In addition, according to U.S. Department of the Treasury (1994, pp. 442-3),

> It is common for senior government officials to sit on boards of both domestic and foreign banks and for officials of both foreign and domestic banks to be on the boards of government linked companies and the government's statutory boards. Currently, the Ministry of Finance's highest ranking civil servant is also the chairman of DBS bank. The current highest ranking civil servant in the Ministry of Trade and industry serves as one of DBS Bank's directors.... The MAS must approve the appointment of directors of locally incorporated banks.

JAPAN[19]

THE JAPANESE BANKING SYSTEM OPERATED under heavy and comprehensive regulation from the end of the war through to the early 1980s. Regulation enforced the segmentation of the supply of different financial services and limited the scope for entry into financial activities. Long-term credit suppliers were kept separate from short-term suppliers, and each had its own sources of funds. Both the level and structure of interest rates were controlled to prevent interest-rate competition from eroding the profitability of the banks. The Bank of Japan also determined total bank lending through its so-called "window guidance," a form of moral suasion. Banks complied to avoid jeopardizing access to this source of cheap funding.

The first moves towards financial liberalization were in response to pressures from the U.S. government in 1983 to raise the value of the yen. While the pace has been deliberate, considerable progress had been achieved by the early 1990s. The Ministry of Finance had ended all controls over interest paid on time deposits by 1993 and on "non-time" deposits by 1994. Only rates on demand deposits remain under control.

The reforms have also gradually reduced the segmentation of financial markets. In the late 1980s, commercial banks could engage in the security business on a limited basis, beginning with the sale of government bonds and then with the power to deal in bonds and commercial paper. This breakdown of segmentation has gained further momentum from the introduction of the *Financial System Reform Act*, which will allow cross-entry between banking and the investment business through subsidiaries.

The commercial banking system in Japan consists mainly of privately owned banks. As is well known, many of these banks share cross-ownership with industrial and commercial organizations.[20] The principal government-owned financial institution is the post office savings system, a major competitor to the commercial banks that accounts for approximately 15 percent of savings deposits. It does not compete with commercial banks in its investments, transferring funds to the government for special purposes and to the other two government-owned institutions, the Japan Development Bank and the Export-Import Bank.

South Korea[21]

South Korean banking was tightly controlled by government at the beginning of the 1980s. The government owned substantial shares of the major commercial banks, specified preferential interest rates for priority credit, and established uniform interest rates on other loans. Financial reform began in 1981 and 1982 with the government's divestment of its ownership in the five large city banks. Soon afterwards, the government abolished preferential interest rates on priority credits and gave banks greater leeway to set interest rates on other lending. Banks could also expand into activities such as trust banking, leasing, and dealing in government securities.

Despite the liberalizing measures in the 1980s, the South Korean banking system remains tightly controlled and segmented. Most observers suggest that it has remained under direct government control despite the change in ownership. The government controlled the appointment of the presidents of the major banks right up to 1993. In addition, loans to priority sectors at preferential rates still dominate the banks' portfolios, accounting for almost half of all bank lending. In recognition of the slow progress of liberalization, the government recently announced further reform in its "blueprint," which calls for the elimination of some of the remaining controls on the banking system and for the liberalization of foreign exchange and the capital account.

Malaysia[22]

The process of financial reform had begun in Malaysia before the beginning of the 1980s. The government announced measures to reduce its control over deposit and loan rates and to reduce the liquidity requirements imposed on the banks in 1978. Nevertheless, it still maintained its ability to influence domestic banks through its substantial ownership interests. In addition, reform measures were not directed towards the system of priority credits that the central bank developed in the mid-1970s. Guidelines from the central bank still require commercial banks to direct up to 20 percent of their outstanding loans to designated sectors or groups.

The pace of financial reform in Malaysia has been uneven. As a reaction to the developing pattern of interest rates, the central bank in 1983 required each bank to tie its lending rates to a base lending rate (BLR) based on its cost of funds and other expenses. Later, controls on deposit rates were reintroduced temporarily to deal with a liquidity crisis from late 1985 through early 1987. Subsequently, the control of lending rates was tightened further by the requirement that the BLRs of two lead banks serve as references for all other banks. Only since 1991 has the BLR again been determined by a bank's own expenses. The maximum lending rate for a bank is set by its BLR plus four percentage points.

Despite the 15 years of reform, the system of priority credits to small businesses, low-cost real estate, and *Bumiputras* (ethnic Malay) remains a feature of the banking system. Malaysia also maintains a system of exchange control "to ensure that exports receipts are received promptly in Malaysia, to assist the Central Bank in monitoring the settlement of international payments and receipts and to encourage the use of financial resources for productive purposes" (Yusof et al., 1994, p. 292).

In addition to its role as regulator, the Malaysian government maintains a substantial ownership interest directly and indirectly through quasi-government entities in major commercial banks that dominate the system. The two largest banks, which together account for one-third of bank assets, are either owned or effectively controlled by the government.

TAIWAN[23]

PRIOR TO 1975, THE GOVERNMENT STRICTLY CONTROLLED the banking system of Taiwan. It set interest rates on loans and deposits, prohibited entry of new banks, restricted branching, and influenced the allocation of credit through its ownership of all the major banks. Since then, considerable liberalization has taken place. Beginning in 1976, controlled interest rates were tied to money market rates; in 1980 rates on negotiable debentures and certificates of deposit were decontrolled, and in 1986 many deposit rates were removed from control when the number of deposit categories was reduced from 13 to 4. The end of interest-rate controls came with the *Banking Law* of 1989, which removed all deposit and loan rates from control. This law also liberalized the rules on branching and removed the prohibition on the entry of new banks. While the authorities also proposed the privatization of the state-owned banks, the plan met with severe opposition, which has held up any progress towards this goal.

The reforms have considerably reduced the state's control over the banking system. Nevertheless, the government maintains a signficiant degree of control through its ownership of 12 of the 16 domestic banks, which between them hold 90 percent of banking assets. Shea (1994) describes the government's influence as follows (pp. 249-50):

> In addition to financial regulations, government banks are subject to the same administrative control as other government agencies. All their directors and senior management are appointed by the government. Numerous regulations and restrictions on personnel, accounting, budgeting and auditing have turned government banks into bureaucracies which operate inefficiently and are sluggish in bringing in financial innovations. Pressuring or lobbying for loans from representatives or government officials has distorted credit policy.... MOF and CBC [Central Bank of China] have often ordered financial institutions to expand or restrict loans to certain economic activities, industries, or borrower groups.

CHINA[24]

THE BANKING SYSTEM OF CHINA was established in the 1950s with the People's Bank of China as a monobank that served as a central bank and as an agency for channelling funds to finance the activities of state enterprises according to the central credit plan. Specialized banks – such as the Agricultural Bank, the Bank of China, and the People's Construction Bank – served as departments of the People's Bank. In 1984, the People's Bank assumed the responsibilities of a central bank and the new Industrial and Commercial Bank shared commercial banking activity with the specialized banks.

The banking system remains dominated by four specialized banks that account for over 90 percent of the funds of the banking system. These banks concentrate their business on their assigned sectors. Fewer than 20 percent of their loans go to the private sector. The *National Treatment Study* (U.S. Department of the Treasury, 1994, p. 248) describes the place of the banks in the planning process: "The concentration of China's financial resources in the state-owned banks has ensured that state-owned enterprises have enjoyed preferential access to formal credit. As subsidies and other government expenditures have moved off the budget in the past decade, the banking system has become an increasingly important conduit for quasi-fiscal financing to the state sector."

Yi (1994) supports this view and suggests that banks are under pressure from local government in their credit allocation, are required to lend to government to support its deficits and must respond to "imperative 'recommendations' ... to provide loans under certain conditions to 'high priority projects'." Similarly, Li concludes (1994, pp. 33-4), "since 1986, the central bank has tried to exercise both direct and indirect control measures, such as setting quotas for fixed asset loans. The central bank indirectly controls credit, mainly by its credits to the various specialized banks."

The government has recently announced plans to convert the specialized banks into commercial banks. Three new "policy banks" will fulfil the present role of the specialized banks. While these measures provide a possible framework for reform, there has been little progress toward changing the lending philosophy of the banks.

INDONESIA[25]

AT THE BEGINNING OF THE 1980S, the Indonesian banking system was tightly controlled through the government's ownership of the dominant commercial banks, through controls on interest rates on deposits and loans, and through directed lending programs. No new banks, either domestic or foreign, had been established since the 1960s, and branching was subject to strict controls.

Reform of the banking system took place in two distinct phases. The first, in 1983, encouraged the banks, especially the dominant state-owned banks, to operate to a greater degree on commercial principles. The main elements of the reform were the removal of controls from interest rates on loans and deposits, the elimination of credit ceilings through which the central bank directed credit by sector and bank, and a reduction of the categories of credit eligible for refinance at the central bank on favourable terms.

The second phase, which began in 1988, was directed towards fostering greater competition in the banking sector. The initial measures included the opening of entry to new domestic banks and to joint-venture banks with foreign participation. Branching was made easier for domestic banks that could demonstrate their soundness and was extended to additional centres for foreign banks. Private banks were able to compete for the deposits of state enterprises for the first time and could gain foreign-exchange powers more easily. Finally, all banks gained through a substantial reduction in reserve requirements.

The reform process continued after the first 1988 package, but the emphasis shifted to other areas of the financial sector. Nevertheless, banks were subsequently permitted to take part in leasing, venture capital, factoring, consumer credit, and credit cards. Other reforms strengthened the system of prudential regulation by restricting the loans that could be made to single borrowers and associated groups. In 1990, further steps were taken to reduce the central bank's role in the allocation of credit and to move the interest rates on the remaining credits closer to market levels. Some reversal of the movement to market allocation of credit occurred in 1990 when the government introduced the requirement that domestic banks allocate 20 percent of their loans to small business.

Vietnam[26]

Until 1988, the banking system in Vietnam consisted of a monobank that served as the central bank and also allocated credit according to the central plan. That year, State Bank of Vietnam was assigned the function of the central bank, and several of its former departments were established as separate commercial banks, each with its own specialized area of lending. In addition, several quasi-private banks have been established since then, generally with the participation of the State Bank of Vietnam. Many features of the Vietnam banking system reflect its incomplete transition from a planned credit system. Interest rates on loans and deposits are still set by the central bank. In addition, the central bank still sets credit guidelines that determine the credit allocations of the commercial banks.

Philippines

The Philippine banking system has been through a number of swings with respect to liberalization. While the entry of foreign banks was prohibited in 1948, many new banks were formed through 1972. Beginning in the early 1970s, the government increased its degree of intervention, "setting relatively low ceilings on both deposit and loan rates and dictating preferential interest rates for favoured industries and projects" (U.S. Department of the Treasury, 1994, p. 422). The government also imposed a moratorium on new banking licenses in 1977.

Like many of the other countries surveyed, a movement towards liberalization began in the 1980s. The first step – the removal of interest-rate ceilings – was taken in 1981. The moratorium on new entry was removed in 1989 and existing banks that met prudential standards were free to establish new branches at their own judgment.

Endnotes

1. The author is indebted to R. Glick, R. Harris, B. Sutton, and J. Helliwell for their useful comments and suggestions.
2. The growth in this region has been documented and analyzed in the World Bank's *The East Asian Miracle* (1993).
3. Sutton (1995) discusses the success of insurance companies in entering the Chinese market. The initial experience of Canadian underwriters in China was described in "The Strange Saga of Noble China," *Globe and Mail*, September 26, 1995. Developments in the regulatory treatment of

4   these sectors are discussed in U.S. Department of the Treasury (1994).
4   The measurement of the output of the financial services industry for national income accounting purposes remains a contentious issue among economists; see Chant (1988), pp. 31-46.
5   This broad money measure is no more than an indication of the demand for banking services. Different amounts of intermediary services may be associated with the same level of intermediation. A bank that offers deposits matched in maturity to a portfolio of marketable securities provides fewer services to its customers than a bank that offers demand deposits against a portfolio of non-marketable loans from a varied group of borrowers.
6   The initial sample included all countries for which data were available in International Monetary Fund, *International Financial Statistics*. International financial centres were excluded from the sample because their broad-money measures reflected substantial financial claims held by non-residents.
7   The use of this benchmark assumes that financial liberalization has been independent of per capita income. If, as is likely, financial liberalization is associated with development, the estimates understate the effect of liberalization on M/Y ratios.
8   The estimates exclude a number of countries for which there are no adequate data, namely China, Vietnam, and Taiwan.
9   There is some question whether South Korea should be classed in this group. *The National Treatment Study* (U.S. Department of the Treasury, 1994) suggests (p. 363) that "foreign banks have experienced difficulties... in gaining market access to Korea," but does not elaborate on the point.
10 See Saunders and Walter (1994).
11 See Bank of Montreal (1994).
12 The sources of data used for this section are discussed in Appendix 1.
13 These tables use different data for Canadian and other international banks. See Appendix 1 for discussion.
14 The Bank of Canada defines "local" activities as those claims or liabilities of an office or a bank made with residents of the country in which the office booking the claim or liability is located and which are denominated in the domestic currency of the country (see Bank of Canada, 1995).
15 This category is part of the larger category "finance and insurance," which is divided into "insurance" and "other financial services." These data measure the value of exports of financial services. To the extent that these exports do not in turn use imports, they represent the contribution of the trade in financial services to GDP.
16 This appendix draws on U.S. Department of the Treasury (1994).
17 See Greenwood (1986).
18 See Greenwood (1986).

19 See Horiuchi (1992), Kitigawa and Kurosawa (1994), and Suzuki and Yomo (1986).
20 See Connor (1994, p. 2) for details.
21 See Nam (1994), Greenwood (1986), and Park and Kim (1994).
22 See Greenwood (1986), Yusof et al. (1994), and Sheng (1992).
23 See Shea (1994) and Yang (1994)
24 See Li (1994), Yi (1994), and Sutton (1995).
25 See Chant and Pangestu (1994).
26 This section is based on interviews conducted by the author in Vietnam in October 1994.

Bibliography

Aliber, Robert J. "International Banking: A Survey." *Journal of Money, Credit and Banking*, XVI (November 1984, part 2):661-78.

Bank of Canada. *Bank of Canada Review: Notes to the Tables* (January 1995).

Bank of Montreal. *Annual Report, 1994*.

Bryant, Ralph. *International Intermediation*. Washington: The Brookings Institution, 1987.

Caprio, Gerard. "Introduction," in *Financial Reform: Theory and Experience*. Edited by Gerard Caprio, Izaak Atiyas and James A. Hanson. Cambridge: Cambridge University Press, 1994, pp. 1-9.

Caprio, Gerard, Izaak Atiyas and James A. Hanson (eds). *Financial Reform: Theory and Experience*. Cambridge: Cambridge University Press, 1994.

Caves, Richard E. "Discussion," in Federal Reserve Bank of Boston, *Key Issues in International Banking*. Conference Series no. 18. Boston, 1977, pp. 87-90.

Chant, John F. *The Market for Financial Services: Deposit-Taking Institutions*. Vancouver: The Fraser Institute, 1988.

Chant, John and Mari Pangestu. "An Assessment of Financial Reform in Indonesia," in *Financial Reform: Theory and Experience*. Edited by Gerard Caprio, Izaak Atiyas and James A. Hanson. Cambridge: Cambridge University Press, 1994, pp. 223-75.

Clark, Jeffrey. "Economies of Scale and Scope at Depository Financial Institutions: A Review of the Literature." Kansas City, Mo.: Federal Reserve Board of Kansas City, 1988.

Darroch, James L. *Canadian Banks and Global Competitiveness*. Montreal: McGill-Queen's University Press, 1994.

Federal Reserve Bank of Boston. *Key Issues in International Banking*. Conference Series no. 18. Boston, 1977.

Fieleke, Norman S. "The Growth of Banking Abroad: An Analytical Survey," in Federal Reserve Bank of Boston, *Key Issues in International Banking*. Conference Series no. 18. Boston, 1977, pp. 9-40.

Goodman, Laurie J. "Comment," *Journal of Money, Credit and Banking*, 16 (November 1984, part 2):678-84.

Greenwood, John G. "Financial Liberalization and Innovation in Seven East Asian Economies," in *Financial Innovation and Monetary Policy*. Edited by Yoshio Suzuki and Hiroshi Yomo. Tokyo: University of Tokyo Press, 1986, pp. 79-105.

Grubel, Herbert G. "A Theory of Multinational Banking." *Banca Nazionale del Lavoro Quarterly Review*, 123 (December 1977):349-63.

_____. *Conceptual Issues in Service Sector Research: A Symposium*. Vancouver: The Fraser Institute, 1987.

Horiuchi, Akiyoshi. "Financial Liberalization: The Case of Japan," in *Financial Regulation: Changing the Rules of the Game*. Edited by Dimitri Vittas. Washington: The World Bank, 1992, pp. 45-119.

Houthakker, Hendrik S. "Comment." *Journal of Money, Credit and Banking*, 16 (November 1984, part 2):684-90.

International Monetary Fund. *International Financial Statistics*. Washington: IMF.

Kitagawa, Hiroshi and Yoshikata Kurosawa. "Japan: Development and Structural Change of the Banking System," in *The Financial Development of Japan, Korea and Taiwan: Growth, Repression and Liberalization*. Edited by Hugh Patrick and Yung Chul Park. New York: Oxford University Press, 1994, pp. 81-128.

Lewis, M.K. and K.T. Davis. *Domestic and International Banking*. London: Philip Allan, 1987.

Li, Kuo-Wai. *Financial Repression and Economic Reform in China*. Westport, Conn.: Praeger, 1994.

Murray, John D. "Comment." *Journal of Money, Credit and Banking*, 16 (November 1984, part 2):690-95.

Nam, Sang-Woo. "Korea's financial reform since the early 1980s," in *Financial Reform: Theory and Exprerience*. Edited by Gerard Caprio, Izaak Atiyas and James A. Hanson. Cambridge: Cambridge University Press, 1994, pp. 184-222.

Park, Yung Chul and Dong Won Kim. "Korea: Development and Structural Change of the Banking System," in *The Financial Development of Japan, Korea and Taiwan: Growth, Repression and Liberalization*. Edited by Hugh Patrick and Yung Chul Park. New York: Oxford University Press, 1994, pp. 188-221.

Patrick, Hugh and Yung Chul Park (eds). *The Financial Development of Japan, Korea and Taiwan: Growth, Repression and Liberalization*. New York: Oxford University Press, 1994.

Pecchioli, R.M. *The Internationalization of Banking: The Policy Issues*. Paris: Organisation for Economic Co-operation and Development, 1983.

Pigeon, Lorraine and Brent Sutton. *The Canadian Financial Services Industry: The Year in Review, 1995*. Ottawa: The Conference Board of Canada, 1995.

Rutenberg, David P. and Edwin H. Neave. "International Competitiveness of Canadian Financial Institutions." Paper prepared for the conference on the reform of the Canadian financial services industry, 1996.

Saunders, Anthony and Ingo Walter. *Universal Banking in the United States: What Could We Gain? What Could We Lose?* New York: Oxford University Press, 1994.

Shea, Jia-Dong. "Taiwan: Development and Structural Change of the Financial System," in *The Financial Development of Japan, Korea and Taiwan: Growth, Repression and Liberalization*. Edited by Hugh Patrick and Yung Chul Park. New York: Oxford University Press, 1994, p. 222-87.

Sheng, Andrew. "Bank Restructuring in Malaysia," in *Financial Regulation: Changing the Rules of the Game*. Edited by Dimitri Vittas. Washington: The World Bank, 1992, pp. 195-236.

Sutton, Brent. *Out from Behind the Great Wall: Emerging Opportunities for Financial Institutions in China*. Report 146-95. Ottawa: The Conference Board of Canada, 1995.

Suzuki, Yoshio and Hiroshi Yomo. *Financial Innovation and Monetary Policy.* Tokyo: University of Tokyo Press, 1986.

Terrell, Henry S. and Sydney J. Key. "The Growth of Foreign Banking in the United States," in Federal Reserve Bank of Boston. *Key Issues in International Banking.* Conference Series no. 18. Boston, 1977, pp. 54-86.

United States, Department of the Treasury. *National Treatment Study.* Washington: 1994.

Vittas, Dimitri (ed). *Financial Regulation: Changing the Rules of the Game.* Washington: The World Bank, 1992.

Walter, Ingo. *Barriers to Trade in Banking and Financial Services.* London: Trade Policy Research Centre, 1985.

White, William R. "International Agreements in the Area of Banking and Finance: Accomplishments and Outstanding Issues." Paper presented at the Conference on Monetary and Financial Integration in an Expanding (N)AFTA: Organization and Consequences, University of Toronto, Centre for International Studies, 15-17 May 1996.

World Bank. *The East Asian Miracle.* New York: Oxford University Press, 1993.

Yang, Ya-Hwei. "Taiwan: Development and Structural Change of the Banking System," in *The Financial Development of Japan, Korea and Taiwan: Growth, Repression and Liberalization.* Edited by Hugh Patrick and Yung Chul Park. New York: Oxford University Press, 1994, pp. 288-324.

Yi, Gang. *Money, Banking and Financial Markets in China.* Boulder: Westview Press, 1994.

Yumoto, Maashi, Kinzo Shima, Hajime Koike and Hiroo Taguchi. "Financial Innovation in Major industrial Countries," *Financial Innovation and Monetary Policy.* Edited by Yoshio Suzuki and Kiroshi Yomo. Tokyo: University of Tokyo Press, 1986, pp. 45-78.

Yusof, Zainal Aznam, Awang Adek Hussin, Ismail Alowi, Lim Chee Sing and Sukhdave Singh. "Financial reform in Malaysia," in *Financial Reform: Theory and Experience.* Edited by Gerard Caprio, Izaak Atiyas and James A. Hanson. Cambridge: Cambridge University Press, 1994, pp. 227-320.

Comment

Reuven Glick
Federal Reserve Bank of San Francisco
San Francisco

AS A U.S. RESIDENT, I would like to commend Industry Canada for organizing this conference on Asia Pacific economic issues. Contrary to what one might think, the United States is not much further ahead in thinking about these issues. While there has been much press and political attention devoted to issues such as the U.S.-Japan bilateral trade balance, there has been much less academic attention. Asia Pacific academic conferences in the United States still seem to lack the cachet attached to discussions of the future of a European monetary union or of Latin American economic reform programs.

This is despite the fact that Asia accounts for roughly 35 percent of U.S. international trade – a percentage that has exceeded that of U.S.-European trade for almost 10 years. This relative paucity of academic attention to Pacific Basin issues is beginning to change in the United States. I am glad to see that the same thing is happening in Canada.

Professor Chant has provided an accessible paper about how economic developments in Asia might affect the demand for the financial services of Canadian banks. The paper has three parts. The first reviews the financial liberalization experiences of Pacific Basin economies, in particular their treatment of foreign banks. Banking systems in the region are categorized into three groups. At one end, there are *predominantly market-based* systems (such as Japan, Hong Kong, and Singapore), where interest rates and credit allocation are primarily market determined. At the other end are the still heavily *state-directed* banking systems (China and Vietnam). In the middle are the *mixed-market* systems (Indonesia, Thailand, Malaysia, Taiwan, and South Korea), where governments still directly or indirectly interfere with interest rates and credit allocation decisions.

The second part of the paper formulates simple hypotheses about the determinants of the international flows of bank services. The "follow-the-leader" approach suggests that domestic banks go abroad to handle the business of home multinationals operating abroad. The "banking advantage" approach suggests that domestic banks go abroad to exploit special bank advantages they might have – such as superior management skills or innovative financial products.

The last part then loosely "tests" whether these hypotheses are supported by Canadian experience in Asia. It concludes that the pattern of Canadian international banking activity in Asia most closely conforms to the follow-the-leader approach because Canadian banks operating in East Asia tend to be concentrated in Japan, Hong Kong, and Singapore, where Canadian foreign direct investment is concentrated.

The paper provides a good starting point for a discussion of the future of Canadian banks in Asia. My comments on how the paper can be improved fall into two broad areas. First, the paper needs to better identify the determinants of Canadian banks going abroad. Second, it needs to put the discussion of Canadian banking opportunities in the context of the larger issue of the future of the banking industry in increasingly more competitive financial markets, involving not just banks, but also securities firms, mutual funds, and so on.

Determinants of Canadian Banking Abroad

AT FIRST GLANCE, it is hard to quibble with the paper's main conclusion, since most econometric studies of international lending suggest that overseas banking is linked to the home country's volume of international trade and investment. But before one totally accepts the follow-the-leader approach and dismisses the banking-advantage approach, one can play devil's advocate by raising several possibilities:

1 Canadian banks may indeed have an advantage in retail branch banking, but because of adverse foreign-country treatment, this advantage may not be exploitable abroad, implying a constrained equilibrium;
2 Canadian banks may *not* have an advantage in retail branch banking but rather in some area, such as wholesale banking or off-balance sheet banking, that is also correlated with the volume of Canadian multinational operations abroad; or
3 Canadian banks may have advantages in both retail branch banking and other wholesale banking services because the two can go hand in hand, making them observationally equivalent.

Let me address these three possibilities in turn. As Professor Chant has highlighted in the paper, many Asian countries still formally or informally limit competition in retail banking activities. National treatment is not the norm in Asian banking markets. Many countries formally limit entry by foreign banks. For example, Malaysia and Singapore have formal moratoriums on new onshore banking licenses. Thailand has had no new foreign banks since 1985, though it recently announced that it would grant five additional branch licenses to foreign banks by 1997. Taiwan prevented the entry of new domestic banks until 1989; it did not really loosen up on foreign-bank entry until 1994. Even Australia did not grant licences to foreign banks until 1985. Indonesia only allows foreign-bank entry through joint ventures with Indonesian firms. (Foreign banks were not allowed to operate in Canada, until 1980, except as representative offices.)

Many countries – Indonesia, Malaysia, Singapore, Thailand, Hong Kong, and until recently, the Philippines – limit the number of foreign-bank branches, including off-site ATM (automatic teller machines). In Singapore and Malaysia, foreign banks cannot join existing local ATM networks. Even after establishment, foreign banks often face other impediments, such as higher capital requirements than local firms, tighter limits on lending, and foreign-exchange positions tied to local capital.

Even where equal national treatment of foreign banks is the law, foreign banks face other barriers. For example, the foreign-bank share of Japan's retail deposit market is limited not by law but by high real estate and operating costs associated with setting up branches in Japan, as well as by anti-competitive institutional arrangements (such as *keiretsu*) that involve close links between Japanese banks and other Japanese corporate groups. There are restrictions which, while they are applied equally to local and foreign banks, still put foreign banks at a disadvantage. Examples of this include (ironically) restrictions on ATM usage to daytime hours when ATM were first introduced in Japan and restrictions on the introduction of new financial products, in which foreign banks specialize,

The implication of these government regulations, exclusionary practices, high costs, etc. is that Canada's branch-banking expertise cannot be implemented in retail banking abroad. As a result, it is not surprising that most foreign banks, including Canadian banks, have made only limited inroads into foreign-bank markets, particularly in retail banking. A cursory examination of the available data indicates that the foreign banks account for less than 10 percent of total deposits and other assets in most Asian countries. (Foreign penetration of the banking systems in Singapore and Hong Kong is relatively high compared to most other countries because of the extent of offshore banking; but Singapore limits the activities of foreign banks in its domestic banking sector. Foreign penetration of Malaysia's banking system is also high, but this is mainly an artifact of its former colonial status with the United Kingdom.)

In light of this existing constrained equilibrium, the future of Canadian and other foreign banks in Asia depends on whether further financial reforms in Asia will be accompanied by a relaxation of those constraints. The answer is uncertain. Despite references to an Asian model of development and growth, there is no single model of financial reform in the region. Some countries (the Philippines, for example) have recently purposely opened up the domestic banking sector to foreign banks, specifically to encourage competition and shake up an inefficient financial market. With 38 banks now, the Philippines is permitting 10 new foreign banks to enter. Many of the world's largest banks have applied.

Other countries (Malaysia, Thailand, and even more so South Korea and Taiwan) are proceeding very cautiously with bank reform. But because there is also a growing recognition that their still relatively undeveloped financial markets have become a bottleneck, hurting growth, these countries appear to be focusing on the development of equity and corporate bond markets to improve financial efficiency. Thus further liberalization of financial markets provides no guarantee that Canadian or other banks will be able to expand, particularly in retail banking.

What if Canadian banks do *not* have a particular banking advantage abroad in managing retail-branch networks, but rather their advantage is in something else, such as wholesale business or the provision of other specialized financial services, and that this other advantage is also correlated with the volume of Canadian multinational operations abroad? As of roughly 1992, Canada had 65 commercial banks with 7,749 (full-service) branches, giving it the highest ratio of bank branches to population of any industrial country. It is not at all clear to me, however, that Canadian banks necessarily have an advantage over the largest U.S. banks in this area, even though U.S. law has historically limited interstate branch banking, except through bank holding companies. For example, the populations of Canada and the state of California are roughly the same. California may provide a sufficiently large market to give Bank of America, headquartered in San Francisco, the same degree of retail branch banking expertise as Canadian banks have. It would be interesting to know how successful Canadian banks are in the United States, as well as in other parts of the world where fewer limits have been imposed on the retail activities of commercial banks. Specifically, how successful have Canadian banks been in countries that allow freer retail activities?

What about the possibility that Canadian banks have some other advantage that also happens to be correlated with Canadian foreign direct investment? Although Professor Chant would be the first to admit that he has not carried out any sophisticated econometric tests himself, it is worth mentioning that he really has only two data points: Japan, Hong Kong, and Singapore have lots of Canadian banking assets and lots of Canadian direct investment. The other countries have fewer banking assets and fewer Canadian bank assets.

U.S. banking activities are also heavily concentrated in Japan, Singapore, and Hong Kong. Why? Because these countries are the international financial centres of the region. In fact, Japan and Singapore are the third and fourth largest foreign-exchange trading centres in the world, after the United Kingdom and the United States. Hong Kong is fifth, just behind Switzerland but ahead of Germany.

What made these countries international financial centres? The answer is: a good infrastructure, a stable economic environment, and a low degree of regulatory control. It is not surprising that international banks that provide wholesale and other specialized financial services will be drawn to countries that offer these conditions. Nor is it surprising, therefore, that Canadian banks, if they indeed have some advantage in wholesale banking, will first locate in the international financial centres when seeking to penetrate the Asian market. So it is not clear that Canadian banks are playing follow-the-leader, because the same conditions that attract banks to Japan, Hong Kong, and Singapore may also have independently attracted Canadian multinational firms as well.

This leads to the third possibility – namely, that retail banking skills and other banking advantages, such as wholesale banking or the provision of other sophisticated banking services, may be observationally equivalent; that is, the two can go hand in hand. Specifically, having a lot of branches can create an advantage not just in retail banking but in providing wholesale banking and other banking services as well. For example, having a lot of branches spread out geographically, as in Canada, may allow a bank to become adept at centrally coordinating Treasury operations, netting out transactions and matching counter parties, trading across time zones, etc. All of these skills can prove useful in conducting wholesale-level transactions, as well as off-balance sheet activities, such as foreign-exchange trading, interest-rate and currency swaps, and providing trade credit. Therefore, again it is unclear whether Canadian bank operations in Asia have been motivated by the desire to follow the lead of multinational corporations or by particular bank advantages. I suppose this could be resolved if one had balance sheet data on the assets and liabilities of Canadian banks in Asia, broken down by the nationality of creditors and debtors, but such data are hard to come by.

Future of the Banking Industry

THUS FAR I HAVE TALKED ABOUT PROBLEMS in properly distinguishing between the follow-the-leader and banking-advantage approaches when understanding the motivation for Canadian banks operating in Asia. A second area that merits greater discussion (perhaps in another paper) is the place of Canadian banking experiences in Asia within the larger context of the changing competitiveness of banks in the world's financial markets. As is well known, in industrial countries, in particular, bank competitiveness has been affected by deregulation measures and technological innovations that have allowed the growth of open capital markets and non-bank intermediation, through mutual funds, securitization, derivatives, and so on. As a result, banks' traditional activities, such as bank deposit taking and lending, have been losing ground.

In the United States, the commercial bank and savings-and-loan bank combined share of total (commercial and mortgage) loans has fallen over the last 25 years from 55 to 30 percent. Domestic banks in Japan have also lost ground to open capital markets and non-bank competitors. The share of Japanese commercial-bank lending out of total funds raised by non-financial firms has declined from 80 percent to less than 50 percent over the same period.

This trend has not been confined to industrial countries. In South Korea, the growth of non-bank financial intermediaries has caused the share of demand and time deposits held by commercial banks to drop from 70 percent in the 1970s to 36 percent in 1992. (This is one reason why remaining government control of South Korea's commercial-banking system has not been as harmful as one might have thought: the relatively unregulated non-bank financial institutions facilitated more efficient allocation of credit across the economy.) Taiwan has long had an unregulated curb market involving financial transactions among small borrowers and lenders.

How have commercial banks responded to this competition for balance sheet lending? They have sought to adapt in two broad ways: 1) they have expanded their fee-based, off-balance sheet products and services, such as letters of credit, loan commitments, derivatives, etc.; and 2) when possible, they have expanded their investment-related activities into underwriting, advising, brokering, mutual funds, etc.

In fact, there is some evidence that the rise in fee-based services and products at U.S. banks has offset a large part of the decline in their share of on-balance sheet financing. This suggests that banks are adapting by carrying out some of their more traditional functions in a different form. This greater emphasis on fee-based income and other income-generating activities is also found in the overseas activities of U.S. and other foreign banks in Asia. Foreign banks in Taiwan held less than 1 percent of deposits but accounted for 20 percent of foreign-exchange trading, 30 percent of export loans, and 92 percent of advances on imports. Foreign banks account for only 3 percent of deposits in Thailand but handle half of the country's foreign-exchange business. Some of the largest U.S. banks earn more than 50 percent of their profits overseas.

Professor Chant's paper does not discuss how Canadian banks are reacting to changing financial market conditions. How are they doing in fee-based activities? How aggressive are they in competing for investment-banking services – i.e., underwriting the growing bond and stock issuance in the region? One needs to know how Canadian banks are doing in these activities in the United States and other markets before one knows how competitive they will be in Asia.

I understand that since 1987 Canadian banks have been allowed to own securities firms, through which they can underwrite and broker in addition to providing traditional retail banking and financial services. This potentially puts Canadian banks on a par with "universal" banks in some European countries, such as Germany.

In conclusion, this paper provides a good starting point in discussing what opportunities Asia might offer to Canadian banks. It does not yet fully resolve the question whether Canadian banks can be successful or not in seeking to exploit these opportunities.

Richard G. Harris & Stephen T. Easton
Department of Economics   Department of Economics
Simon Fraser University   Simon Fraser University
Canadian Institute for Advanced Research

International Tourism in the Asia Pacific Region and its Implications for Canada[1]

THE EXPANSION OF INTERNATIONAL TOURISM has been a striking characteristic of the modern period of economic growth, and it is frequently cited as one of the major manifestations of globalization. The remarkable growth performance of the Asian economies has led to interest in the possibilities for increased economic activity in the tourist sector and in exports of tourism services generally. The purpose of this paper is to review some of these trends and to put them in the larger perspective provided by the past and prospective growth experience in the Asia Pacific region and in the context of the tourism industry viewed in relation to other sectors that produce tradeable goods and services. As in the case of any international comparison involving a particular sector, the questions that one can address are limited by data constraints. Broadly speaking, the focus is on Canadian tourism export services in the region. To address that question properly, a number of salient facts and theoretical perspectives are used.

The economics of tourism is a subject in its own right, about which a great deal has been written.[2] The literature largely divides along two lines. First, there are a number of demand studies dealing with the determinants of tourism, many of which are developed in the context of a closed economy. Second, tourism is described from a partial-equilibrium industry perspective, with various factors being identified as affecting demand and supply – taxation policies, tourism promotion, and so on. What makes the Asia Pacific region unique is the combination of the high rates of economic growth in the Asia region and the emergence of a number of important macroeconomic and socioeconomic trends in demography and relative prices that are certain to have a strong influence on the pattern of economic development.

In analyzing tourism in the Asia Pacific region and its implications for Canada, we first review some basic facts on world tourism, tourism in the Asia Pacific region, and the role of this tourism in the Canadian economy. We also review some of the salient partial- and general-equilibrium theories that address the relevant endogenous and exogenous variables. Some attention is paid to long-run dynamics and to the evolution of comparative advantage. We then turn to an empirical evaluation of the growth of international tourism in Japan. This country was selected for two reasons – because of the availability of good panel data, and because Japan provides an excellent case study of what may occur in other Asian economies with respect to tourism as they climb the ladder of economic development. We also look at Canada as a destination country for tourists from the Asia Pacific region, using Canadian arrivals data. This is done primarily to isolate some important relative-price and income effects since the data available do not enable us to estimate a complete model of international tourism flows. Finally, we undertake a simulation analysis regarding future international tourism export services from Canada to China, taking as given some basic trends in growth, demography, productivity, and comparative advantage.

INTERNATIONAL TOURISM: A REVIEW OF TRENDS AND FACTS

THE WORLD TOURISM INDUSTRY

IT HAS BEEN POINTED OUT BY NUMEROUS WRITERS on economic globalization that tourism, along with transport and telecommunications, is one of the prime forces driving the internationalization of the world economy. The tourism industry itself is arguably the largest industry in the world.[3] It employs 204 million people worldwide (10.6 percent of the global workforce) and produces 10.2 percent of world gross domestic product (GDP). Tourism is the

TABLE 1

WORLD INTERNATIONAL TOURISM TRENDS, 1950-92

	AVERAGE ANNUAL INCREASE	
PERIOD	ARRIVALS	RECEIPTS[1]
	(Percent)	
1950-59	11.7	13.9
1960-69	8.4	10.4
1970-79	6.2	18.6
1980-89	4.5	8.4
1980-84	2.7	1.8
1985-90	6.6	17.0
1986-92	5.8	12.2

1 Data reported in U.S. dollars.
Source: World Tourism Organization.

leading producer of tax revenue across all industries at US$655 billion (1993) and at US$3.4 trillion (1993) is the largest industry in terms of gross output. From an aggregate demand perspective, tourism has become increasingly important. It accounts for 10.9 percent of all consumer spending, 10.7 percent of all capital investment, and 6.9 percent of all government spending. It is not

TABLE 2

WORLD TOURISM GROWTH, 1950-92

YEAR	INTERNATIONAL ARRIVALS (THOUSANDS)	INTERNATIONAL RECEIPTS[1] (US$ MILLIONS)
1950	25,282	2,100
1960	69,296	6,867
1961	75,281	7,284
1962	81,329	8,029
1963	89,999	8,887
1964	104,506	10,073
1965	112,729	11,604
1966	119,797	13,340
1967	129,529	14,458
1968	130,899	14,990
1969	143,140	16,800
1970	159,690	17,900
1971	172,239	20,850
1972	181,851	24,621
1973	190,622	31,054
1974	197,117	33,822
1975	214,357	40,702
1976	220,719	44,436
1977	239,122	55,631
1978	257,366	68,837
1979	273,999	83,332
1980[2]	287,906	102,372
1981[2]	289,784	103,750
1982[2]	289,177	97,880
1983[2]	292,177	98,695
1984[2]	320,142	109,004
1985[2]	329,636	115,424
1986[2]	340,808	139,811
1987[2]	366,758	171,577
1988[2]	393,865	197,743
1989[2]	427,884	210,837
1990[2]	455,594	255,074
1991[2]	455,100	261,070
1992[2]	475,580	278,705

1 Excluding international fare receipts.
2 Revised figures.
Source: *World Travel Yearbook, 1994.*

surprising that tourism is effectively represented by countless industry and trade associations, and that it receives government support through innumerable agencies at virtually all levels. Industrial strategies based on the promotion of tourism are increasingly popular, particularly at state and local government levels. The focus of this paper is on international, as opposed to domestic, tourism in the Asia Pacific region.

Growth rates for international tourism are shown in Table 1. Growth in all subperiods is substantially in excess of the rates of growth of national income. Of particular note is the rapid increase in the growth of tourism receipts during the second half of the 1980s. Evidence on the levels of international tourism is provided in Table 2, which shows that nearly 500 million people participated by 1992. The high rates of growth in the industry have produced forecasts of even greater growth in the future by groups such as the World Tourism Organization (WTO). It has been forecast, for example, that over the 1992-2002 decade tourism will grow at an annual rate of 6.1 percent – 23 percent faster than world economic growth.

TABLE 3

INTERNATIONAL TOURIST ARRIVALS AND RECEIPTS[1]
WORDWIDE AND BY REGIONS

		1986		1992	
REGIONS	SERIES	NUMBERS[2]	SHARE OF THE WORLD TOTAL (PERCENT)	NUMBERS[3]	SHARE OF THE WORLD TOTAL (PERCENT)
World	A	340,891	100	476,000	100
	R	140,023	100	279,000	100
Africa	A	9,458	3	17,000	4
	R	2,970	2	5,167	2
Americas	A	70,972	21	102,100	21
	R	37,652	27	76,567	27
East Asia Pacific	A	32,539	10	58,300	12
	R	16,671	12	43,291	16
Europe	A	217,218	64	287,500	60
	R	76,057	54	147,205	53
Middle East	A	7,973	2	7,200	2
	R	5,003	4	4,356	2
South Asia	A	2,731	1	3,500	1
	R	1,670	1	2,119	1

Note: A = arrivals (thousands); R = receipts (US$ millions).
1 Figures on receipts exclude international transport.
2 Totals for 1986 are revised figures.
3 Totals for 1992 are provisional revised estimates.
Source: World Tourism Organization.

The tourism industry has attracted considerable attention from the perspective of the balance of payments – that is, as a source of export services. The general world picture of the relative importance of various regions is provided in Table 3. Europe continues to dominate in terms of both arrivals and receipts, with over 50 percent of the world totals in both cases. The Western Hemisphere is second, with about one-quarter of the world totals. East Asia and the Pacific are a distant third, at about 10 percent of the world total, although this has grown considerably in the past decade.

Tourism in the Asia Pacific Region

Unfortunately data on the East Asian part of this region is much less comprehensive than for the OECD countries. Tourism output within the Asia Pacific region – spending by travelers plus capital investment – is forecast by the WTO to rise from $800 billion in 1995 to $2 trillion by 2005.[4] At the same time, the region's share of world travel is expected to rise from 23 to 27 percent. The international implications of this growth depend considerably upon whether that tourism occurs within or outside the region. Intra-Asia tourism accounted for 66 percent of total trips made in the region in 1993. This is comparable to most other regions in the world with the exception of Europe (Table 4). The high intra-regional European share is no doubt an artifact of both the number of countries in Europe and their geographical proximity.

Not surprisingly, participation rates in international tourism vary considerably across East Asia, reflecting both levels of economic development and legal restrictions on travel that exist in a number of countries. For example, in 1994 the participation rates were approximately 13 percent for Japan, 5 percent for South Korea, 2 percent for Thailand, and 0.2 percent for China.

Looking at the entire Asia Pacific area as a destination region, we get a slightly different view of matters. In terms of the origin of total visitors within

Table 4

Share of Intra-regional Tourism in Total Tourism of Region

	Percent
Europe	87
East Asia	70
Americas	67
Africa	62
Middle East	60
South Asia	21

Source: WTO figures cited in *The Economist* (29 July 1995).

the region, Asia clearly dominates both North America and the Pacific and is growing in terms of market share. From 1984 to 1994, visitors originating in Asia had increased from approximately 50 to 60 percent of the market (Table 5).

The distribution of destinations in the Asia Pacific region is given in Table 6, together with growth rates. All regions are holding their relative attractiveness as destinations, but there is clear evidence that North-East and South-East Asia are experiencing slightly better than average growth. This could reflect the increased intra-regional share of tourism in East Asia or an increase in out-of-region arrivals.

Some further evidence on the pattern of travel is provided by recent Japanese data. Japanese departures by country of destination are reported in

TABLE 5

ORIGIN OF VISITORS IN THE ASIA PACIFIC REGION, 1984-94

	ANNUAL GROWTH RATE	MARKET SHARE	
	1984-94	1984	1994
		(Percent)	
Asia	11.3	49.4	59.9
Pacific	9.0	8.3	8.2
North America	3.6	22.2	13.2
Europe	10.6	12.2	13.9

Source: Pacific-Asia Travel Association, *Annual Statistical Report, 1994*.

TABLE 6

TOURIST ARRIVALS IN THE ASIA PACIFIC REGION, BY REGION OF ORIGIN, 1984-94

	INCREASE	ARRIVALS	MARKET SHARE	
	1984-94	1984-94	1984	1994
	(Percent)		(Percent)	
North America	10.8	7,512,912	9.0	10.4
South Asia	11.0	2,875,116	–	4.0
North-East Asia	9.8	25,433,934	33.3	35.3
South-East Asia	11.0	22,516,334	26.4	31.3
Australia and New Zealand	11.5	4,684,265	5.3	6.5
South and Central Pacific	4.6	2,544,585	–	3.5
Hawaii	–	6,455,160	–	9.0
Total	9.2	72,022,306	–	100.0

Source: Pacific-Asia Travel Association, *Annual Statistical Report, 1994*.

Table 7 for three months in early 1995. It is clear that countries such as Hong Kong, South Korea, Singapore, and China play a much greater role for Japanese travellers than do North American destinations.

Additional evidence on the pattern of arrivals in the region can be gleaned by looking at one country in the region for which some recent data are available. Table 8 reports arrivals data for Hong Kong. We see here the

TABLE 7

DEPARTURES OF JAPANESE NATIONALS BY DESTINATION, JANUARY TO MARCH 1995

DESTINATION	JANUARY	FEBRUARY	MARCH
Cambodia	684	740	1,064
China	39,240	50,222	67,136
Taiwan	57,157	60,439	71,371
Hong Kong	84,789	71,365	83,463
India	3,553	5,315	5,480
Indonesia	33,277	31,215	21,522
South Korea	105,112	108,095	130,854
Malaysia	16,151	15,200	19,708
Nepal	1,112	2,029	2,296
Philippines	23,740	20,657	23,918
Singapore	58,626	52,446	67,831
Thailand	46,324	47,313	57,320
Vietnam	3,632	5,707	5,033
Canada	13,243	12,929	17,276
Mexico	1,954	2,352	2,024
United States	348,409	311,920	387,893

Source: Japan National Tourist Organisation, "Visitor Arrivals and Japanese Overseas," *Travellers Report* (July 1995).

TABLE 8

VISITOR ARRIVALS AT HONG KONG, BY REGION OF ORIGIN, 1993-94

	1993	1994
South-East Asia	1,239,458	1,196,835
Taiwan	1,777,310	1,665,330
China	1,732,978	1,943,678
Japan	1,280,905	1,440,632
United States/Canada	945,098	961,329
West Europe	1,046,080	1,126,079
Australia/New Zealand	312,552	316,338
Other	603,119	680,935

Source: Hong Kong Tourist Association, *Statistical Report of Tourism 1994*.

TABLE 9

TOURISM RECEIPTS, WORLDWIDE AND EAST ASIAN MARKETS

	1992	1993	1994
	(US$ billions)		
Worldwide	303.9	305.8	321.5
East Asia and Pacific	46.6	51.7	58.9
South-East Asia	16.9	18.9	23.1

Source: World Tourism Organization.

overwhelming importance of intra-regional visitors relative to visitors from other regions. The relative importance of China is obvious, but so are visitors from South-East Asia, Taiwan, and Japan.

In terms of receipts in the balances of payments, tourism was expected to yield US$321.5 billion worldwide in 1994, representing an increase of 5 percent over the 1993 total of US$305.8 billion (Table 9). In 1992, US$303.9 billion was recorded for tourism receipts worldwide. East Asia and the Pacific accounted for 18.35 percent of world tourism receipts or an estimated total of US$58.9 billion in 1994. South-East Asia was expected to record a total of US$23.1 billion in terms of total receipts in 1994. While these figures may seem quite large, they should be put in perspective. International tourism is dwarfed relative to spending on domestic tourism in the developed economies. For example, in 1990 total travel receipts in the United States, both domestic and international, were estimated at $617 billion; this compares with total domestic receipts of $575 billion. Total world tourism receipts (both domestic and international) are in the range of 13 percent of world GDP.[5]

CANADIAN TOURISM DATA

INTERNATIONALLY, CANADA RANKS WELL DOWN THE LIST in terms of either destinations or receipts at the aggregate level. In 1992, Canada was the ninth most popular destination in the world, just behind Germany and just ahead of Mexico, with 3.43 percent of total world international arrivals, or just over 16 million visitors. A large fraction of that total was accounted for by visits from the United States. In terms of international tourism receipts, Canada ranked tenth, just behind Singapore and just ahead of Hong Kong. Receipts were US$5.75 billion or 2.06 percent of the world's total. At 8 percent, Canada's growth rate in tourism receipts from 1980 to 1992 was just below the world average.

Tables 10 and 11 provide data on total Canadian arrivals by region of origin, in both cases excluding the United States. The most remarkable feature of these data is the growth in Asia's share in total arrivals. From 1972 to 1994, that share grew from about 13.5 percent to almost 31 percent. Adding Oceania, whose share is roughly constant, the Asia Pacific basin (excluding Latin America) accounts for nearly 40 percent of the non-U.S. Canadian tourist market.

In Table 12 we document the distribution of these arrivals by country from 1972 to 1994. In general, we see here the relative importance of Japan, followed by "other Asia," Hong Kong, and "other Pacific." Table 13, which shows the same distribution as shares of all arrivals in 1994, displays broadly similar patterns, with "other Pacific" moving ahead of Hong Kong. In Table 14 the "other Asia" component is broken down for more recent years. Here we see the relative importance of the newer East Asia markets, in particular South Korea and Taiwan.

TABLE 10

TOURIST ARRIVALS AT CANADIAN FRONTIERS BY REGIONS OF ORIGIN (OTHER THAN THE UNITED STATES), 1980-94

	NON-U.S.	EUROPE	AFRICA	ASIA	OCEANIA[1]
1980	2,163,013	1,377,852	45,359	384,829	176,104
1981	2,144,732	1,350,651	49,091	391,494	169,930
1982	1,974,661	1,231,754	50,626	361,440	172,570
1983	1,775,739	1,060,157	50,603	382,396	158,974
1984	1,887,222	1,110,306	48,342	426,743	188,743
1985	1,808,038	1,042,188	41,835	420,592	188,476
1986	2,259,827	1,304,810	44,730	564,234	231,137
1987	2,642,636	1,564,236	42,726	648,310	258,379
1988	3,105,860	1,770,751	47,096	823,591	284,956
1989	3,276,838	1,804,784	48,461	948,488	330,623
1990	3,256,391	1,779,325	51,019	962,062	327,476
1991	3,240,529	1,792,773	48,001	966,180	284,625
1992	3,303,479	1,854,394	49,869	978,114	291,761
1993	3,477,721	2,011,273	50,386	991,914	271,335
1994	3,513,812	1,947,426	48,651	1,083,463	299,723

1 In addition to Australia and New Zealand, Oceania includes "other Pacific islands."
Source: Statistics Canada.

Table 11

Distribution of Tourist Arrivals at Canadian Frontiers by Regions of Origin, 1972-94

	Europe	Africa	Asia	Oceania
		(Percent)		
1972	70.42	1.58	13.47	6.38
1973	71.48	1.81	12.31	7.34
1974	68.52	2.08	13.67	8.50
1975	67.22	1.97	14.80	8.19
1976	66.52	2.14	15.19	9.02
1977	66.27	2.33	15.32	9.04
1978	65.48	2.14	16.53	9.21
1979	65.43	1.93	17.14	8.69
1980	63.70	2.10	17.79	8.14
1981	62.98	2.29	18.25	7.92
1982	62.38	2.56	18.30	8.74
1983	59.70	2.85	21.53	8.95
1984	58.83	2.56	22.61	10.00
1985	57.64	2.31	23.26	10.42
1986	57.74	1.98	24.97	10.23
1987	59.19	1.62	24.53	9.78
1988	57.01	1.52	26.52	9.17
1989	55.08	1.48	28.95	10.09
1990	54.64	1.57	29.54	10.06
1991	55.32	1.48	29.82	8.78
1992	56.13	1.51	29.61	8.83
1993	57.83	1.45	28.52	7.80
1994	55.42	1.38	30.83	8.53

Tourism: Comparative Advantage and the Balance of Payments

A COMPARISON OF A COUNTRY'S EXPORT SHARE of a particular good or service relative to its total exports with the corresponding figures for other countries provides a very rough measure of that country's comparative advantage. In Figure 1 we display a graph of international tourism receipts as a proportion of total Canadian exports. We can see that receipts have been relatively stable over the last 20 years.

These numbers can be put in perspective by comparing them with a similar set of revealed competitiveness figures for a number of member countries of the Organisation for Economic Co-operation and Development (OECD) in Table 15. Canada's ratio of travel exports to total merchandise exports hovers

TABLE 12
ARRIVALS AT CANADIAN FRONTIERS FROM THE ASIA PACIFIC REGION, 1972-94

	OTHER ASIA	HONG KONG	INDIA	JAPAN	AUSTRALIA	NEW ZEALAND	OTHER PACIFIC	TOTAL
1972	26,430	12,021	19,855	52,438	22,393	5,793	29,352	168,282
1973	25,245	13,826	13,788	71,095	31,046	8,857	41,200	205,057
1974	33,802	17,371	18,091	77,543	40,151	10,694	52,816	250,468
1975	42,583	20,778	22,023	90,411	41,487	11,701	55,508	284,491
1976	55,980	23,065	31,420	106,783	57,049	12,816	72,409	359,522
1977	54,084	21,204	21,742	97,532	51,348	11,572	65,847	323,329
1978	69,087	20,430	28,241	127,827	59,153	15,839	79,100	399,677
1979	79,678	23,833	36,700	158,582	67,243	18,302	89,106	473,444
1980	97,370	30,839	48,547	162,253	65,967	20,117	90,020	515,113
1981	105,529	36,924	50,527	146,461	64,651	18,114	87,165	509,371
1982	110,989	39,559	27,330	139,447	67,589	16,628	88,353	489,895
1983	119,719	46,371	29,949	138,716	62,737	14,594	81,643	493,729
1984	133,676	51,563	35,143	162,246	74,902	16,925	96,916	571,371
1985	130,718	43,272	34,553	174,503	76,028	15,682	96,766	571,522
1986	160,153	64,545	46,199	235,185	85,021	27,381	118,735	737,219
1987	151,723	66,694	53,031	311,687	89,686	35,067	133,626	841,514
1988	192,909	87,785	66,739	404,592	100,841	39,416	144,699	1,036,981
1989	230,182	123,080	63,076	462,699	120,316	42,555	167,752	1,209,660
1990	239,983	129,609	54,621	474,132	122,294	39,529	165,653	1,225,821
1991	243,774	127,104	49,080	480,308	109,369	31,264	143,992	1,184,891
1992	251,020	126,398	45,794	495,823	113,329	30,799	147,633	1,210,796
1993	261,590	123,145	43,805	505,812	102,692	31,188	137,455	1,205,687
1994	331,695	127,443	43,912	515,667	117,742	30,580	151,401	1,318,440

TABLE 13

1994 SHARES OF ASIA-PACIFIC CANADIAN TOURISM ARRIVALS

OTHER ASIA	HONG KONG	INDIA	JAPAN	AUSTRALIA	NEW ZEALAND	OTHER PACIFIC
25.16	9.67	3.33	39.11	8.93	2.32	11.48

Source: Statistics Canada (1995).

TABLE 14

CUMULATIVE MONTHLY ARRIVALS FROM "OTHER ASIA" TO CANADA, JANUARY 1990-AUGUST 1994

CHINA	INDONESIA	SOUTH KOREA	PHILIPPINES	SINGAPORE	TAIWAN	THAILAND
152,279	47,436	256,390	102,054	97,675	225,535	58,808

Source: Statistics Canada (1995).

FIGURE 1

TOURISM RECEIPTS AS A PROPORTION OF EXPORTS, CANADA, 1928-92

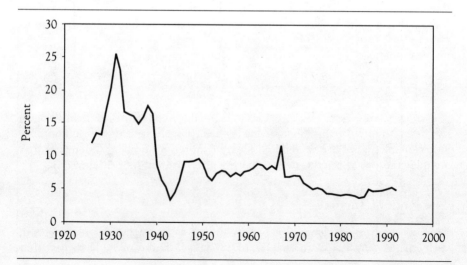

TABLE 15

REVEALED ADVANTAGE IN TOURISM OF SELECTED OECD COUNTRIES, 1978-85

	RATIO OF TRAVEL EXPORTS TO MERCHANDISE EXPORTS
United States	5.26
United Kingdom	6.47
France	7.77
West Germany	3.42
Italy	11.43
Japan	0.55
Netherlands	2.51
Belgium and Luxembourg	3.46
Spain	34.35
Greece	36.89
Australia	4.73

near 5 percent. This puts Canada at the low end of the range within the OECD, suggesting that its comparative advantage remains in the area of natural resources and manufactured goods.[6]

THE ECONOMIC THEORY OF INTERNATIONAL TOURISM

THE ECONOMIC THEORY OF TOURISM is largely based on a partial-equilibrium model and focuses on the demand side. In this section we briefly review that theory. The discussion then turns to international tourism, and what the theory of international trade suggests with respect to the determinants of international tourism flows. The last section deals with issues of growth in the international economy and how it might affect international tourism flows.

THE DETERMINANTS OF TOURISM DEMAND

THERE IS A LARGE AND WELL-DEVELOPED EMPIRICAL LITERATURE dealing with the determinants of the demand for tourism, largely in a domestic context.[7] The general theory of the demand for tourism has a number of elements that are relevant for the purposes of evaluating international tourism flows.

The Role of Income

Estimates of tourism demand generally find it to be an income-elastic good, with income elasticities in the 1.2 to 3.0 range. These strongly suggest that tourism is a luxury good and that tourism budget shares depend on the level of individual income. This has a number of implications with respect to the determination of comparative advantage in tourism and the impact of economic growth on tourism receipts and expenditures.

The literature has been rather poor at distinguishing between permanent and transitory income effects. If tourism services are important in total consumption smoothing, then we should expect to see less volatility in these expenditures than in employment or income movements associated with business cycle movements. A loose reading of the literature would suggest that the "luxury" status of tourism, and in particular international tourism, is synonymous with the proposition that tourism is a buffer-type good that reacts strongly to short-term movements in income. In other words, tourism is the first budget item that consumers scale down during bad periods and that they augment during booms. If that is the case, then it would suggest that the demand for tourism is strongly procyclical, and that it is subject to buffer-stock effects, which means that backlogs of demand tend to be triggered at some point during an upturn in an economic cycle. For small, open economies that depend heavily on tourism, this raises a set of conventional concerns about the "export instability" typically associated with commodity exporting.

Wages and the Value of Time

Tourism is complementary with leisure and, via the usual leisure-demand models, interacts with the determination of labour supply as in the Becker model of the allocation of time. That model suggests that changes in wage rates affect both full income and the opportunity cost of time. Rising wage rates therefore induce both a positive income effect on the demand for tourism and a negative substitution effect. This analysis can be used, for example, to rationalize the observation that the average time for a typical tourist trip has been falling dramatically. It also has some implications regarding long-haul versus short-haul trips.

Transport Costs

For long-distance international tourism, the cost of travel is a major expenditure, and it is a fixed cost in the purchase of an "international tourism experience." Being a fixed cost, it should be an important determinant of an individual's decision to travel internationally but should be less important for decisions on margins such as the number of nights to spend abroad or other trip-related spending. A major feature of the time series data has been the reduction in the real cost of air transport; how important this is in explaining international travel and whether this trend will continue are important research issues.

The Demand for Variety

Tourism is generally regarded as an important life experience for most people and in addition is characterized by a demand for variety in this experience – in other words, for seeing different places and having different tourist experiences. Conventional models of product differentiation easily accommodate this type of effect, although the implementation of these models has so far been incomplete. If tourism is a highly differentiated good, this suggests that from the supply-side perspective, market structure in the tourism industry may be more properly viewed as monopolistically competitive rather than competitive. Deliberate product differentiation strategies, as opposed to price competition, are therefore likely to be important.

Experience and Hysteresis

Tourism is one of many goods that may be subject to some type of dynamic learning or experience effect; economists use other words such as "habit persistence," "rational addiction," or "hysteresis" – all suggestive of the same effect. That is, once tourism is experienced, it raises the marginal utility of that good. Disentangling empirically the experience effect from the income effect is a tricky matter. In the case of international tourism, however, the argument is potentially quite important. It could mean, for example, that policies that promote the international tourism experience can lead to a permanent increase in the demand for that good (hysteresis).

Demography

Demography influences tourism demand in different ways. First, there is the demand for tourism by retired individuals, for whom tourism is generally thought to represent a much larger budget share than for younger, working-age individuals. Depending upon the demographic evolution of the population, this would suggest that certain types of tourism demand will grow relative to others. At the same time, younger individuals demand different types of experiences and generally find long-distance international travel less stressful. This can have important implications for the demographic composition of international travel to certain destinations.

Demographic trends interact importantly with the life-cycle model of consumption, which strongly suggests a hump effect in the consumption pattern of an individual occurring between the ages of 40 and 55. The model unfortunately has little to say about the composition of those expenditures. A society with a strong middle-age "baby boom" can expect to have falling saving rates and strong consumption growth. Much of this growth may fall on tourism, depending upon other factors such as the role of relative prices. In an open economy, these life-cycle effects lead naturally to deficits and surpluses on the current account, which in turn can lead to sector-specific effects that are consumption-sensitive. Tourism may be such a sector.

Closely related to demographic trends are changes in the role of women and the family. The strong increase in the labour force participation rate of women in North America has placed additional coordination constraints on families, which makes long vacations more problematic. If these trends emerge in Asia, they may have important effects on the demand for international tourism.

Tourism as an International Tradeable Service

WHAT DOES ECONOMIC THEORY SUGGEST regarding the international pattern of trade in tourism services? There are three conventional models here: the Ricardian model of comparative advantage; the factor-endowments model of trade; and the "new trade theory" approach to intra-industry trade in differentiated goods.[8]

Ricardian Theory

Ricardian theory suggests that the observed exports of a country's good or service are a function of that country's cost advantage in that good relative to other countries. All goods are presumed to be homogeneous across countries and produced under conditions of constant cost. For example, in the case of Japan, a net importer of tourism services, a Ricardian explanation would hinge on demonstrating that Japan has relatively higher costs in the provision of tourism relative to manufactured goods. The cost argument, as will be seen below, is closely associated with the issue of long-term productivity growth and real exchange rates.

Factor Endowment Theory

The simplest theory of tourism is one based on natural endowments, be they mountains or museums. Tourism exports are driven by the unique factor "services" that those endowments generate. The theory predicts that countries that have certain advantageous natural endowments will be net exporters of tourism services. Environmentalists correctly point out that many of these natural "endowments" are public goods that are subject to congestion effects. For example, the benefits of the national parks diminish if they become too crowded. This has a number of implications that are not explored in this paper but that have important implications for how the rents on these natural endowments are captured through the pricing of international tourism services.[9]

The factor-endowments theory, while obviously suggestive, is difficult to quantify. We could all agree, however, that Canada is well endowed with mountains and Italy is well endowed with museums. These endowments are largely permanent, compared with other forms of comparative advantage arising from the accumulation of physical or human capital. One implication of this theory is that if a country has a unique factor endowment that gives it world monopoly power with respect to that factor, some form of export tax on that service would have a welfare-improving impact on the exporting country.

Intra-Industry Trade and Product Differentiation

If one assumes that tourism is a highly differentiated product and that the degree of differentiation is determined by market forces, then the standard models of trade in differentiated products suggest that we should see countries both exporting and importing large volumes of tourism services. Moreover, these trade volumes will be larger the closer countries are in terms of factor endowments and the lower are transport costs. It is clear that this theory rings true in the context of tourism. The gains from trade are reflected in the total variety of experiences offered internationally and domestically. Trade restrictions in such a world can be highly damaging because of the negative effect this has on the total world variety available.

An important issue in a multi-country world is the extent to which trade is intra- versus extra-regional. Some data were presented earlier on this question. The intra-industry trade model suggests the important role that transport costs play in this respect. Low transport costs – i.e., geographic proximity – will lead to large volumes of intra-regional, intra-industry trade. That is, the model would predict that volumes of intra-regional tourism should grow faster than long-distance international tourism. This contrasts with the factor-endowments theory, which makes no such predictions.

Summary

Each of these theories seems plausible in its own right, and most trade analysts would accept that each is capable of providing a partial explanation of observed tourism patterns. In the Canadian case, the obvious natural endowments and low population density mean that Canada offers both a unique destination in the Asia Pacific region but one that competes for Asian visitors in some measure with similar destinations such New Zealand or Australia. On the other hand, the long distances separating Canada from Asia would suggest that cross-Pacific tourism is likely to grow less quickly than intra-regional tourism in East Asia. The proximity of Canada to the United States may serve to generate some tourism due to multiple destination trips or what is referred to as "circuit tourism."[10]

TOURISM AND UNBALANCED ECONOMIC GROWTH

AMONG THE MOST INTERESTING and potentially important factors affecting tourism in the Asia Pacific region are the significant differences in levels of income across the region and the high rates of economic growth in a number of countries, in particular China and the newly industrialized countries (NIC) of East Asia. Assuming that these trends remain intact, tourism will be a major imported service as incomes rise in the region. Canada, in particular, could be a strong beneficiary of this development. The purpose of this section is to propose an open-economy version of an unbalanced, two-country growth model and highlight its implications for trade in services.

Real Exchange Rates

To put this issue in perspective, it is important to recognize that tourism services are unlike manufactured goods in that they are produced by using inputs that we traditionally think of as being associated with the service industries and that we treat as non-traded. The ratio of non-traded service prices to traded good prices is referred to as the *real exchange rate*. Tourism is different from other services in that it is tradeable. An increase in the real exchange rate typically induces an economy to produce less tradeables and more services. In the case of tourism services, however, that interpretation is not necessarily correct in that an increase in tourism prices may induce a supply response that increases the net export of tourism services.

Coupled with this observation is what is known as the Balassa-Samuelson effect: rich countries tend to have higher real exchange rates – that is, services cost more relative to manufactured goods in richer countries. Therefore, there is a close, empirically well-documented relationship between economic growth and the price of services. Helliwell (1994) reports a regression of relative price levels for 1990, using 100 countries in which the coefficient on relative GDP per capita is 0.5042. This can be interpreted as an approximate elasticity; in other words, a 1 percent increase in relative income per head raises the relative price level by 0.5 percent. This can be viewed as a corollary of the convergence hypothesis: as countries' income levels converge, their prices of services also converge.

The implications of this observation for tourism trade in the Asia Pacific region are extremely important. As these countries grow, their tourism prices will rise relative both to those of other goods and to tourism prices in high-income countries, including Canada.

A Model

We now sketch a two-country model of tourism between two regions in a state of unbalanced growth. We will call one region Asia, with subscript $i = a$, which has both lower initial income and a higher income growth rate, and the other region $i = o$ ("other") – a region with high income but a lower rate of growth. Income per capita in each region is denoted by y, population by P. All prices are measured relative to a common tradeable manufactured good whose price is set equal to unity. Growth and GDP in real terms are expressed in units of the manufactured good. Consumption of within-region tourism is denoted by X, out-of-region tourism by Y, and consumption of the manufactured good by C. Demand functions for X and Y for region a on a per capita basis are given by

$$X_a = X(p_a, p_o, y_a)$$
$$Y_a = Y(p_o, p_a, y_a) \tag{1}$$

There is a similar pair of demand functions for region o, and also demand functions for manufactured goods. Production possibilities in each country are given by a Ricardian "production possibility frontier" (PPF) between manufactured and tourism goods. Let aggregate tourism be denoted by T_i and aggregate manufactured goods by M_i. Thus the supply side in country a is described by

$$A_a M_a + B_a T_a = P_a$$

where A_i and B_i are productivity coefficients.

Market clearing requires that

$$T_o = P_a Y_a + P_o X_o,$$

with a similar equation for Asia. Economic growth is reflected in the fact that the coefficients A and B are diminishing over time, shifting the PPF out for a given population. The PPF, however, shifts out and rotates at the same time, giving rise to the "unbalanced" nature of growth.

Productivity growth is given by

$$\hat{A}_a = -\theta_a, \; \theta_a > 0$$
$$\hat{B}_a = -\lambda_a, \; \lambda_a > 0 \tag{2}$$

The "hat" notation denotes proportionate rates of change of the same variable without a "hat." The consequence of this specification, assuming that θ is greater than λ, is that productivity growth is unbalanced; holding resources constant – i.e., population – the economy is able to produce more manufactured goods over time relative to tourism services. Real GDP measured in units of manufactured goods is given by P/A, and thus real GDP per capita is given by $y = 1/A$. We assume the growth rate of manufactures is higher in Asia than in "other," hence $\theta_a > \theta_o$.

The dynamics of this model are given by a profile of population growth rates and per capita income growth rates:

$$\hat{P}_a = n_a$$
$$\hat{y}_a = \theta_a \qquad (3)$$

The model is closed with a pair of equations for the evolution of the price of tourism in each country. The relative price of tourism T in country a and its rate of change are given by

$$p_a = B_a/A_a$$
$$\hat{p}_a = \theta_a - \lambda_a \qquad (4)$$

Consider now the value of international tourism exports of country o – i.e., $Z = p_o Y_a$. Calculating the rate of change of Z we get

$$\hat{Z} = n_a + \varepsilon_a^{cross}(\theta_a - \lambda_a) + (\varepsilon_a^{own} + 1)(\theta_o - \lambda_o) + \eta_a \theta_a \qquad (5)$$

In this formulation, "own" denotes own-price elasticity of international tourism demand, "cross" denotes cross-price elasticity between international and domestic tourism, and η denotes the income elasticity of demand. This formulation emphasizes the importance played by relative productivity growth in tourism in both countries. The Balassa-Samuelson effect in dynamic form refers to the fact that in comparing countries a and o we have

$$\theta_a - \lambda_a > \theta_o - \lambda_o$$

so that tourism prices are rising faster in a than in o. This is true at the same time that income growth is higher in a in o. One can get an idea of the orders of magnitude here by plugging in some representative elasticities. Suppose the own-price elasticity is -0.8, the cross-price elasticity is -0.5, and the income elasticity is 1.5. We also use the Helliwell number that sets the growth rate of p_a at one-half the growth rate of θ_a. This gives us the result that

$$\hat{Z} = n_a + 1.75\theta_a + 0.2(\theta_o - \lambda_o)$$

This indicates the sensitivity of tourism exports to the growth rate of population, the growth rate of per capita income, and relative productivity growth in the destination region. In general, for this set of elasticity values one can expect that tourism exports will be quite sensitive to income growth rates in the lower-income region.

TOURISM FLOW MODELS

WE DEVELOP TWO ECONOMETRIC MODELS of tourism flows. Ideally, one would model and estimate a complete structural model of flows, using both demand and supply conditions in all originating and destination countries. Unfortunately, the data for such an exercise are not available. One approach is to model Japanese tourism outflows – that is, the choice by Japanese to travel abroad and their choice of destination. This model is best viewed as describing demand determinants from the perspective of a single country, which reflects the variation in those determinants over time in the originating country. We chose Japan because of the availability of time series data, but Japan also provides a good case study of an East Asian country that has gone through a period of rapid growth. To that extent, the evolution of its international tourism may provide useful predictions of what other Asian countries could experience in the future.

The second model, developed later on, looks at arrivals to Canada from 1972 to 1991. This model captures a combination of demand determinants in the originating countries and variation in supply over time within Canada. It is perhaps most useful at capturing cross-country differences in income and other parameters, and it provides a Canadian perspective on the international dimensions of arrivals.

A MODEL OF JAPANESE INTERNATIONAL TOURISM

THE MOST EXTENSIVE TRAVEL/TOURISM DATA are reported with respect to three units of measurement – arrivals at frontiers, nights spent in hotels, and arrivals at all accommodations. Few countries collect all three series of data. We explore the relationship between the number of arrivals at frontiers, nights spent, and arrivals at all accommodations from Japan to 12 (or fewer) countries for which the relevant data are available, over a 20-year period (1972-91). Frontier arrivals are one measure of tourism, but it has a number of deficiencies. In Europe, for example, it is easy to pass from one country to the other. It is less easy to do so from, say, Australia or Iceland.

Given the theory discussed above and the data available, we explore the effects of the cost of travel as a function of fares and distance, relative prices between Japan and other countries, and the individual characteristics of different countries as a measure of their potential attractiveness. We estimate both the international tourism participation elasticities, and the expenditure elasticities conditional on choosing to travel overseas. The growth in both

Table 16

Tourist Arrivals from Japan, Various Markets, 1972-91 (Average Annual Growth)

Country	Arrivals at Frontiers	Nights Spent in Hotels	All Accommodations
		(Percent)	
Australia	18.6	–	–
Austria	–	8.6	8.5
Canada	10.7	–	–
Denmark	–	1.2	–
Finland	–	7.9	–
France	6.6	8.0	–
Germany	–	6.0	6.8
Greece	5.5	–	–
Iceland	8.4	–	–
Italy	5.7	5.1	5.6
Netherlands	–	2.1	2.1
New Zealand	16.0	–	–
Portugal	5.0	5.8	5.7
Spain	6.6	6.2	–
Switzerland	–	5.1	5.3
Turkey	12.1	11.1	13.2
United Kingdom	5.6	4.0	–
United States	10.8	–	–
Average	9.2	5.5	6.0
Sample Total (1991)	1,058,469	12,455,823	2,136,558

Source: Organisation for Economic Co-operation and Development (1991).

population and income fosters greater outward-bound tourism, and the size of certain age cohorts also plays a central role in what outbound travel has taken place. The technique used is a pooled cross-section time series analysis, with the destination countries identified as being "fixed effects."[11]

Between 1972 and 1989 Japanese departures increased at an average rate of 11.5 percent per year. In the sample of countries for which arrivals-at-frontier data are available, the average increase in arrivals was 9.8 percent per year. Over the longer period described in Table 16, arrivals at frontiers from Japan increased at an average annual rate of 9.2 percent, while nights spent in hotels and arrivals at all accommodations rose by 5.5 and 6 percent, respectively.

Arrivals at Frontiers

Our discussion of the arrivals-at-frontier model begins with what would be the usual approach if there were but a single destination for Japanese tourists. We simply regress Japanese arrivals at the frontier against the variables of interest. In effect, this assumes that all countries are randomly likely to have Japanese tourists drop in after accounting for the variables of interest. Table 17 reports some results of the estimating equation.

This regression is the basic equation from which we develop the more detailed statistical model. The features of interest are the cost of transportation to each country, LJD; the real exchange rate, LRJX; the size of the Japanese population, LJPOP; the composition of the Japanese population, reflected in the share of the population between the ages of 45 and 75, SOLD; and the population of each of the destination countries, LPOP. The latter captures what might be either relative size, which would come out of an intra-industry trade model, or factor-endowments effects if these are proportional to population. The dependent variable is the logarithm of arrivals at the frontier of each country. Each of the variables prefixed by an "L" reflects the log of that variable. Thus the coefficients are the relevant elasticities (except for that on SOLD since it is a share).

Of interest in the simplified model summarized in Table 17 is the fact that the signs of most of the coefficients are in the expected direction. The transportation cost is measured as the real yen cost per mile muliplied by the number of miles from Tokyo to a major air destination in the foreign country. The point estimate of the coefficient on LJD suggests that a 10 percent increase in the cost of travel leads to a 2 percent fall in visits (measured by arrivals at the frontier). The effect of the size of the Japanese population is to increase tourism, as is the effect of the destination country's size. A higher real per capita income in Japan is also associated with an increase in tourism. The only anomaly is the real exchange rate. In this model, in which the exchange rate is measured in foreign currency units per Japanese yen, a higher exchange rate, LRJX, should lead to more tourism as the relative-price and income effects are both positive. Note that relative to the theoretical discussion above

TABLE 17

ARRIVALS-AT-FRONTIER MODEL RESULTS

VARIABLE	ESTIMATED COEFFICIENT	STANDARD ERROR	T-RATIO 233 DF
LJD	−0.20211	0.1090	−1.855
LRJX	−0.11302	0.3070E-01	−3.682
LJPCY	3.4877	0.4014	8.689
LJPOP	8.7227	1.641	5.314
LPOP	1.0063	0.2143E-01	46.97
SOLD	−14.220	3.593	−3.957
Constant	−196.92	32.54	−6.052

Table 18

Arrivals-at-Frontier Model with Fixed Country Effects

Variable	Estimated Coefficient	Standard Error	T-Ratio 222 DF
LJD	0.10134	0.1039	0.9754
LRJX	0.64112	0.9351E-01	6.856
LJPCY	3.2866	0.3732	8.806
LJPOP	7.2937	1.395	5.228
LPOP	3.3124	0.4335	7.642
SOLD	−17.373	3.372	−5.152
United Kindgom	3.1312	0.6069	5.159
France	2.3226	0.7309	3.178
Greece	4.3584	1.614	2.701
Iceland	12.655	3.146	4.023
Italy	−1.3123	1.095	−1.198
Portugal	2.9784	1.645	1.811
Spain	0.34660	1.078	0.3216
Turkey	−3.8135	1.086	−3.511
Australia	6.1835	1.238	4.993
New Zealand	10.178	1.883	5.405
Canada	5.1706	0.9878	5.234
Constant	−211.46	28.44	−7.435

there is no variable capturing the within-country change in the price of tourism relative to other goods. This shortcoming is unfortunate and should be addressed in future research.

This model blends all of our sample countries together without taking into account country-specific effects. We need to account for these by introducing dummy variables for each country; this is done in Table 18.[12]

When statistically significant, in each case the coefficient has the expected sign. The dummies for each country are relative to the United States. It is a little disappointing that the coefficient on travel costs is not significant in the expected direction.

Real Exchange Rate

The real exchange rate reflects two components – price levels in the countries of departure and destination, and the nominal exchange rate. Their behaviour over the past 20 years is displayed in Table 19. The average rate of growth in the consumer price index (CPI) across the 20 countries is 11 percent. At 5.2 percent,

TABLE 19

COMPONENTS OF THE REAL EXCHANGE RATE, 1972-91
(AVERAGE ANNUAL RATE OF CHANGE)

	CONSUMER PRICE INDEX	EXCHANGE RATE DEPRECIATION VIS-À-VIS JAPANESE YEN	
		NOMINAL	REAL
		(Percent)	
Australia	8.9	6.5	2.8
Austria	4.6	0.9	1.4
Canada	7.0	5.5	3.6
Denmark	7.4	4.0	1.8
Finland	8.6	4.4	1.1
France	7.6	5.1	2.6
Germany	3.6	1.2	2.8
Greece	16.7	14.6	3.0
Iceland	29.8	26.6	2.0
Italy	11.6	8.7	2.4
Japan	5.2	–	–
Netherlands	4.4	1.5	2.3
New Zealand	10.8	7.9	2.3
Portugal	16.8	13.9	2.2
Spain	11.6	7.2	0.8
Switzerland	3.9	-0.7	0.5
Turkey	35.4	34.5	4.4
United Kingdom	9.6	6.6	2.2
United States	6.2	4.8	3.7
Average	11.0	8.8	2.9

Japan is among the lower-inflation countries. Against the yen, the average nominal exchange rate depreciated by 8.8 percent, with only Switzerland registering an appreciation. The average real exchange rate, expressed as the number of real foreign currency units per real yen, depreciated at an annual rate of 2.9 percent against the yen, and none of the countries displayed a real appreciation against the Japanese currency.

The expected sign on the real exchange rate is positive as a higher real exchange rate means that foreign goods are cheaper for Japanese tourists, thus encouraging tourism. The magnitude of the regression coefficient suggests that a 10 percent depreciation of a country's bilateral real exchange rate relative to the yen will encourage a 6 percent increase in tourism as measured by arrivals. This can be interpreted as an own-price elasticity of 0.6 in Japanese participation rates in international tourism.

Real Income

The growth in Japanese per capita real income (measured in real yen) during the 1970-91 period was substantial. The average annual increase was 3.1 percent, compared with about 1.2 percent for the United States over the same period. At 3.3, the point estimate of the income elasticity of tourism demand, with demand measured as "frontier visits," is very high. This strongly suggests that participation in international tourism is closely linked to per capita income.

Japanese Population Growth and Demography

The coefficient on the size of the Japanese population suggests an increasing participation rate with the growth in the population. However, given the collinearity with income there are reasons to be cautious about this coefficient. There is some subtlety in the population elasticity. In this regression, the variable SOLD is also included, defined as the share of the population that is over the age of 45. In effect, the simple elasticity on population holds the fraction of the population over 45 constant while it increases the total population. If the demographic structure of the population is changing so that the share of the population that is older is increasing, this diminishes the tourism to the countries visited by the Japanese. The effect of an increase in the age of the population is substantial. An increase of 0.01 (from 0.24 to 0.25) means a reduction in frontier arrivals of 1.7 percent. Over the period 1972 to 1991, the share of the population in this cohort rose from 0.24 to 0.33. Of course, since the population also increased during this period and since the elasticity is high, the net effect depends on the appropriately weighted increases in the population and its structure. In the event, despite the high elasticities, the effect of the average annual change of the population and its composition is to *reduce* the amount of Japanese tourism to our sample of countries by about 3 percent per year!

Our interpretation of the population variable may also reflect the dispersion of income as well as a simple scale effect. In the absence of data reflecting income-specific population cohorts, the growth of segments of the population with much higher levels of income will lead to highly non-linear associations between total population growth and tourism. The idea of a threshold effect between income and population is difficult to model at this level of aggregation, but it may add to the story as well. A shift in total income that is a shift towards higher-income groups will increase demand to the extent that the marginal propensity to spend on tourism is higher for high-income groups.

Size of the Destination Country

The variable LPOP reflects the (log of the population) size of the destination country for Japanese tourists. The underlying assumption is that, other things being equal, there will be a greater tendency to visit larger countries than smaller ones. Although we could have considered physical size, we felt that it was at least as likely that population would be the more important characteristic capturing unmeasured attractiveness of a country. Since the physical size of the country is fixed and site-specific physical effects would be captured in the country-specific dummies, we let the destination-country population act as a proxy for the "people" characteristics of tourism such as cultural attractions.

Fixed Effects

The country dummies reflect the country-specific behaviour associated with unidentified tourist magnetism. There are undoubtedly a host of site-specific characteristics for each country that we cannot identify. Our assumption is implicitly that these will be captured in the coefficient of the dummy variable. Naturally, this assumption leaves a lot to be desired. If the data were adequate, we would also check for the development of new sites within each country. But with only a dozen countries in our cross-section and 20 observations in time series, we are already stretching the limits of the data. Given these limitations, however, the fixed effects are interesting. There are four countries in the Asia Pacific basin for which we have data – Australia, Canada, New Zealand, and the United States. The fixed effects suggest that New Zealand is the most "attractive," with Canada and Australia about even. The coefficients of 5 on both countries indicate that Japanese tourism to both countries is about five times that of travel to the United States after having corrected for distance and country-size effects. To this extent, all three countries have a revealed bilateral comparative advantage in tourism exports to Japan relative to the United States and to virtually all European destinations. Unfortunately, these data do not allow us to compare these countries with other parts of Asia as competing Japanese tourist destinations.

Cost of Travel

The cost-of-travel variable, LJD, is the real cost of an airline trip to the destination country, measured in real yen. The coefficient in the regression is insignificant and of the "wrong" sign. The average rate of growth in the cost of travel from Japan is –6.1 percent per year. In Japanese terms, travel costs were falling. (Of course, the costs of going to each country differs as the cost per mile is multiplied by the distance traveled.)

Nights Spent in Hotels

If we look at the number of nights spent in hotels, the results are similar to those associated with arrivals at frontiers, but with a few differences.[13]

TABLE 20

NIGHTS SPENT IN HOTELS: DEPENDENT VARIABLES

VARIABLE	ESTIMATED COEFFICIENT	STANDARD ERROR	T-RATIO 222 DF
LJD	0.26674	0.9925E-01	2.688
LPOP	0.14631	0.3457E-01	4.233
LJDEPT	0.41150	0.8530E-01	4.824
LJPCY	0.30677	0.1007	3.045
LRJX	0.52930	0.1097	4.825
SOLD	2.4233	1.244	1.947
Austria	−1.6675	0.8535	−1.954
Denmark	−2.1279	0.8271	−2.573
Finland	−2.7644	0.7679	−3.600
France	−1.2568	0.3232	−3.889
Germany	−0.81653	0.2544	−3.210
Italy	−4.9236	0.8669	−5.679
Netherlands	−2.4365	0.3507	−6.949
Portugal	−6.4560	0.6438	−10.03
Spain	−4.4454	0.6060	−7.335
Switzerland	0.45687	0.6588	0.6935
Turkey	−7.5303	0.7830	−9.617
Constant	3.7081	2.374	1.562

Coefficient estimates are reported in Table 20. We include in the regression the number of departures from Japan, LJDEPT. In effect, the regression asks the question, "Once you have made the decision to take a trip to a particular country, how many nights do you spend in a hotel?" Unfortunately, we do not have data on any destinations in the Asia Pacific region for this measure of demand.

First, as would be expected, the number of departures is related to the number of nights spent in hotels. Second, the number of nights spent tends to increase with age. The distance measure is now significantly positive. This is interesting and consistent with theory, insofar as it suggests that once you are in the country, the farther you have had to travel, the more nights you will tend to spend in that country. A 10 percent increase in distance costs implies a 2.7 percent longer stay, based on a point estimate of the coefficient. Third, as expected, a higher real exchange rate is consistent with more nights spent abroad. A 10 percent higher real exchange rate means a 5.3 percent increase in stay durations. Higher real income also means longer stays. A 10 percent increase in income leads to a 3 percent increase in the number of nights spent in hotels. Each of these results suggests the importance of making the distinction between intensive versus extensive decision margins. Clearly Canada, for example, relative to East Asian destinations has a comparative disadvantage with respect to distance. If, however, the decision is made to travel to Canada, we have a relative advantage in attracting visitors for longer stays. The results suggest that both the real-exchange-rate effect and the age effect work in favor of a long-distance destination such as Canada.

Finally, an increase in the age of the population also leads to an increase in the number of nights spent at hotels. Recall that this result is conditional on having made the departure from Japan. It is consistent with the observation above that the number of tourist arrivals at frontiers declines when a larger fraction of the population is older.

As might be expected, tourists from Japan tend to spend more nights in larger countries. This is revealed by the positive coefficient on LPOP, the (log) of the population in the country in which the nights are spent.

ARRIVALS AT ALL ACCOMMODATIONS

A DIFFERENT MEASURE OF DEMAND IS ARRIVALS of Japanese at all accommodations. This is not the same as nights spent at hotels as it does not total the number of nights, but records arrivals. As such, it may be subject to the same limitations as frontier arrivals. The regression results in Table 21 are very similar to those for frontier arrivals. In particular, even though the sample of countries is quite different (see Table 16 for the list of countries for which data are available), all the economic variables have the same signs and significance as for frontier arrivals. The data are consistent with high income and population

TABLE 21

ARRIVALS AT ALL ACCOMMODATIONS: DEPENDENT VARIABLES (WITHOUT LJDEPT)

VARIABLE	ESTIMATED COEFFICIENT	STANDARD ERROR	T-RATIO 127 DF
LJD	0.23150	0.1257	1.841
LJPCY	3.3319	0.4989	6.678
LPOP	2.6681	0.5759	4.633
LRJX	0.58701	0.1481	3.963
LJPOP	4.6189	1.848	2.499
SOLD	−11.899	4.292	−2.772
Austria	45.797	8.848	5.176
Germany	995.02	213.0	4.671
Italy	2.0208	0.2043	9.893
Netherlands	993.15	213.0	4.662
Portugal	5.1065	0.8588	5.946
Switzerland	48.530	8.831	5.495
Constant	−157.07	38.38	−4.092

Table 22

Arrivals at All Accommodations: Dependent Variables (With LJDEPT)

Variable	Estimated Coefficient	Standard Error	T-Ratio 127 DF
LJD	0.25775	0.1419	1.816
LJPCY	2.0807	0.4131	5.036
LPOP	2.8137	0.5860	4.802
LRJX	0.55574	0.1591	3.492
LJDEPT	0.28504	0.1206	2.364
SOLD	−3.9730	2.058	−1.930
Austria	47.968	8.961	5.353
Germany	1048.8	216.7	4.841
Italy	2.0114	0.2037	9.875
Netherlands	1047.0	216.7	4.832
Portugal	5.2883	0.8608	6.144
Switzerland	50.654	8.923	5.677
Constant	−68.023	11.50	−5.913

elasticities, a positive response to relative prices (through the real exchange rate), and a decrease in arrivals associated with age. The distance variable is of the wrong sign although not statistically significant at the usual levels.

In Table 22, the arrival data are treated slightly differently. In this case, we include the departure variable, LJDEPT. That is, we pose the question of the effect of different variables on the number of arrivals at all types of accommodation in a country conditional on the aggregate number of departures from Japan. Again the evidence is consistent with the earlier regressions. The distance measure is positive but statistically insignificant at the usual levels. There is a negative sign associated with the fraction of the population that is older. Again, this is not unreasonable, and it stands in contrast to, and support of, the number-of-nights data discussed above. Although a tourist will spend more nights at a hotel, the number of arrivals tends to decline with age.

Conclusion

Combining the price and income effects on both nights and arrivals data leads to some interesting insights. Participation in international tourism, as measured by arrivals, is highly sensitive to income, with an elasticity close to 3; this should be placed in context, however, by the observations showing that total Japanese participation in international tourism remains well below European rates. Given the decision to travel abroad, the additional income effect on the number of nights spent in hotels, which can be viewed as a proxy for additional demand at the extensive margin, is much lower, with an elasticity around 0.3. The combined effect implies that per capita income growth, at least in the intial stages, has its most powerful effect on the participation decision. This suggests, for example, the important role that a reduction in long-distance transport costs might have, given their weight relative to total income. The effects of the evolution of the real exchange rate are similar although not as dramatic. Both participation and nights are about equally sensitive to changes in the bilateral real exchange rate but less so than income. Note that this result is consistent with the model described earlier, which suggests a positive effect of the real exchange rate but one that is quantitatively less strong than that of income.

THE ARRIVALS OF TOURISTS TO CANADA IN 1972-92: A 17-COUNTRY ANALYSIS

TABLE 23 DISPLAYS THE RESULTS of a set of pooled regressions that examine the inflow of tourists to Canada from 17 different countries, with the number of tourists acting as the dependent variable. All variables are in natural logs so that the coefficients are elasticities, and there are 21 observations for each country – one for each year from 1972 to 1992. The first regression model is a function only of the price and income variables that are expected to influence tourism patterns. The second regression includes the population of the sending countries as an additional scale variable, and the third model adds a fixed effect for each of the 17 countries sending tourists to Canada. The behaviour of the coefficients gives an indication of the forces acting on them as different explanatory variables are introduced.

VARIABLES

Real Exchange Rate Indexes

LREXI represents the Canadian CPI relative to that of each of the 17 countries over the period. The relative CPIs are adjusted by the exchange rate between Canada and each of those countries, giving a real bilateral exchange rate. It is expressed as an index with a value of 100 in 1972, and there are variations over time associated with each country from that base. In our model, a higher relative CPI in Canada implies less tourism. Thus we expect a negative sign on the real exchange rate.[14]

The variable LUSREXI is an index of relative prices between the CPI in the United States and each of the other countries, adjusted by the U.S./other

Table 23

Inflows of Tourism to Canada from 17 Countries, 1972-92

Variable	Estimated Coefficient	Standard Error	T-ratio[1]
Model 1			
LRYI	1.2603	0.2029	6.212
LTCRY	−0.38073	0.7941E-01	−4.794
LREXI	−0.89025	0.1562	−5.701
LUSREXI	0.35933	0.1902	1.890
Constant	11.056	0.1946	56.81
Model 2			
LRYI	0.92207	0.1645	5.604
LTCRY	−0.63732	0.3659E-01	−17.42
LREXI	−0.92315	0.1514	−6.097
LUSREXI	0.63418	0.1704	3.722
LPOP	0.64725	0.4387E-01	14.75
Constant	0.73951	0.7164	1.032
Model 3			
LRYI	0.73278	0.2114	3.466
LTCRY	−0.51673	0.9251E-01	−5.586
LREXI	−0.94903	0.1415	−6.706
LUSREXI	0.58478	0.1720	3.401
LPOP	1.7454	0.3470	5.030
Australia	0.70199	0.4694	1.495
Belgium	−0.44962E-01	0.6229	−0.7219E-01
Denmark	0.75827	0.8468	0.8954
France	−1.1723	0.1672	−7.010
Germany	−1.5016	0.1716	−8.749
Greece	0.30443	0.6248	0.4873
Hong Kong	2.1027	0.8812	2.386
India	−4.6667	1.009	−4.625
Italy	−1.7622	0.1327	−13.28
Japan	−2.3576	0.3001	−7.857
Netherlands	0.51623	0.5045	1.023
New Zealand	2.2901	1.010	2.268
Portugal	0.17308	0.6347	0.2727
Spain	−2.3344	0.2148	−10.87
Sweden	0.31788	0.6853	0.4639
Switzerland	1.1793	0.7654	1.541
Constant	−17.621	6.213	−2.836

1 Model 2 = 351 DF; Model 3 = 335 DF.

country exchange rate. As previously, this real exchange rate is indexed at 100 in 1972.

We would expect a rise in the U.S. CPI relative to the Canadian CPI to induce more tourists to visit Canada. That is, we expect Canadian tourism is a substitute for U.S. tourism from third countries. Thus the coeffient on LUSREXI is expected to be positive.

Real Income

The variable LRYI is the log of an index of real per capita income in each country. As before, we have a "units" problem. Again, we take 1972 as the base year and look at variations over time.

Travel Costs

The variable LTCRY is the log of travel costs relative to domestic income. It is the fraction of per capita income expended in each country to get to Canada. For example, for Indians it is about three times the average income; for Swedes, it is about 5 percent. By measuring travel costs as a fraction of per capita income, we do not need to be concerned about the units as we were for real income and exchange rates.

RESULTS

MODEL 1 CONTAINS ONLY THESE RELATIVE-PRICE and quantity variables as explanatory forces in Canadian tourism. The regression is consistent with our understanding of the coefficients. Although the data are pooled, there are no fixed effects in this model. The estimates suggest an income elasticity of about 1.2; a cost of transportation elasticity of –0.38 with respect to the share of income absorbed by the trip to Canada (in measuring distance, we have taken whichever port, Vancouver or Toronto, is closer to the supplier country); and real-exchange-rate effects that are consistent with the United States' being a substitute for tourism to Canada. The own-price elasticity is strong relative to the cross-price effect from the United States.

Model 2 adds the population of the sending countries to the regression, thus adding a scale effect. As is apparent from Table 23, by introducing the population the value of the elasticity on per capita real income falls to 0.9 and the coefficients on each of the other terms are strengthened in both absolute size and significance. The coefficient/elasticity on the (log of) population is about 0.6. In other words, an increase in the size of the country of origin will increase the number of tourists but less than proportionally.

The final model, Model 3, adds a fixed effect to each country to capture individual variations not found in the other variables. One can imagine circumstances such as historical associations playing a role in tourism with respect to the United Kingdom or to other Commonwealth countries, for example. Indeed, the assumption of the fixed-effect approach is that there is at least some element of individuality not modeled in each country's demand for Canadian tourism services. In the model, the United Kingdom is the country whose dummy variable is not displayed, so the coefficients are relative to that country.

The effect of the introduction of individual country parameters is to depress the income elasticity slightly and to raise the population elasticity, but to leave the rest of the coefficients relatively unchanged. The tripling in the size of the population elasticity is interesting. Once we have identified the countries' "fixed" contribution to Canadian tourism, the growth of their populations generates a very substantial increase. If we view these 17 countries as special, then we might have more confidence in the Model 2 elasticity of about 0.6, but if we think that they are typical of those countries supplying Canadian tourism, then population growth is a powerful force in generating tourism.

Chinese Tourism in Canada: Some Illustrative Calculations

IN THIS SECTION WE REPORT some simulations on growth in both arrivals and number of nights for tourists from China coming to Canada. The model uses the base parameter values derived in the discussion of Japanese tourism. The central idea is to assume that the growth and tourism experiences found in the data for Japan will be replicated for China – that is, the same parameter values will apply. It is important to remember, however, that China has both a much lower income and a much larger population than did Japan in 1972. Current tourism arrivals in Canada from China are about 20,000 to 25,000 per year – one of Canada's smaller markets. However, it is the prospect for future growth in this market that is the issue here.

The baseline parameter values used in the model are as follows:

Parameter	Arrivals	Nights
Income elasticity	1.5	0.3
Own-price elasticity	−0.9	−0.3
Age effect	−10.0	2.0
Population elasticity	1.3	

The real exchange rate is assumed to evolve at one-half the rate at which per capita growth rates between China and Canada differ. The per capita growth rate for China is driven by a convergence equation: growth starts at 11 percent in 1990 and falls to 8 percent by 2020 under the high-growth scenario. Under the low-growth scenario, it starts at about 6 percent and falls to 5.5 percent.

The Chinese income is about one-third the Canadian level by 2020 under the high-growth scenario, but barely moves under the low-growth scenario. The per capita growth rate in Canada is taken to be 1.2 percent.

The results of these simulations are shown in Figures 2 to 8 for a variety of assumptions that depart from these central-case parameter values. Figure 2 shows the effect on arrivals of both high- and low-growth scenarios. The remaining figures explore a variety of assumptions in the model under the high-growth scenario. In almost all cases, growth rates of both arrivals and nights produced by the model are in the double digits – even getting close to 40 percent in a high-income scenario.[15]

In Figures 3 and 4 we see the effect of varying the age sensitivity of the model. Notice the much wider variation between arrivals versus nights in the low age sensitivity model. The aging population in China under the high age sensitivity assumption has quite a dampening effect on the growth in arrivals.

In Figures 5 and 6 the sensitivity of growth rates to variations in the price elasticity are shown. The key elasticity here is that on the real exchange rate between China and Canada. Under the assumed growth hypothesis, the real exchange rate between the two countries is expected to appreciate (for China) by a factor of 3.7 over the 30-year period. We see that under a high price elasticity scenario, growth is about 10 percent higher in both demand variables – arrivals and total nights spent in hotels.

In Figures 7 and 8 the effects of varying income elasticities are shown. Here we see that lowering income elasticities produces a corresponding strong

FIGURE 2

ARRIVALS AT FRONTIERS:
HIGH- AND LOW-GROWTH SCENARIOS

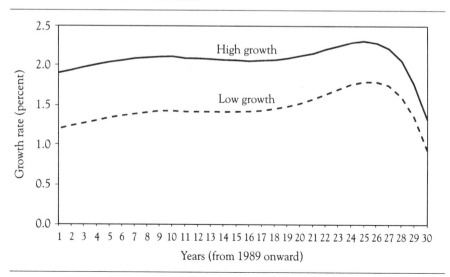

FIGURE 3

HIGH AGE SENSITIVITY OF ARRIVALS AT FRONTIERS AND NIGHTS SPENT AT HOTELS

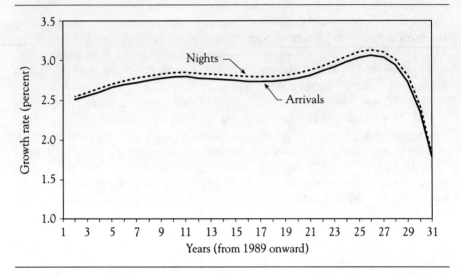

FIGURE 4

LOW AGE SENSITIVITY OF ARRIVALS AT FRONTIERS AND NIGHTS SPENT AT HOTELS

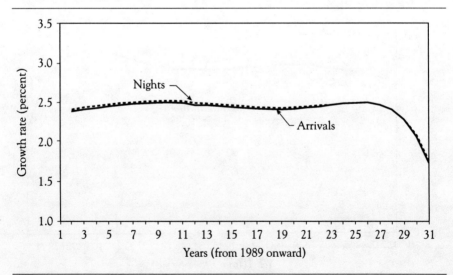

Figure 5

High Price Elasticities of Arrivals at Frontiers and Nights Spent at Hotels

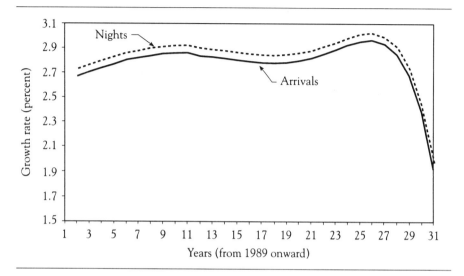

Figure 6

Low Price Elasticities of Arrivals at Frontiers and Nights Spent at Hotels

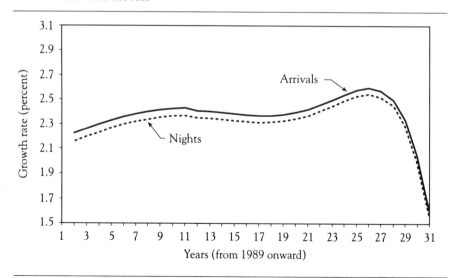

FIGURE 7

HIGH INCOME SENSITIVITY OF ARRIVALS AT FRONTIERS AND NIGHTS SPENT AT HOTELS

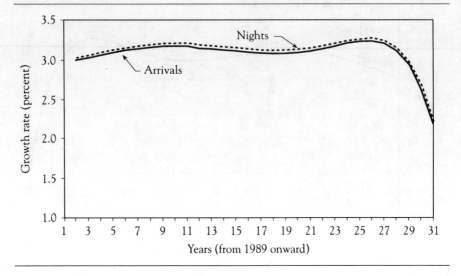

FIGURE 8

LOW INCOME SENSITIVITY OF ARRIVALS AT FRONTIERS AND NIGHTS SPENT AT HOTELS

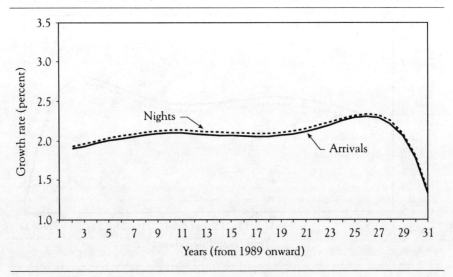

reduction in growth rates and in the growth of arrivals relative to the central-case assumptions.

Conclusion

THE TOURISM MARKET IN THE ASIA PACIFIC REGION is bound to be one of the most dynamic in the region, assuming growth maintains its recent pace. From the Canadian perspective, tourism exports to the Asia Pacific region may help eventually to offset the traditional bilateral trade deficit in merchandise trade with that area. The evolution of this market will be sensitive to a number of important developments. These include: the rate of economic growth in the region; the evolution of real exchange rates bilaterally between Canada and the economies of East Asia; potential developments in trans-Pacific transport costs; and demographic trends in the region. It is clear that intra-regional tourism in the western Pacific will be a major competitor for long-distance tourist destinations, including Canada. The two models of tourism flows discussed in this paper – the Japanese departures model and the Canadian arrivals model – suggest that if economic growth in East Asia replicates the postwar Japanese experience, Canada's tourism sector will be a major beneficiary of this growth.

Endnotes

1. We are grateful to our discussant Tae Oum for comments and to conference participants. Please send all correspondence related to paper to authors at the Department of Economics, Simon Fraser University, Burnaby, B.C. V5A 1S6; our e-mail addresses are *rharris@sfu.ca* and *easton@sfu.ca*.
2. Useful surveys include Crouch (1995), and Pearce (1995).
3. These figures are from Naisbitt (1994), who cites the World Tourism Organization.
4. "Asia Goes on Holiday." *The Economist* (20 May 1995).
5. These estimates are contained in *Travel Industry World Yearbook* (1994), pp. 8-10.
6. For confirmation of this, see the Hejazi and Trefler paper in this volume.
7. See Crouch (1992, 1995).
8. The "new trade theory" of intra-industry trade in differentiated products is described in Helpman and Krugman (1985). The more traditional approaches are well known but tend rarely to be used in the tourism literature, which remains largely based on partial-equilibrium models.

9   See Copeland (1991).
10 Pearce (1995), ch. 3, provides some data on circuit tourism. He suggests that for long-haul (cross-oceanic) travel, circuit tourism is fairly common. Accounting for this effect in empirical work is almost never done.
11 The advantage of this technique is that it enables the coefficients to be estimated by taking heteroscedasticity, autoregression and contemporaneous error structures into account.
12 In Table 18, the United States is the residual country and is not identified explicitly with a dummy. There is, of course, an additional limitation to the analysis: we do not have an exhaustive list of countries to which Japanese tourists go for their vacations.
13 The sample is different in this case as we do not have frontier arrivals data and nights spent in hotels data for the same countries. A number of experience-hysteresis variables were tried (such as cumulative visits), but all with insignificant results. On a straight time series approach, however, including lagged dependent variables is suggestive of strong persistence effects, but in the absence of tight theoretical constraints and given the short time series available, it would be unwise to use such a model to make inferences regarding the possible presence or absence of hysteresis.
14 The mechanical reason for choosing an index is that different CPIs expressed in home-currency units actually correspond to different commodity baskets. For example, the Canadian CPI relative to the French CPI is adjusted by the exchange rate so that we have a magnitude of the French franc/Canadian dollar relationship. If we were looking at Japanese tourism, then the units would be yen/Canadian dollar. Although both are correct, unless we stack the regression and use a different relative-price term for every country, we need a common dimensionality even though we are looking at the log of the index. Another alternative would be to construct true purchasing-power-parity indexes for the CPI. In any case, by choosing an index we lose the cross-sectional variation associated with the base year. This scaling problem among countries and their currencies clearly poses a dilemma for anyone examining cross country data, and we have chosen this mechanism to deal with it.
15 The high- and low-growth scenarios involve changing the relevant elasticity parameter from its central value by one standard error in the appropriate direction. The convergence coefficient was set at 0.1 and 0.05 for the high- and low-growth scenarios. In Figures 3 and 4, the age elasticity on arrivals was respectively set at 13 and 7, and the elasticity on nights was set at 3 and 1; in Figures 5 and 6, the price elasticity on arrivals was respectively set at 1.35 and 0.45, and the elasticity on nights at 0.45 and 0.15; in Figures 7 and 8, the income elasticity on arrivals was respectively 2.0 and 1.0, and the elasticity on nights was 0.4 and 0.2.

BIBLIOGRAPHY

Copeland, Brian R. "Tourism, Welfare and De-industrialization in a Small Open Economy." *Economica*, 58, 232 (1991):515-29.

Crouch, G.I. "The Effect of Income and Price on International Tourism." *Journal of Tourism Research*, 19 (1992):643-64.

_____. "The Study of International Tourism Demand: A Review of the Findings." *Journal of Travel Research* (summer 1995):12-23.

Dwyer, L. and P. Forsyth. "Foreign Tourism Investment: Motivation and Impact." *Annals of Tourism Research*, 21, 3 (1994).

Helliwell, J.F. "Asian Economic Growth," in *Pacifc Trade and Investment: Options for the 90s*. Edited by W. Dobson and F. Flatters. International Development Studies Program, John Deutsch Institute for the Study of Economic Policy. Kingston, Ont.: Queen's University, 1994.

Helpman, E. and P. Krugman. *Market Structure and Foreign Trade*. Cambridge, Mass.: MIT Press, 1985.

Japan. Statistics Bureau Management and Coordination Agency. *Japan Statistical Yearbook*. Various years.

Lufthansa. *OAG Desktop Flight Guide Worldwide Edition*, 17, 2 (April 1992).

Mak, J. and K. White. "Comparative Tourism Development in Asia and the Pacific." *Journal of Travel Research* (summer 1992):14-23.

Naisbitt, J. *Global Paradox*. New York: Avon Books, 1994.

Organisation for Economic Co-operation and Development. *Tourism Policy and International Tourism in OECD Member Countries, 1990-91*. Paris: OECD, 1991.

Pacific Asia Travel Association. *Annual Statistical Report, 1994*.

Pearce, Douglas. *Tourism Today: A Geographical Analysis*. Second edition. Essex, U.K.: Longman Scientific and Technical, 1995.

Peterson, John. "Export Shares and Revealed Comparative Advantage: A Study of International Travel." *Applied Economics*, 20 (1988):351-65.

Statistics Canada. CANSIM Data Base. Monthly arrivals into Canada, series numbers D145807 Belgium, D145808 Denmark, D145810 Germany, D145811 Greece, D145813 Netherlands, D145814 Portugal, D145815 Spain, D145816 Sweden, D145817 Switzerland, D145818 UK, D145822 Hong Kong, D145823 India, D145825 Japan, D145836 Australia, D145837 New Zealand, D164016 Finland, D164021 Iceland, D164022 Ireland, D164033 Norway, D164042 Turkey, D164116 China, D164119 Indonesia, D164124 South Korea, D164136 Philippines, D164139 Singapore, D164142 Taiwan, D164143 Thailand.

Travel Industry World Yearbook: The Big Picture, 1993-1994. Volume 37. New York, NY: Child and Waters Inc., 1995.

United Nations. Department of International Economic and Social Affairs. *The Sex and Age Distributions of the World Populations*. 1994 revisions. New York: United Nations, 1994.

Walton, J. "Tourism and Economic Development in the ASEAN," in *Tourism in South-East Asia*. Edited by M. Hitchcock, V. King and M. Parnwell. London: Routledge, 1993.

Witt, S., M. Brooke and P. Buckley. *The Management of International Tourism*. London: Unwin Hyman, 1991.

World Bank. "World Tables," in World Bank Data on Diskette, 1991, 1993.

Comment

Tae H. Oum
Faculty of Commerce
University of British Columbia

The paper:

- reviews key trends in tourism – worldwide, by continent with a focus on Asia Pacific tourism, and tourism exports by Canada;
- develops a model of international tourism in which tourism is treated as an internationally tradeable service; and
- estimates several interesting reduced-form models on Japanese tourist destinations and tourist arrivals in Canada.

The key findings are as follows:

- Asian countries' per capita income and real exchange rates vis-à-vis the Canadian and U.S. currencies are main drivers for Canada's international tourism exports to Asia.
- Transport costs and changes in the population and its age structure also impact on Asia's international tourism imports.
- Intra-regional tourism in the eastern Pacific – including Canada – will be a major competitor for long-distance tourist destinations of East Asians.

The implications for Canada's tourism exports are as follows:

- The tourism market in the Asia Pacific region is bound to be the fastest-growing market because economic growth in the region is likely to continue at the pace observed in recent years.
- However, Canadian tourism exports to Asian markets will be sensitive to: a) the rate of economic growth in the region; b) the evolution of real exchange rates between Canada and Asian countries; c) potential changes in transpacific transport costs; and d) demographic trends in Asian countries.

Overall comment on the paper:

- Overall, this is an excellent work. Particularly, I would like to commend the authors for developing a theoretical model of international tourism explicitly from the theory of international trade. Their model is based on the unbalanced economic growth between the manufacturing sector and the service sectors. In particular, they find that tourism prices (relative to prices of manufactured goods) are rising faster in developing countries (Asia) than in developed countries (Europe, Canada, the United States, etc.) because of the Balassa-Samuelson effect – i.e., there is a larger differential in productivity growth between manufacturing and services in developing countries than in developed countries.

- Their two-country model ("a" denotes an Asian developing country, and "o" denotes other developed country) shows that the rate of change in the value of international tourism exports of country "a" depends on: 1) relative productivity growth in tourism in both countries; 2) population growth in country "a"; and 3) the income elasticity of demand and the own- and cross-price elasticities (between international and domestic tourism) of tourism demand in country "a."

Comments on the empirical models:

- Need for more explicit linkages between theoretical results and empirical models.

  The authors estimate Japanese tourist destination models (arrivals at frontiers, nights spent in hotels, and all accommodations) and the tourist origin models to Canada (from 17 countries). Although the authors tried to include some variables implied in the theoretical parts of the paper, the empirical models lack explicit linkage with the theoretical results. For example, the relative prices of tourism services in alternative destination countries (including home countries) are not included in any of the models. It is possible that the data did not allow them to create a multilateral tourism price index, but I am sure that data on the relative prices of tourism and manufactured goods in countries of origin and destination are available. As implied in their theoretical model, the inclusion of such variables could improve empirical results. Their models include transport costs, exchange rates, incomes, and population – variables that are usually included in traditional models estimated without theory. It would be highly desirable to establish a clearer linkage between the theoretical model and the empirical models.

- Japanese destination model: arrivals at frontiers.

  There appears to be a minor problem in the regression model reported in Table 18. I do not see the reason why $LPOP$ (population of destination country) is relevant for Japanese destination choice. The *LPOP* variable appears to make the country effects behave strangely. For example, Iceland's country effect is the highest of all. I do not see why people would want to go to Iceland for tourism more than any other country if economic conditions (variables) are identical. This strange result may have been caused by the very small population of Iceland, which can explain only a tiny fraction of the variation in the dependent variable while *LPOP* accounts for a large variation in the case of countries with large populations (e.g., the United States, Spain). The high country effects for Canada, New Zealand, and Australia and the negative effect for Italy are also suspect. Removal of *LPOP* is likely to produce more reasonable results.

- Use of Japan model for prediction on Asia.

The authors use the Japan model to predict future Chinese tourism to Canada. Although Japan was chosen mainly for data reasons, that country is in many ways quite different from the rest of Asia. The Japanese people are far more careful and selective in spending than the nationals of most other countries at a similar stage of economic development. Japan still has a much lower intensity of international tourism than most other advanced countries. The intensity of international tourism in other Asian countries is likely to be far higher than Japan's was at a similar economic development stage.

Michael Baker & Dwayne Benjamin
Department of Economics
University of Toronto

Asia Pacific Immigration and the Canadian Economy[1]

AS IS EVIDENT IN THE DIVERSITY OF PAPERS PRESENTED at this conference, the importance of the Asia Pacific economies relative to the world economy has grown over the past 20 years. From a Canadian perspective, the Asia Pacific region has become a major source of imports, as well as a significant export market. Investment flows increasingly in both directions between Canada and the Pacific Rim countries. In addition to these arm's-length economic transactions, however, and in parallel with the growth of Asia Pacific economies, the region has become a significant source of new immigrants to Canada. While the impact of this phenomenon extends well beyond the economic realm, our objective here is to examine some elements of their relationship with the Canadian economy.

We explore this relationship from two perspectives. First, we examine the degree to which the performance of the Canadian economy affects immigrant inflows from the Asia Pacific region, and then assess the contribution that these immigrants make to the Canadian economy. We focus on three areas in which these contributions take place. The most direct impact of immigrants from the Asia Pacific region occurs in the labour market, through the skills and human capital they bring to Canada. The best measure of this contribution is the value placed on those skills by the market – the earnings they command. Second, as suggested by recent research, immigrants provide links between their countries of origin and their adopted countries. These links may arise through a variety of channels – through knowledge of source-country goods markets that facilitates exports from Canada, and through preferences for source-country goods that increase imports. This information can lead to increased trade between Canada and the Asia Pacific region, or it may extend beyond trade and affect the patterns of foreign investment in both areas.

Clearly, these areas do not exhaust the ways in which Asia Pacific immigration may have affected the Canadian economy. We cannot estimate what cannot be measured, so that possible externalities, such as the contribution of immigrants to the entrepreneurial talent pool, cannot be assessed. Similarly, we cannot easily measure the impact of immigration on the outcomes experienced by Canadian-born residents. We also ignore the impact of immigrants on consumption patterns, especially with respect to housing or to congestion in cities. Since we are looking for Canada-wide evidence, we miss some of the possibly important more disaggregated effects, such as the impact of Asia Pacific immigration on Vancouver or Toronto. Indeed, one obvious conclusion from our study is that a broad topic such as the links between Asia Pacific immigration and the Canadian economy will obscure considerable detail. Nevertheless, as a first approach, the broad picture outlined here represents a significant improvement over the little that is known at present about this important group of immigrants.

The paper begins with an overview of recent immigration from the Asia Pacific region and shows that there are significant differences in the patterns of immigration from the countries that comprise the region. A similar view emerges from the analysis of the profile of the stock of Asia Pacific immigrants in Canada. Then, using labour market outcomes as a measure of assimilation into the Canadian economy, we find that Asia Pacific immigrants assimilate at the same rate as those from other parts of the world but, perhaps surprisingly, start at somewhat lower earnings levels. Again, this generalization masks substantial heterogeneity. Finally, we look at international linkages that are correlated with the stock of immigrants in Canada. We find that there is a strong correlation between both imports and exports and the stock of immigrants from the Asia Pacific region. On the other hand, we find no evidence that foreign investment patterns in either direction are related to past immigration patterns.

Throughout our analysis we use evidence of immigration patterns and outcomes in the United States as a point of comparison. Methodologically, this enables us to disentangle North American/immigrant-source-country effects from a "purer" immigrant effect. Of course, in bringing in U.S. evidence, we occasionally confront serious data comparability problems that place some limitations on the number of Asia Pacific countries that we can include in our analysis. Nevertheless, the U.S. evidence serves an important role in either dismissing or reinforcing results based on Canadian evidence alone.

IMMIGRATION FLOWS TO CANADA

IN DESCRIBING AND CHARACTERIZING Asia Pacific immigration flows to Canada, we focus on their broad determinants, especially the extent to which they are sensitive to economic conditions. Any analysis of immigration flows must address both supply and demand factors. On the demand side, flows of immigrants to Canada are determined by immigration policy, in terms of both the aggregate number of immigrants admitted and the types of immigrants that receive preferential treatment. Of course, these policy parameters themselves are sensitive to Canadian economic conditions, as well as to international political considerations, such as refugee crises. On the supply side, immigration levels from particular countries will depend on the relative economic benefits expected from migrating to Canada versus staying at home or migrating elsewhere. While viewing immigration to Canada through the lens of a supply-and-demand model may be helpful, as we shall see the non-economic (i.e., historical or political) determinants of migration may be more important. One way to help isolate some of these factors is to compare immigration patterns between Canada and the United States. This will enable us to net out the source-country-specific supply conditions, leaving historical, economic, and policy differences between Canada and the United States as the determinants of immigration patterns.

As a starting point, we look at aggregate Asian immigration to Canada and the United States.[2] Administrative statistics go further back for all of Asia than for particular countries within Asia, reflecting the fact that relatively few Asians emigrated to North America in the mid-20th century. The primary difference between the Asia aggregate and the Asia Pacific group is the inclusion of immigrants from South Asia (India, Pakistan, etc.) and the Middle East — two substantial groups. Nevertheless, the aggregate data do provide an approximation of the longer-term trends in immigration flows (Figure 1). The most obvious pattern is the upward trend in Asian immigration to both Canada and the United States. While the overall number of immigrants is higher in the United States (currently just over 300,000 versus over 100,000 in Canada), Canadian levels are larger as a percentage of the population. Asian immigration to Canada accelerated in the mid-1960s, primarily due to new selection criteria that ended discrimination against these immigrants (Green, 1976, 1995). As a share of total immigration, Asians now represent almost 50 percent for Canada and about 30 percent for the United States. The U.S.-Canada difference holds even after dropping Mexican immigrants from the U.S. total, so that at least since the late 1980s, Asian immigrants have comprised a larger share of all immigrants to Canada.

Asia Pacific immigration flows as a percentage of Asian immigration are reported in Figure 2 for the shorter time period for which the data are more detailed. While that share is generally higher in the United States, it has been trending downward over the period. In Canada, the Asia Pacific share displays considerable variation over time (between 55 and 65 percent) and rises sharply in the late 1980s. This latest trend can largely be explained by immigration from one source country — Hong Kong. Immigrants from the British colony now represent 20 percent of all Asian immigrants to Canada, a level much higher than the historical average. The level of immigration from Hong Kong, however, has always been higher in Canada than in the United States, where it represents about 3 percent of the flow of Asian immigrants. China

FIGURE 1

ASIAN IMMIGRATION TO CANADA AND THE UNITED STATES, 1955-92

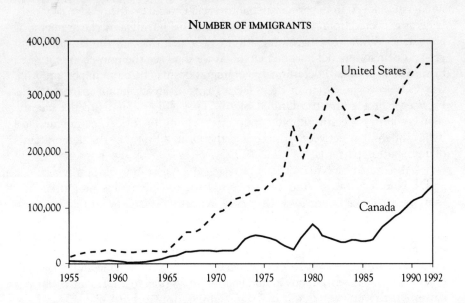

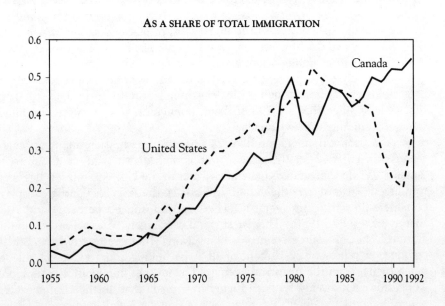

FIGURE 2

ASIA PACIFIC IMMIGRATION AS A SHARE OF TOTAL ASIAN IMMIGRATION TO CANADA AND THE UNITED STATES, 1974-92

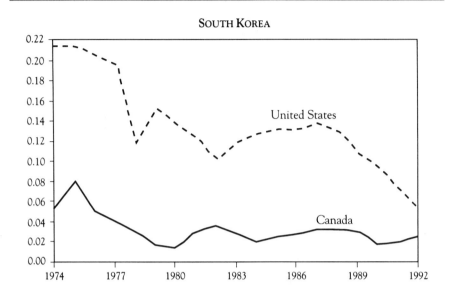

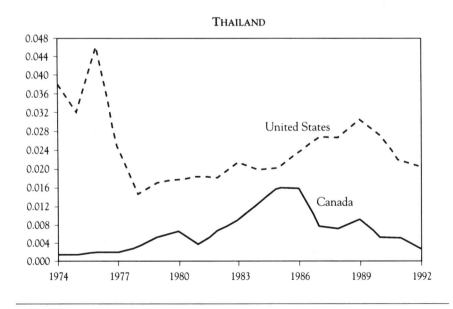

FIGURE 2 (CONT'D)

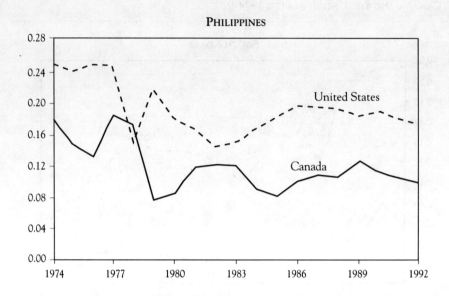

PHILIPPINES

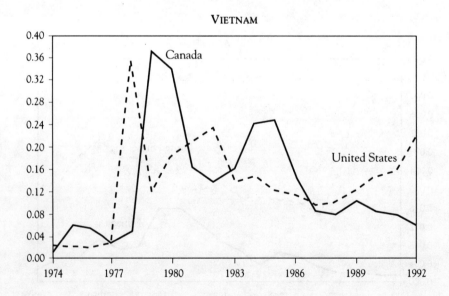

VIETNAM

Figure 2 (cont'd)

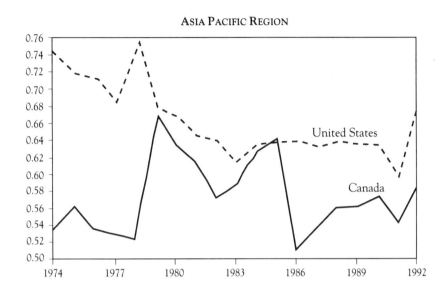

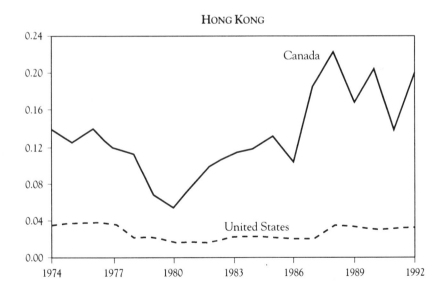

Figure 2 (cont'd)

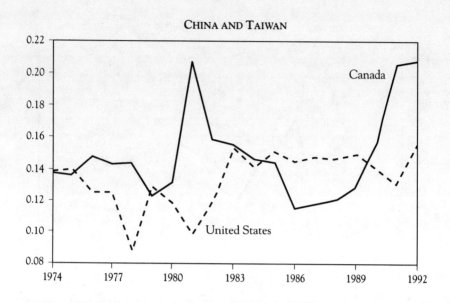

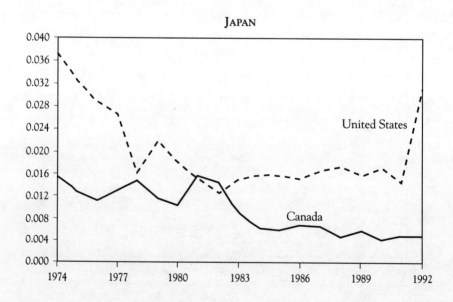

and Taiwan together are important source countries for Canada and the United States, especially in recent years: in 1993, they represented about 20 and 15 percent, respectively, of total Asian immigration to each country.[3]

The remaining Asia Pacific countries have historically provided a larger fraction of U.S. immigrants. Both Japan and Thailand are negligible source countries, together accounting for less than 5 percent of Asian immigrants to either Canada or the United States. South Korea has been more important, at one time providing 20 percent of Asian immigrants in the United States and 5 percent in Canada. These proportions are now around 7 percent in the United States and between 2 and 3 percent in Canada. The Philippines has also been an important source country, especially in the United States. By 1993, 18 percent of Asian immigrants to the United States, and 9 percent to Canada, were from the Philippines. Finally, while immigration from Vietnam was negligible in the early 1970s, after the conclusion of the war there was a considerable refugee flow to North America. The timing is different in the two countries, but both Canada and the United States saw Vietnamese immigration rise to about one-third of the Asian total in the late 1970s and early 1980s. At present, Vietnamese immigrants comprise about 10 percent of Asian immigrants to Canada and 20 percent of those who choose the United States. In summary, then, Asia Pacific immigration to Canada has primarily been from Hong Kong and China, whereas the other countries of South-East Asia have been more important for the United States.

The variations in the flows of Asia Pacific immigration to North America, both over time and across countries, may enable us to identify the effect of economic variables on these flows. To this end, we estimate a simple (but prototypic) supply-side migration model in which migration is an individual (or household) decision. We assume that migrants compare the expected utility of migration versus non-migration. If utility is otherwise independent of location, the expected utility should depend on the expected level of income of the individual as well as the risk of economic success or failure represented by the distribution of income.

When moving to a model based on aggregate data, we must make compromises in mapping theoretical variables into measured proxies. Not only are there aggregation assumptions, but data limitations are paramount, especially in dealing with international data that include developing countries. Our base specification is derived from the standard Harris-Todaro style model:

$$\ln IMM_{St} = \beta_0 + \beta_1 \ln PCGDP_{St} + \beta_2 \ln PCGDP_{Ct} + \beta_3 URATEC_t + \beta_4 \ln POPDEN_{St} + \varepsilon_{St}$$

where IMM_{St} is the number of immigrants to Canada from source country S at time t, $PCGDP_{St}$ is per capita gross domestic product (GDP) in the source country, $PCGDP_{Ct}$ is per capita GDP in Canada, $URATEC_t$ is the unemployment rate in Canada, and $POPDEN_{St}$ is the population density of the source country. The $PCGDP$ variables represent the expected income levels in the two countries and are approximations to the true income variables that affect individual migration decisions. Similarly, $URATE_C$ is a proxy for the risk of migration and is correlated with the probability of employment (success) upon arrival in Canada. $POPDEN$ will capture some of the congestion factors that may affect immigration decisions and, in some specifications, will control for the size of the potential immigrant pool. We restrict β to be equal across the Asia Pacific countries. While this may not be entirely valid, it does enable us to exploit cross-country variation in income levels and provides a summary of the overall sensitivity of Asia Pacific immigration to economic conditions.[4]

Estimates of this simple model are presented in column 1 of Table 1. The results are consistent with the intuitive view suggested by the supply-side model: immigration to Canada is strongly related to GDP differentials – i.e., to a decrease in source-country GDP and an increase in Canadian GDP. The Canadian unemployment rate has the anticipated negative sign but is insignificantly different from zero. Population density has its expected positive sign. We note, however, that Canadian GDP may capture both supply and demand effects. Indeed, our interpretation of these reduced-form coefficients as supply parameters is more by assumption than by evidence. Immigration levels are capped by the Canadian government, and these ceilings are sensitive to Canadian economic conditions. Therefore, when the economy is growing, immigration increases to some extent because the maximum increases. The easiest way to account for this effect and thus gauge more accurately supply-side sensitivity to Canadian economic conditions is to include total immigration to Canada, IMM_{Ct}, as an additional control variable. This should help absorb the demand-side factors that are correlated with the level of economic activity.

Adding $\ln IMM_{Ct}$ to the regression is analogous to estimating a log share equation with $\ln \left(\dfrac{IMM_{St}}{IMM_{Ct}} \right)$ as the dependent variable, except that the coefficient on $\ln IMM_{Ct}$ is not restricted to equal 1. If the coefficient turns out to be 1, this indicates that Asia Pacific immigration moves proportionately with total immigration to Canada. In column 2 we present the results from this specification. Somewhat surprisingly, the coefficient on $\ln IMM_{Ct}$ is very close to 1. The parameter on $PCGDP_{St}$ of the source country is virtually unaffected, and that on the $PCGDP_{Ct}$ is slightly smaller, as expected. Nevertheless, the effect is still large, suggesting an elasticity of immigrant supply of 3 with respect to changes in Canadian $PCGDP$; in other words, a 1 percent increase in Canadian GDP is associated with a 3 percent increase in Asia Pacific immigration.

As noted earlier, this regression relies on both time-series and cross-section variation to identify the economic effects. Our results could, however, be biased by political or historical factors – or indeed by other unobserved factors – that are correlated with both the $PCGDP$ of the source country and migration.

TABLE 1

DETERMINANTS OF ASIA PACIFIC IMMIGRATION TO CANADA
DEPENDENT VARIABLE: LOG IMMIGRATION TO CANADA
(STANDARD ERROR IN PARENTHESES)

	(1)	(2)	(3)	(4)
Intercept	−27.440*	−29.906*	−4.362	−70.156*
	(8.270)	(8.029)	(5.046)	(34.090)
Log total immigration		1.091*	0.691*	1.303*
		(0.297)	(0.124)	(0.337)
Log GDP source country	−1.007*	−1.043*	0.509	0.646*
	(0.159)	(0.154)	(0.267)	(0.274)
Log GDP Canada	4.178*	3.013*	−1.426	4.334
	(0.899)	(0.926)	(0.929)	(3.093)
Unemployment rate Canada	−0.072	0.037	−0.091*	0.0806
	(0.055)	(0.061)	(0.033)	(0.094)
Log population density source country	0.574*	0.558*	3.404*	3.893*
	(0.088)	(0.085)	(0.993)	(1.107)
Hong Kong			−14.243*	−16.431*
			(4.017)	(4.139)
Indonesia			−2.585*	−2.509*
			(0.270)	(0.270)
Japan			−7.493*	−8.328*
			(1.354)	(1.410)
South Korea			−6.496*	−7.306*
			(1.416)	(1.465)
Malaysia			0.179	0.433
			(0.872)	(0.875)
Philippines			−1.896*	−2.187*
			(0.520)	(0.537)
Singapore			−14.603*	−16.386*
			(3.239)	(3.341)
Thailand			−3.273*	−3.323*
			(0.264)	(0.263)
Taiwan			−8.102*	−9.076*
			(1.693)	(1.752)
Trend				−1.168
				(0.086)
p-value for economic variables	0.0001	0.0001	0.0372	0.0134
R-squared	0.25	0.30	0.93	0.93

* Statistically significant at the 5 percent level.
Note: The omitted country category is China. Sample size: 187. See the appendix for the explanation of variables.

In the third column of Table 1 we include dummy variables for each source country (China being the excluded category) in an attempt to control for these factors. The results for this country fixed-effect specification are strikingly different. First, the coefficient on $PCGDP_S$ becomes positive and insignificant; and second, that on $PCGDP_C$ is negative and insignificant. The patterns of the dummy variable coefficients themselves capture permanent differences in immigration patterns across the Asia Pacific countries and are similar to what we would expect on the basis of Figures 1 and 2.

The elimination of the source-country $PCGDP$ coefficient suggests that poor countries send more immigrants to Canada for reasons other than their low income. Of course, fixed-effects estimation can be quite brutal in discarding permanent income differences across countries as "admissible" evidence. Nevertheless, the lack of robustness of the economic coefficients to the inclusion of country dummies leads one to be skeptical about their interpretation. Certainly, using the within-country time-series variation is not enough to yield a coefficient on source-country $PCGDP$ that is consistent with the theory. Once one admits the possibility of other sources of country heterogeneity that are correlated with income, the seemingly simple task of isolating the income effect in a simple migration model becomes more complicated. For example, holding initial conditions constant, as countries grow and their work force becomes more educated, their potential emigrants become a better match for the more developed Canadian economy. Thus the $PCGDP_{St}$ variable may be capturing unobserved changes in the quality of human capital in source countries. Similarly, the importance of other political factors that affect migration decisions may diminish as countries grow wealthier. In a cross-section it may be the poor countries, for example, that are the leading sources of refugees, and it would be inappropriate to interpret these immigrant flows as a response to an income differential.

Finally, in the last column, we add a linear time trend to capture any other unobserved changes in the propensity of individuals to migrate to Canada. In this specification, we control for both the permanent and trend differences in migration and the economic variables, leaving only within-country deviations from a common trend as the source of identification. Once we discard that information, there is clearly little evidence that economic variables explain much of the variation in immigration to Canada from the Asia Pacific countries, though $PCGDP_S$ becomes marginally significant. There are two possible interpretations: 1) that we are "throwing the baby out with the bath water" and discarding all the relevant exogenous variation in economic conditions; or 2) that there is no genuine relationship between migration levels and the economic variables. In the latter case, most of the immigration flows would be explained by unobservable historical or political factors, and perhaps by the "permanent" income differential that exists between Canada and most of the Asia Pacific countries. The data are not rich enough to distinguish between these hypotheses.

In any simplified, aggregate model, there is always the possibility of bias caused by omitted variables. For example, over the sample period the Canadian economy experienced considerable change in the composition of labour demand. Such changes could easily be correlated with both $PCGDP_{Ct}$ and immigrant flows, leading to bias in ordinary-least-squares (OLS) equations. Similarly, as already mentioned, source-country immigrant supply shocks could be correlated with source-country $PCGDP$ and immigrant levels. A fruitful way to address these possibilities is to compare Asia Pacific immigration flows for Canada and the United States. As we have already seen, the timing and composition of these flows have been quite different. To what extent can these differences be accounted for by differences in economic conditions between the two countries? By focusing on the differences in immigration levels as a function of differences in economic conditions, we can assess the degree to which Canada and the United States compete in the market for Asia Pacific immigrants.

To see the link between the Canada-U.S. exercise and the Canada-only exercise, assume the following immigration equations hold for the two countries:

$$\ln IMM_{CSt} = \beta_{C0} + \beta_{C1}\ln PCGDP_{St} + \beta_{C2}\ln PCGDP_{Ct} + \beta_{C3}URATEC_t + \beta_{C4}\ln POPDEN_{St} + \varepsilon_{CSt}$$

$$\ln IMM_{USt} = \beta_{U0} + \beta_{U1}\ln PCGDP_{St} + \beta_{U2}\ln PCGDP_{Ut} + \beta_{U3}URATEU_t + \beta_{U4}\ln POPDEN_{St} + \varepsilon_{USt}$$

In our interpretation of this model, we restrict the "pure" supply-side variables to have the same effect on immigration to either country: from the individual migrant's perspective, a change in his local income has the same effect on his propensity to migrate to either the United States or Canada, holding U.S. and Canadian opportunities constant. If we maintain the assumption that utility is independent of location, we can similarly restrict the coefficients on the other variables to be equal in Canada and the United States. Of course, the restriction of the coefficients would also be problematic if the supply-side interpretation itself were invalid. For example, if immigrants to Canada are more "demand"-determined than those to the United States, then the coefficients on these variables could differ across countries.[5]

While the response coefficients are restricted to be equal in Canada and the United States, the error terms and intercepts are allowed to vary across destination countries. As in the previous exercise, we place more structure on the error term, such as adding trends and source-country effects, as we proceed. For now, assume that the error has the following form:

$$\varepsilon_{jst} = \phi_{NAt} + \sum_{s} \xi_{St} + u_{jst}$$

where ϕ_{NAt} is a "North America" effect that may or may not be fixed. This could represent common changes in technology or labour demand that affect immigrant demand, or the attractiveness of North America to Asia Pacific immigrants. ξ_{St} represents source country-specific supply shocks that have a common effect on immigration flows to Canada and the United States (such as a refugee crisis), and u_{jst} represents the residual. Clearly, ϕ_{NAt} and ξ_{St} may contaminate the estimation of individual destination-country equations. Differencing the Canada and U.S. immigration equations, however, yields:

$$\Delta \ln IMM_{St} = \beta'_0 + \beta'_2 \Delta \ln PCGDP_t + \beta'_3 \Delta URATE_t + v_t$$

where $v_t = u_{cst} - u_{ust}$. The supply-side variables and common demand-side variables, are eliminated by the differencing. In this way, then, this specification addresses some of the concerns with the Canada-only estimation. Of course, if the implicit cross-country parameter restrictions are invalid, then this interpretation of the differenced equations would also be invalid. Under those circumstances, the regression would have to be interpreted at face value; in other words, it would have to be seen as a model in which the differences in immigration patterns between Canada and the United States are a function of differences in economic conditions in the two countries, as well as of permanent historical factors.

Differences in the data available for Canada and the United States force some compromises in the implementation of this specification. First, as previously noted, China and Taiwan must be pooled, since the U.S. data do not separate immigration from the two countries until the early 1980s. Second, data on immigration from Malaysia, Singapore, and Indonesia were not available for the United States over the early part of the sample period. Finally, since the source-country PCGDP differences out, Vietnam can be included in our sample.

The results of the estimation of the differenced model are presented in Table 2. Column 1 contains the coefficients for the simplest version of the model. As can be seen, the only significant coefficient is the intercept, which captures the lower average level of Asia Pacific immigration to Canada relative to the United States. The PCGDP variable has the "correct" sign, as does the difference in the unemployment rate, but neither is significant. In column 2, we add a control for the total number of immigrants admitted. As in Table 1, this should absorb some of the immigration policy effects, at least as far as total admissions are concerned. The coefficient on total admissions is positive but insignificant. None of the other coefficients change significantly. As noted in Figure 2, there appear to be permanent differences between Canada and the United States in immigration patterns across Asia Pacific source countries. These may be due to differences in history (wars, Commonwealth ties, etc.) or to locational preferences on the supply side. In column 3 we add country fixed effects to account for these differences. Note as well that because we are looking at differenced data, these country effects do not represent levels of supply-side

TABLE 2

DETERMINANTS OF ASIA PACIFIC IMMIGRATION TO CANADA
DEPENDENT VARIABLE: DIFFERENCE BETWEEN LOG IMMIGRANTS TO
CANADA AND THE UNITED STATES
(STANDARD ERROR IN PARENTHESES)

	(1)	(2)	(3)	(4)	(5)
Intercept	−1.354*	−0.497	0.001	−1.469	−0.972
	(0.374)	(0.648)	(0.343)	(0.938)	(0.466)
Δ Log total immigration		0.627	0.627*	0.526	0.526*
		(0.388)	(0.192)	(0.393)	(0.189)
Δ Log GDP	4.178	5.520	5.520*	−0.671	−0.670
	(3.608)	(3.681)	(1.816)	(5.677)	(2.726)
Δ Unemployment rate	−0.082	0.030	0.030	−0.140	−0.140
	(0.075)	(0.102)	(0.050)	(0.156)	(0.080)
Hong Kong			1.521*		1.526*
			(0.188)		(0.183)
Japan			−0.932*		−0.932*
			(0.188)		(0.183)*
South Korea			−1.567*		−1.567*
			(0.188)		(0.183)
Philippines			−0.600*		−0.600*
			(0.188)		(0.183)
Thailand			−1.709*		−1.709*
			(0.188)		(0.183)
Vietnam			−0.192		−0.192
			(0.188)		(0.183)
Trend				0.070	0.070*
				(0.049)	(0.023)
p-value for economic variables	0.410	0.287	0.007	0.619	0.129
R^2	0.014	0.033	0.776	0.048	0.790

* Statistically significant at the 5 percent level.
Δ Difference between the Canadian and U.S. variables.
Note: China and Taiwan (pooled) are the omitted country category. Sample size: 133. See the appendix for the explanation of variables.

conditions in the source countries but supply factors that are correlated with permanent differences in attachment to Canada or the United States.

The results are quite interesting. The difference in PCGDP now has a large significant effect. Indeed, the elasticity is quite large, at approximately 5. This suggests that once we account for permanent differences in immigration patterns between the United States and Canada, increases of Canadian GDP relative to the United States attract immigrants to Canada. The unemployment rate does not seem to have much effect. This may be due to the fact that it is unclear how to interpret differences in Canadian and U.S. unemployment rates (see Card and Riddell, 1993), and their interpretation as a measure of "risk" may be inappropriate. The coefficient on total immigration is positive and significant. This indicates that as immigration to both countries has increased, a significantly greater share of the Canadian total has been accounted for by the Asia Pacific region. The country effects themselves are as expected: Canada receives many more immigrants from Hong Kong, and as many from Vietnam and China, but fewer from the other Asia Pacific countries. The joint significance of the country effects suggests that these non-economic supply-side variables are the most important determinants of immigrant flows between the Asia Pacific countries and the United States and Canada. Finally, in the last two columns we add a time trend. As in the previous table, inclusion of the time trend obliterates the significance of the PCGDP variable. This suggests that if we are unwilling to "explain" trends in immigration patterns with trends in PCGDP, then the remaining covariation provides no evidence of a link between economic conditions in destination countries and immigration flows. As before, this may be too harsh a conclusion to draw from the available data.

A Profile of the Immigrants

USING DATA FROM THE 1991 CANADIAN CENSUS, we next construct a sketch of the stock of immigrants as of 1990. These data provide a view of immigrants from four source countries in the Asia Pacific region – Hong Kong, China, the Philippines, and Vietnam. Other Asia Pacific immigrants are aggregated into a single category.[6] We compare these individuals to immigrants from all other source countries (the category "other"), as well as Canadian-born residents.

Our initial working sample includes men and women between the ages of 16 and 64. We exclude respondents from the Atlantic provinces and the two territories, as information on place of birth and immigrant status is coded differently for these regions, making it impossible to identify residents of those areas who were born in the Asia Pacific region.

In Table 3 we present some (average) characteristics of the men in each group; the corresponding statistics for women are reported in Table 4. The trends in the source-country composition of Canada's immigrants, outlined earlier, are clearly evident here for both sexes. The majority of each category of Asia Pacific immigrants arrived in Canada after 1970 (IM7175 or later). On the other hand, "other" immigrants are more evenly distributed across the cohorts. Given that there is a higher incidence of migration among the young, it is not surprising that Asia Pacific immigrants also tend to be younger than individuals in the comparison groups. The exceptions are immigrants of both sexes from China and women from the Philippines. Note that immigrants from China are more likely to have arrived in the 1950s and 1960s.

TABLE 3

SAMPLE CHARACTERISTICS: MEN

	CANADIAN-BORN RESIDENTS	ASIA PACIFIC IMMIGRANTS					
		OTHER IMMIGRANTS	HONG KONG	CHINA	PHILIPPINES	VIETNAM	EAST AND SOUTH-EAST ASIA
IM55P		0.13	0.00	0.11	0.00	0.00	0.01
IM5660		0.12	0.01	0.06	0.00	0.00	0.01
IM6165		0.09	0.02	0.03	0.01	0.00	0.01
IM6670		0.17	0.08	0.09	0.06	0.01	0.08
IM7175		0.16	0.19	0.13	0.26	0.06	0.18
IM7680		0.10	0.14	0.13	0.21	0.39	0.21
IM8185		0.08	0.13	0.15	0.14	0.30	0.16
IM8690		0.15	0.42	0.30	0.32	0.25	0.34
Age	36.38	42.19	34.19	44.97	35.89	32.24	36.51
Married	0.61	0.74	0.65	0.82	0.66	0.49	0.65
Quebec	0.32	0.15	0.03	0.07	0.04	0.18	0.16
Ontario	0.37	0.58	0.52	0.43	0.50	0.45	0.45
Prairies	0.20	0.12	0.13	0.15	0.26	0.25	0.14
British Columbia	0.12	0.15	0.32	0.34	0.21	0.12	0.25
City	0.59	0.84	0.97	0.94	0.95	0.95	0.93
Weeks worked	38.34	39.31	35.86	37.42	38.67	33.77	33.92
"Participate"	0.81	0.79	0.75	0.70	0.84	0.75	0.67
Employed	0.79	0.79	0.78	0.77	0.81	0.68	0.74
Unemployed	0.08	0.08	0.05	0.07	0.08	0.13	0.07
Schooling	13.48	13.84	15.98	13.24	15.35	12.63	14.87
UI receipt	0.15	0.13	0.07	0.11	0.14	0.18	0.12
SA receipt	0.08	0.07	0.02	0.05	0.04	0.09	0.05
Poor	0.12	0.15	0.20	0.23	0.12	0.25	0.26
Citizen	1.00	0.72	0.61	0.67	0.65	0.71	0.60
Self-employed	0.11	0.14	0.13	0.17	0.05	0.06	0.18
Wages	24,277.18	26,703.68	22,342.64	18,714.40	20,570.54	15,873.97	18,721.54

Note: The sample includes all individuals 16 to 64 years of age. "Other immigrants" are immigrants from countries other than the Asia Pacific. "SA receipt" is based on the census category "other government transfer programs." "Poor" is based on the Statistics Canada low-income cutoff. "Weeks worked" is the number of weeks worked in the reference year. "Participate" is the proportion of individuals with both weeks worked and wage earnings positive in the previous year.

Source: 1991 Canadian census.

Asia Pacific immigrants also display different settlement patterns than "other" immigrants and Canadian-born residents. All immigrants are more likely to live in Ontario or British Columbia. As is well known, the greatest proportion of immigrants from Hong Kong and China are in these provinces. Also, Asia Pacific immigrants are more urban: 95 percent or so of most groups live in an urban area. The proportion of "other" immigrants living in a city is

TABLE 4

SAMPLE CHARACTERISTICS: WOMEN

	CANADIAN-BORN RESIDENTS	OTHER IMMIGRANTS	ASIA PACIFIC IMMIGRANTS				
			HONG KONG	CHINA	PHILIPPINES	VIETNAM	EAST AND SOUTH-EAST ASIA
IM55P		0.13	0.00	0.03	0.00	0.00	0.01
IM5660		0.11	0.01	0.06	0.00	0.00	0.01
IM6165		0.09	0.01	0.05	0.02	0.00	0.01
IM6670		0.16	0.07	0.09	0.09	0.01	0.08
IM7175		0.16	0.17	0.13	0.22	0.06	0.18
IM7680		0.11	0.14	0.14	0.15	0.35	0.22
IM8185		0.09	0.14	0.16	0.15	0.33	0.16
IM8690		0.15	0.46	0.34	0.36	0.26	0.33
Age	36.68	41.92	34.18	44.39	38.35	33.93	36.37
Married	0.63	0.72	0.66	0.80	0.61	0.57	0.66
Quebec	0.32	0.15	0.04	0.06	0.06	0.18	0.14
Ontario	0.37	0.59	0.52	0.44	0.51	0.45	0.45
Prairies	0.19	0.12	0.13	0.14	0.23	0.23	0.15
British Columbia	0.12	0.15	0.31	0.36	0.21	0.14	0.26
City	0.61	0.85	0.97	0.95	0.94	0.95	0.93
Weeks worked	30.90	30.08	29.65	28.63	37.48	26.67	27.69
"Participate"	0.70	0.66	0.67	0.59	0.80	0.63	0.61
Employed	0.65	0.62	0.63	0.58	0.80	0.57	0.59
Unemployed	0.07	0.08	0.06	0.07	0.05	0.12	0.08
Schooling	13.50	13.12	14.78	11.62	15.34	11.66	13.68
UI receipt	0.14	0.12	0.11	0.12	0.12	0.18	0.11
SA receipt	0.09	0.08	0.02	0.06	0.04	0.11	0.05
Poor	0.15	0.17	0.22	0.22	0.19	0.29	0.26
Citizen	1.00	0.72	0.61	0.67	0.65	0.71	0.60
Self-employed	0.05	0.06	0.06	0.08	0.02	0.04	0.11
Wages	12,980.55	13,024.58	13,468.09	10,275.47	16,637.88	8,703.92	10,078.06

Note: The sample includes all individuals 16 to 64 years of age. "Other immigrants" are immigrants from countries other than the Asia Pacific. "SA receipt" is based on the census category "other government transfer programs." "Poor" is based on the Statistics Canada low-income cutoff. "Weeks worked" is the number of weeks worked in the reference year. "Participate" is the proportion of individuals with both weeks worked and wage earnings positive in the previous year.
Source: 1991 Canadian census.

about 85 percent, while fewer than 60 percent of Canadian-born residents are in these areas.[7]

Individuals' participation in the labour market is captured by variables that measure weeks worked and employment for positive wages in the reference year, as well as by employment/population and unemployment/population rates in the reference week. Among men (Table 3), Asia Pacific immigrants, except those from the Philippines, are less likely to have worked in the reference year. Given the previous discussion, we might expect that this is just an age or assimilation effect. Most of the differences are significant, however, in regressions that control for age and period of arrival in Canada (not shown). Activity in the reference week provides a view of how labour market participation is divided between employment and unemployment. Immigrants from Hong Kong and China have the same employment rates as "other" immigrants but lower unemployment rates. In contrast, immigrants from Vietnam and the rest of East and South-East Asia have lower employment rates. Finally, consistent with the evidence on employment in the previous year, Filipino immigrants have the highest labour market participation of all groups.[8]

The results for women (Table 4) display similar patterns. First, controlling for age and period of arrival, women from the Philippines are more likely to have worked in the reference year, while those from other areas of East and South-East Asia have lower employment rates. Second, all Asia Pacific immigrants except those from Vietnam have lower unemployment rates in the reference week. Finally, Filipino women have the highest labour market participation rate of all groups.

Measures of formal schooling provide an indication of the levels of skills these immigrants bring to Canada.[9] Immigrants of both sexes from Hong Kong and the Philippines make up the most highly educated groups. Their educational level is roughly two years higher than that of "other" immigrants, and these differences remain when controlling for age and period of arrival. Immigrants from other East and South-East Asian countries also have higher education levels in the raw means, although this difference disappears in the conditional means. Finally, immigrants from both China and Vietnam have less education than "other" immigrants. Conditioning on age and period of arrival, the difference is significant for men from China and for women from both China and Vietnam.

To provide various views of the economic status of Asia Pacific immigrants to Canada, we examine their participation in the unemployment insurance (UI) and other government transfer programs – labelled "social assistance (SA) receipt" – as well as the incidence of poverty at the household level.[10] Among men (Table 3), immigrants from Hong Kong, China, and other countries in East and South-East Asia are less likely than "other" immigrants to have received UI benefits in the reference year, both in the raw and adjusted means. On the other hand, immigrants from Vietnam have higher UI participation. There is less variation in the unadjusted rates for women (Table 4). Adjusting for age and period of arrival, the results are very similar to those for men – lower rates for immigrants from Hong Kong, the Philippines, and other East and South-East Asia countries, and a higher rate for those from Vietnam.

A similar story emerges in the analysis of SA receipt.[11] Again, it is the Vietnamese who have the highest rate of receipt of these types of benefits among both men and women. Finally, immigrants from many of the Asia

Pacific countries live in households in which income is below the Statistics Canada low-income cutoff. Filipino men are the primary exception. We also note that while Hong Kong immigrants have a higher poverty rate than "other" immigrants in the raw means for both sexes, they have significantly lower rates when age and period of arrival are controlled for.

Next we examine characteristics that capture some non-economic dimensions of assimilation. The assumption of Canadian citizenship is an external signal of re-affirmation of the decision to migrate. While in the raw means for both sexes a lower percentage of Asia Pacific immigrants are citizens, this appears to be due to their recent arrival and to the fact that a period of residence in Canada is required before citizenship can be obtained. Controlling for age and period of arrival, all immigrants from this region have a significantly higher citizenship rate than immigrants from other countries. The other aspect of assimilation (not shown in the tables) that we examine is the adoption of one of Canada's official languages as the home language. While knowledge of English or French have obvious implications for success in the labour market, it may also indicate the degree of assimilation into Canadian society. Among immigrants whose mother tongue is neither English nor French, and applying the usual controls, all Asia Pacific immigrants except those from the Philippines have significantly lower rates of adoption of these languages at home.

In the last two rows of each table, we report the self-employment rates and raw wage and salary earnings of each group. There are common patterns in the relative self-employment rates of men and women. Higher rates are observed for Chinese and other East and South-East Asian immigrants, while Filipinos tend to have lower rates. For men, immigrants from the Pacific Rim have lower earnings than both "other" immigrants and Canadian-born residents. Within the Asia Pacific immigrant group, those from Hong Kong have the highest earnings, and those from Vietnam the lowest. Given that many of these immigrants arrived in Canada quite recently, we might expect that at least part of these differences would be accounted for by the stage of assimilation. Among women, immigrants from Hong Kong and the Philippines have higher raw earnings than "other" immigrants and Canadian-born residents.

If nothing else, this analysis suggests that it would be inappropriate to refer to Asia Pacific immigrants as a homogeneous group. Across the limited number of distinct groups that can be identified in the 1991 census, we observe considerable heterogeneity among immigrants from that region with respect to the skills they bring to Canada and to their participation in the labour market and in transfer programs. What they do share is relatively recent arrival in Canada.

THE HUMAN CAPITAL CONTRIBUTION

WE NEXT EXAMINE THE ASSIMILATION of Asia Pacific immigrants in Canada, drawing comparisons with immigrants from other countries and with Asia Pacific immigrants to the United States. The evidence in the previous section provided two views of this assimilation: Asia Pacific immigrants assumed Canadian citizenship more quickly than immigrants from other countries, but were slower to assume one of Canada's official languages at home. Economists, however, typically use different standards to define assimilation. They examine the rates of immigrants' earnings growth in the labour market of their adopted country, relative to the rates for native-born residents or for immigrants who arrived in an earlier period. By this definition, if immigrants' earnings grow faster than those of one of these comparison groups, we say that they are assimilating. Within the context of the standard human-capital earnings function that is typically assumed, assimilation takes the form of excess returns to labour market experience, which are correlated with the length of time an immigrant cohort has been in Canada.

Success in the labour market might be expected to depend, in part, on how well an immigrant's human capital matches opportunities in the Canadian labor market. Canada uses a point system to evaluate some immigrants, to flag those applicants whose skills appear to be a particularly good match. The point system applies to immigrants admitted as independent applicants under the business class, and to a lesser extent to those admitted as assisted relatives. Therefore, one might expect that the proportion of immigrants from a particular source country admitted under the point system would have some bearing on their eventual labour market success. In Table 5 we present the proportions of IM8690 Asia Pacific immigrants from various destinations by admission classes.[12] These data are for immigrants by country of last residence, which does not exactly match our division of immigrants by place of birth. For example, refugees who stop over in a third country on their way to Canada will be coded differently by place of last residence. It is difficult to determine the error introduced into the analysis by this difference in definition.

There are some striking differences in class composition across the Asia Pacific countries. Immigrants from China are mostly admitted under the family class, while those from Vietnam tend to be refugees. The proportions evaluated under the point system range from a low of 0 percent for Laos to 87 percent for Taiwan. These differences might account for the variation in skills that can be observed across groups. For example, Hong Kong immigrants, who are mostly admitted under the point system, have higher levels of education than most other immigrants. They also reveal that our residual group of "other East and South-East Asians" is very heterogeneous.

In performing the analysis we limit our sample to individuals who worked "full year/full time" in the reference year – i.e., who worked 40 weeks or more at mostly full-time work – in order to make our comparisons within a group that has a fairly homogeneous labour supply. We also add data from the 1986 Canadian census to the analysis. This enables us to follow cohorts of immigrants over time, from 1985 to 1990. The addition of these data is important because we want to separate any cohort effects – any permanent effects

Table 5

Class of Admission IM8690

	Family	Refugee	Assisted Relative	Entrepreneur	Self-Employed	Investor	Independent	Point System
China	0.58	0.01	0.14	0.01	0.00	0.00	0.26	0.41
Hong Kong	0.17	0.00	0.05	0.19	0.01	0.04	0.54	0.82
Philippines	0.45	0.00	0.16	0.03	0.00	0.01	0.36	0.56
Vietnam	0.27	0.65	0.08	0.00	0.00	0.00	0.00	0.08
Brunei	0.17	0.00	0.11	0.05	0.01	0.00	0.66	0.83
Indonesia	0.33	0.01	0.07	0.24	0.01	0.03	0.31	0.66
Japan	0.34	0.01	0.02	0.08	0.03	0.02	0.50	0.65
Kampuchea	0.03	0.96	0.01	0.00	0.00	0.00	0.00	0.02
South Korea	0.31	0.00	0.11	0.43	0.02	0.00	0.13	0.69
Laos	0.03	0.97	0.00	0.00	0.00	0.00	0.00	0.00
Macao	0.26	0.01	0.04	0.30	0.00	0.12	0.27	0.73
Malaysia	0.28	0.00	0.05	0.19	0.01	0.01	0.46	0.72
Mongolia	0.33	0.33	0.00	0.00	0.00	0.00	0.33	0.34
Singapore	0.15	0.01	0.09	0.07	0.02	0.18	0.48	0.84
Taiwan	0.13	0.00	0.03	0.41	0.01	0.21	0.21	0.87
Thailand	0.75	0.11	0.03	0.02	0.00	0.00	0.09	0.14
Rest of Asia Pacific	0.19	0.20	0.05	0.23	0.01	0.07	0.25	0.61
Total Asia Pacific	0.27	0.14	0.08	0.13	0.01	0.03	0.34	0.59
Rest of the world	0.37	0.19	0.11	0.03	0.02	0.00	0.28	0.44

Note: Classes are defined as follows: family class, refugee, assisted relative, entrepreneur, self-employed, investor, and independent. Immigrants who were not in the family class or refugees were considered as admitted under the point system.

Source: Employment and Immigration Canada, *Immigration Statistics*, various issues.

on earnings from being in a particular arrival cohort – from the effects of assimilation. This is only possible if we exploit the quasi-panel information provided by multiple cross-sections of data.[13]

The difference between cohort effects and assimilation effects is demonstrated in Figures 3 and 4. In Figure 3 we graph the average earnings of immigrants from the four most recent cohorts (IM8185 through IM6670), as measured in the 1986 census. Earnings are plotted against the length of time that members of these cohorts had been in Canada as of 1985. The graph suggests that immigrants' earnings rise with the length of their residence in Canada: those who arrived earlier make more money than recent arrivals. Ignoring the issues of making comparisons relative to Canadian-born residents or to older immigrants and controlling for observable characteristics for the moment, this graph provides one measure of economic assimilation. For example, the graph suggests that over the first five years in Canada, an immigrant's earnings rise by roughly 18 percent (32,700/27,719), which is calculated by taking the ratio

FIGURE 3

ASIA PACIFIC IMMIGRANTS: AVERAGE ANNUAL EARNINGS BY COHORT, 1986

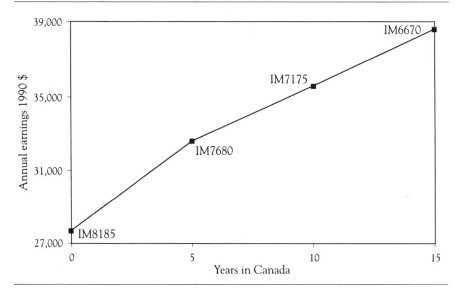

Note: The sample is comprised of immigrant paid workers who worked 40 weeks or more at mostly full-time work. All earnings are converted to 1990 dollars.
Source: 1986 Canadian census.

of IM7680's average earnings to IM8185's average earnings. Likewise, the return to the second and third five-year periods in Canada are roughly 9 percent (35,659/32,700) and 8 percent (38,686/35,659), respectively.

Information from the 1991 census enables us to check the accuracy of these predictions by observing the actual earnings growth experienced by these cohorts over their first, second, and third five-year periods in Canada. In Figure 4 we graph the annual earnings of each cohort in 1985 in a different format. We also graph their average earnings in 1990, based on 1991 census data. By following a cohort over time, we can measure the actual earnings growth it experienced over the period. The solid lines for each cohort map this progress. The dotted lines provide a view of the accuracy of our predictions based on the 1986 cross-section, as well as the measure of any cohort effects. For example, the dotted line starting at the 1985 earnings for IM7175 reveals that IM7680 did not achieve our prediction for the second five-year period in Canada – that is, the dotted line does not meet the solid line for IM7680 on the right side (1990) of the graph. The cross-section estimate over-predicted the earnings growth that this cohort would experience. The difference between the dotted line and IM7680's 1990 earnings is a measure of a cohort effect. In 1990, IM7680 had the same years in Canada as IM7175 did in 1985. Therefore,

Figure 4

Asia Pacific Immigrants: Assimilation and Cohort Effects, 1985 and 1990

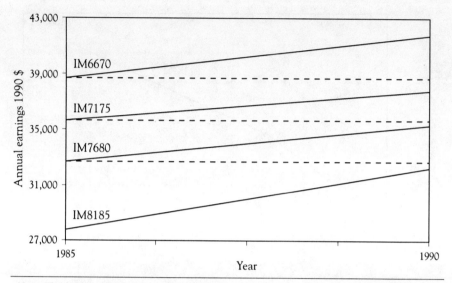

Note: The sample is comprised of immigrant paid workers who worked 40 weeks or more at mostly full-time work. All earnings are converted to 1990 dollars.
Source: 1986 and 1991 Canadian censuses.

IM7680 did not match the outcome achieved by the preceding cohort, holding years in Canada constant. This is evidence that IM7680's earnings are permanently lower than IM7175's, and is what people refer to when they speak of the "declining quality" of recent immigrants. Each of the dotted lines in Figure 4 reveals that the cross-section predictions overestimated the return to each of the five-year periods in Canada. They also provide evidence of cohort effects, the largest being in the comparison between IM6670 and IM7175.

When we estimate the assimilation and cohort effects, we also take account of earnings trends in the economy, normalizing by the earnings growth of a control group – Canadian-born residents or earlier immigrants, for example. This idea is demonstrated in Figure 5. Here we graph the average earnings for each cohort and for Canadian-born residents in each year. Our measure of assimilation is the difference in the bars for a given cohort (their earnings growth between the two censuses), less the difference in the two bars for Canadian-born residents. By the measure of unadjusted earnings it appears that all cohorts experienced some assimilation – excess returns to experience – over this period. The cohort effects are measured by the position of cohort i in 1985 relative to Canadian-born residents, compared with the relative position of cohort $i+5$ in the 1990 census. For example, in 1990 IM8185 (cohort $i+5$)

FIGURE 5

ASIA PACIFIC IMMIGRANTS: EARNINGS BY COHORT, 1985 AND 1990

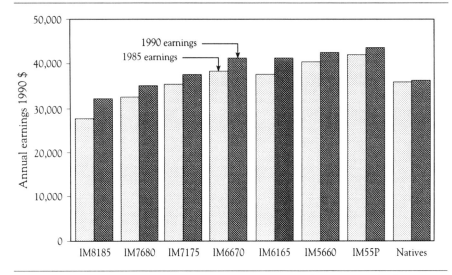

Note: The sample is comprised of immigrant paid workers who worked 40 weeks or more at mostly full-time work. All earnings are converted to 1990 dollars.
Source: 1986 and 1991 Canadian censuses.

has the same number of years of residence in Canada as IM7680 (cohort i) did in 1985. The comparison for these cohorts reveals that in 1990, IM8185 did not quite achieve the relative position of IM7680 in 1985, consistent with the evidence in Figure 4.

We make similar comparisons in a regression context in order to control for observable characteristics among the different cohorts. We use a pooled sample of Canadian-born residents and immigrants from the two censuses. The empirical specification is:

$$lny_{it} = X_{it}\beta + \gamma_1 YSM_{it} + \gamma_2 YSM_{it}^2 + \sum \delta_D IM(COH_D)_{it} + \alpha t + \varepsilon_{it}$$

where YSM is a measure of the years since migration to Canada for an individual, $IM(COH_D)$ are indicators for cohorts IM8690 through IM55P, X is our set of control variables, and t is a time effect that is assumed to be the same for Canadian-born residents and for immigrants.[14] The parameters on the YSM terms provide the estimates of assimilation. The cohort dummy variables provide estimates of the cohort effects, at the point of entry to Canada (YSM = 0). Our control variables include a quadratic in experience, education, and marital status, as well as a dummy variable for individuals who work between 40 and 48 weeks.

We focus on men in this section, since the changing labour market participation of women over the period poses non-trivial econometric problems. The estimates are reported in Table 6. The first column contains results that are comparable with those in other studies in this area. The returns to assimilation – the excess returns to experience for immigrants – are a little over 2 percent a year in the first years of residence in Canada. This is larger than estimates from previous periods (Bloom and Gunderson, 1991; Baker and Benjamin, 1994; and Bloom, Grenier and Gunderson, 1995), but it is consistent with recent evidence that the return to assimilation over the first years in Canada – i.e., for IM8185 – increased during the second half of the 1980s (Grant, 1995). Also starting with IM6670, the estimates of the cohort effects suggest that the entry earnings of successive immigrant cohorts have been declining, although the differences are not very large.

In the second column we present the results of allowing Asia Pacific immigrants to have different assimilation and cohort effects than immigrants from other countries. While it might have been preferable to allow these effects to differ by country of origin, detailed place-of-birth information is not available in the 1986 census. The results suggest that Asia Pacific immigrants have slightly higher assimilation rates in their first years in Canada, although the difference is not significant at conventional levels. There is evidence, however, that more recent immigrants in this group start at a greater initial-earnings disadvantage than their counterparts from other countries. For example, Asia Pacific members of IM8185 had a roughly 14.5 percent earnings disadvantage at entry relative to other members in this group.

Finally, in the third column we modify this specification to allow Canadian-born residents and immigrants to have different parameters on the control variables (X). While we do find that the returns to education and experience are different for the two groups, the basic message about Asia Pacific immigrants does not change: they experience the same rate of assimilation as immigrants from other countries, but start at lower entry-earnings levels.

One way to discover whether the difference in entry earnings between Asia Pacific immigrants and those from other countries is a phenomenon peculiar to Canada or peculiar to those immigrants themselves, is to perform a comparable analysis of this group in the United States. If we find similar results for Asia Pacific immigrants in that country, this suggests either an Asia Pacific effect or a phenomenon peculiar to North America.

To follow up on this suggestion, we examine data from the 1980 and 1990 U.S. censuses. These data cover a different time period (the reference years are 1979 and 1989) than that used for Canada, but they are sufficient to provide a view of Asia Pacific immigrants relative to other groups. We have selected the sample so that it would be as comparable to our Canadian samples as possible. One point worth noting is that the period of arrival is coded slightly differently in the U.S. censuses, so that cohort definitions do not match perfectly.

TABLE 6

ASSIMILATION PROFILES OF IMMIGRANTS TO CANADA
DEPENDENT VARIABLE: LOG EARNINGS (MEN)
(STANDARD ERRORS IN PARENTHESES)

	(1)	(2)	(3)
$Ysm \times 100$	2.131* (0.362)	2.024* (0.395)	2.403* (0.395)
$Ysm - Sq \times 100$	−0.022* (0.007)	−0.020* (0.007)	−0.021* (0.007)
Asia Pacific $Ysm \times 100$		0.448 (0.907)	0.497 (0.905)
Asia Pacific $Ysm - Sq \times 100$		−0.006 (0.027)	−0.007 (0.027)
IM55P	−0.456* (0.081)	−0.433* (0.085)	−0.045 (0.088)
IM5660	−0.412* (0.070)	−0.396* (0.074)	0.006 (0.078)
IM6165	−0.403* (0.063)	−0.384* (0.067)	0.023 (0.071)
IM6670	−0.379* (0.054)	−0.361* (0.058)	0.064 (0.064)
IM7175	−0.389* (0.045)	−0.365* (0.048)	0.066 (0.055)
IM7680	−0.354* (0.034)	−0.299* (0.037)	0.144* (0.046)
IM8185	−0.384* (0.022)	−0.346* (0.024)	0.106* (0.038)
IM8690	−0.408* (0.015)	−0.376* (0.017)	0.084* (0.034)
Asia Pacific IM55P		−0.456* (0.197)	−0.468* (0.196)
Asia Pacific IM5660		−0.136 (0.148)	−0.145 (0.148)
Asia Pacific IM6165		−0.221 (0.133)	−0.231 (0.132)
Asia Pacific IM6670		−0.129 (0.108)	−0.132 (0.107)
Asia Pacific IM7175		−0.123 (0.092)	−0.122 (0.092)
Asia Pacific IM7680		−0.197* (0.074)	−0.211* (0.073)
Asia Pacific IM8185		−0.145* (0.052)	−0.158* (0.052)
Asia Pacific IM8690		−0.133* (0.034)	−0.111* (0.034)

* Statistically significant at the 5 percent level.
Note: The sample comprises paid workers employed 40 weeks or more in mostly full-time work during the reference year. The coefficients are from a regression of ln earnings on the variables reported in the text. In column (3), we allow immigrant specific parameters on the control variables. The f-test for equality of the cohort effects are 2.39 ($p = 0.0192$) in column (1). The f-test for joint significance of immigrant interactions in column (3) is 62.57 ($p = 0.0001$).

Source: 1986 and 1991 Canadian censuses.

Before proceeding, we note that any differences between the two countries may be due to differences in the source-country composition of Asia Pacific immigrants, which we discussed earlier. In Table 7 we present the compositions of the stock of Asia Pacific immigrants in the United States in 1990 and in Canada in 1991. The proportions of immigrants from the Asia Pacific region in the two countries are roughly similar, at 16 or 17 percent. Within this group, however, the source-country compositions differ. Canada has a far higher proportion of immigrants from Hong Kong and, to a lesser extent, China. The proportions from the Philippines and Vietnam are roughly similar. On the other hand, the United States has a much larger proportion from other countries in the region, at nearly 43 percent.[15]

The U.S. regression results are reported in Table 8. We again present a base set of estimates and then allow differential effects for Asia Pacific immigrants, and separate immigrant/Canadian-born parameters on the control variables. The estimates of assimilation in the first column suggest returns of just under 2 percent in the first years in the United States. The estimates of the cohort effects suggest that more recent cohorts have slightly higher entry earnings than their predecessors, although the differences are not large. The results of allowing the Asia Pacific interactions are reported in column 2. First, Asia Pacific immigrants have roughly the same rate of assimilation as those from other countries, a result similar to that observed in Table 6. Second, while the U.S. Asia Pacific cohort interactions are negative, they are generally smaller than in the Canadian results. Asia Pacific immigrants start at a larger initial earnings deficit than other immigrants, but the difference is smaller in the United States than in Canada. Finally, allowing separate immigrant parameters on the control variables (column 3) does not alter the basic results. This analysis, therefore, confirms the evidence in Table 6 that Asia Pacific immigrants start out at lower earnings levels than other immigrants.

TABLE 7

ASIA PACIFIC IMMIGRANTS TO CANADA AND THE UNITED STATES AS A PROPORTION OF ALL IMMIGRANTS, BY SOURCE COUNTRY (MEN ONLY)

	CANADA	UNITED STATES
Hong Kong	0.22	0.04
China	0.22	0.14
Vietnam	0.16	0.15
Philippines	0.19	0.24
Other South-East Asia	0.21	0.43
All Asia Pacific immigrants as a proportion of all immigrants	0.16	0.17

Note: The sample is comprised of paid workers employed 40 weeks or more at mostly full-time employment.
Source: 1990 U.S. and 1990 Canadian censuses.

TABLE 8

ASSIMILATION PROFILES OF IMMIGRANTS TO THE UNITED STATES
DEPENDENT VARIABLE: LOG EARNINGS (MEN)
(STANDARD ERRORS IN PARENTHESES)

	(1)	(2)	(3)
$Ysm \times 100$	1.909*	1.775*	2.014*
	(0.171)	(0.186)	(0.186)
$Ysm - Sq \times 100$	−0.026*	−0.025*	−0.025*
	(0.004)	(0.005)	(0.005)
Asia Pacific $Ysm \times 100$		0.342	0.435
		(0.461)	(0.460)
Asia Pacific $Ysm - Sq \times 100$		0.012	0.010
		(0.013)	(0.013)
IM59P	−0.265*	−0.229*	0.133*
	(0.025)	(0.027)	(0.032)
IM6064	−0.259*	−0.229*	0.135*
	(0.023)	(0.025)	(0.030)
IM6569	−0.255*	−0.232*	0.134*
	(0.020)	(0.022)	(0.027)
IM7074	−0.250*	−0.226*	0.138*
	((0.016)	(0.017)	(0.024)
IM7579	−0.235*	−0.214*	0.151*
	(0.011)	(0.013)	(0.021)
IM8084	−0.272*	−0.263*	0.102*
	(0.015)	(0.016)	(0.023)
IM8589	−0.216*	−0.211*	0.153*
	(0.012)	(0.013)	(0.021)
Asia Pacific IM59P		−0.305*	−0.294*
		(0.065)	(0.065)
Asia Pacific IM6064		−0.206*	−0.191*
		(0.060)	(0.060)
Asia Pacific IM6569		−0.116	−0.095
		(0.050)	(0.050)
Asia Pacific IM7074		−0.123*	−0.098*
		(0.041)	(0.041)
Asia Pacific IM7579		−0.088*	−0.061*
		(0.028)	(0.028)
Asia Pacific IM8084		−0.024	0.002
		(0.037)	(0.037)
Asia Pacific IM8589		−0.017	0.022
		(0.030)	(0.030)

* Statistically significant at the 5 percent level.
Note: The f-test in column (1) for the equality of the cohort effects is 2.38 (p = 0.027). The f-test in column (3) for the joint significance of the immigration interactions with the control variables is 121.16 (p = 0.0001). For other notes, see Table 6.
Source: 1980 and 1990 U.S. censuses.

Immigrants' success in Canada may be related to the types of jobs they obtain. In turn, differences in outcomes across immigrant groups might be accounted for by corresponding differences in their distribution of employment across "good" and "bad" jobs. We provide a view of this issue by examining the

TABLE 9

ADJUSTED DIFFERENCES IN THE OCCUPATIONAL DISTRIBUTION OF ASIA PACIFIC IMMIGRANTS
(STANDARD ERRORS IN PARENTHESES)

	HONG KONG	CHINA	PHILIPPINES	VIETNAM	OTHER EAST AND SOUTH-EAST ASIA
Managerial	0.042*	−0.041*	−0.080*	−0.085*	−0.016
	(0.013)	(0.013)	(0.014)	(0.014)	(0.014)
Teaching	−0.030*	−0.023*	−0.012	−0.046*	−0.029*
	(0.007)	(0.007)	(0.007)	(0.007)	(0.008)
Natural sciences	0.093*	0.063*	0.054*	−0.009	0.053*
	(0.010)	(0.010)	(0.011)	(0.011)	(0.011)
Medicine and health	0.011	0.001	0.007	0.041*	0.001
	(0.005)	(0.005)	(0.005)	(0.005)	(0.006)
Social sciences	−0.004	−0.005*	−0.008*	−0.011*	0.007
	(0.004)	(0.003)	(0.004)	(0.004)	(0.004)
Arts	0.006	0.001	−0.005	−0.006	−0.003
	(0.004)	(0.004)	(0.004)	(0.004)	(0.004)
Construction	−0.052*	−0.075*	−0.058*	−0.049*	−0.044*
	(0.009)	(0.009)	(0.010)	(0.010)	(0.010)
Sales	0.029*	−0.021*	−0.049*	−0.018	0.015
	(0.009)	(0.009)	(0.010)	(0.010)	(0.010)
Other primary	−0.006*	−0.002	−0.002	−0.006*	−0.005*
	(0.003)	(0.003)	(0.003)	(0.003)	(0.003)
Processing	−0.024*	−0.004	0.054*	0.009	−0.002
	(0.007)	(0.007)	(0.008)	(0.008)	(0.008)
Machining	−0.067*	−0.078*	0.111*	0.084*	0.013
	(0.013)	(0.013)	(0.014)	(0.014)	(0.015)
Transport	−0.026*	−0.032*	−0.035*	−0.023*	−0.018*
	(0.007)	(0.007)	(0.007)	(0.007)	(0.008)
Other occupations	−0.029*	−0.018*	0.008	0.010	0.012
	(0.009)	(0.009)	(0.010)	(0.010)	(0.010)
Agriculture	−0.005*	−0.006*	−0.003	−0.005*	−0.003
	(0.003)	(0.003)	(0.003)	(0.003)	(0.004)
Service	0.028*	0.210*	0.032*	0.061*	0.024*
	(0.010)	(0.010)	(0.011)	(0.011)	(0.012)
Clerical	0.034*	0.028*	−0.014	0.054*	−0.004
	(0.009)	(0.009)	(0.010)	(0.010)	(0.010)

* Statistically significant at the 5 percent level.
Note: The sample is comprised of immigrant workers who worked 40 weeks or more at mostly full-time work. The coefficients are from regressions of an indicator variable of employment in the indicated occupation on education, experience, region, and dummy variables for immigration from the indicated countries.
Source: 1991 Census of Canada.

occupational distribution of Canadian immigrants in Table 9. Here we only use the 1991 census so we can compare the distribution across different source countries in the Asia Pacific region. The reported statistics are the parameters on dummy variables for immigrants from the indicated countries, in a regression of an indicator variable for employment in the relevant occupation on education, experience, region, and the place-of-birth dummies. The equation is estimated for the sample of immigrants working full year/full time in 1990. Finally, we order the occupations by their relative earnings from a regression of log earnings on education, experience, region, and dummy variables for the occupations and 16 industries within this same sample.

If a certain group was overrepresented in good jobs, we would expect to find a predominance of positive coefficients in the first occupations listed in Table 9, as well as a predominance of negative coefficients for the lower-ranked occupations. This pattern is roughly discernable for immigrants from Hong Kong. The majority of significant negative coefficients are found towards the bottom of the table. Just the opposite pattern is apparent for Vietnamese immigrants. They are relatively overrepresented in services and clerical occupations, and underrepresented in managerial and teaching occupations. The good job/bad job difference between these two groups may be explained by the information in Table 5 on class of admission. Hong Kong immigrants are primarily admitted under the point system so their skills may be matched with good opportunities in Canada. On the other hand, the Vietnamese enter Canada primarily as refugees, so they are not selected on the basis of their skills.

This analysis also provides yet another view of the considerable heterogeneity within the Asia Pacific group that may be masked in our earnings regressions (Tables 6 and 8).[16] While we find that Asia Pacific immigrants start at an earnings disadvantage relative to other immigrants, this average result may not characterize the outcomes of immigrants from Hong Kong or Vietnam very well. For example, given their high relative levels of observable skills and favourable occupational distribution, we might expect Hong Kong immigrants to actually enter Canada at a relative earnings advantage or to enjoy higher rates of assimilation than other immigrants. Investigation of such heterogeneity within the Asia Pacific group should be possible with the release of the 1996 census, which will provide another cross-section of data with detailed information on place of birth.

IMMIGRATION AND TRADE LINKS

IN THIS SECTION WE CHANGE OUR FOCUS to examine some of the indirect effects of Asia Pacific immigration on the Canadian economy, in particular as it affects trade patterns. This brings our discussion closer to some of the questions addressed in other papers presented at this conference.

There are a number of mechanisms by which immigrants from the Asia Pacific region may expand trade. These are discussed, for example, in Head and Ries (1995), Globerman (1995), and Gould (1994). The most important means by which trade is expanded is through the reduction of transaction costs, brought about by increased information. Immigrants from the Asia Pacific region bring with them knowledge of their home-country markets, which provides both opportunities for export and new sources of imports. Immigrants may also have preferences for goods produced in their countries of origin. While this is almost certainly the case, the magnitude of that effect remains an open question. As these mechanisms are inherently unobservable (or at least non-quantifiable, or unmeasured), our emphasis will be on establishing whether there is an empirically robust link between immigration from and trade with the Asia Pacific region. For more thorough discussions of the trade theory, or indeed for descriptions of trade between Canada and the Asia Pacific region, readers are directed towards the more trade-oriented conference papers.

Our analysis borrows considerably from Head and Ries (1995), who estimate a formal gravity model of trade between Canada and each of its trading partners, estimating the impact of the cumulative stock of immigrants from a given country on imports from and exports to that country. Our base specification is a variation of their gravity

$$lnX_{jt} = \beta_0 + \beta_1 lnGDPS_{jt} + \beta_2 lnGDPC_t + \beta_3 ln\frac{P_S}{P_C} + \beta_4 OPENC_t + \beta_5 OPENS_{jt} + \beta_6 lnSIMM_{jt} + v_{jt}$$

$$lnM_{jt} = \gamma_0 + \gamma_1 lnGDPS_{jt} + \gamma_2 lnGDPC_t + \gamma_3 ln\frac{P_S}{P_C} + \gamma_4 OPENC_t + \gamma_5 OPENS_{jt} + \gamma_6 lnSIMM_{jt} + \varepsilon_{jt}$$

for exports X_{jt} to country j and imports M_{jt} from country j. GDPS and GDPC are the GDPs of the source country j and Canada respectively, P_S / P_C are the real terms of trade, OPENS and OPENC are measures of the openness of the economies (the sum of imports and exports as a fraction of GDP), and $SIMM_{jt}$ is the cumulative sum of immigrants to Canada from country j since 1974. Our specification differs from Head and Ries' in that we do not include controls for distance to Canada or for whether the country is adjacent to Canada, and we have less restrictive controls for GDPS and GDPC.

Head and Ries, who estimate their equations on a sample that includes all countries, found (approximately) that β_6, the effect of immigrants on exports, was 0.1 and that γ_6, the effect of immigrants on imports, was 0.3. The results of interacting an East Asian (similar to Asia Pacific) indicator with the immigration variable yielded higher effects, on the order of $\beta_6 = 0.3$ and $\gamma_6 = 0.7$. Our sample differs from theirs most significantly in that we only include the Asia Pacific countries in our sample.[17] In our empirical analysis, we extend their approach to take into account other sources of heterogeneity that may affect the analysis.[18]

The results of estimating these equations are reported in Table 10. We report a base specification in column 1 for exports and imports, respectively. We are less interested in the coefficients on the trade-motivated control variables, but simply note that they are similar to those of Head and Ries. Our discussion will focus on the immigration coefficients. The export equation coefficient is very similar to the finding in Head and Ries (for East Asia): we find an elasticity of 0.301. However, we find a much smaller import elasticity of 0.2. Nevertheless, both of these results suggest that there is a strong correlation between trade and immigration: countries with which Canada has more trade are also countries from which it receives more immigrants.

This correlation raises a very important question: In which direction does the causality work? Do strong trade links provide information to potential immigrants and facilitate immigration flows? Or alternatively, are there common unobserved factors at work, increasing connections between Canada and the Asia Pacific regions and independently increasing both immigration and trade? There is a possibility, then, that these correlations are purely spurious. While we have no credible means of addressing the simultaneity problem (we doubt whether plausible instrumental variables exist), we take a number of approaches in addressing the heterogeneity problem.

The simplest way to control for some of the heterogeneity is to include country fixed effects in the regression.[19] This will control for permanent characteristics of the trading partners that may be correlated with both immigration and trade. These results are presented in column 2. As can be seen, the fixed effects are generally significant. Their inclusion also significantly switches the sign of the terms-of-trade coefficient. For exports, however, the immigration coefficient is virtually unaffected (at 0.305). For the import equation, the immigration coefficient has actually increased. This suggests that some of the unobserved heterogeneity was negatively correlated with immigration. The new estimate is 0.832, and closer to Head and Ries' estimate for East Asia.

Another way to control for country-specific heterogeneity is to include a lagged dependent variable (LDV) in the specification. The lagged import and export variables should control for some of the heterogeneity that moves slowly over time. Inclusion of the LDV significantly reduces the immigration coefficients in both trade equations (column 3), so that the immigration coefficient is insignificant in the import equation. Taking the coefficients at face value, however, we can compute long-run effects of immigration on trade.[20] They are 0.659 for exports and 0.933 for imports, both quite large. However, we do not place much credence in these estimates since the standard errors on the immigration variables are also very large. As well, if there is serial correlation in the trade equations, then the lagged dependent variable coefficients will be biased. Our motivation for including the LDV is to absorb heterogeneity, not to characterize the dynamics of the impact of immigrants on trade, and we

TABLE 10

LINKS BETWEEN IMMIGRATION AND TRADE
(STANDARD ERRORS IN PARENTHESES)

	LOG EXPORTS			LOG IMPORTS		
	(1)	(2)	(3)	(1)	(2)	(3)
Lagged dependent variable			0.727* (0.052)			0.686* (0.047)
Ln GDPC	1.412* (0.426)	1.649 (0.869)	1.666* (0.537)	3.552* (0.846)	0.258 (2.362)	1.051 (1.445)
Ln GDPS	0.618* (0.065)	0.503 (0.258)	−0.015 (0.167)	0.223 (0.130)	0.368 (0.700)	0.054 (0.444)
Ln terms of trade	2.085* (0.132)	−0.672* (0.175)	−0.341* (0.114)	2.652* (0.263)	−2.635* (0.476)	−0.758* (0.320)
Open Canada	−0.059* (0.021)	0.002 (0.012)	−0.002 (0.007)	−0.016 (0.042)	0.106* (0.032)	0.026 (0.021)
Open source country	0.002* (0.001)	−0.003 (0.002)	−0.001 (0.001)	−0.001 (0.002)	−0.005 (0.004)	0.000 (0.003)
Ln cumulative immigrants	0.301* (0.037)	0.305* (0.087)	0.180* (0.073)	0.219* (0.073)	0.832* (0.236)	0.293 (0.190)
China		−4.179* (0.357)	−1.564* (0.330)		−9.979* (0.970)	−2.737* (0.805)
Hong Kong		−0.752 (1.003)	−1.010 (0.635)		−2.877 (2.725)	−1.312 (1.713)
Malaysia		−2.217* (0.866)	−1.244* (0.554)		−4.346 (2.353)	−1.621 (1.483)
Thailand		−3.310* (0.656)	−1.241* (0.440)		−5.181* (1.783)	−1.600 (1.142)
Indonesia		−3.879* (0.504)	−1.335* (0.364)		−6.081* (1.370)	−1.835* (0.894)
South Korea		−1.266* (0.618)	−0.862 (0.396)		−3.364* (1.680)	−1.196 (1.063)
Philippines		−3.920* (0.796)	−1.853* (0.551)		−7.791* (2.164)	−2.588 (1.424)
Singapore		0.104 (1.311)	−0.598 (0.824)		0.087 (3.562)	−0.562 (2.224)
Trend		−0.023 (0.030)	−0.043* (0.020)		−0.047 (0.083)	−0.047 (0.053)

* Statistically significant at the 5 percent level.

Note: See the appendix for the explanation of variables. Regressions also include an intercept term. The sample covers the years 1974-92. The omitted country category is Japan.

do not place so strict an interpretation on the results. In fact, we view these results as suggestive of a need to control for time-changing heterogeneity in the trade equations.

If the common factors that affect both immigration and trade are changing over time, then fixed effects will not fully address the heterogeneity bias. As witnessed by the theme of this conference, the Asia Pacific region has, in many respects, become more important in the world economy over the past 20 years. Immigration and trade with Canada may be the joint product of this specific manifestation of globalization. If that is true, it would certainly affect U.S. trade and immigration patterns as well. Thus, as with our earlier analysis of immigration patterns in both countries, we can use differences between Canada and the United States to net out the correlation between immigration and trade, controlling for the common factors that may contaminate estimates based on the Canadian experience alone. This exercise has particular promise because of the differing patterns of immigration to Canada and the United States from the Asia Pacific region.

As with our earlier analysis, bringing the United States into the analysis forces us to make compromises in our sample because of data availability. In particular, we must drop Singapore, Malaysia, and Indonesia from our sample, and pool China and Taiwan. We then estimate the following equations:

$$\ln X_{jt} = \beta_0 + \beta_1 \Delta \ln GDP_t + \beta_2 \Delta \ln P_t + \beta_3 \Delta OPEN_t + \beta_4 \Delta \ln SIMM_t + v_{jt}$$

$$\ln M_{jt} = \gamma_0 + \gamma_1 \Delta \ln GDP_t + \gamma_2 \Delta \ln P_t + \gamma_3 \Delta OPEN_t + \gamma_4 \Delta \ln SIMM_t + v_{jt}$$

where $\Delta \ln GDP$ is the difference in GDP between Canada and the United States; $\Delta \ln P$ is the difference in price levels; $\Delta OPEN$ is the difference in openness; and $\Delta \ln SIMM$ is the difference in the cumulative stock of immigrants from source country j. The source-country characteristics fall out with differencing (restricting coefficients to be equal for Canada and the United States). Thus the Canada-U.S. comparison has the benefit of netting out world or North American common factors that relate to both trade and immigration, but also Asia Pacific factors common to both Canada and the United States. The results of the estimates of this equation are presented in Table 11.

As suspected, the coefficients on immigration are smaller than in the previous table, with the elasticity of immigration in both import and export equations being about 0.2. However, once we control for source-country fixed effects, the coefficients rise. The export elasticity is 0.45, higher than all of the previous estimates. The import elasticity is 0.58, also quite high. For completeness, we provide estimates that include a lagged dependent variable. The immigration terms remain significant, and the estimated long-run elasticities are 0.57 and 0.77 for exports and imports. Thus, after comparing Canada and the United States, we find that the conclusions to be drawn from Table 10 were not especially misleading. It appears that Canada's trade patterns within

TABLE 11

LINKS BETWEEN TRADE AND IMMIGRATION: DIFFERENCING CANADA AND THE UNITED STATES
(STANDARD ERRORS IN PARENTHESES)

	LOG EXPORTS			LOG IMPORTS		
	(1)	(2)	(3)	(1)	(2)	(3)
Δ Lagged dependent variable			0.704* (0.046)			0.739* (0.047)
Δ Log GDP	2.619 (2.588)	1.544 (2.014)	1.941* (0.970)	0.905 (3.199)	1.695 (2.503)	1.705 (1.192)
Δ Log terms of trade	1.891 (1.278)	1.876* (0.875)	1.298* (0.441)	1.943 (1.580)	1.307 (1.088)	0.832 (0.541)
Δ Open economy	0.028 (0.038)	0.033 (0.025)	0.028* (0.013)	0.037 (0.047)	0.027 (0.032)	0.019 (0.016)
Δ Log cumulative immigrants	0.185* (0.043)	0.448* (0.108)	0.168* (0.059)	0.193* (0.053)	0.583* (0.134)	0.202* (0.073)
China and Taiwan		0.095 (0.137)	−0.084 (0.067)		0.416* (0.171)	−0.030 (0.085)
Hong Kong		−0.863* (0.257)	−0.312* (0.136)		−1.213* (0.319)	−0.402* (0.170)
South Korea		0.454* (0.132)	0.177* (0.069)		0.542* (0.164)	0.193* (0.084)
Philippines		−0.894* (0.112)	−0.190* (0.068)		−0.890* (0.139)	−0.132 (0.078)
Thailand		−0.014 (0.179)	0.073 (0.086)		0.265 (0.223)	0.178 (0.107)
Trend		0.009 (0.012)	0.004 (0.006)		−0.010 (0.014)	−0.004 (0.007)

* Statistically significant at the 5 percent level.
Note: See the appendix for the explanation of variables. The omitted country category is Japan.

the Asia Pacific region are tilted towards those countries from which it has received more immigrants. The same holds true for the United States, which has a different mix of Asia Pacific immigrants. Therefore, we are left with a reasonably robust conclusion that immigration is correlated with trade, with the correlation with imports being slightly higher than with exports. We must still exercise caution, however, since we have not controlled for country-specific (Canada or United States) factors or addressed possible simultaneity bias. Similarly, the restriction of the equality of the Canadian and U.S. trade parameters may not be valid. Nevertheless, if these factors were important, we would expect greater differences between the results in Tables 10 and 11.

Immigration and Foreign Direct Investment

AS A FINAL EXERCISE, we explore linkages between immigration from the Asia Pacific region and investment flows between that region and Canada. Our focus will be on foreign direct investment (FDI), as opposed to broader investment in Canadian bonds, etc. The motivation for looking at these linkages is the same as for trade: transaction costs may be reduced with the presence of an immigrant community, as immigrants facilitate information flows between regions. As with the analysis of trade, our objective is quite limited. We are interested in whether a robust correlation exists between immigration and FDI, not in constructing a complete model of international financial flows. Nevertheless, it is important to control for any factors affecting FDI that may be correlated with immigration. We follow the same methodology as in previous sections. We estimate a base model, then augment it with country fixed effects and time trends. We then use the United States as a control sample and compare Canada-U.S. differences in investment patterns. As it turns out, the data comparability problems are more severe in this exercise than in the previous analyses.

We estimate the following base model. Foreign investment flows in both directions between Canada and the Asia Pacific countries are modelled as a function of economic variables that should affect the level of foreign investment. We also include a measure of the cumulative stock of immigrants from the Asia Pacific source countries to assess the impact of immigration. The base regressions are given by

$$FDI_{jt} = \beta_0 + \beta_1 GDP_{Ct} + \beta_2 GDP_{St} + \beta_3 \Delta ln Y_{Ct} + \beta_4 \Delta ln Y_{St}$$
$$+ \beta_5 \Delta ln P_{Ct} + \beta_6 \Delta ln P_{St} + \beta_7 ln PI_{Ct} + \beta_8 ln PI_{St}$$
$$+ \beta_9 \Delta ln TSE_t + \beta_{10} SIMM_{St} + \varepsilon_{St}$$

and are estimated for investment between Canada and the Asia Pacific region in both directions. Unlike the earlier equations, these are estimated in levels rather than logs, since it is possible for investment values to be positive, zero, or negative. FDI is measured as the change in the book value of investments held by shareholders in the source country. It should be correlated with the actual flow of investments, though it will also reflect changes in the capital value of existing investments. We control for the levels of GDP in both the source countries and Canada, as well as for percentage changes in GDP ($\Delta ln Y$), in real price levels ($\Delta ln P$), and in the price indices for investment goods ($\Delta ln PI$). We also include a measure in the change of value of the Toronto Stock Exchange (TSE) stock price index. All variables are measured in real terms. There are a variety of theoretical reasons for including these variables, but we do not wish to give this model more theoretical emphasis than it deserves. Most of the variables are designed to reflect changes in investment opportunities,

Table 12

Links Between Asia Pacific Immigration to Canada and Foreign Direct Investment (FDI)
(Standard Errors in Parentheses)

	FDI from Asia Pacific region to Canada		FDI from Canada to Asia Pacific region	
	(1)	(2)	(3)	(4)
GDP Canada	−43 (260)	1,673 (1,400)	363 (250)	−1,819 (1,210)
GDP source country	134* (30)	244 (130)	59 (30)	430* (110)
$\Delta \ln$ GDP Canada	7.216 (8.407)	0.481 (10.276)	−3.832 (7.914)	6.390 (8.908)
$\Delta \ln$ GDP source country	−1.026 (3.900)	−1.335 (4.090)	0.110 (3.671)	−0.074 (2.545)
$\Delta \ln$ prices Canada	16.288 (9.951)	11.268 (11.112)	−1.140 (9.368)	7.547 (9.632)
$\Delta \ln$ prices source country	−1.670 (4.676)	0.850 (5.105)	−2.482 (4.402)	−1.119 (4.425)
$\Delta \ln$ investment prices Canada	−7.400 (9.255)	−3.467 (10.041)	1.357 (8.713)	−4.874 (8.704)
$\Delta \ln$ investment prices source country	0.690 (4.460)	−1.794 (4.853)	2.010 (4.198)	1.433 (4.206)
$\Delta \ln$ TSE index	0.307 (0.933)	0.564 (0.963)	0.262 (0.878)	0.157 (0.835)
Sum immigrants (source country)	1.286* (0.471)	1.408 (0.821)	−0.070 (0.444)	0.313 (0.712)
Hong Kong		150.504 (174.270)		488.238* (151.063)
Taiwan		133.042 (174.283)		472.197* (147.607)
South Korea		113.117 (160.897)		432.794* (139.471)
Malaysia		145.968 (176.629)		481.122* (153.108)
Singapore		168.647 (183.073)		567.183* (158.693)
Trend		−21.848 (16.067)		18.710 (13.927)

* Statistically significant at the 5 percent level.
Note: Units: the dependent variable is the level of foreign investment in millions of dollars. GDP is measured in billions of dollars, and the changes in log values are multiplied by 100 (approximately corresponding to percentage changes). The omitted country category is Japan.

which ought to be correlated with changes in the expectation of returns for foreign investors.

The results of estimating the base model are presented in columns 1 and 3 of Table 12. In this specification we find a significant positive effect of immigration on FDI. The estimated coefficient suggests that an increase in the number of Asia Pacific immigrants is associated with an increase in FDI in Canada from the region of $1.286 million dollars, or about $1,000 per immigrant. Very few of the other coefficients are significant. Also, there does not appear to be a relationship between the stock of immigrants and Canadian FDI in the Asia Pacific region. In the next specification (columns 2 and 4) we add country fixed effects and a time trend. Here, the immigration effect ceases to be significant. On the basis of these results, it would be very difficult to claim that there was a robust statistical relationship between immigration and investment.

Next, we compare investment and immigration flows between Canada and the United States. This exercise enables us to control for time-varying North America (or world) effects, as well as time-varying source-country effects that are common to both the United States and Canada. Unfortunately, the U.S. data only permit including three countries in the sample – Japan, South Korea, and Hong Kong. This places real limitations on the analysis, but it should at least enable us to evaluate the robustness of the results based on the Canadian data alone. The results of this exercise, presented in Table 13, clearly indicate that there is no statistically significant relationship between immigration and FDI in any of the specifications. While this may be a consequence of the small sample, more likely it confirms the results based on the Canadian data: there is no stable correlation between FDI and immigration.

Conclusions

THIS CONFERENCE IS JUST ONE OF A GROWING NUMBER OF FORUMS in which the importance of the Asia Pacific region to the world economy is debated. While we adopt a particular Canadian perspective, the various papers presented here draw on issues that many countries face. This region is increasingly a centre for world trade, a major reservoir of investment capital, and the source of many of the world's migrants. In Canada, the specific manifestations of these trends can be viewed abstractly in our international trade and finance accounts, and more concretely in the attention that our citizens who hail from this region command.

In this paper, we view the consequences of the emergence of the Asia Pacific region through the lens of Canadian immigration. Asia Pacific immigrants, who as a group are often so prominent in the media, are somewhat less so in the data. They are currently the largest single group of immigrants to Canada and thus, by their numbers, demand our attention. Nevertheless, their

TABLE 13

IMMIGRATION AND FOREIGN DIRECT INVESTMENT: DIFFERENCE BETWEEN CANADA AND THE UNITED STATES
(STANDARD ERRORS IN PARENTHESES)

	FDI FROM THE ASIA PACIFIC REGION		FDI TO THE ASIA PACIFIC REGION	
	(1)	(2)	(3)	(4)
Δ GDP	2180 (2,040)	8,579 (6,650)	464 (410)	−1,364 (1,890)
Δ Δ ln GDP	−26,393 (51,722)	−1,102 (43,203)	−20,411 (10,499)	−27,826* (12,257)
Δ Δ ln investment prices	−12,063 (31,617)	−4,676 (22,464)	−9,290 (6,418)	−11,479 (6,373)
Δ Δ ln prices	−5,570 (36,077)	−12,018 (25,014)	6,612 (7,323)	8,509 (7,097)
Δ Δ ln stock market indices (TSE − DOW)	3,555 (10,668)	1,970 (7,320)	3,680 (2,165)	4,158* (2,077)
Δ sum of immigrants	−5.400 (3.104)	−5.849 (7.404)	−1.064 (0.630)	−1.694 (2.101)
Hong Kong		5,928* (1,079)		715* (306)
South Korea		3,495 (2,773)		187 (787)
Trend		647 (667)		−193 (189)

* Statistically significant at the 5 percent level.

Note: The first Δ refer to differences between Canada and the United States, while the second Δ are time differences. Units are as follows: the dependent variable is measured in millions of dollars; GDP, in billions of dollars; and immigrants, in thousands. Unlike in Table 12, the changes in log values are not scaled by 100 (i.e., they represent changes in proportions, not percentage changes). The omitted country category is Japan.

activities in the labour market are very similar to those of immigrants from other parts of the world. At the level of detail available in the current data sets, they assimilate at the same rate as other immigrants and enter Canada with somewhat larger initial earnings deficits. Another point that our analysis makes clear, however, is that Asia Pacific immigrants are a very heterogeneous group. Income levels and earnings characteristics vary widely across source countries, as do the administrative criteria applied to their admission to Canada. Similarly, while flows of immigration from all countries are commonly sensitive to economic conditions in Canada, the patterns of these flows differ significantly, at least partially reflecting different historical relationships between Canada and the source countries. Therefore, it is possible that more detailed analysis would provide a sharper distinction among immigrants from specific Asia Pacific countries and reveal greater variation in their respective contributions to the Canadian labour market.

A similar story emerges from our analysis of linkages between immigration, trade, and investment. We do find that the levels of imports and exports between Canada and the Asia Pacific countries are significantly correlated with the stock of immigrants from these countries in Canada. This finding is also robust to controls for many of the factors related to globalization, which may lead to an overstatement of this relationship. The causality in this relationship remains unclear, however, and a cautious researcher would be reluctant to view the results as confirming that these immigrants open up new markets for Canada. Our analysis of foreign investment leads to more robust conclusions: we find no evidence of a correlation between investment from different regions of the Asia Pacific region and the stock of immigrants it sends to Canada.

If there is something peculiar to Asia Pacific immigrants in the specific dimensions investigated here, the data only whisper of its existence. Future analysis based on more detailed data may provide more pronounced confirmation. Alternatively, the outstanding features of the Asia Pacific region may lie in other avenues of its economic interactions with Canada.

Appendix

Data Sources and Explanations of Variables

Immigration Flows

Canada

a) Asian Immigration (aggregate): CANSIM, series D36
b) Asia Pacific immigration by country: Employment and Immigration Canada, *Immigration Statistics*, various issues (Ottawa: Supply and Services Canada)
c) Class of immigration by source country: Employment and Immigration Canada, *Immigration Statistics*, various issues (Ottawa: Supply and Services Canada)

United States

a) Asian immigration (aggregate): *Historical Statistics of the United States*, Part 1 (1975); and U.S. Department of Commerce, Bureau of the Census, *Statistical Abstract of the United States*, various issues.

b) Asia Pacific immigration (by country): U.S. Department of Commerce, Bureau of the Census, *Statistical Abstract of the United States*, various issues; and Department of Justice, Immigration and Naturalization Service, *Annual Report*, 1994.

Variables for the Immigration Flow Equations

a) Immigration flows as described above.
b) *PCGDP* is from the Penn World Tables (PWT), 5.6 (Alan Heston and Robert Summers), real GDP per capita using chain index (1985 international price), POPULATION is also from the PWT.
c) Unemployment rate in Canada is from CANSIM (series D767611, annual average), the unemployment rate of all individuals 15 years old and over.
d) Unemployment rate in the United States is from CITIBASE (series LHUR, annual average), the unemployment rate of all individuals 16 years old and over.

Variables for the Trade Equations

a) Immigration flows as described above, used to form cumulative sums of immigrants from each country since 1974.
b) GDP is real GDP based on population (POP) and *PCGDP* variables from the Penn World Tables (see above).
c) Openess is from PWT and is the sum of imports and exports as a share of GDP.
d) Prices are defined as the price of GDP expressed in constant U.S. dollars.
e) Imports and exports are from the International Monetary Fund's Direction of Trade Statistics (IMF/DOTS) database. Nominal values are converted to real using the price indices in the PWT.

Variables for the Investment Equations

a) Immigration flows and the cumulative stock as described above.
b) Real GDP as described above.
c) FDI in Canada from CANSIM, "Canada's International Direct Investment Position." Investment in the Asia Pacific region is based on "Canada's Direct Investment Abroad," and investment in Canada on "Foreign Direct Investment in Canada," for various countries. All series give the nominal book value of the investments. These are converted to real data using the price indices from PWT.

d) FDI in the United States: both FDI in the United States and investment abroad come from the Bureau of Economic Analysis, *Survey of Current Business*, various issues. Nominal book values are converted to constant 1985 dollars using the price indices from PWT.
e) The price of investment is the price index for the investment component of GDP in the PWT.
f) The stock market indices are: TSE, retrieved from the CANSIM; and DOW, retrieved from CITIBASE.

Census-Based Data

Canada

The Canadian data are drawn from the 1986 and 1991 censuses. Samples are for individuals between the ages of 16 and 64. Observations from Newfoundland, Nova Scotia, New Brunswick, Prince Edward Island, and the Yukon and Northwest Territories are excluded. The full-year/full-time samples include paid workers who worked more than 39 weeks in the reference year in mostly full-time employment.

United States

The U.S. data are drawn from the 1980 and 1990 U.S. censuses, 5 percent Public Use Microdata Samples (PUMS). All samples are for individuals between the ages of 16 and 64. The full-year/full-time samples include paid workers who worked more than 39 weeks in the reference year, usually for more than 30 hours a week (corresponding to the definition of full-time in the Canadian census).

Endnotes

1 We thank Don DeVoretz, Richard Harris, Elizabeth Ruddick, and conference participants for helpful suggestions. Terence Yuen provided excellent assistance in assembling our data sets. This project was partially funded by the Social Sciences and Humanities Research Council.
2 For descriptions of immigrant flows from other source countries, see Green (1995) or Baker and Benjamin (1994).
3 China and Taiwan are pooled in the U.S. immigration figures until 1982, which made a separate treatment impossible.

4   The *PCGDP* data are taken from the Penn World Tables, while *URATEC* is from CANSIM. The data are unavailable for all countries. In particular, GDP data are unavailable for Vietnam in the PWT.
5   See Akbari and DeVoretz (1992) for an example of a demand-side treatment of immigration flows to Canada.
6   The census codes this category as "other East/Southeast Asian countries." They include Japan, North and South Korea, Macao, Taiwan, Indonesia, Malaysia, Singapore, and Thailand.
7   We define urban residence by the 19 cities identified in the census files.
8   The labour market participation is the sum of the employment rate and the unemployment rate in the reference week.
9   We note that some immigrants in the older cohorts may have immigrated at a young age, and thus would have been educated in Canada.
10 For a more complete discussion of transfer program participation of immigrants to Canada, see Baker and Benjamin (1995a and 1995b).
11 This variable is coded from the census variable "other income from government sources." It includes social assistance, disability benefits, and veterans' allowances and pensions, and excludes UI benefits and retirement pensions and benefits.
12 Data on admission class by source country is available from 1983 through 1992. We choose to examine the IM8690 cohort as it matches the census definition and provides a fairly accurate view of the class decomposition for recent entrants.
13 For a more detailed exposition of the need for the quasi-panel approach, see Borjas (1985) or Baker and Benjamin (1994).
14 The assumption of a common time effect on native and immigrant earnings is necessary for identification.
15 The largest constituents of this group are immigrants from Japan and Taiwan.
16 In Baker and Benjamin (1995c), we explore this heterogeneity with respect to earnings in greater detail, though our focus is broader, both in terms of looking at immigrants from all source regions, as well as Canadian-born residents of the matching ethnicity. In that paper, the distinction between "Chinese" and "other South-East Asian" immigrants is also noted.
17 We also use slightly different data sources. As they do, we use the Penn World tables for the key macroeconomic variables, but we use the IMF Direction of Trade database instead of Statistics Canada trade data. We use the IMF data, in part, because we need comparable data for the United States and Canada. Our sample includes Japan, China, Hong Kong, Malaysia, Thailand, Indonesia, South Korea, the Philippines, and Singapore. Because Taiwan was not a member of the IMF, no data were available.

18 We use the OLS method to estimate our equations while they use a tobit because of the large number of "zeroes" in their data. Except for four observations (1974-77 for China), we have no zeroes. Instead of using the tobit, we replace the zeroes with a small number. While this procedure is generally unadvisable, it has no impact on our results, insofar as the conclusions are insensitive to the choice of small number used to replace the missing observations for China. Indeed, it also makes no difference if we drop these observations altogether.

19 Head and Ries include a region dummy for East Asia. This would be comparable to our specification without fixed effects, since the intercept in our base specification captures the common Asia Pacific "effect."

20 If λ is the coefficient on the lagged dependent variable and β is the coefficient on immigration, the long-run impact of immigration is given by $\beta/(1-\lambda)$.

BIBLIOGRAPHY

Akbari, Ather and Don DeVoretz. "The Substitutability of Foreign-Born Labour in Canadian Production, Circa 1980." *Canadian Journal of Economics*, 25, 3 (August 1992): 604-14.

Baker, Michael and Dwayne Benjamin. "The Performance of Immigrants in the Canadian Labor Market." *Journal of Labor Economics*, 12 (1994): 369-405.

_____. "The Receipt of Transfer Payments by Immigrants to Canada." *Journal of Human Resources*, 30, 4 (Fall 1995a): 650-76.

_____. "Labor Market Outcomes and the Participation of Immigrant Women in Canadian Transfer Programs," in *Diminishing Returns: The Economics of Canada's Recent Immigration Policy*. Edited by Don DeVoretz. Toronto: C.D. Howe Institute and the Laurier Institute, 1995b.

_____. "Ethnicity, Foreign Birth and Earnings: A Canada/ U.S. Comparison." Manuscript, University of Toronto, 1995c.

Bloom, David, Gilles Grenier and M. Gunderson. "The Changing Labor Market Position of Canadian Immigrants." *Canadian Journal of Economics*, 27, 4b (November 1995):987-1005.

Bloom, David and Morley Gunderson. "An Analysis of the Earnings of Canadian Immigrants," in *Immigration, Trade, and the Labor Market*. Edited by John Abowd and Richard Freeman. Chicago: University of Chicago Press, 1991.

Borjas, George. "Assimilation, Changes in Cohort Quality, and the Earnings of Immigrants." *Journal of Labor Economics*, 3 (1985): 463-89.

Card, David and Craig Riddell. "A Comparative Analysis of Unemployment in Canada and the United States," in *Small Differences that Matter: Labor Markets and Income Maintenance in Canada and the United States*. Edited by D. Card and R. Freeman. Chicago: University of Chicago Press, 1993.

DeVoretz, Don (ed). *Diminishing Returns: The Economics of Canada's Recent Immigration Policy*. Toronto: C.D. Howe Institute and the Laurier Institute, 1995.

Globerman, Steven. "Immigration and Trade," in *Diminishing Returns: The Economics of Canada's Recent Immigration Policy*. Edited by Don DeVoretz. Toronto: C.D. Howe Institute and the Laurier Institute, 1995.

Gould, David M. "Immigration Links to the Home Country: Empirical Implications for U.S. Bilateral Trade Flows." *The Review of Economics and Statistics*, 76 (1994): 302-16.

Grant, Mary. "New Evidence on Immigrant Assimilation in Canada." Unpublished paper, University of Toronto, 1995.

Green, Alan C. *Immigration and the Postwar Canadian Economy*. Macmillan of Canada, 1976.

⸻⸻⸻. "A Comparison of Canadian and US Immigration Policy in the Twentieth Century," in *Diminishing Returns: The Economics of Canada's Recent Immigration Policy*. Edited by Don DeVoretz. Toronto: C.D. Howe Institute and the Laurier Institute, 1995.

Head, Keith and John Ries. "Immigration and Trade Creation: Econometric Evidence from Canada." Manuscript, University of British Columbia, 1995.

Comment

Donald J. DeVoretz
Department of Economics
Simon Fraser University

BAKER AND BENJAMIN HAVE CONFRONTED a very ambitious task – namely, to sort out the economic impacts of Asian, and presumably South Asian, immigration on the Canadian economy. The logic of their argument is, as usual, clear and straightforward. They initially attempt to measure the rationale for Asian immigrant entry into Canada – in this case, the supply conditions for Asian immigrants. Then they successively measure Asian immigrant earnings and their trade and investment impacts on the Canadian economy. In addition, following Globerman (1995), they employ U.S. immigrant experience and data to control for many common unobservables (such as globlization) that would have a joint effect on both immigration levels and trade. The results of their exercise are stated in their conclusions: Asians differ little in their earnings experience from other foreign-born earners; Asian investment is not correlated with the stock of Asian immigrants; and there is no uni-causal relationship between immigration and trade. These modest results, which are based on aggregated Canada-wide data, are to be expected, given the concentrated regional nature of Canada's immigration flows. For example, British Columbia since 1990 has received over 200,000 immigrants, with 70 percent residing in the greater Vancouver area. In contrast to the Baker and Benjamin analysis, how could a 20 percent increase in Vancouver's economically active population go undetected?

The major weakness of the Baker and Benjamin analysis is that it looks at aggregate data. Many have fallen prey to the aggregation error in investigating the economic impact of immigration (see Akbari and DeVoretz, 1992). Immigrants to Canada are urbanized, and three cities – Vancouver, Toronto, and until recently Montreal – received over 50 percent of the yearly inflow. This is where the impact should be measured. I will return to this theme at the end.

My criticisms of this essay are few but crucial. In this commentary, I will argue that:

- Canada's immigrant movement is explained by immigrant demand functions rather than by a supply function, and the Baker and Benjamin analysis based on a Borjas self-selection supply sorting function is misplaced.
- Earnings assimilation and labour market experience, although interesting to immigrants, are not the relevant impacts of immigrants on the Canadian economy. It is their impact on the size and direction of congestion externalities, on public finances, and on the long-term restructuring of urban consumer demand that are the important impacts.
- Controlling for fixed effects with a heterogeneous sample by introducing the U.S. immigrant experience only complicates the understanding of any impact of Asian immigrants in Canada.
- Finally, I would argue that if one analyzed the impact of immigrants at the correct aggregation level – the urban core, where immigrants reside – and used the correct indicators and removed confusing cross-border references, one would find that Asian immigrants cause significant impacts in local economies.

Why do Asian immigrants come to Canada? Contrary to the supply-side arguments originally offered by Borjas for the United States, where they are appropriate, and now by Baker and Benjamin for Canada, there exists a substantial literature to support the hypothesis developed in Figure 1 regarding the demand argument for immigrants. DeVoretz and Maki (1978, 1983) and Akbari and DeVoretz (1992) argue that e-b, or occupational immigrant demand, is a product of excess demand for a particular skill. Points are awarded for an occupation if insufficient graduates are available or if wages are rising. Under this view, it is true that immigrants self-select but, in addition, the initial wave of immigrants – i.e., the 15 to 30 percent who are point-selected – are also selected by the Canadian government. In other words Canadian immigrants under the "point system" are doubly selected. This is particularly true with Asian immigrants, who were point-selected in a proportion ranging between 30 and 50 percent during the post-1990 period. The remaining 50 percent of post-1990 Asian movers under the family reunification class are not a self-selected supply but are a delayed multiple of the initial entrants. This alternative demand-based

Figure 1
Canada's Demand and Rest of World Supply of Immigrants

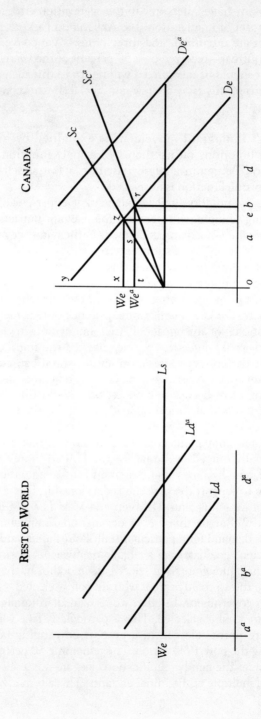

interpretation of Canadian immigration flows is crucial to most of the Baker and Benjamin analysis: Canadian immigrants do not just singly self-select and arrive on a "first come, first serve" basis as in the United States, but they are subject to crude excess labour market demand indicators in the first instance. Moreover, unlike in the United States, the yearly Canadian totals change, ranging from 80,000 to 250,000 per annum over the last 10 years, as excess demand alters in the economy (Green, 1995).

Given this economic scrutiny of Canada's immigrants, the presumption is that some economic impact would be detected somewhere and that any comparison to the hemispheric quota system of the United States must be regarded with suspicion. Moreover, given the Canadian points system, immigrant policy options abound in the Canadian context to raise the economic contribution of immigrants. For example, let us argue that language is a key condition for the rate of Asian earnings assimilation. If Asians or any cohort are not assimilating in wages under the Canadian system, you just raise the points for English/French. Canada just did this. This tinkering with the excess demand function via raising or lowering points and altering the mix allows a great deal of policy options, which were totally unavailable to the United States during the study period. For this reason alone, using the United States to standardize for global fixed effects in the later regression analysis by Baker and Benjamin is incorrect.

The next step in the Baker and Benjamin argument is to review the income growth of Asian immigrants. They argue that the model they use to explain earnings assimilation is based on human-capital theory and that the assimilation returns lead to eventual Asian immigrant "catch up." Their findings are optimistic: "The returns to assimilation are a little over 2 percent in the first years in Canada."

On the contrary, I and others have found a collapse of human-capital transfers during the 1980s (Coulson and DeVoretz, 1993), leading to an age/earning profile for Asians that lies everywhere below the Canadian-born control group.

What is the importance of this conflicting evidence? Does it really matter to the Canadian economy if the "catch-up" point is delayed? The answer is yes. If foreign-born earnings growth rates actually slow down in the future, then this implies lower taxes and higher use of public services. In fact, Akbari (1995) reports a minor positive tax transfer for the 1985-90 immigrant cohort due to the lower initial earnings of that cohort. A lack of foreign-born earnings growth could quickly turn this transfer negative in the future.

In sum, what is at stake in this earnings analysis is not the rate of earnings assimilation per se but its impact on the Canadian economy through public finance transfers. I am much less optimistic about the positive public-finance impact of the Asian 1986-90 cohort than that implied by the Baker and Benjamin earnings analysis.

Figure 2

Crossover Points for United Kingdom, Eastern European and Asian Immigrants

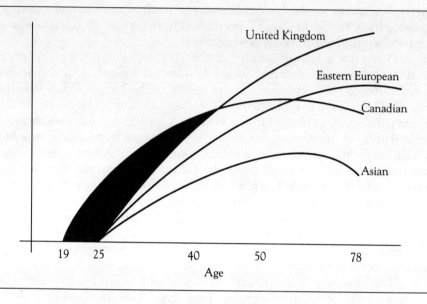

Trade and Investment

BAKER AND BENJAMIN, following others (Globerman, 1995; Head and Reis, 1995) attempt to isolate the Canadian relationship between exports, imports, and immigration. Their results are similar to those of Head and Reis – namely, positive import and export elasticities with respect to immigration. Baker and Benjamin then worry about spurious correlation – namely, a third force commonly affecting both Asian trade and immigration. I am suspicious of using the U.S. trade and immigration patterns in first differences to control for one major common third force – i.e., globalization – which in turn raise the significance of their findings of trade and immigration. Specifically, the key variable – immigration levels – cannot be put in a differenced form in their export or import equations, given the vastly different nature of the two countries' immigration policies. Thus, at a minimum, any trade and immigration effect found in the Canadian context must reveal itself in changing patterns to be convincing. And to date, as Globerman notes, this has not occurred.

Baker and Benjamin find no significant relationship between foreign direct investment (FDI) and Asian investment in Canada and vice versa under a country fixed-effects model. This is to be expected, given their macro approach with limited observations. Only a portion of FDI may be sensitive to "transaction costs" that are reduced with more immigration. For example, some investment in countries with culturally specific rules (China) and some types of Canadian-bound investment (e.g., real estate) may have substantial transaction costs. But FDI, as broadly treated in the Baker and Benjamin analysis, is unable to capture any of these micro inducements to redirect FDI via immigration.

In sum, a critical reading of the Baker and Benjamin contribution reveals that Asian immigration has had a minor impact on the Canadian economy, which is limited possibly to changing levels of Asian exports and immigration.

What if we limited our analysis to the unit that matters in an immigration context, and focused on impacts that affect the structure of the urban economy? Would we get more robust results when we measure the economic impacts of Asian immigrants? The answer is yes. I produce these below for Vancouver.

IMMIGRANT-INDUCED GROWTH: VANCOUVER'S PROJECTED EXPENDITURE PATTERNS

THE FOLLOWING SIMULATION EXERCISE in the Vancouver context more appropriately measures the impact of recent Asian immigrants. Figure 3 reports the outcome or the final stage of a view of immigrant-induced growth for one urban area – namely, Vancouver. Given that Vancouver's immigrants enter with divergent expenditure patterns and have differences in expenditure response to income changes and differential lifetime earning performances, then Figure 3 can be constructed.

Figure 3 reports a 10-year simulated growth path (1991-2001) for the expenditures in the 14 major commodity groups by foreign-birth status.[1] While it is true that without immigration Vancouver's expenditure patterns would modestly change (Canadian-born line), it is the Asian immigrant group that will lead the change in Vancouver's demand structure. The future growth in consumption patterns is predictable: shelter, household operations, furnishings, and transport will lead the way in restructuring Vancouver's economy.

Now let us look to a counterfactual world, with zero net immigration – i.e., Vancouver's situation circa 1984-86. The stock of immigrants who entered before 1981 would still be residents in the 1990s and as they age, their earnings and expenditures would decline. Modest growth in expenditure patterns would emerge from an aging Canadian-born population, as noted.

This simulation model no doubt has limitations. But what it does note is that if one limits the analysis to the economy that immigrants enter and one ferrets out the appropriate differences in Asian economic characteristics, significant economic impacts will emerge. If these impacts are diluted economy-wide and are sought in areas dominated by other mainstream economic actors

FIGURE 3

VANCOUVER EXPENDITURE PATTERNS, 1991-99

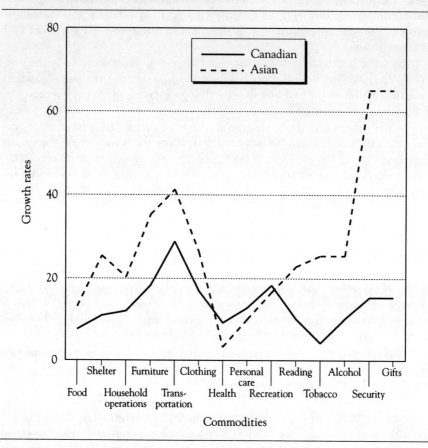

– for example, trade and foreign investment – then there will be little immigrant impact.

It is the old joke again. When the drunk was asked why he was looking for his wallet by the curb and not out in the field where he dropped it, he replied, "The light is by the curb." We just need more lights.

Endote

1. For example, the 7.5 percent increase in food expenditures by Canadian-born households between 1991 and 2001 is a product of the following multiplication: projected growth rate of Vancouver-based, Canadian-born households (i.e., 0.6 percent per annum) × 10-year growth rate of the average Canadian-born household income (2.2 percent) × the income elasticity of demand for food (0.5).

Brian R. Copeland
Department of Economics
University of British Columbia

Trade, Environment and Growth in the Asia Pacific Region: Implications for Canada

INDUSTRIAL PRODUCTION IN EAST ASIA GREW at the rate of 9 percent between 1965 and 1990. Several countries, such as Taiwan, South Korea, and Indonesia, grew at much higher rates, and China is currently experiencing a very high rate of growth. While this has led to large increases in incomes, it has also been accompanied by serious environmental problems. Several recent reviews of East Asian environmental quality have documented large increases in pollution emissions, declines in air and water quality, and depletion of natural resources such as forests and wildlife habitat (Brandon and Ramankutty, 1993; Edmonds, 1994; and Tisdell, 1995). Most countries in the region responded by enacting environmental legislation during the 1980s. However, in many countries implementation and enforcement of environmental standards has been hampered, in part because they lack the necessary institutions to deal with large-scale strains on the environment.

There are three important ways in which environmental problems in Asia can affect Canada. First, some pollutants have a global impact and can worsen Canadian environmental quality directly. CFC releases can contribute to the deterioration of the ozone layer; carbon emissions can affect the global climate; and habitat destruction can reduce biodiversity. In these cases, all countries, including Canada, have an interest in reaching international agreements to control international pollution externalities. A major obstacle to such agreements, however, is that different countries attach different priorities to global environmental problems. In many developing countries, access to clean drinking water, urban sewage systems, and improving urban air quality is of far more pressing concern. Moreover, current levels of global emissions are very uneven across countries. One would therefore expect agreements to reduce pollution to be linked to other issues, such as international transfers of income or possibly international trade agreements.

Second, Canadians may be concerned about environmental quality in other countries even if it does not directly affect Canadian environmental quality. This concern may be based on humanitarian grounds, if pollution causes illness and death in other countries. It may also arise from a belief that the current generation holds the earth's environment in trust for future generations, and that the responsibility of any individual for this task does not stop at national borders. Finally, there may be concerns that environmental degradation and population pressures in foreign countries may increase pressure on Canada to accept higher levels of immigration.

The third mechanism that links environmental quality in Asia to Canada is international trade. This linkage is the main focus of this paper. A good such as a plastic cup can be thought of as the embodiment of the inputs that were used to create it. In addition to labour and capital services, the production of plastic generates pollution, which consumes environmental services. Hence when we import a plastic cup, we are implicitly importing foreign environmental services. This has two consequences. First, Canadian demand for Asian pollution-intensive goods partially contributes to the demands placed on the Asian environment. This has led to a great deal of recent research on whether or not trade liberalization leads to increased pollution. Second, because production tends to locate in the country or region that allows firms to generate the highest profits, there is some concern that weak environmental regulations in Asia may be used to lure industry away from Canada, or that increased competition from countries with weak environmental policies may weaken the political resolve in Canada to enact and enforce strong environmental regulations. This has led to some discussion of amending international trade agreements to allow countries to respond to "environmental dumping" by imposing countervailing tariffs on imports from countries with weak environmental regulations.

The purpose of this paper is to review the linkages between economic growth, international trade, and environmental quality, and discuss their implications for Canadian policy. I begin with a brief overview of some indicators of pollution emissions in Asia Pacific countries in the second section. In the third section, I introduce a simple model that provides a framework to analyze the effects of trade liberalization and growth on environmental quality. This model predicts that trade liberalization can alter the distribution of pollution across countries and may lead to an increase in global pollution. I then review the empirical evidence, which suggests that economic growth in developing countries tends to lead to an increase in many (but not all) types of pollution, but that several indicators of environmental quality also eventually improve with income. There is clear evidence that environmental quality in many East Asian countries is worsening. There is also evidence that pollution-intensive industries have been migrating from rich to poor countries, but there is no consensus on whether or not this is due to differences in environmental policies. Studies that have attempted to measure the effect of differences in pollution policy on migration of industry have failed to find a significant causal relationship.

In the fourth and fifth sections I consider the implications for policy. One of the key issues is whether trade and environmental policy should be linked. This has been the subject of heated debate between environmentalists, economists, and others with an interest in policy. In the final section I present some conclusions.

AIR POLLUTION IN SELECTED ASIA PACIFIC COUNTRIES

ASIAN COUNTRIES FACE A VARIETY OF ENVIRONMENTAL PROBLEMS. In some countries, access to clean water and soil erosion are among the most serious of these problems. In other countries, air quality in urban areas is extremely poor.[1] In this section, I provide a brief overview of air pollution in selected Asia Pacific countries and include Canada for comparison. I focus on air pollution because there are relatively good cross-country estimates of both emissions and ambient air quality, and also because two of the major air pollutants (carbon emissions and CFCs) have global consequences that will affect both Canada and Asia directly.

In Tables 1 and 2, I present estimates of emissions of carbon dioxide and sulphur dioxide, while in Table 3 are estimates of consumption of CFCs and halons. Carbon emissions are thought to be a major contributor to global warming, and the burning of fuels that generate carbon dioxide (CO_2) also

TABLE 1

CARBON DIOXIDE EMISSIONS FROM INDUSTRIAL SOURCES, CANADA AND SELECTED ASIA PACIFIC COUNTRIES, 1960-90 (THOUSANDS OF TONS)

COUNTRY	EMISSIONS				PER CAPITA 1990	PER GDP 1990[a]
	1960	1970	1980	1990		
Canada	52,700	90,610	115,820	115,260	4.35	0.20
China	215,295	211,607	406,440	678,016	0.61	1.86
Hong Kong	807	2,263	4,489	7,401	1.26	0.12
Indonesia	5,844	9,050	25,825	38,506	0.21	0.36
Japan	63,997	202,973	254,881	289,288	2.34	0.10
Malaysia	1,112	4,092	7,764	16,216	0.90	0.38
Philippines	2,303	6,733	9,971	11,752	0.19	0.27
South Korea	3,455	14,230	34,312	65,884	1.54	0.28
Thailand	1,012	4,190	10,921	25,535	0.46	0.32
Vietnam	2,061	7,696	4,633	6,337	0.10	–

a   Kg/year.
Source: United Nations Environment Programme, 1993.

TABLE 2

SULPHUR DIOXIDE EMISSIONS FROM HUMAN-MADE SOURCES,
CANADA AND SELECTED ASIA PACIFIC COUNTRIES, 1975-90
(THOUSANDS OF TONS)

	EMISSIONS					
COUNTRY	1975	1980	1985	1990	PER CAPITA 1990	PER GDP 1990[a]
Canada	5,319	4,643	3,704	3,800	143.3	6.7
China	10,180	13,370	17,260	19,990	17.7	54.8
Hong Kong	109	166	144	150	25.9	2.5
Indonesia	201	329	435	485	2.7	4.5
Japan	2,570	1,600	1,180	1,140	9.2	0.4
Malaysia	193	272	271	263	14.7	6.2
Philippines	807	1,040	510	370	6.0	8.4
South Korea	234	271	324	333	7.9	1.4
Taiwan	609	1,040	693	605	29.8	7.0
Thailand	224	420	507	612	11.2	7.6
Vietnam	40	34	38	39	0.6	4.2

a   Kg/year.
Source: United Nations Environment Programme, 1993.

TABLE 3

CONSUMPTION OF CFCS AND HALONS,
CANADA AND SELECTED ASIA PACIFIC COUNTRIES, 1986-91 (TONS)

	CONSUMPTION			
COUNTRY	1986	1990	1991	PER CAPITA 1990[a]
Canada	23,176	15,302	10,461	0.58
China	31,668	53,268	62,821	0.05
Indonesia	2,494			0.01[b]
Japan	135,089	120,074	88,436	0.97
Malaysia	3,840	4,194	4,098	0.24
Philippines	4,359	3,142	2,112	0.05
South Korea	9,394			0.23[b]
Thailand	2,360	6,984	8,324	0.16

a   Kg/year.
b   1986.
Source: United Nations Environment Programme, 1993.

lead to emissions of other pollutants that reduce air quality. Sulphur dioxide (SO_2) can cause acid rain and respiratory problems. CFCs and halons are ozone-depleting gases. In each case, I have included data on Canada for comparison.

Perhaps the most striking fact that emerges from these tables is that Canada is a major polluter. On a per capita basis, Canada produces more CO_2 and SO_2 than any Asian country. On the other hand, as Table 4 indicates, air quality in major Canadian urban centres is better than in countries such as China or Thailand. Thus, while Canada pollutes a lot, it also manages to insulate its citizens from the effects of the pollution to a greater extent than many of the lower-income Asian countries are able to. This is partly because of Canada's wealth, and partly because of its low population density. Canada is not much different from other rich countries in this respect (although its sulphur dioxide emissions are rather high). A high level of consumption is typically accompanied by a high level of pollution.

The second important implication of this data comes from the time trends in pollution. While carbon emissions in Canada appear to have levelled off somewhat (and indeed fell between 1980 and 1990 in European countries such as France, West Germany, and Sweden), they are increasing quite dramatically in Asian countries, such as China, South Korea, and Thailand. Sulphur dioxide emissions and CFC consumption have been falling in most OECD countries (including Canada), but they are increasing in China and Thailand.

TABLE 4

CONCENTRATION OF SUSPENDED PARTICULATE MATTER IN
MAJOR CANADIAN AND ASIA PACIFIC REGION URBAN CENTRES, 1981 AND 1989
(MEAN ANNUAL VALUES, $\mu g/m^3$)[a]

COUNTRY	CITY	1981	1989
Canada	Vancouver	73	51
	Toronto	59	66
China	Beijing	422	407
	Guangzhou	260	159
	Shanghai	205	231
	Shenyang	406	422
	Xian	350	526
Hong Kong	Hong Kong		95[b]
Japan	Tokyo	63	47
Thailand	Bangkok[c]	118	115

a Measured in city centre commercial districts.
b 1988.
c Measured in suburban residential areas.
Source: United Nations Environment Programme, 1993.

The scope for further increases in pollution emissions in the Asia Pacific is reflected in the difference in per capita pollution rates. As the Asia Pacific region's economies grow, per capita air pollution emissions will creep up toward the much higher levels of the OECD countries. Because of the very high population of this region, the potential for global environmental impacts is significant. China is already among the top three emitters of carbon dioxide, nitrogen dioxide, and sulphur dioxide worldwide (the United States is the largest emitter). It is close to surpassing U.S. emissions of the first two gases, and at its current rate of economic growth will soon become the most important single contributor to the stock of major global air pollutants. Moreover, this will occur even if China were to adopt U.S. emission intensities, and even if its per capita contribution to pollution were to remain much lower than those of the OECD countries.

Effects of Growth and Trade on the Environment

Theory

A COUNTRY'S ENVIRONMENTAL QUALITY is the result of a complex interaction between the natural ecosystem and the economy. In the face of this complexity, it is useful to begin with a very simple model to try to highlight some of the forces determining environmental quality.[2] Much insight can be obtained by thinking about the demand and supply for pollution, or more generally, *environmental services*. Pollution is a special case of the use of environmental services – when one pollutes, one releases waste into the environment and exploits services provided to us by the natural environment. The environment provides us with many other services in addition to waste disposal, and this analysis can be extended to them also. But for concreteness, it is useful to focus on pollution.

I begin with a country that has not yet opened up to international trade (that is, the country is in autarky). For simplicity, I focus on the case where pollution does not spill over international borders. Transboundary pollution is treated using a similar framework in Copeland and Taylor (1995a). In the left panel of Figure 1, I have sketched the pollution demand and supply in a country I call "Home." The demand for pollution is derived from producer and consumer behaviour. Let τ be the implicit cost to a firm or individual of releasing a unit of pollution. If pollution taxes are charged explicitly, then τ is the pollution tax. If regulations are imposed, then τ is the shadow price of pollution induced by the regulations (the addition to the marginal cost of production or consumption that is induced by the need to comply with pollution regulations). If pollution is unregulated, then $\tau = 0$. The derived demand for pollution is then simply the amount of pollution that producers and consumers will generate when facing a pollution "price." In other words, it measures the marginal benefit of polluting.

Figure 1

Pollution Demand and Supply, "Home" and "Foreign"

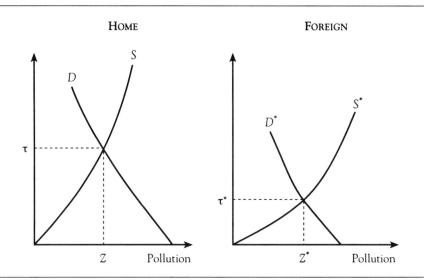

The derived demand for pollution slopes down in autarky for two reasons. First, as the cost of polluting (τ) rises, the prices of pollution-intensive goods rise and consumers shift away from them. This leads to less production of these goods, and hence less demand for pollution. Second, for a given amount of production, an increase in τ gives firms incentives to choose less pollution-intensive production techniques – they may install abatement equipment, switch to different inputs (e.g., use a coal with a lower sulphur content), or adopt a different production technology. Finally, the position of the derived demand curve depends on income. Higher-income countries have a higher demand for consumption, which typically translates into a higher demand for pollution.

The supply curve for pollution represents the trade-off between income and pollution that society is willing to make at the aggregate level. The shadow price of pollution ("τ") on the vertical axis measures the extra income that the country could generate if it allowed an extra unit of pollution. Thus the supply curve represents the amount of pollution that a country is willing to tolerate in return for compensation in the amount τ. Under this interpretation, a pollution supply curve is much like a labour supply curve: higher output and income compensate for increased pollution, just as higher wages compensate for the disutility of effort. However, unlike a labour supply curve, the shape and position of the pollution supply curve is critically dependent upon government regulation. If the government does not regulate pollution at all, and if there are no other institutions to do so, then the pollution supply curve is perfectly elastic and is a horizontal straight line through $\tau = 0$. In this case, the country allows unlimited pollution, regardless of the marginal damage that it causes. If instead the government uses a fixed pollution tax τ, then the supply curve is a horizontal straight line at $\tau = \bar{\tau}$. If an aggregate pollution quota is enforced, so that the government ensures that total pollution satisifies $z \leq \bar{Z}$, then the pollution supply curve is a vertical straight line at $z = \bar{Z}$.

As a benchmark, it is useful to consider the case where pollution regulation is perfect and reflects underlying preferences for environmental quality. In this case, an alternative and more familiar interpretation is that the supply curve measures the *marginal damage* from pollution. For any given level of pollution, it measures the amount that citizens of the country would be willing to pay to have pollution reduced by a unit. The position of the supply curve then depends on many factors, but the two most important are income and the fragility of the local environment. Since the demand for environmental quality rises with income, an increase in income should cause the supply curve to shift up and to the left. A richer country is willing to pay more for a reduction in pollution. Moreover, for a given level of income, we would expect that a country with a fragile environment would be less willing to tolerate pollution than one with a very forgiving environment. Thus, with all else equal, a country with a fragile environment would have a supply curve up and to the left of that of countries with a resilient environment.

Finally, it is important to keep in mind that the case of perfect regulation is only a hypothetical benchmark, and that few if any countries have achieved this standard. Consequently, perhaps the most critical factor affecting pollution supply is the responsiveness of government regulators to societal preferences. Because pollution control suffers from public goods problems, an increased demand for environmental quality will not be translated into actual improvements in the environment unless institutions are in place to control pollution, and these institutions are responsive to the preferences of the citizens.

Equilibrium is determined by the intersection of demand and supply. This yields both the level of pollution (Z) generated by the country and the shadow price of pollution (τ).

With our simple model in hand, let us consider the effects of economic growth on pollution. It is useful to think of the effects of growth in terms of what Grossman and Krueger (1993) and Copeland and Taylor (1994) call scale, composition, and technique effects. Growth increases the productive capacity of the economy. Provided consumption patterns and pollution regulations stay the same, one would expect the derived demand for pollution to rise. This is called the *scale effect*. In Figure 1, this would correspond to an outward shift of the demand curve. Next, the composition of goods produced may change. If growth is biased toward pollution-intensive industries, then the *composition effect* may further increase the derived demand for pollution (reinforcing the outward shift of the curve). On the other hand, if higher-income consumers switch to less pollution-intensive goods (such as services) then the composition of production may well shift in favour of cleaner goods, and the composition effect may work against the scale effect, reducing the demand for pollution. Let us suppose that, on balance, the scale and composition effects tend to increase the demand for pollution in autarky, and thus the demand curve shifts out. Finally, we have the *technique effect*. Because environmental quality is a normal good, growth should lead to an increase in the demand for environmental quality and hence a reduction in the willingness to allow pollution. Thus, provided that policy responds to the collective wishes of consumers, one would expect the supply curve to shift in.

Because growth leads to an increase in the derived demand for pollution and a reduction in supply, the shadow price of pollution (or environmental services) tends to rise with growth. However, the net effect of growth on pollution is ambiguous. If the scale and composition effects are stronger than the technique effect, then pollution will rise, but if the income elasticity of demand for environmental quality is high, then the technique effect can dominate and pollution can fall. Notice, however, that because environmental quality is a public good, the mere fact that its income elasticity of demand is large will not guarantee an improvement in environmental quality. Rather, the demand for environmental quality will only be translated into an actual improvement in environmental quality if a country has a political system that responds to popular demands and has the institutions in place to design and enforce environmental regulations.

Thus our simple model predicts that the effects of growth on environmental quality (still in the absence of international trade) depends on the strength of the income elasticity of demand for environmental quality and on the responsiveness and effectiveness of the political and regulatory systems.

Let us now consider the effects of international trade on environmental quality. First, international trade can increase the scale of production (which tends to increase pollution) and can increase incomes (which tends to reduce pollution through the technique effect). Potentially much more important, however, are the large composition effects that may accompany trade liberalization. Suppose that a country has a comparative advantage in pollution-intensive industries. Then trade liberalization will cause these industries to expand, increasing the derived demand for pollution. In such cases, pollution will rise with trade. On the other hand, if the country has a comparative advantage in relatively clean industries, then the derived demand for pollution falls and pollution may fall with trade.

Whether or not a country has a comparative advantage in clean or dirty industries depends on the interaction between the relative supply of factors (such as capital, labour, and land), how factor intensities of production correlate with pollution intensities (e.g., are capital-intensive industries more or less pollution intensive than labour-intensive industries), and on absolute income levels (which influence pollution policies). For example, if capital-intensive industries tend to be pollution intensive, then (as long as pollution policies across the two countries are similar) one would expect pollution to rise with trade in the capital-abundant country and to fall with trade in the capital-poor country.

Income differences can also play a role in determining the effects of trade on pollution. Copeland and Taylor (1994) consider the case where countries differ *only* in income levels, and where all countries have perfect regulatory regimes. In this case, a high-income country has relatively strict pollution regulations (high τ) and hence a comparative disadvantage in pollution-intensive industries. This is illustrated by comparing the left and right panels of Figure 1. Home has a high income and hence a high τ in autarky, and Foreign has a low income and low τ^* in autarky. This difference in the effective price of using environmental services creates a motive for trade, and pollution-intensive industries relocate to the low-income country. Trade thus creates a "pollution haven" in the low-income country. However, provided that pollution supply curves accurately reflect preferences, such trade benefits both countries. The rich country purchases the environmental services of the poor country, in return for which it sells the services of its human capital. While this type of trade is sometimes criticized by environmentalists as being exploitive, it reflects much the same type of exchange as occurs when high-income countries sell human capital-intensive goods to poor countries in return for goods intensive in unskilled labour. The poor country is worse off than the rich country before and after trade, but trade raises the absolute welfare levels of each.

However, the result that this type of trade is mutually beneficial depends critically on the assumption that the supply of pollution reflects the true social marginal damage from pollution. To see what can go wrong, suppose that regulatory institutions do not adequately respond to liberalized trade. Consider a country that does not regulate pollution at all. In this case, the pollution supply curve is flat at $\tau = 0$ before and after trade. This may be a reasonable response in a closed economy if the demand for pollution-intensive goods is not too strong. But once this country opens up to trade, its lack of pollution regulations could make it a very attractive location for pollution-intensive industries, leading to a large deterioration in environmental quality. Without a mechanism to dampen the increase in pollution in a way that reflects the community's preferences, there is no reason to expect that trade will be beneficial. Although the country's income rises (at least in the short run), the damage from increased pollution may more than offset the income increase and make the country worse off. In such cases, trade liberalization can lead to large discrete drops in welfare.

The effects of trade on a fishery provide a simple example of this mechanism (this takes us beyond "pollution," but the principles are the same). If there is no trade (and provided foreign fishing vessels are excluded from the fishing grounds), then the domestic market can provide a natural check on overfishing (as long as the domestic demand for fish is not too large relative to the stock of fish). If there is a large increase in fishing, then the price of fish drops, which renders fishing less attractive and leads to exit from the industry. Without regulation, fishing effort may be excessive, but the limited size of the domestic market provides partial protection for the fish stock. However, once the country is opened up to trade, this natural dampening mechanism is eliminated. A large increase in fishing effort need not lead to a decline in the price of fish, since surplus fish may be exported. Thus trade liberalization that is not accompanied by an institution to regulate the fishery can lead to the collapse of the fish stocks.[3]

Even if governments adequately respond to "community" preferences, one might still argue that such trade is undesirable if one believes that community preferences do not properly reflect higher ethical principles. In the model of Copeland and Taylor (1994), trade based solely on income-induced differences in environmental regulations lowers average global environmental quality. The increase in pollution in the poor country is larger than the reduction in pollution in the rich country. If we believe that we have a moral obligation to maintain the quality of the earth's environment on behalf of future generations of humans or current and future generations of non-human wildlife, then such trade could be deemed undesirable.

Finally, if we allow for transboundary pollution, we obtain the same qualitative effects of trade on the environment (Copeland and Taylor, 1995). However, because environmental quality is now a global public good, theory would suggest that there will be excessive pollution both before and after trade liberalization, even if each country has good internal control over its own environment. In this case, trade liberalization need not raise global welfare if it leads to a large increase in global pollution. Moreover, global income distribution can play an important role in determining who gains and loses from trade. Copeland and Taylor provide a simple example where high-income countries lose from trade liberalization while low-income countries gain. This is because lower-income countries can credibly commit to increase their pollution upon the opening of trade, and therefore trade liberalization gives lower-income countries a strategic advantage in their interaction with richer countries over global pollution levels.

One of the critical issues in dealing with global pollution is the implicit distribution of global property rights over the environment. If polluters have the right to pollute, then high-income countries have to bribe or coerce lower-income countries into controlling their pollution levels. The international trade and investment regime alters the *de facto* system of property rights and affects the trade-offs between bribes and coercion. One of the results of Copeland and Taylor (1995) is that a commitment to free trade can increase the level of the transfer from North to South needed to maintain global environmental quality.

In summary, theory suggests the following. First, opening up a country to trade removes the natural ceilings on production in any one sector imposed by the size of the domestic market. This means that if a country without institutions in place to control pollution is opened up to trade, then trade may potentially result in a large deterioration in environmental quality in countries with a comparative advantage in pollution-intensive goods. There should be no presumption that trade is welfare-improving in such cases – the trade-induced benefits of higher income may be more than offset by the costs of increased pollution. Second, even if institutions to control pollution are in place, trade between countries with large differences in income can lead to a relocation of pollution-intensive industries to lower-income countries. That is, in theory, trade can create "pollution havens" in low-income countries. This would suggest that the pollution content of exports (the amount of pollution released per unit of output) from a poor country would (all else being equal) be higher than the pollution content of exports from a rich country. Finally, trade can lead to an increase in average worldwide pollution.

In theory the solution to all of these problems seems simple – externalities should be internalized. In practice, however, the case of imperfect regulation is perhaps the closest to reality. There are often very long lags between recognizing environmental problems and developing and adapting institutions to control them. Canada's economic history is replete with examples of institutional failure in the area of environment and conservation. Dewees and Halewood (1992) document how the Ontario government failed to avert (and in some ways was complicit in) the environmental devastation around Sudbury, despite the availability of knowledge about pollution control methods used in other jurisdictions. Allardyce (1972) and Gillis (1986) review the case of sawdust pollution in rivers in the Maritimes and Ontario. The routine dumping of sawdust into rivers was responsible for reducing fish populations and likely wiped out the shad fishery in the Bay of Fundy. Moreover, while a law was passed in 1865 that explicitly prohibited such dumping, it was not until after 1900 that it was enforced. Nor can one dismiss these examples as relics of our past. The recent collapse of the fishery off the East Coast of Canada suggests that Canadian conservation policy is still inadequate.

The reality is that trade liberalization and foreign capital inflows will take place in a policy regime that does not fully internalize environmental externalities. While in theory the problem is not trade but lack of regulation, in practice, institutions to deal with such problems take time to evolve and do not always have the political support required to be effective. Consequently, in practice, trade and capital flows can be a major contributor to environmental problems if they stimulate pollution-intensive industry.

Empirical Evidence

There has recently been considerable interest in the empirical relationship among economic growth, trade, and environmental quality. This literature is still, however, in its early stages, and the conclusions must be viewed as tentative.

One of the major problems confronting researchers in this area is the absence of good data. Ideally, what is needed is a cross-country sample containing direct measures of environmental quality, pollution intensities of production, and measures of the marginal cost of environmental regulations. In practice, good cross-country measures of environmental quality are available only for some types of pollutants, such as air pollutants. Cross-country measures for other types of pollution are more difficult to construct. This is an important limitation, because, as Shafik (1994) shows, different types of pollutants respond very differently to income changes. Measures of pollution intensities of production by industry are available for the United States, but not for most other countries. Most researchers simply apply these U.S. measures to other countries to obtain rough cross-country estimates of the pollution intensity of production. Finally, it is very difficult to construct measures of the marginal cost of environmental regulation. Some studies use survey data to form an estimate of the total expenditure by firms on pollution control. However, at best, this yields only a measure of the average cost of compliance. Others use ordinal rankings of the stringency of countries' pollution regulations.

Let us now turn to studies of the effects of growth on pollution. Grossman and Krueger (1993) use data on urban air quality and find an inverted U-shaped relationship between air pollution (sulphur dioxide and suspended particulates) and per capita income. Starting at low levels of income, pollution first rises with growth, peaks at a per capita income of about $5,000 per year, and then falls. Shafik (1994) uses similar data and finds that pollution peaks at a slightly lower income. Seldon and Song (1994) report similar results using data on aggregate national emissions, although they find that aggregate national emissions tend to peak at a higher level of income ($8,000 per year).

Studies of this type are often used to support the idea that economic growth is an important part of the solution to the world's environmental problems. However this interpretation is potentially misleading. First, many countries have incomes far below the estimated turning point. Seldon and Song (1994) project large increases in emissions over the next half-century, particularly in light of projected rapid economic growth in Asia. Thus, if the historical relation between income and air pollution persists, economic growth will lead to *lower* worldwide average air quality throughout the lifetimes of people alive today.

Second, the relationships obtained for air quality do not apply to all forms of pollution. Beghin et al. (1994) note that no consistent relationship has been found between income and concentrations of heavy metals (mercury, arsenic, etc.). Shafik (1994) finds that availability of clean drinking water and urban sanitation tends to rise with income. However, water quality in rivers tends to fall with income (dissolved oxygen tends to fall and fecal coliform tends to rise). Also, municipal waste per capita and carbon emissions per capita both tend to rise. This suggests that many pollution problems are not mitigated as income rises.

One weakness of these studies is that they do not explicitly control for the effect of international trade. This is important for the interpretation of the studies. If rich countries mainly solve their pollution problems by encouraging polluting industries to locate elsewhere, then it is by no means clear that developing countries will experience the same income/pollution profile as did the developed countries – eventually we will run out of new places to shift the polluting industries. That is, if the composition effect of trade is driving the results more than the technique effect, then one must be very cautious in interpreting results like Grossman and Krueger's as evidence that higher income can allow us to solve pollution problems.

There is indeed some evidence that the compositional effect of international trade may be very important. Lucas et al. (1992) find that although many high-income countries are experiencing a fall in the pollution intensity of national product, this appears to be due to a change in the composition of output rather than a movement toward cleaner production methods. Effluents per unit of manufacturing output tend to increase over time, but the composition of output has been changing as well, with manufacturing tending to shift out of high-income countries. Low and Yeats (1992) find that while high-income countries produce a large fraction of the world's pollution-intensive goods, there is evidence that pollution-intensive industries have been migrating to lower-income countries over the past 25 years. The share of such products exported from Southeast Asia rose between 1965 and 1988 (from 3.4 to 8.4 percent), while the share exported from North America fell. Robison (1988) also finds that the United States has increased its imports of pollution-intensive goods.

Lee and Roland-Holst (1994) provide a direct estimate of the implicit trade in environmental services between a high- and a low income country. Our theory predicts that the pollution content of exports from a low-income country (i.e., the pollution generated per unit of output) should be higher than the pollution content of exports from a high-income country. Lee and Roland-Holst construct estimates of the pollution content of trade between Japan and Indonesia and find that Indonesia's exports to Japan are about six times as pollution intensive as are Japan's exports to Indonesia. Their calculations are hampered by their use of U.S. pollution intensity coefficients to estimate the pollution content of trade. That is, they implicitly assume that the production techniques are the same in both countries because there are no good cross-country data on the pollution intensity of production techniques. However, if pollution intensities were higher in Indonesia, then the results would be strengthened.

Lee and Roland-Holst also attempt to determine how the pollution content of trade between Japan and Indonesia would be affected by further trade liberalization. They construct a computable general equilibrium model and simulate the effects of reductions in trade barriers in Indonesia. Their results indicate that trade liberalization would increase pollution in Indonesia.

Theory predicts that pollution intensities of production should differ across countries. Because direct measures are difficult to obtain, Dessus et al. (1994) attempt to construct them by correlating emission intensities in the United States with input use. That is, they use U.S. data to estimate an equation that predicts emission intensities from input use. They then collect data on input use in different countries to arrive at cross-country estimates of pollution intensities of production. They find that there is considerable variation in pollution intensities of production across countries, for example, China appears to use production techniques that are more pollution intensive than average.

These results are tentative. They are consistent with the theoretical predictions that differences in pollution policy can induce improvements in environmental quality in high-income countries (and a worsening of environmental quality in low-income countries) as a result of composition effects. However, they are also consistent with other interpretations (such as shifting comparative advantage based on factor endowments). More work is needed to increase our understanding of these issues.

Although the evidence indicates that pollution-intensive production may be (slowly) shifting away from OECD countries, there is little evidence available to confirm the *cause* of this shift. Most studies (e.g., OECD, 1993) find that pollution control costs are relatively small (less than 3 percent of total production costs). Grossman and Krueger (1993) find that measures of pollution control costs are not a significant variable when explaining imports into the United States from the *maquiladora* sector in Mexico. Tobey (1990) uses an ordinal measure of the stringency of pollution regulations across countries and finds that this variable does not appear to affect the pattern of trade. Jaffe et al. (1993) review the evidence on the effects of pollution control costs on plant relocation, and find little to demonstrate that policy differences explain investment flows.

More work is needed to reconcile the evidence of the shifting location of pollution-intensive industries with the relatively low measures of pollution abatement costs. One possibility is that rapidly developing countries tend to have a comparative advantage in industries that happen to be pollution intensive for reasons other than policy differences. However, it is also possible that the data on the costs of complying with environmental regulations may understate the true costs. Most measures calculate average abatement costs, and one would expect that marginal abatement costs would be higher.

Finally, it is not inconsistent with the theory to observe large changes in location coming from small differences in regulation-induced costs. Markusen et al. (1993) develop a simple model of multiplant firms and find that small changes in policy can lead to large discrete changes in market structure as firms relocate plants. Also, Copeland and Taylor (1995a) develop a model in which differences in pollution taxes in autarky shift pollution-intensive industries to lower-income countries, but these differences in pollution taxes are eliminated by free trade. In equilibrium, although production has relocated, one would not be able to infer the reason for this relocation by looking at equilibrium marginal abatement costs (since they are equalized by the very movements their differences caused). As is standard in trade models, international trade tends to mitigate international differences in factor prices, and in some cases may equalize them. In our context, this means that trade tends to reduce international differences in the price of environmental services. Thus our model predicts differences in ambient environmental quality (which we indeed observe), international migration of pollution-intensive industries (which we observe), but not necessarily large differences in marginal abatement costs across countries after the migration has occurred.

Policy

ONE OF THE KEY ISSUES IN TRADE AND ENVIRONMENT LITERATURE is whether or not trade and environmental policy should be linked or treated separately. An important idea from the field of public economics is the notion that good policies should focus narrowly on the problem at hand. Using this approach, the case for linkage would seem to be weak. Pollution problems arise because consumers and producers do not pay the full social cost of the environmental services they use. The ideal solution is to introduce good pollution regulations or taxes to internalize these costs. Trade-policy problems arise because governments interfere with the free flow of goods and services across borders, and the solution to these problems is to negotiate a set of rules that restrict their ability to raise trade barriers. In this view, trade and environmental policy are aimed at two separate problems, so two separate sets of instruments are required. If trade liberalization leads to more pollution, it is not because of bad trade policy, but rather because of bad pollution policy, and so the solution is to fix the pollution policy.

Moreover, many of those who work in the field of trade policy argue that linking trade and environmental policy could open up a large loophole in trade agreements that could be exploited by protectionists. In this view, freer trade is thought to be a desirable norm to promote economic growth throughout the world. The current regime of relatively low trade barriers is the result of a long struggle aimed at progressively reducing overt trade barriers and restricting the ability of governments to succumb to special interest groups who lobby for trade protection. Advocates argue that to preserve this regime and achieve further trade liberalization, the focus should remain narrowly on trade, and that loopholes should be closed rather than opened.

Upon closer examination, the case for keeping trade and environmental policy separate is not as strong as either of these arguments would suggest. First, recall Figure 1, in which the pollution supply curve is drawn as upward sloping. This reflects the view that increases in pollution levels are increasingly harmful. In practice, it means that regulatory authorities have to adjust pollution regulations in response to increases in the demand for the use of environmental services. Under virtually all pollution regulatory systems in place in Canada, this adjustment is not automatic but requires active intervention in markets. Moreover, many of the arguments that increases in income could lead to a reduction in pollution rely on an inward shift of the pollution supply curve in response to higher income levels. Once again, this is not an automatic response but is policy-induced. This means that there is a compelling argument for coordinating environmental policy with changes in trade policy. That is, as trade is liberalized, the optimal environmental policy changes (see Copeland, 1994). This is not an argument for explicit linkage in formal trade agreements. However, it does suggest that unless the environmental policy regime is sufficiently flexible and effective, then there is no presumption that a trade liberalization will be welfare-improving.

Moreover, there is no obvious reason why trade liberalization should have primacy. The current approach is to agree on a trade regime, and hope that the environmental policy regime will be adequate. Why should one not think of accepting the realities of the environmental policy regime, and then asking what sort of trade regime is appropriate? As Chichilnisky (1994) argues, if environmental externalities in a given country are not properly internalized, then one can make a case that free trade is an inappropriate regime for that country at that point in time.

The second argument for keeping trade and environmental policies separate is based on political economy – the concern that such a linkage would create more loopholes for protectionist lobbies. In this view, an international trade agreement fulfills a very important domestic function in that it provides a mechanism through which governments can credibly commit to say "no" to protectionist lobbies. The history of trade protection underscores the importance of such a commitment device. In eras of weak and loophole-ridden trade agreements, governments typically succumb to protectionist pressures and undermine the free trade regime.

This interpretation sheds some light on the success of international trade policy in reducing protectionism as compared with the rather lacklustre record of environmental policy in dealing with pollution. Trade policy is backed up with a good commitment device (a trade treaty), while environmental policy has no such commitment device. A government that is thinking of caving in to protectionists has to think not only about angering those who will be harmed in its own country, it also has to consider the ramifications of violating an international treaty. This has served as a powerful deterrent to some of the more blatant forms of protectionism. But because most environmental policy is treated as an internal responsibility of each country, there is no similar deterrent effect to help (or force) governments to stand up to lobbyists who oppose tightening up environmental policy. This suggests that while linking trade agreements to environmental agreements may indeed weaken trade agreements and lead to more protection, it could also lead to better environmental policy by providing a stronger incentive to implement and enforce policy that may be opposed by local lobbies.

One's view on such matters is no doubt influenced by the importance that one places on the costs of an inefficient trade regime versus the costs of a weak environmental policy regime. If environmental services are abundant, then the costs of linkage (more protectionism) will outweigh the benefits. But if environmental services are scarce, and pollution and conservation problems are growing in severity, then the costs of ignoring the indirect effects of trade liberalization on the environment can be very high and the argument against linkage is not so compelling.

Moreover, the strengths and weaknesses of the arguments for linkage vary with the type of environmental problem. In what follows, I examine the linkage between trade and environmental policy and focus on four main concerns: 1) the effects of trade liberalization on Canada's environmental policy, 2) transboundary pollution, 3) extraterritoriality, and 4) environmental dumping.

Canada's Environmental Policy

THE FIRST ISSUE IS ONE OF COORDINATION, rather than linkage. While most of the (environment-related) discussion of rapid economic growth has focused on emerging environmental problems in Asia, it is important to keep in mind that there will be increased pressure on the Canadian environment as well. High economic growth rates in Asia, coupled with high deforestation rates and rapid population growth will lead to large increases in demand for energy, raw materials, forest products, and food. As these are goods in which Canada has a comparative advantage and which are also environmentally sensitive, it is important that Canada's pollution and conservation policies be responsive to increased demands for the use of environmental services.

Transboundary Pollution

TRANSBOUNDARY POLLUTION OCCURS when pollution generated by one country directly harms another country. Uncoordinated environmental policies at the national level are unlikely to be fully effective in controlling this type of pollution, because while the polluting country has to bear the full costs of pollution abatement, the benefits are spread among all affected countries. As a result, countries have an incentive to pollute too much.

Transboundary pollution is perhaps one of the most serious environmental issues that will affect Canada-Asia relations in the years to come. Carbon and CFC emissions are good examples. In 1990 China released in total about six times as much carbon dioxide from industrial sources into the atmosphere as did Canada (see Table 1). However, on a per capita basis, Canada's emission rate is seven times as high as China's. Similarly, Canada's per capita consumption of ozone-depleting gases (CFCs and halons) is more than 10 times as high as China's (Table 3). This means that as China's economy continues to expand, there is huge potential for increases in emissions of gases that contribute to global warming and depletion of the ozone layer.

Most observers agree that the only sensible solution to problems of transboundary pollution is international cooperation and negotiation. However, international trade policy instruments are relevant in three ways. First, if negotiations fail, then the trade policy regime will have some influence on the amount of transboundary pollution generated. But although trade sanctions may be effective in certain bilateral cases of pollution, it is unlikely that any trade instrument at Canada's disposal would have an effect on pollution levels in Asia.

Second, an environmental agreement needs to be enforced and needs incentives for all major polluters to become signatories. The use of trade sanctions tied to environmental agreements is one of the few enforcement devices available in an international context. This was the approach adopted in the Montreal Protocol on ozone-depleting gases, and Canada should be a strong supporter of this practice, provided the procedures under which trade instruments would be invoked are clearly defined and dispute settlement mechanisms are in place.

Third, income distribution is the critical factor underlying any agreement on global environmental quality. Essentially, a global environmental agreement has to determine 1) how much pollution to allow, and 2) how to divide up the right to pollute across countries. Because rich countries such as Canada are currently large polluters (on a per capita basis), any agreement would involve cuts in emissions by rich countries combined with restrictions on the growth of emissions in poor countries. But why should poorer countries accept a regime in which the rich countries have unilaterally laid a claim to a large share of the right to pollute? As coercion does not seem to be a viable option, rich countries will have to pay poor countries not to pollute. The trade regime can play a role in affecting the size of the transfers from rich to poor countries by affecting the threat point of the negotiations. For example, in Copeland and Taylor (1995), rich countries prefer to link an environmental agreement to a trade agreement, while poor countries prefer to leave negotiations over global environmental quality until after a free trade regime is established. This suggests that the U.S. strategy of linking NATFA to an environmental accord had merit from the U.S. perspective, and that the benefits (and costs, depending on one's point of view) of such linkage should not be ignored in future trade negotiations involving other countries.

EXTRATERRITORIAL APPLICATION OF ENVIRONMENTAL NORMS

CERTAIN TYPES OF ENVIRONMENTAL PROBLEMS attract international concern even when the environmental damage is confined to one country and there is no threat to the competitiveness of industries in other countries. Wilderness preservation and biodiversity are examples of such problems. Many Canadians care about the fate of rain forests in tropical countries, just as many Europeans care about forests or seal-harvesting practices in Canada. In principle, these issues are very similar to transboundary pollution: formally, both types of environmental problem can be modeled as an activity by people in one country that harms people (subjectively or physically) in another country. Consequently, the solution is potentially the same, and negotiation is called for. In some cases (such as wilderness preservation), it may be possible for foreigners to purchase land for the purpose of preserving it. If negotiations or purchases are not feasible, then trade sanctions could be used to influence foreign activity. However, as with transboundary pollution, it is probably not a wise option for Canada to consider. Unilateral actions by Canada would be ineffective and would be viewed as attempts to infringe on foreign sovereignty. Moreover, by legitimizing this practice, Canada would likely increase the frequency with which it is the target of such actions.

Environmental Dumping

ENVIRONMENTAL DUMPING IS SAID TO OCCUR if an exporting firm benefits from access to environmental services that are underpriced; that is, if a firm sells its product in a foreign market for less than its full social cost because it does not have to pay for the environmental damage that it causes. While the phrase "environmental dumping" is very catchy, it is more accurate to think of the firm as the recipient of an implicit subsidy. Access to environmental services at a price less than their opportunity cost is equivalent to receiving an explicit subsidy in a regime where environmental services are priced. It is well established in international trade law that countries may in some cases respond with import tariffs to *explicit* subsidies given by foreign governments to firms to promote exports. The notion of environmental dumping is simply that the right to levy countervailing duties should be extended to *implicit* subsidies that arise because environmental regulations are not in place or enforced.

Most international trade economists tend to strongly oppose the extension of countervail law to include environmental dumping. First, they argue that the countervail process is often captured by protectionist interests and widening the scope of such actions would create another loophole in trade agreements. Second, they argue that typically it is not in the national interest to try to discourage foreign export subsidies of any type because they improve the terms of trade of the importing country. If someone wants to sell us something at a price below cost, Canada's aggregate consumption opportunities expand. Finally, a policy of imposing "eco-duties" may not work if the intention is not simply protectionist but also to create incentives for improved environmental quality in the exporting country. The argument here is that the effectiveness of environmental policy is increasing in the level of per capita income. Since trade sanctions reduce the income of the target country, they make it that much more difficult to implement environmental policy reforms.

On the other hand, many environmentalists believe that future trade agreements should be amended to allow countries to use countervailing duties to offset implicit subsidies given to firms in countries with weak pollution regulations. This rule would ensure that trade is in accordance with comparative advantage and not distorted by subsidies. Moreover, countries would be free to impose strict environmental regulations without fear of industries fleeing to pollution havens. Since concern about potential job loss drives much of the lobbying against improved environmental protection, such a policy would also diffuse some of the opposition to reform of domestic pollution policy. Daly and Goodland (1994) put the case very bluntly:

> National protection of a basic policy of internalization of environmental costs constitutes a clear justification for tariffs on imports from a country which does not internalize its environmental costs of production. This is not "protectionism" in the usual sense of protecting an inefficient industry, but rather it is the protection of an efficient national policy of internalization of environmental costs (p. 401).

Because the issue of environmental dumping is much more controversial than the other issues, I discuss it at length in the next section of the paper. I proceed by first identifying the conditions under which environmental dumping may occur, and then discussing the benefits and costs of adopting a regime under which retaliation is permitted.[4]

ENVIRONMENTAL DUMPING EXAMINED

THE CONCEPT OF ENVIRONMENTAL DUMPING seems simple enough at first – it occurs if environmental regulations are too weak. In practice, the definition of dumping is not straightforward. In standard countervail cases, one can infer the presence of a subsidy to inputs if a firm's costs are less than the market value of inputs. The key problem in pollution regulation, however, is that environmental services are typically not traded on open markets, so the extent of the subsidy is difficult to gauge. A variety of institutional arrangements – ranging from tribal traditions to government regulations backed up by criminal law – have been developed to regulate the use of environmental services. Many countries have institutions that are not able to reconcile the competing demands on the environment, and as a result environmental services are underpriced. However, because there is no market, there is no simple benchmark for the "correct" price of pollution, and arriving at a practical definition of environmental dumping can be controversial.

Differences in Regulations

THERE ARE SEVERAL POSSIBLE METHODS of detecting environmental dumping. Canada and the United States (and most other countries) tend to impose a variety of regulations on firms aimed at enforcing an environmental standard that is determined by science ("safe" levels of exposure to toxins) and the political process. Firms may, for example, face design standards for their factories, emissions ceilings measured either per unit of output or input or at the plant level, or restrictions on allowable inputs or production processes. One potential area for the detection of environmental dumping might then be to look for cases where regulations are too weak.

In practice, however, this strategy is unlikely to be fruitful. In the context of a sometimes bewildering array of regulations, what does it mean for a country to have environmental regulations that are too weak? There are many ways that regulations may differ across countries for a given pollutant in a given industry. Countries may specify different design standards, have different monitoring practices, have different restrictions on production processes or allowable inputs, have different emission standards at the firm or industry level, and so on. However, different regulatory strategies can lead to similar environmental outcomes, so differences in regulations across countries cannot be taken as an indication of dumping.

Moreover, efficient pollution abatement procedures and equipment may vary across firms within a given industry in the same country. In the United States, the Environmental Protection Agency (EPA) has given firms some flexibility in meeting pollution targets by allowing pollution permits to be partially transferable in some cases. Under this scenario, firms would choose their own techniques, provided aggregate ambient targets are met. This would mean that in equilibrium, different firms may adopt different strategies to meet the targets and may have different emissions intensities. For example, under the SO_2 marketable permit system, firms have the option of buying pollution permits rather than installing new abatement equipment in order to meet the new targets.

This makes it difficult to define weak standards based on production techniques. There may be some cases where there are no gray areas. For example, if there is little scope for substitution, and if a given process or toxic emission is so harmful that it is effectively banned in one country, then evidence of *any* emissions of the toxin in the other country may be sufficient to determine that standards are relatively weak. However, more generally, one cannot infer that a country is not properly regulating the environment merely because firms use different techniques, or because a given firm abates more than another.

Implicit Pollution Taxes

A POTENTIALLY FRUITFUL APPROACH is to avoid focusing specifically on firm behavior, and instead consider the implicit pollution tax. The use of environmental services by a polluting firm is not very different from using any other input such as labour or capital. In a competitive market, the wage represents the opportunity cost of labour services. An efficient pollution regulatory system would ensure that firms paid a price for polluting that reflected the opportunity cost of environmental services. While firms do not actually pay a tax under most regulatory systems, the implicit pollution tax is given by the equilibrium marginal abatement costs. This is the cost to the firm of emitting the last unit of pollution. It also measures the implicit value attached to the use of environmental services by the authorities.

One might be tempted to conclude that environmental dumping occurs if the implicit pollution tax is too low. Unfortunately, even this procedure is not as simple as it might seem. In practice, marginal abatement costs vary widely across firms in both Canada and the United States. There are two reasons for this. First, an efficient policy targets pollution damage, not just pollution emissions. A unit of sewage dumped into Juan de Fuca Strait off Victoria, British Columbia, causes less environmental damage than a unit of sewage dumped into Lake Ontario. Consequently, the same pollution generated at different locations may require different treatment to attain the same level of environment quality. Differences across firms are efficient in this case.

A second reason why marginal abatement costs differ across firms is that regulations are often inefficient. Consider two firms for which a given unit of emissions causes the same amount of environmental damage. Regulators often impose blanket regulations on all firms without regard to differences in marginal abatement costs. In the United States, firms are sometimes permitted to trade pollution permits among themselves (subject to EPA approval) to offset the inefficiencies of blanket regulations. In Canada, this rarely occurs. Thus, in practice, measuring an industry's marginal abatement cost requires calculating an average. Some firms will be above the average and some firms below. This is an indication that a regulatory policy is inefficient, not necessarily that environmental dumping at the industry level is occurring.

A Simple Trade Model

CAN THE IMPLICIT POLLUTION TAX BE USED as a measure of the strictness of a country's pollution policy? Let us use the simple trade model presented in section two. Consider two countries, Home and Foreign, and for simplicity, focus on the case of one pollutant, Z. Because I am considering environmental dumping, and not transboundary pollution, I assume that the harmful effects of pollution do not spill across international borders.

First, suppose there is no trade. The derived demand for pollution is denoted by D^a and D^{a*} in Figure 2, and is the marginal benefit of polluting the environment. The pollution supply curves are denoted by S and S^*. In autarky, equilibrium pollution levels are determined by the intersection of demand and

Figure 2
Pollution Demand and Supply, "Home," "Foreign" and "World"

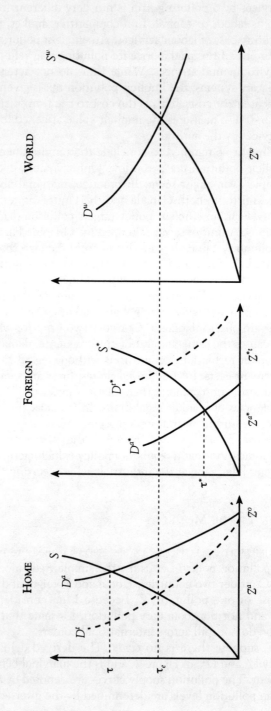

supply. As illustrated, the implicit pollution tax (the marginal abatement cost) is higher in Home than in Foreign country. Foreign has weaker environmental policy in autarky, and this is observable since $\tau^* < \tau$.

Now consider the effects of trade. Free trade usually tends to reduce differences in factor prices across countries. That is, although workers cannot move freely across countries, free trade in goods means that workers in trading countries indirectly compete with each other via competition in the goods market. This process can also be relevant in the market for environmental services. In the example in Figure 2, pollution-intensive goods are cheap in Foreign and expensive in Home. Consequently, Home imports these goods from Foreign. This reduces the demand for pollution at Home, and increases it in Foreign. Figure 2 illustrates the case where factor prices are fully equalized by trade. In free trade, the World derived demand for pollution equals the World supply. This leads to an equilibrium price of pollution services τ_e. The new post-trade demands for pollution in each country (D^t and D^{t*}) are indicated by the dashed lines.[5] In line with comparative advantage, pollution-intensive firms have relocated to Foreign. This puts increased pressure on Foreign's environmental services, but also leads to a tightening of pollution regulations there. In equilibrium, the implicit or explicit price of pollution (τ_e) is the same across countries. Trade has eliminated differences in implicit pollution taxes across countries, but at the same time has redistributed the burden of pollution.

However, perhaps more realistic is the case where differences in pollution taxes remain after trade is opened. Consider Figure 1 again, but now suppose that it illustrates a free trade equilibrium. In this case, $\tau > \tau^*$. Trade in goods is not sufficient to equalize the implicit price for pollution in this case. But does this mean that there is environmental dumping?

The key problem in trying to detect environmental dumping from differences in implicit pollution taxes is that it is very difficult to know what lies behind the pollution supply curve. There are two possibilities: the pollution supply curve may fully reflect the marginal damage from pollution (in which case pollution regulation is optimal and there is no environmental dumping), or pollution supply may be excessive because pollution regulations are inadequate. It is impossible to distinguish between these two cases simply by looking at differences in implicit pollution taxes across countries.

Why might the willingness to accept pollution differ across countries? One reason may be that the foreign environment has a greater absorptive capacity for pollution. This means that environmental quality may be just as good in Foreign as in Home, despite more pollution emissions. Trade in this case can be good for the environment, as it shifts pollution away from the more delicate ecosystem of Home. However, it may also be that the absorptive capacities of the environments are identical across countries, but that Foreign's marginal damage is lower because Foreign is a poor country. If environmental quality is a normal-good, then poor countries are willing to accept more pollution in return for higher incomes than are rich countries.

In this case, different pollution taxes across countries simply reflect the fact that, in general, trade does not fully equalize factor prices (even in theory). If countries have very different factor endowments, they tend to produce different types of goods, and arbitrage in goods does not bring the factors into close enough competition to eliminate price differentials across countries. In practice, wages have clearly not been equalized by trade between rich and poor countries. Yet we do not levy countervailing duties against countries that have low wages. Rather, low wages are a reflection of comparative advantage, and trade is the mechanism through which gaps in wages across countries are supposed to be reduced. Moreover, when there are fixed factors that generate rents (such as land), there is no expectation that factor prices will be equalized. This is true even within a single country: there is no reason to expect that a residential lot in Vancouver will fetch the same rent as a similar lot in Saskatoon. Analogously, there is no reason to expect that trade will equalize the price of environmental services. Differences in factor prices across countries are not sufficient to infer the existence of implicit subsidies.

Moreover, differences in implicit pollution taxes are not a necessary consequence of environmental dumping. To see this, suppose that countries use aggregate pollution quotas to control pollution (and that these are allocated efficiently across firms). As illustrated in Figure 3, Home has a very tight quota, and Foreign has a weak quota. Then, if labour and environmental services are the only inputs, the model with fixed pollution quotas is simply a standard Heckscher-Ohlin model. Provided that the differences in the quota/worker ratio are not too large, trade will equalize the implicit price of pollution across countries. Hence $\tau = \tau^* = \tau_e$. However, the initial permit supply may be completely arbitrary and need not bear any relation to marginal damage. Thus, as illustrated, suppose the Foreign supply of permits is too high. Then, although permit prices are equalized by trade in goods, the equilibrium permit price is nevertheless below foreign marginal damage ($\tau_e < MD^*$). In this case, most economists would probably agree that Foreign's firms are receiving an implicit subsidy because its environmental services are underpriced.[6]

The final major weakness in using implicit pollution taxes to detect environmental dumping is illustrated in Figure 4. In this case, attitudes toward the environment are identical across countries (the marginal damage curves are the same), but in equilibrium, the shadow price of environmental services is higher in Home than in Foreign, because the derived demand for pollution is higher in Home – perhaps Home has a comparative advantage in pollution-intensive industries. In this case, Home has both stricter environmental regulations, and more pollution. Foreign does not need tight pollution regulations because pollution-intensive industries do not find it profitable to locate there, so a comparison of marginal abatement costs across countries would reveal $\tau^* < \tau$. However, it is hard to make the case that a country with relatively little pollution is engaged in environmental dumping.

FIGURE 3

POLLUTION QUOTAS, "HOME," "FOREIGN" AND "WORLD"

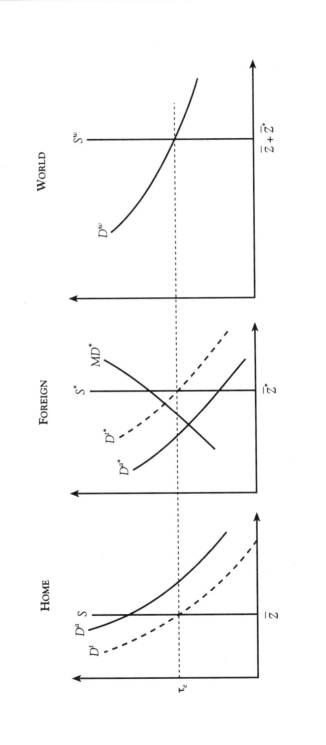

FIGURE 4

EFFECTS OF DEMAND ON THE PRICE OF ENVIRONMENTAL SERVICES

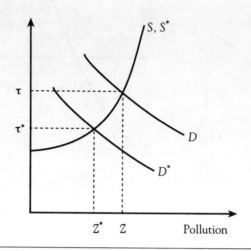

This analysis reveals serious problems in attempting to use the implicit pollution tax as an indicator of environmental dumping. I have shown that dumping may occur even when implicit pollution taxes are equal across countries, and that dumping (in the sense that economists think of it) need not occur when the foreign pollution tax is lower than at home. Moreover, it is possible for environmental dumping to occur even when a country's implicit pollution tax is higher than its trading partner's tax. (To see this in Figure 4, suppose that Home's pollution tax were lower than illustrated, but still higher than τ^*. Then Home would be providing a subsidy to its industry, since the pollution tax is less than marginal damage.)

Moreover, the economist's case for identifying an implicit subsidy for environmental services rests upon comparing the implicit price of pollution with marginal damage. In practice, no one knows what marginal damage is. Economists continue to debate how to measure it, but in practice, environmental standards are set by scientists, administrators, or politicians, and formal cost-benefit analysis plays a very minor role. Consequently, it is difficult to make the case that Canadian environmental regulations fully reflect the marginal damage of pollution here. So how could we expect to make such an inference for other countries?

Ambient Standards

IF WE ABANDON AS IMPRACTICAL the use of cross-country norms for regulatory standards or implicit pollution taxes, what is left? Perhaps the most promising possibility is to define environmental dumping based on ambient environmental quality. That is, parties to a trade agreement could agree on standards for environmental quality. These are standards for air and water quality measured as concentrations of pollutants at various receptor points. A country would be guilty of subsidizing pollution-intensive industries if it allowed the pollution concentration in the environment to rise above the agreed upon standard. The advantage of this approach is that it allows for trade between different countries whose environments have different carrying capacities. Moreover, the relative levels of pollution taxes or production techniques are irrelevant, provided that countries hit the ambient targets. Finally, information on marginal damage is not required.

This is illustrated in Figure 5. Pollution concentration in the environment is measured on the horizontal axis. Since ambient environmental quality is the target, it is natural to think of firms as being assessed an explicit or implicit charge based on their contribution to pollution concentration. That is, a tax system would charge firms based on their emissions and adjusted according to the impact of the emissions on ambient quality.[7] The derived

FIGURE 5

AMBIENT STANDARDS IN "HOME" AND "FOREIGN"

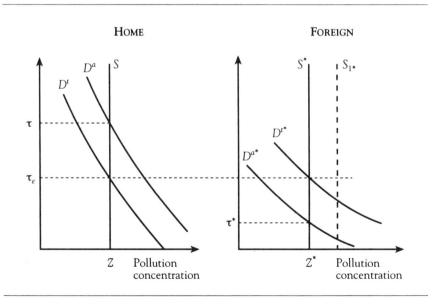

demand for the right to cause environmental damage is downward sloping, as usual. It reflects the marginal cost of attaining a given level of environmental quality. I have illustrated the demand for two countries. The demands may differ because of differences in factor endowments or the absorptive capacity of the environment. For concreteness, suppose that the costs of abating pollution emissions are identical across countries, but that Home's environment is more fragile than Foreign's. That is, to attain a given level of environmental quality, more pollution must be abated in Home than in Foreign. Consequently, the marginal cost of attaining a given level of environmental quality is higher in Home than in Foreign. Suppose that each country implements the same ambient standard ($S = S^*$), then the equilibrium pollution tax is higher in Home than in Foreign ($\tau > \tau^*$).

Now consider the effects of trade. Trade tends to reduce the gap between the implicit price of environmental damage across countries. Suppose that the gap is eliminated, and that τ_e is the equilibrium price. Then pollution-intensive industries will shift from Home to Foreign, taking advantage of the less sensitive foreign environment. This shifts the demand for environmental damage up in Foreign from D^{a*} to D^{t*} and down in Home from D^a to D^t (relatively cleaner industries relocate to Home). In equilibrium, both countries gain from trade through more efficient allocation of resources across countries. However, environmental quality is not adversely affected by trade, because both countries adjust their policies to maintain the ambient targets. There is no dumping in this case.

Now suppose that Foreign instead allows ambient environmental quality to deteriorate to $S_1^* > S^*$. Then the implicit tax on environmental damage may still be equalized across countries by trade. However, the foreign country would be guilty of environmental dumping by this definition, since pollution concentration in the environment exceeds the standard S^*. The appropriate countervailing duty would be equal to the level of the additional tax necessary to reduce pollution to the level at which the ambient standard is met.

Notice that this definition of dumping does not require a measure of marginal damage, nor does it require information on implicit or explicit pollution taxes or on regulatory procedures. One simply samples environmental quality, and compares it to the standard. Calculation of the countervailing duty would require information on abatement costs but not on marginal damage.

Another advantage of this system is that it is flexible enough to allow for differences in attitudes toward pollution across countries. For example, one could allow different countries to adopt different ambient standards and define dumping to occur when they violate their own standards. Of course, this might make countries reluctant to adopt strict standards. Alternatively, different standards could be negotiated for different countries, and countervailing duties would be applied only when the negotiated standards were violated.

Finally, note that the pollution regulation regime in the United States bears some resemblance to the ambient scheme discussed here. The EPA sets national standards, and states are required to design regulatory systems to meet these standards. One of the rationales behind the EPA approach was explicitly to prevent states from competing against each other based on differences in pollution regulations.

Response to Environmental Dumping

SUPPOSE THAT WE AGREE ON HOW TO DETECT environmental dumping. Should Canada support a regime that allows for the application of countervailing duties to offset the trade-distorting effects of weak environmental policy? The answer to this question depends on the weight that Canada places on the wellbeing of its own citizens relative to those of its trading partners. But the issue here is not whether Canada should unilaterally start levying countervailing eco-duties – clearly it should not. Rather, the issue is whether in future trade agreements Canada should support a regime that 1) ties acceptance and implementation of the agreement to achieving certain environmental goals, and 2) allows the use of trade instruments as a response to foreign non-compliance. In what follows, I therefore focus on Canada's interests in discussing the merits of such a regime.

Provided that one has determined that a subsidy exists, and provided that pollution does not spill over international boundaries, then, in principle, the optimal response to implicit environmental subsidies should not be different from the optimal response to any other subsidy. If the subsidized goods are imported by Home, then, in a competitive market, Home's terms of trade improve, and Home is better off.[8] In this scenario, retaliation against environmental dumping may seem counter to the national interest. However, the benefits and costs of foreign environmental subsidies will be spread unevenly among Canadians – some will gain and some will lose. If the Canadian industry affected contracts, then those who have strong ties to it will suffer losses. Countervailing duties in such cases may protect the incomes of those in the industries affected.

Once we depart from the assumption of perfect markets, the analysis is more problematic. Depending on the type of market failure, foreign export subsidies may benefit or harm Home. If there is unemployment in Home, then foreign export subsidies may exacerbate it. If Home is not properly regulating its pollution, then the lack of foreign regulation may in fact be helpful, since it prevents pollution-intensive industry from entering Home's market, so the lax foreign pollution laws act as a second-best method of pollution control in Home.

If there is learning by doing or other forms of dynamic external economies, then Foreign's strategy may be to endure excessive pollution in the short run to build up the industry. Once the industry is successful and can survive on its own without the environmental subsidy, Foreign can toughen up its pollution regulation. However, by then the industry may have left Home for good. In such a case, Home's policy may be to impose subsidies of its own. However, the optimal subsidy is typically a subsidy to output, training, or learning, and not to match weak foreign environmental regulations. The danger is that in an era of budget restraints, Home's regulators may be tempted to avoid tough enforcement of local environmental regulations. A retaliatory tariff may be attractive in this case, since it generates revenue and protects the domestic industry.

If there is imperfect competition, then, as Markusen, Morey and Olewiler (1993) have argued, firm and plant location may be very sensitive to differences in environmental policy. In this case, tariffs on imports from foreign countries can be welfare improving, and retaliation against foreign export subsidies can provide a convenient excuse for such actions.

Barrett (1994) considers the use of environmental policy in cases where domestic and foreign firms have market power and earn economic profits. In such markets, environmental dumping can reduce Foreign's costs and thus give its firms a strategic advantage over Home's firms. This can shift profits away from Home's firms to Foreign and reduce Home's welfare. In this case, retaliation against Foreign's environmental dumping can be in Home's interest. Both countries might be better off if they agreed to not subsidize their firms. However, such an agreement may not be feasible, since it intrudes on Foreign's environmental policy. Thus an option to retaliate increases Home's bargaining power with respect to Foreign.

In summary, there are a number of potential scenarios in which a country may find it worthwhile to engage in environmental dumping and a trading partner may find it worthwhile to retaliate. In principle, the issues do not seem to be fundamentally different from those arising from explicit export subsidies. In cases of (non-environmental) export subsidies, the GATT allows countries to respond with countervail laws. Moreover, these laws have been used and duties have been imposed. This suggests that losses to specific factors in directly affected industries often carry more political weight than the more widely diffused terms of trade benefits.

The GATT thus appears to be inconsistent on this issue. It is not clear that environmental subsidies are fundamentally different from any other export subsidy. Moreover, in treating implicit environmental subsidies as different from other types of subsidies, the GATT appears biased against proactive intervention to correct market failure. The GATT supports the legitimacy of trying to force a foreign government to remove subsidies that distort prices away from market prices (and true opportunity costs). But it does not support the legitimacy of trying to force a country to impose a pollution tax to move prices toward their true opportunity costs. Thus the message of GATT is that the market outcome is the norm. This is not tenable in the presence of pollution externalities.

The inconsistency in the GATT rules on subsidies need not, however, be an argument for widening the scope of countervail laws. Rather, we should consider the merits of any kind of countervail laws. There are two standard arguments for the use of countervail laws. The first is that they are a necessary loophole in trade agreements to buy off protectionist forces. The second is that they are necessary to enforce trade agreements – governments have incentives to cave in to political pressure to protect domestic industry with subsidies, and the threat of a foreign countervail action can act as a deterrent. If countervail laws are mainly in place to appease protectionists, then the argument for extending them to environmental subsidies appears at first quite weak, since doing so would further undermine the liberal trading regime. However, the appeasement argument can cut both ways. Extending countervail laws to environmental subsidies may appease domestic interests who oppose stronger pollution regulations. Moreover, the enforcement argument behind countervail laws may also apply to environmental issues. Just as agreeing to a countervail regime is a form of political commitment to not use export subsidies, an explicit international agreement on environmental policy can act as a form of commitment to avoid caving in to political pressure to weaken enforcement of environmental standards, and can result in a cleaner environment. Such agreements have to be enforced, and revoking trade concessions to countries that do not abide by environmental agreements is one way to do so.

To conclude, there is no clean distinction between traditional arguments for countervail and their extension to environmental duties. Perhaps there should be no real distinction in Canada's position on the two issues. This is perhaps the strongest argument against extending countervail to include environmental dumping. Canada's experience with countervail laws has not been a happy one. Countervail laws in the United States have provided a convenient vehicle for protectionists there to harass Canadian exporters. Because the definition of environmental dumping is fuzzy, the administrative process can be captured by protectionists. Moreover, environmentalists in Europe may well target Canada for various environmental transgressions.

Conclusion

PART OF THE DISAGREEMENT OVER THE ISSUE of linking trade and environmental policy no doubt reflects differences over how free trade affects the environment and in the valuation of environmental quality. If one puts a high value on environmental quality and believes that freer trade tends to worsen environmental quality, then it is difficult to be persuaded by the argument that extending countervail laws to cover environmental dumping may weaken the liberal trading regime and raise average trade barriers. Alternatively, if one believes that freer trade may either enhance or have a limited effect on environmental quality, then one is likely to be more sympathetic to the concern that green countervail laws may simply add another loophole for protectionists to exploit.

Better empirical evidence could be helpful here. As noted earlier, there is some evidence that pollution will rise with global economic growth, at least for the next couple of generations. And there is evidence that the pollution intensity of exports from lower-income countries with relatively weak enforcement of environmental standards has increased over time. This suggests that trade may well be contributing to an increase in global pollution. But there is still much that we do not know.

There is little convincing evidence that differences in environmental policies have a significant influence on trade patterns. Perhaps, however, the test has been too strong. There is also little evidence that many of the policies that are treated as export subsidies by the GATT have a significant effect on the pattern of trade. Nevertheless, the use of countervail laws is officially sanctioned in such cases.

The empirical issues may not be satisfactorily resolved for some time, if ever. Meanwhile, we are left with the conclusion that the GATT is inconsistent in its treatment of subsidies. The current rules appear to be biased in favour of certain types of producer interests and against environmental interests. Retaliation is sanctioned in cases where governments introduce distorting subsidies, but not when they fail to eliminate implicit subsidies arising from pollution distortions. It is difficult to justify this inconsistency, particularly at a time when global environmental quality is worsening.

However, rather than resolving the inconsistency by simply expanding the range of potential countervail actions, it would probably be in Canada's interests to try to narrow the scope for *all* types of countervail actions. That is, countries could agree on common definitions of illegal subsidies of all types, including subsidies for the use of the environment. With clearer definitions of illegal subsidies, the potential for the use of countervail as a harassment device would be reduced. But at the same time, trade policy instruments would be available in cases where countries do not adhere to agreements they have made with respect to either explicit or implicit subsidies.

ENDNOTES

1   For a very detailed discussion of China's environmental problems, see Edmonds (1994). As well, the United Nations Environment Programme (1993) has brought together data on various types of environmental problems (such as water quality, deforestation rates, soil quality, etc.), for countries throughout the world, including some Asian countries.
2   The model is based on Copeland and Taylor (1994,1995b).
3   The interpretation of the size of the domestic market in autarky as providing a natural dampening mechanism on pollution is discussed at

4 length in Copeland and Taylor (1995c) with the aid of a model with production externalities. Chichilnisky (1994) and Brander and Taylor (1995) analyze the effects of trade liberalization on a fishery.
4 There has been much discussion of environmental dumping at the policy level. Esty (1995), McCaskill (1994), and Wilkinson (1994) provide useful discussions of the issues. See also Rauscher (1994).
5 For simplicity, we assume that the supply curves are not affected by trade. This is true, for example, in the simple model of Copeland and Taylor (1994). More generally, supply curves may move with trade, but that will not affect the points I want to make here.
6 This argument does not hinge on the use of pollution quotas. Referring to Figure 2, suppose that the pollution supply curve is determined by politics and is everywhere to the right of the true marginal damage curve in Foreign. Then trade can equalize the implicit price of pollution (however the policies are implemented), but in equilibrium, $\tau = \tau^* < MD^*$.
7 Montgomery (1972) shows that a permit system based on this principle would implement the ambient quality target at minimum cost.
8 On the other hand, if the foreign goods compete with Home's exports to third countries, or if they compete with Home's exports to Foreign, then Home's terms of trade worsen, and Home as a whole is worse off.

Bibliography

Alardyce, G. "'The Vexed Question of Sawdust': River Pollution in Nineteenth Century New Brunswick." *Dalhousie Review*, 52 (1972):177-90.
Barrett, S. "Strategic Environmental Policy and International Trade." *Journal of Public Economics*, 54 (1994):325-38.
Beghin, J., D. Roland-Holst and D. van der Mensbrugghe. "A Survey of the Trade and Environment Nexus: Global Dimensions." *OECD Economic Studies*, 23 (1994):167-92.
Brander, J. and M. S. Taylor. "International Trade and Open Access to Renewable Resources: The Small Open Economy Case." NBER Working Paper No. 5021. Cambridge, Mass.: National Bureau of Economic Research, 1995.
Brandon, C. and R. Ramankutty. "Toward and Environmental Strategy for Asia." World Bank Discussion Paper No. 224. Washington, DC: World Bank, 1993.
Chichilnisky, G. "Global Environment and North-South Trade." *American Economic Review* (1994).
Copeland, B. R. "International Trade and the Environment: Policy Reform in a Polluted Small Open Economy." *Journal of Environmental Economics and Management*, 26 (1994): 44-65.
Copeland, B. R. and M. S. Taylor. "North-South Trade and the Environment." *Quarterly Journal of Economics*, 109 (1994): 755-87.
_____."Trade and Transboundary Pollution." *American Economic Review*, 85(1995a): 716-37.

———. "Trade and the Environment: A Partial Synthesis." *American Journal of Agricultural Economics*, 77 (1995b): 765-771.

———. "Trade, Spatial Separation, and the Environment." NBER Working Paper 5242. Cambridge, Mass.: National Bureau of Economic Research, 1995c.

Daly, H. and R. Goodland. "An Ecological-Economic Assessment of Deregulation of International Commerce Under GATT." *Population and Environment*, 15 (1994):395-427.

Dessus, S. et al. "Input-Based Pollution Estimates for Environmental Assessment in Developing Countries." Technical Paper No. 101. Paris: OECD Development Centre, October 1994.

Dewees, D. N. and M. Halewood. "The Efficiency of the Common Law: Sulphur Dioxide Emissions in Sudbury." *University of Toronto Law Journal*, 18 (1992).

Edmonds, R. L. *Patterns of China's Lost Harmony: A Survey of the Country's Environmental Degradation and Protection*. London: Routledge, 1994.

Esty, D. C. *Greening the GATT: Trade, Environment and the Future*. Washington: Institute for Internation Economics, 1994.

Gillis, R. P. "Rivers of Sawdust: The Battle over Industrial Pollution in Canada, 1865-1903." *Journal of Canadian Studies*, 21 (1986):84-103.

Grossman, Gene M. and Alan B. Krueger. "Environmental Impacts of a North American Free Trade Agreement." In *The Mexico-U.S. Free Trade Agreement*. Edited by Peter Garber. Cambridge, Mass.: MIT Press, 1993.

Jaffe, A. B. et al. "The Effect of Environmental Regulation on International Competitiveness: What the Evidence Tells Us." Unpublished paper. Harvard University, 1993.

Lee, H. and D. Roland-Holst. "International Trade and the Transfer of Environmental Costs and Benefits." In *Applied Trade Policy Modeling*. Edited by J. Francois and K. Reinert. Cambridge: Cambridge University Press, 1994.

Low, Patrick and Alexander Yeats. "Do 'Dirty' Industries Migrate?" In *International Trade and the Environment: World Bank Discussion Papers*. Edited by Patrick Low. Washington, DC: World Bank, 1992.

Lucas, R. et al. "Economic Development, Environmental Regulation and the International Migration of Toxic Industrial Pollution: 1960-1988." In *International Trade and the Environment: World Bank Discussion Papers*. Edited by P. Low. Washington, DC: World Bank, 1992.

Markusen, J., E. Morey and N. Olewiler. "Environmental Policy When Market Structure and Plant Locations Are Endogenous." *Journal of Environmental Economics and Management*, 24 (1993):69-86.

McCaskill, K. A. "Dangerous Liaisons: The World Trade Organization and the Environmental Agenda." Policy Staff Paper No. 94/14. Ottawa: Foreign Affairs and International Trade Canada, June 1994.

Montgomery, W. E. "Markets in Licenses and Efficient Pollution Control Programs." *Journal of Economic Theory*, 5 (1972):395-418.

Rauscher, M. "On Ecological Dumping." *Oxford Economic Papers*, 46 (1994):822-40.

Robison, H. D. "Industrial Pollution Abatement: The Impact on the Balance of Trade." *Canadian Journal of Economics*, 21 (1988):187-99.

Seldon, T. M. and D. Song. "Environmental Quality and Development: Is There a Kuznets Curve for Air Pollution Emissions?" *Journal of Environmental Economics and Management*, 27 (1994):147-62.

Shafik, N. "Economic Development and Environmental Quality: An Econometric Analysis." *Oxford Economic Papers*, 46 (1994):757-73.

Tisdell, C. "Asian Development and Environmental Dilemmas." *Contemporary Economic Policy*, 13 (1995):38-49.

Tobey, J. A. "The Effects of Domestic Environmental Policies on the Pattern of World Trade: An Empirical Test." *Kyklos*, 43 (1990):191-210.

United Nations Environment Programme. *Environmental Data Report*. Oxford: Blackwell, 1993.

Wilkinson, D. G. "NAFTA and the Environment: Some Lessons for the Next Round of GATT Negotiations." *The World Economy*, (1994):395-412.

Someshwar Rao & Ashfaq Ahmad
Micro-Economic Policy Analysis
Industry Canada

Canadian Small and Medium-Sized Enterprises: Opportunities and Challenges in the Asia Pacific Region[1]

WHILE CANADA, LIKE MANY OTHER INDUSTRIALIZED COUNTRIES, has been experiencing serious economic difficulties since the mid-1970s, its current problems appear to be more critical than those of many other OECD countries. Very high and persistent levels of unemployment, poor productivity growth, stagnant real incomes, large and growing government and external debt levels, and a weak Canadian dollar are the symptoms of poor adjustment to profound structural changes in the global economy. These global trends include dramatic and rapid changes in product and process technologies; the information revolution; large and continuous shifts in the comparative-advantage position of countries around the globe; closer economic integration among nations; increasingly fierce competition for markets, investment, technology, and skilled employees; and a trend decline in commodity prices.

International commerce, the lifeblood of the Canadian economy, could play a potentially pivotal role in addressing the serious economic problems confronting this country. But to realize fully the benefits of international trade, foreign direct investment (FDI), and technology, Canada must gradually overcome the four major structural weaknesses in its international commercial relations – the heavy concentration of its international trade, direct investment, and innovation activities in the hands of a few very large firms; excessive dependence on the U.S. market; the competitiveness problems of Canadian exports with respect to prices, quality, and variety; and a heavy reliance on resource and resource-based products for export revenue.

A deepening of Canada's commercial linkages with the dynamic and fastest-growing Asia Pacific markets could be very helpful in overcoming those structural difficulties. These countries – especially China, the newly industrialized economies (NIE), and other members of the Association of South-East Asian Nations (ASEAN)[2] – offer immense potential as a target for Canadian exports and investment in the next century. Increased participation in the Asian markets by Canadian businesses, especially small and medium-sized enterprises (SME), could be one of the main vehicles for stimulating economic growth and job creation in this country.

It is recognized that in today's environment, the health of small businesses is crucial for increased economic activity, innovation, and job creation. SME represent over 98 percent of all business enterprises in Canada, and their shares of total sales and employment have grown considerably in the 1980s and 1990s. However, the job- and wealth-creating capabilities of SME will not be fully utilized until they become more actively engaged in exporting. The dynamic Asian markets could enable Canadian SME to sustain and improve their growth record and overcome the disadvantages associated with size, and to improve their absolute as well as relative economic performance.

The main objective of this study is to analyze future opportunities and challenges of the Asia Pacific region for Canadian business, with a special focus on SME. It seeks to address the following important research and policy issues:

- How big is the market potential for Canada in the Asia Pacific region?
- How well has Canada performed in the region until now in terms of exports and direct investment activity?
- What are some of the major challenges posed by the need to deepen Canada's commercial linkages with the region?
- Does firm size play an important role in the overall participation of Canadian firms in Asian and other international markets?
- How does the competitiveness of Canadian SME compare with that of large firms in Canada?
- Do SME and large firms have different motivations in their outward thrust?
- How does the economic performance of Canadian SME compare with that of their large Canadian counterparts?
- Do outward-oriented Canadian firms, especially SME doing business in Asia, perform better than firms that are absent from foreign markets?
- How do the characteristics, strategies, and performances of Canadian firms – large as well as small or medium-sized – compare with those of their American counterparts?
- What are some of the major problems that Canadian firms, especially SME, face in doing business in Asia and elsewhere?
- What could governments do to help Canadian SME to increase their participation in Asian and other international markets?

In an effort to answer these questions, our research makes use of three types of information – macro-level data on the gross domestic product (GDP), trade, and FDI; firm-level financial data; and company survey results. We collected financial information on about 1,300 Canadian and 2,600 American companies.

For analytical purposes, they are disaggregated into three size classes (SME, large, and very large), 12 major industry groupings, and three types of market orientation – firms doing business in Asia and elsewhere; firms doing business in other international markets but not in Asia; and firms doing business solely in the domestic market.[3]

Our research findings show that the economic performance of outwardly oriented Canadian SME is superior not only to that of their domestically oriented counterparts but also to that of large and very large firms. Moreover, outwardly oriented Canadian SME are also very competitive vis-à-vis American firms. However, the international participation rate of Canadian SME is very low, compared with that of large and very large Canadian firms. Their participation rate, especially in the Asia Pacific region, could be increased by effective government policies in three key areas – liberalization of trade and investment flows; market intelligence and market awareness initiatives; and incentives for exporting and investing.

CANADA AND THE ASIA PACIFIC[4]

OVER THE PAST 25 YEARS, THE ASIA PACIFIC ECONOMIES have expanded at a much faster pace than those of any other country or region in the world. The emergence of the Asia Pacific region as a major centre of commercial activity in the world, especially in manufacturing production, presents Canada with many opportunities to expand and diversify its international trade and investment linkages, and to significantly improve the growth prospects and dynamism of its economy.

THE ASIA PACIFIC MARKETS

WITH OVER 1.7 BILLION PEOPLE, the Asia Pacific region is the most populous in the world (Figure 1). While the region has experienced faster rates of growth in output, living standards, trade, and investment during the postwar period than any other, it is important to bear in mind that there are enormous differences among Asian countries with respect to population size and growth, economic development, natural resource endowment, and the industrial structure of output, employment, and trade.

In 1992, the Asia Pacific economies had a combined GDP in excess of US$5 trillion, compared with nearly US$7 trillion each for North America and the European Union. Per capita income levels (measured in market exchange rates), however, varied a great deal across the region, ranging from a low of US$374 in China to more than US$29,000 in Japan, and averaging almost US$3,000 overall (Figure 2).

More importantly, the Asia Pacific region has experienced an unprecedented growth in output and living standards over the past two decades.

FIGURE 1

POPULATION AND POPULATION GROWTH, ASIA PACIFIC REGION AND OTHERS

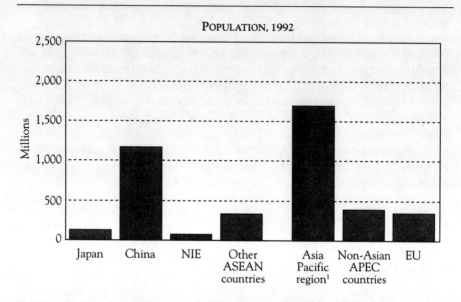

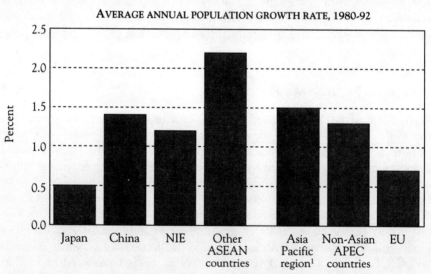

1 The Asia Pacific Region figure includes those for Japan, China, the NIE, and the other ASEAN countries.
Source: Industry Canada compilations using various sources, including APEC Economic Committee (1995).

FIGURE 2

LEVEL AND GROWTH OF GDP PER CAPITA, ASIA PACIFIC REGION AND OTHERS

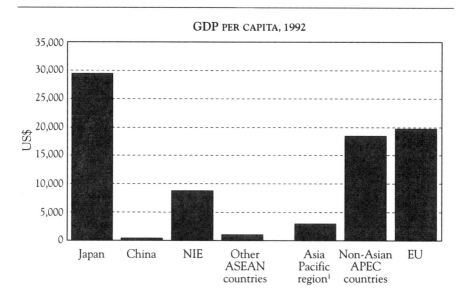

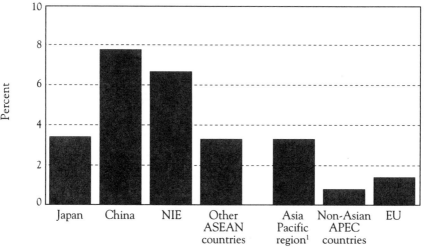

1 The Asia Pacific region figure includes those for Japan, China, the NIE, and the other ASEAN countries.
Source: Industry Canada compilations using various sources, including APEC Economic Committee (1995).

Between 1980 and 1992, real GDP growth among the Asian members of the Asia Pacific Economic Council (APEC)[5] averaged 4.8 percent annually – more than double the rate posted by either non-Asian APEC members or the European Union (EU). Real output in China grew at an exceptional rate of 9.2 percent – almost twice the average growth rate experienced by the Asia Pacific region as a whole.

During this period, real GDP growth in most Asia Pacific economies exceeded population growth by a considerable margin, thus contributing to a significant increase in their living standards. Between 1980 and 1994, real per capita incomes in the Asian APEC group rose at an average annual rate of 3.3 percent, compared with a meagre 0.8 percent in non-Asian APEC countries and 1.4 percent in the EU. Real per capita income growth in China (7.8 percent) and the NIE (6.7 percent) was exceptionally high relative to that of the other Asian APEC countries (Figure 2).

Similarly, exports and imports of the Asia Pacific countries expanded dramatically in the postwar period. Their merchandise exports rose from a mere US$39 billion in 1971 to almost US$900 billion in 1992, almost a twenty-five-fold increase. As with output, the NIE and China posted the fastest growth in trade among the Asia Pacific economies. More importantly, the share of the Asia Pacific countries in the total value of world merchandise exports rose from 11.5 percent in 1972 to about 20 percent in 1992. Likewise, their share of world exports of non-factor services have increased substantially over the last 25 years or so. At the same time, their share of total world merchandise imports also grew dramatically. In short, the Asia Pacific region has become a major and dynamic centre of commercial and economic activity in the world.

A continuation of favourable domestic conditions – high saving and investment rates; a highly skilled, easily adaptable and growing work force; market and outward-oriented economic policies, etc. – and of worldwide economic integration (including continued convergence in technology and productivity levels), coupled with a stable global economic environment, an expanding world economy, and freer world trade, would help the Asian countries, especially China, the NIE, and the ASEAN group, to pursue their climb on the value-added ladder and substantially improve their relative standing in the world economy.

If current trends continue into the future, the Asia Pacific economies could account for about 40 percent of world trade by the year 2010, compared with over 20 percent in 1992. In other words, future prospects for Canadian exports, investment, jobs, and real income could depend critically on how well Canada is able to penetrate these rapidly expanding markets, as well as on its capacity to adjust to the growing competitive challenge from these countries in both domestic and third-country markets, especially in manufactured products trade (World Bank, 1993; Helliwell, 1994; and Rao, 1992).

Trade Linkages

THE IMPORTANCE OF THE ASIA PACIFIC MARKET for Canadian exports and imports has increased significantly over the past 25 years or so. The share of exports to the region in Canada's total merchandise exports rose from about 7 percent in 1971 to over 8 percent in 1994 (Figure 3). That modest advance, however, masks a substantial increase in the importance of the NIE, China, and the ASEAN countries for Canadian exports. For example, the share of the NIE in total Canadian merchandise exports to the Asia Pacific region rose from 16 percent in 1980 to 27 percent in 1994. During the same period, the Japanese share of Canadian exports to the region declined from 66 percent to 53 percent.

On the import side, the importance of the Asia Pacific countries has also grown substantially. Their share of Canada's total merchandise imports rose from about 7 percent in 1971 to 13 percent in 1994 (Figure 3). As with exports, however, Japan's relative importance declined considerably – from 58 percent in 1980 to 43 percent in 1994. Even the share of NIE in Canadian imports from the region declined slightly during the same period. In contrast, the share of China and the ASEAN countries rose dramatically – from 3 to 15 percent for China and from 4 to 12 percent for the ASEAN group. The growing importance of these countries could be attributed to low labour costs as well as a significant real depreciation of their currencies vis-à-vis those of Japan and the NIE.

In short, Canada's trade linkages with the Asia Pacific countries have grown very rapidly, and that region has surpassed the EU as our second largest trading partner.[6] There is, however, considerable scope for Canadian businesses, especially SME, to expand their exports to the Asian countries and to increase investment in these economies. The large and growing technology and infrastructure needs of Asian nations present excellent opportunities for Canadian firms to expand and diversify their export and investment base, in terms of both products and geographic destination.

Investment Linkages

THE RECENT THEORETICAL AND EMPIRICAL LITERATURE strongly suggests that international trade and FDI are complements, not substitutes. Investment and trade patterns go hand in hand: strong investment linkages lead to strong trade linkages, and vice versa. Canada's commercial relations also show a strong positive relationship between these two pillars.

As with trade patterns, Canada's investment linkages with the Asia Pacific Rim have strengthened considerably over the past 25 years, albeit from a very small base. Between 1980 and 1994, the Asia Pacific share of the total FDI stock in Canada increased sixfold, from 1 to 6 percent. Similarly, the

FIGURE 3

DISTRIBUTION OF CANADIAN MERCHANDISE EXPORTS AND IMPORTS, BY REGION, 1980 AND 1994

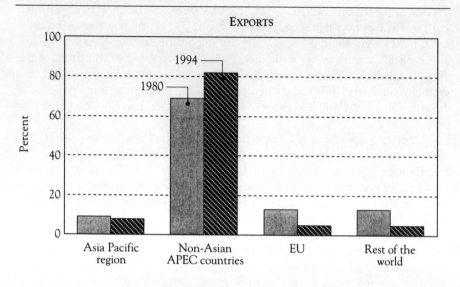

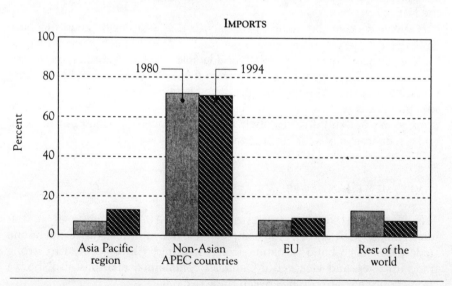

Source: Industry Canada compilations using various sources, including APEC Economic Committee (1995).

region's share of total stock of Canadian direct investment abroad more than doubled during this period – from 3 to 7 percent (Figure 4).

Figure 4

Distribution of Canadian Foreign Direct Investment Stock, by Region, 1980 and 1994[1]

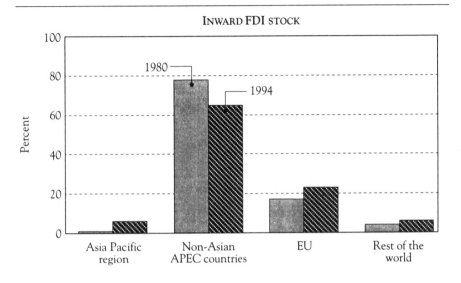

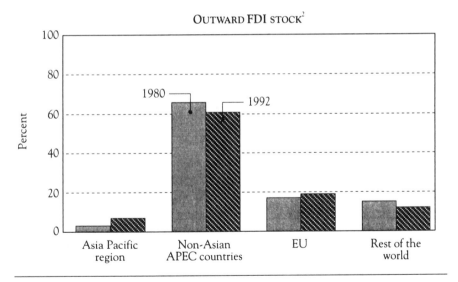

1 Preliminary data.
2 1992 data.
Source: Industry Canada compilations using various sources, including APEC Economic Committee (1995); Statistics Canada, Balance of Payments Division.

The pattern of Canada's investment linkages within the Asia Pacific region has also undergone marked shifts. For example, the shares of China and, in particular, the NIE in the inward and outward stock of Canadian FDI increased dramatically between 1980 and 1994. In 1994, the NIE accounted for 31 percent of the Asia Pacific's inward FDI stock in Canada and for 7 percent of Canada's outward FDI stock in the region – compared with 8 and 7 percent, respectively, in 1980.

In sum, the Asia Pacific region represents a growing share of total Canadian inward and outward FDI stocks. However, Canada's investment linkages with the region are still weak relative to U.S. investment linkages with the Asia Pacific countries. At the same time, as is the case with merchandise trade, the lion's share of Canada's inward and outward FDI stocks is still accounted for by the United States, although its relative importance has been declining somewhat.[7]

Strengthening Canada's Commercial Linkages with the Asia Pacific Region

Resources and resource-based manufactured products account for over 80 percent of Canada's exports to the Asia Pacific region, although they represent only about 50 percent of total Canadian exports. Technology-intensive manufactured exports, on the other hand, also account for nearly 50 percent of all Canadian exports but for only 16 percent of sales to the Asia Pacific countries (Rao, 1992, Table 23).

The importance of primary and resource-based products in the Asia Pacific region's total imports has been declining dramatically – from 56 percent in 1971 to about 40 percent in 1989. By contrast, the share of technology-intensive manufactured imports rose from about 33 to 48 percent (Rao, 1992, Table 6). These trends suggest that Canada needs to broaden its export base in order to strengthen its trade linkages with the Asia Pacific countries.

In attempting to achieve this, Canada will have to overcome strong competitive challenges from the United States, Australia, New Zealand, and the ASEAN countries in resource and resource-based exports to the region and to regain some of its lost market shares in those products. Between 1971 and 1989, for example, Canada's share of China's imports of agriculture, forestry, and fish products, and of resource-intensive manufactured products declined considerably – from about 60 to 18 percent (Rao, 1992, Table 30).

Canada will continue to benefit from the high-quality and competitively priced goods offered by the Asia Pacific countries. Machinery and transportation equipment (52 percent), labour-intensive manufactured products (21 percent), and light manufactured goods (14 percent) account for over 80 percent of Canada's imports from the region (Rao, 1992, Table 24). But Canadian manufacturers of these products, especially labour-intensive goods and light machinery and equipment, will increasingly face fierce competition from the NIE, China, and the ASEAN countries both at home and in Asia Pacific and other foreign markets.

In sum, the Asia Pacific region offers enormous export and investment opportunities for Canada. To take advantage of these opportunities, however, Canada needs to meet four main challenges: to improve the overall competitiveness (costs, variety, quality) of its exports; to broaden its export base; to increase the participation of SME in Asia; and to strengthen its investments linkages with the Asia Pacific economies.

The Growing Importance of SME in Canada

AS A RESULT OF MAJOR TECHNOLOGICAL ADVANCES in production processes, economies of scale and scope associated with the centralization and concentration of economic activity are no longer considered the main sources of competitiveness. Today, flexibility in adapting to the rapidly changing domestic and international economic environment, and innovation have become the mainstays of competitive strength. SME owned and operated by an individual or a small team are able to manage the challenges of structural adjustment alluded to above.[8] As a result, they have become important players in Canada and other industrialized countries in terms of sales, job creation, innovation, and outward orientation (exports and FDI).

What is an SME? While there is no standard definition, in many industrialized countries a manufacturing enterprise employing fewer than 500 employees is considered an SME. Recent research by the United Nations Conference on Trade and Development (UNCTAD) also uses this definition. In the service industries, however, the size threshold is often much lower, being set in most countries at fewer than 50 employees. Clearly, SME are not a homogeneous group.

Numerically, the vast majority (over 98 percent) of all business enterprises in Canada are SME. Conversely, the 2,000 largest corporations represent only 0.1 percent of all business firms in Canada. In all other APEC economies too, SME account for between 89 and 99 percent of all business enterprises (Hall, 1994).

SME make a major contribution to the Canadian economy. It is estimated that they accounted for 57.2 percent of Canada's total private-sector GDP in 1992, up from 52.6 percent in 1982. Three-quarters of the 1992 share was accounted for by small firms, with fewer than 100 paid employees in manufacturing and fewer than 50 paid employees in all other sectors. The rest came from medium-sized firms, with between 100 and 500 employees (Canada, Industry Canada, 1994d).[9]

More importantly, SME augmented their share of total business employment in Canada from 57.7 percent in 1982 to 63.2 percent in 1992 – an

Figure 5

Distribution of Canadian Firms[1] by the Number of Employees, 1982 and 1992

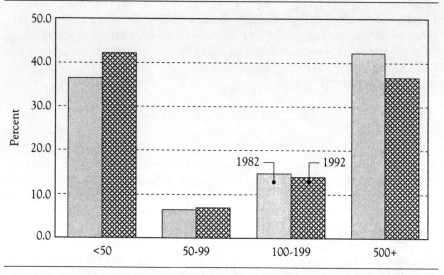

1 Includes self-employment; excludes public administration.
Source: Based on estimates prepared by the Entrepreneurship and Small Business Office, Industry Canada, Ottawa.

increase achieved at the expense of large firms. Moreover, SME accounted for almost all of the net new job creation in Canada in the last 10 years (Figure 5). Similar employment trends are observed in other APEC economies (Hall, 1994).

Sectoral and Regional Distribution

Capital/labour ratios vary from country to country and from industry to industry. However, SME in general tend to be relatively more labour-intensive than large firms across all industries in all countries. Not surprisingly, therefore, small-business activity in Canada is mainly concentrated in non-capital-intensive sectors such as services, and is less prominent in capital-intensive sectors such as communications, manufacturing, or utilities. In 1992, for example, the share of employment accounted for by Canadian SME was in excess of 80 percent in most service-oriented industries such as construction services, wholesale trade, business services, and in accommodation, food, and beverage services. In contrast, in relatively more capital-intensive sectors that proportion ranged from a low of 15 percent in communication and other utilities to 54 percent in manufacturing industries (see Table A2-1).

SME account for the lion's share of employment in all provinces and territories, but their relative importance varies a great deal among them, ranging from a low of 53 percent in Ontario to a high of 85 percent in Yukon in 1994. In all other provinces except Manitoba, the SME share of total employment ranged between 60 and 67 percent, compared to the national average of about 59 percent.[10]

SME AND CANADIAN EXPORTS

TO DATE, SME HAVE PLAYED A VERY SMALL ROLE in Canada's export activity. Although they accounted for 97 percent of all Canadian export firms in 1992, only 9 percent of the value of Canadian exports originated from SME, against over 90 percent from a few large firms.

More importantly, in 1990 only 7.6 percent of Canadian businesses had any exports at all, and the majority of them made only three foreign transactions during the year. This pattern was particularly evident among SME. Nonetheless, Canadian SME are becoming more international, as shown by the increase in the number of SME exporters in recent years – from 33,000 in 1986 to 73,000 in 1992. Yet, Canadian firms with exports represent less than 10 percent of all SME. While reliable estimates for the shares of SME exports in other countries are not available, according to a recent UNCTAD study about one-quarter of manufactured exports in OECD countries were accounted for by SME. Similarly, SME account for one-third of the manufactured exports of the Asia Pacific countries (Hall, 1994).

In Canada, large exporting firms are found mainly in a number of sectors in which Canada has a strong comparative advantage – transportation equipment, pulp and paper, and mining. SME exports tend to be concentrated in industries where new export markets have opened up, especially in fast-growing service sectors. For example, among the top SME export industries – wholesale food and beverages, business service, and plastic products – SME exports accounted for over 40 percent of total exports in 1992. The largest SME export share – close to 70 percent – was in the "miscellaneous" category, representing a wide range of light consumer goods.

SME AND CANADIAN DIRECT INVESTMENT ABROAD

AS WITH TRADE, MUCH OF THE CANADIAN STOCK of direct investment abroad is in the hands of a few very large firms. For example, according to the United Nations Centre on Transnational Corporations (UNCTC), there are about 1,300 Canadian-based multinational enterprises (MNE), both large and small, in Canada, but they represent a meagre 0.2 percent of all Canadian business establishments. In addition, the direct investment activities of Canadian-based firms in foreign countries are highly concentrated. For instance, the top 20 Canadian-based MNE account for about half of the total foreign assets of all Canadian-based MNE (Canada, Industry Canada, 1994c).

SME AND INNOVATION

THE ROLE OF SME IN THE INNOVATION PROCESS is difficult to assess. Recent research, however, suggests that they make a significant contribution to innovation in Canada and elsewhere (Baldwin and Johnson, 1995; UNCTAD, 1993). The Futures Group estimates that American SME account for 55 percent of all manufacturing product innovations and that they produce twice as many innovations (including "significant" innovations) per employed person as large firms (Report of the President, 1995).

In sum, SME are major players in terms of output and employment, and they are the dominant source of net new job creation in Canada and elsewhere. They are an important source of dynamism and innovation, and they make a small positive contribution to Canada's exports and direct investment abroad. As they are increasingly becoming outward-oriented, future SME trade and investment activities in the Asia Pacific region are likely to play a significant role in Canada's success in penetrating Asian markets.

OUTWARD ORIENTATION AND ECONOMIC PERFORMANCE OF CANADIAN SME

OUTWARD ORIENTATION

Rationales

In general, the reasons invoked by SME and large firms for the decision to export or invest abroad are similar. Sales and asset growth, profitability, access to natural resources and technology, government incentives in either or both the home and the host country, broadening of the firm's comparative-advantage position, lower labour and other production costs, availability of skilled labour, adequate physical and technological infrastructure, and the need to overcome formal and informal barriers to trade in host countries are some of the major factors cited by firms in Canada and elsewhere (Caves, 1982; Dunning, 1985; and Rugman, 1987).

In a recent UNCTAD survey (1993), SME mentioned the following six factors as being the most important reasons for undertaking direct investment in foreign countries: expectation of growth in local markets; access to and growth in third-country markets; information gathering; the need to strengthen competitive capacity; low labour costs in host countries; and lower production costs for exporting to third countries and to the home country. These factors were also mentioned by SME as the main reasons for establishing foreign affiliates in Asia, but they were ranked somewhat differently, with the low cost of labour in the host country being rated the third most important reason for investing in Asia (Figure 6).

A recent survey by the Conference Board of Canada (1994) examined the reasons that motivate Canadian firms to expand into the Asia Pacific region. Market diversification, production for export to other markets, favourable tariff/trade regulations, production for local market, and reduced production costs were mentioned by firms of all sizes as the major reasons for expanding into the Asia Pacific markets. The results of another Conference

FIGURE 6

REASONS INVOKED BY TRANSNATIONAL SME FOR INVESTING IN ASIAN AND WORLD MARKETS[1]

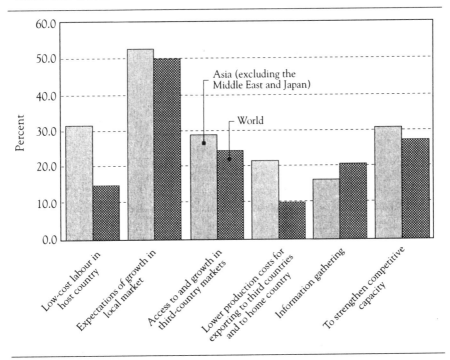

1 The height of the bar measures the number of foreign affiliates that said "yes" to the reason cited, divided by the total number of foreign affiliates.
Source: UNCTAD (1993).

Board survey (1995) of Canadian firms (mostly SME) investing in China are generally similar to the UNCTAD survey findings, with access to the Chinese market, low labour costs, low production costs, and access to other markets being cited.

Firm Size

In our database, 56 percent of all Canadian firms doing business in the Asia Pacific region and other foreign markets are SME, compared to just 12 percent of U.S. firms (Table 1). Similarly, the proportion of SME in the other two categories – firms doing business elsewhere but not in Asia, and domestically oriented firms – is substantially higher in the Canadian sample than in the U.S. sample. The proportion of large and very large firms, on the other hand, is over 75 percent in all three categories of outward orientation in the U.S. sample.

Although SME account for nearly 60 percent of all Canadian firms doing business in Asia and elsewhere, their sales represent a very small fraction (less than 3 percent) of the total sales of Canadian firms exporting abroad. In sharp contrast, the very large firms contributed almost 90 percent to the total sales of all Canadian firms doing business in Asia and elsewhere, compared with 97 percent for the U.S. sample.

In our sample, only 21 percent of all Canadian SME do business in Asia and other international markets, compared with 25 percent of American SME. Similarly, sales of the firms doing business in Asia and elsewhere account for only about 25 percent of the sales of all SME in the two countries (Table 2). As well, the proportion of SME doing business only in other international markets is small (less than 20 percent) in the two countries. On the other hand, roughly 60 to 65 percent of all SME in the two samples are domestically oriented (no exports and no FDI). In sharp contrast, 66 to 80 percent of very large Canadian and American firms are outwardly oriented.

These simple tabulations strongly suggest that outward orientation and the geographic destination of exports and FDI are positively related to firm size. To formally test for the relationship between these three variables, two multiple regression equations were estimated for both Canadian and U.S. firms. The first regresses the outward orientation (a dichotomous variable) on 10 industry and two size dummies, with the constant term capturing the influence of the omitted industry and size categories. The second equation tests for the relationship between the geographic destination of outward orientation (Asia Pacific versus non-Asia Pacific) and firm size, controlling for the influence of industry characteristics.

The coefficients of the two size dummies (SME and large firms) are negative and highly significant in the two equations for both Canada and the United States. Moreover, the coefficient on SME is larger than that on large

TABLE 1
SIZE DISTRIBUTION OF CANADIAN AND U.S. FIRMS, BY MARKET ORIENTATION (PERCENT)

MARKET ORIENTATION	FIRMS DOING BUSINESS IN THE ASIA PACIFIC REGION				FIRMS DOING BUSINESS IN INTERNATIONAL MARKETS BUT NOT THE ASIA PACIFIC REGION				FIRMS DOING BUSINESS ONLY IN DOMESTIC MARKETS			
	NUMBER OF FIRMS		SALES		NUMBER OF FIRMS		SALES		NUMBER OF FIRMS		SALES	
Firm size	Canada	United States	Canada	United States	Canada	United States	Canada	United States	Canada	United States	Canada	United States
Very large	22.3	55.1	88.9	97.2	10.2	40.7	68.8	90.9	3.4	35.1	50.6	86.7
Large	21.4	32.6	8.3	2.6	23.5	40.9	24.4	8.4	11.8	39.2	28.9	11.8
SME	56.3	12.2	2.8	0.2	66.3	18.3	6.9	0.8	84.8	25.7	20.5	1.6
Total	100.0	100.0	100.0	100.0	100.0	100.0	100.0	100.0	100.0	100.0	100.0	100.0

Source: Compiled by the authors using data from various sources.

TABLE 2
MARKET-ORIENTATION DISTRIBUTION OF CANADIAN AND U.S. FIRMS, BY SIZE (PERCENT)

	VERY LARGE FIRMS				LARGE FIRMS				SME			
	NUMBER OF FIRMS		SALES		NUMBER OF FIRMS		SALES		NUMBER OF FIRMS		SALES	
Market orientation	Canada	United States	Canada	United States	Canada	United States	Canada	United States	Canada	United States	Canada	United States
Firms doing business in the Asia Pacific region	64.3	49.4	74.0	70.5	37.0	34.8	36.5	34.9	21.4	25.4	26.0	25.8
Firms doing business in international markets but not the Asia Pacific region	15.9	17.0	12.8	12.1	21.8	20.4	24.0	21.0	13.5	17.7	14.3	17.9
Firms doing business only in domestic markets	19.8	33.6	13.2	17.4	41.2	44.8	39.6	44.1	65.1	57.0	59.7	56.3
Total	100.0	100.0	100.0	100.0	100.0	100.0	100.0	100.0	100.0	100.0	100.0	100.0

Source: Compiled by the authors using data from various sources.

firms in both equations for both countries (see Tables A3-1 and A3-2). These results, in turn, imply that the probability that SME will be active in Asian markets is significantly lower than that for large and very large firms in the two countries.

However, these results do not necessarily imply that SME are inherently disadvantaged in going global, because firm size might be simply capturing the positive influence of other omitted variables. For example, recent research suggests that firm size is positively correlated with firm age, experience in international markets, and managerial ability and vision (Reuber and Fischer, 1995; Ogbuehi and Longfellow, 1994).

Industrial Structure

The distribution by industry of the sales of firms doing business in Asia is similar to the overall distribution for both Canada and the United States. In the Canadian sample, firms in technology- and resource-intensive manufacturing, as well as in the finance, insurance, and real estate industry account for almost 80 percent of the total sales of all firms doing business in Asia and elsewhere, compared with about 70 percent in the U.S. sample (Table 3).

These findings are also confirmed by regression analysis. For example, the coefficient of the technology-intensive manufacturing dummy is positive and statistically significant in the two outward-orientation regression equations for the Canadian and U.S. samples. These results suggest that, other things being equal, firms in technology-intensive manufacturing industries are likely to be much more outwardly oriented and to participate more in the Asia Pacific markets than firms in the control group (mining). Likewise, Canadian firms in resource-intensive manufacturing and in the finance, insurance, and real estate group have a slightly higher propensity to participate in the Asian markets than firms in mining (see Tables A3-1 and A3-2).

ECONOMIC PERFORMANCE

THE COMPETITIVENESS OF SME is crucial to increasing their penetration of Asian markets and meeting the growing challenges of increased competition from other domestic and foreign-based (Asian and other) firms in domestic and international markets. The increased outward orientation in turn would help SME to further improve their competitive position.

How does the past performance of SME compare with that of other firms in Canada? The average annual growth of sales and assets of SME in the past five years was substantially better than that of large and very large firms in all three categories of outward orientation (firms doing business in Asia; firms doing business abroad but not in Asia; and domestically oriented firms). For

TABLE 3
INTERNATIONAL DISTRIBUTION OF SALES OF CANADIAN AND U.S. FIRMS DOING BUSINESS IN THE ASIA PACIFIC REGION (PERCENT)

	AGRICULTURE, FORESTRY, AND FISHING	CONSTRUCTION	FINANCE, INSURANCE, AND REAL ESTATE	LABOUR-INTENSIVE MANUFACTURING	MINING	PUBLIC ADMINISTRATION	RESOURCE-INTENSIVE MANUFACTURING	RETAIL TRADE	SERVICES	TECHNOLOGY-INTENSIVE MANUFACTURING	TRANSPORTATION AND PUBLIC UTILITIES	WHOLESALE TRADE	TOTAL
Canada													
Number of firms	0.8	0.6	8.8	6.0	4.7	0.6	23.6	0.0	6.3	37.9	4.1	6.6	100.0
Sales	0.3	0.0	28.9	2.4	8.2	0.1	18.9	0.0	1.2	30.6	6.1	3.4	100.0
United States													
Number of firms	0.3	0.8	5.2	7.7	3.3	0.0	13.9	2.2	10.2	50.2	3.7	2.4	100.0
Sales	0.1	0.9	10.6	2.6	8.9	0.0	16.4	7.6	3.6	41.0	6.8	1.6	100.0

Source: Compiled by the authors using data from various sources.

example, the average annual five-year growth rates (weighted) of sales and assets of SME in the first group were 11.8 and 12.8 percent, respectively, compared with 4.4 and 8.6 percent for very large firms (Figure 7).

Similarly, in terms of capital productivity (sales/assets) and the average rate of return on assets, SME performed much better than large and very large firms. For example, weighted average rates of return on assets of Canadian companies doing business in the Asia Pacific region were 7 percent for SME, 3.1 percent for large firms, and 1.2 percent for very large firms. At the same time, Canadian SME with business in international markets other than the Asia Pacific region also outperformed their large and very large counterparts in terms of sales growth, asset growth, capital productivity, and return on assets (see Figure A2-1).

On the other hand, the labour productivity (sales/employment) of SME is significantly lower than that of larger firms in all the three categories of outward orientation. In the case of firms doing business in Asia, SME productivity is roughly two thirds that of very large firms (see Figure A2-2).

FIGURE 7

ECONOMIC PERFORMANCE OF CANADIAN SME DOING BUSINESS IN THE ASIA PACIFIC REGION, BY SIZE OF FIRM

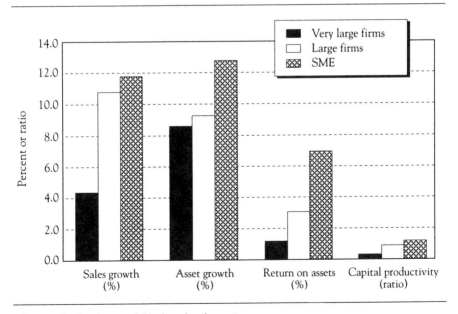

Source: Based on data compiled by the authors from various sources.

The regression results, however, suggest that part of the difference in labour productivity could be attributed to the difference in the capital/labour ratio. The coefficient of this variable in the labour productivity equation, after controlling for size, industry, and outward-orientation variables, is positive and highly significant. Nevertheless, labour productivity is significantly positively correlated with the size variable in the Canadian sample, implying that a large part of the labour productivity gap is the result of the size disadvantage of SME (see Table A3-3). The size variable may be capturing the positive influence of various types of scale and scope economies on output and sales.

In short, Canadian SME performed better than their large and very large Canadian counterparts with respect to four of the five performance indicators.

Relative Performances of Canadian and U.S. Firms

More importantly, the Canadian SME outperformed their U.S. counterparts by a substantial margin in terms of all five indicators in all three categories of outward orientation. For example, the average sales growth and average rate of

Figure 8

Economic Performance of Canadian and American SME Doing Business in the Asia Pacific Region

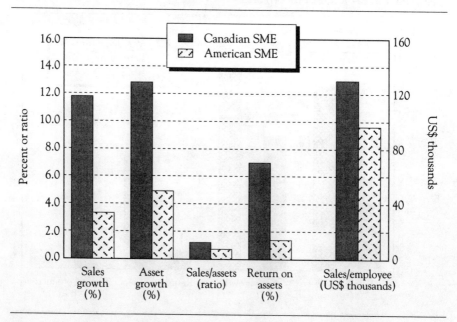

Source: Based on data compiled by the authors from various sources.

return on assets of American SME doing business in the Asia Pacific region were only 3.3 and 1.4 percent, respectively, compared to 11.8 and 7 percent for Canadian SME (Figure 8).

Similarly, very large outwardly oriented Canadian firms generally performed better than their U.S. counterparts (Tables 4 and 5). It is interesting to note that the huge size disadvantage of Canadian firms does not seem to have had an adverse impact on their labour productivity performance relative to that of their American counterparts. For example, the productivity level of very large Canadian firms doing business in the Asia Pacific region is about 9 percent higher, on average, than that of corresponding U.S. firms. For large Canadian firms, the productivity level advantage is almost 25 percent higher.

These findings strongly suggest that Canadian firms, especially SME, are very competitive vis-à-vis their American rivals. Thus they are well positioned to increase their participation in the Asia Pacific region and in other international markets, and to further improve their relative standing.

Links Between Outward Orientation and Economic Performance

The economic performance of Canadian SME that do business in the Asia Pacific region and in other international markets is considerably superior, on average, to that of domestically oriented SME in terms of all five indicators (Table 4 and Figure 9). For example, their sales growth and asset growth is 40 percent or more, on average, higher than that of domestically oriented SME. Similarly, the return on assets of outward-oriented SME is about four times as high as that of domestic-oriented firms. These aggregate results also generally hold for firms in the three major industry groups: technology-intensive manufacturing; resource-intensive manufacturing; and finance, insurance, and real estate.

At the same time, the average performance of very large, outwardly oriented Canadian firms is superior to that of their domestically oriented counterparts with respect to four of the five performance indicators used – labour productivity, capital productivity, the average annual rate of return on assets, and the average annual rate of asset growth. Both categories recorded more or less similar sales growth rates.

The results for large Canadian firms are somewhat mixed. Outwardly oriented firms performed better than domestically oriented ones in terms of the rate of return on assets, capital productivity, and sales growth, while domestically oriented firms outperformed outwardly oriented ones with respect to asset growth. On average, both categories had similar labour productivity levels (Table 4).

In sharp contrast to the Canadian results, there appears to be no systematic positive relationship between outward orientation and economic performance

TABLE 4
ECONOMIC PERFORMANCE OF CANADIAN FIRMS, BY SIZE AND MARKET ORIENTATION

MARKET ORIENTATION	FIRMS DOING BUSINESS IN THE ASIA PACIFIC REGION			FIRMS DOING BUSINESS IN INTERNATIONAL MARKETS BUT NOT THE ASIA PACIFIC REGION				FIRMS DOING BUSINESS ONLY IN DOMESTIC MARKETS		
PERFORMANCE INDICATOR	VERY LARGE	LARGE	SME	VERY LARGE	LARGE	SME	VERY LARGE	LARGE	SME
Sales growth (%)	4.4	10.8	11.8	7.0	10.2	12.5	6.3	8.9	8.6
Asset growth (%)	8.6	9.3	12.8	8.4	9.5	15.4	7.6	14.5	7.6
Sales/assets (US$ thousands)	0.3	0.9	1.2	0.7	1.0	0.9	0.3	0.4	0.4
Sales/employees (US$ thousands)	203.5	138.9	129.7	239.6	152.8	114.4	196.2	149.2	114.5
Return on assets (%)	1.2	3.1	7.0	2.5	3.7	8.4	2.1	1.1	1.8

Source: Compiled by the authors using data from various sources.

TABLE 5
ECONOMIC PERFORMANCE OF U.S. FIRMS, BY SIZE AND MARKET ORIENTATION

MARKET ORIENTATION	FIRMS DOING BUSINESS IN THE ASIA PACIFIC REGION			FIRMS DOING BUSINESS IN INTERNATIONAL MARKETS BUT NOT THE ASIA PACIFIC REGION			FIRMS DOING BUSINESS ONLY IN DOMESTIC MARKETS		
PERFORMANCE INDICATOR	VERY LARGE	LARGE	SME	VERY LARGE	LARGE	SME	VERY LARGE	LARGE	SME
Sales growth (%)	6.7	9.8	3.3	9.4	13.0	7.0	8.4	10.5	5.4
Asset growth (%)	10.3	10.8	4.9	10.8	14.6	6.1	11.2	10.1	8.3
Sales/assets (US$ thousands)	0.5	1.0	0.7	0.4	0.7	0.6	0.3	0.3	0.2
Sales/employee (US$ thousands)	186.7	111.7	96.2	157.9	122.0	107.2	187.0	124.6	107.6
Return on assets (%)	3.9	5.7	1.4	3.1	4.9	3.1	3.6	2.3	2.0

Source: Compiled by the authors using data from various sources.

FIGURE 9

ECONOMIC PERFORMANCE OF CANADIAN SME, BY MARKET ORIENTATION

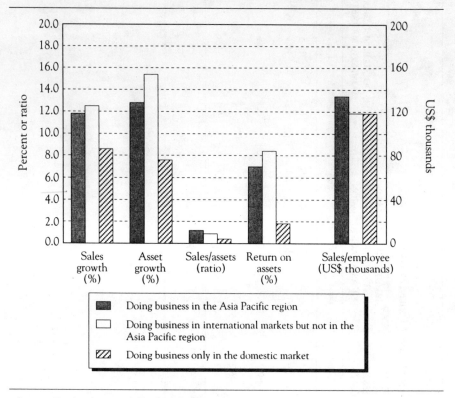

Source: Based on data compiled by the authors from various sources.

among the American firms. Huge differences in the size and dynamism of the two domestic markets may account for the differing results. In other words, outward orientation seems to matter a great deal for improving the performance of firms from a small and slowly growing domestic market.

OBSTACLES TO SME PARTICIPATION IN THE ASIA PACIFIC REGION

CANADIAN TRADE AND FDI IN THE ASIA PACIFIC REGION, particularly in Japan and China, is expected to grow in the coming years. In the Conference Board's 1994 survey, respondents chose Japan more often than any other country in the Asia Pacific region as a priority market for exports and foreign investment, closely followed by China. With respect to FDI intentions, fully 98 percent of respondents in another survey (Conference Board of Canada, 1995) cited China as one of the principal markets for their intended investments. The "rest of Asia," the second most popular choice, was selected by 37 percent of respondents. By contrast, North America and Europe were mentioned by fewer than 20 percent of respondents. These results seem to suggest that the pace of Canadian exports and FDI (via equity joint ventures, cooperative joint ventures, and the establishment of wholly Canadian-owned enterprises) in the Asia Pacific region is likely to accelerate in the near future.

The removal of existing impediments – tariff and non-tariff trade barriers, formal and informal obstacles to FDI – would give additional impetus for Canadian firms to increase their presence in the region. These barriers adversely affect the commercial activities of both SME and large firms. According to the Conference Board (1994), a majority of business firms (53 percent) believe that exports of goods and services offer the best opportunity for expansion into the region. The reduction of impediments to trade and investment in the Asia Pacific region should therefore be a major focus of the Canadian policy agenda.[11]

Trade Barriers

Tariffs

The foregoing discussion is based on estimates of tariffs conducted around 1988 (following the implementation of the Tokyo Round decisions). Since then, a number of Asian countries have unilaterally made significant reductions in the level of tariff protection, and the recently concluded Uruguay Round made further progress in lowering tariff levels among members of the new World Trade Organization (WTO). Unfortunately, more recent and comprehensive estimates of tariff levels are not available for either the Asian countries or Canada, and thus we are compelled to use the 1988 estimates. As our discussion is mainly concerned with the structure of tariff protection – which, it is safe to assume, has remained more or less intact in the post-Tokyo Round era – the lack of more recent data is not expected to have a major impact on our findings.

The average rate of tariff protection is substantially higher in the Asia Pacific countries than in Canada (Rao, 1992, Tables 35 and 36). While it is expected to decline in the wake of the Uruguay Round Agreements, a gap is likely to remain.

The industrial structure of tariffs in Canada and the Asia Pacific countries shows a strong bias for protection in favour of industries in which individual countries have a strong comparative disadvantage. Thus the agri-food sector and resource and resource-based industries in the Asia Pacific region tend to receive a much higher level of tariff protection than their Canadian counterparts. Average post-Tokyo Round tariff rates on agricultural, forestry, and fishing products and on food, beverages, and tobacco, for example, are substantially higher in Japan (ranging between 22 and 29 percent) than in Canada (between 2 and 6 percent). Similarly, the average tariff rate on mineral fuels ranges from a low of 7 percent in Malaysia and Thailand to 17 percent in the Philippines.

In Canada, the average tariff protection for manufactured goods is significantly higher than that for primary products. Within manufacturing, tariffs on labour-intensive manufactured goods such as textiles, clothing, footwear, and furniture and fixtures are substantially higher than on other manufactured imports, ranging from 14.3 percent for furniture and fixtures to 24.2 percent on clothing.

According to the Conference Board's 1994 survey, however, Canadian companies are more concerned with the way tariffs are applied in the Asia Pacific countries than with levels of tariff protection. They indicated that they could compete well with their rivals from other countries so long as they faced the same type of tariff protection, but that they faced a competitive disadvantage if competitors received preferential tariff treatment. In some Asia Pacific countries, tariffs tend to vary for the same product, depending on whether or not the latter is eligible for an exemption from the published tariff. In China, for example, high-tech items whose purchase falls within a state or sectoral plan can be imported at a tariff rate that is significantly lower than the listed level. These variations and the resulting unpredictability of tariff rates make it difficult for companies to export to the Chinese market (U.S. Trade Representative, 1993).

Non-Tariff Barriers

Most Asia Pacific countries have various types of formal and informal non-tariff barriers (NTB) that impede or restrict foreign imports of specific products. Formal NTB range from outright import bans to quantitative restrictions imposed through quotas and import licensing systems. In addition, qualitative restrictions such as standards, testing, labelling, and certification procedures are quite stringent in some countries. Inadequate intellectual-property protection; the lack of transparency in the laws, regulations, and decrees that govern trade; discriminatory government procurement policies; and smuggling are additional barriers to exporting.

Reliable data on non-tariff barriers are not readily available for all countries, especially for the Asia Pacific countries. Nevertheless, using the tariff equivalents of NTB, it has been estimated that non-tariff protection in Japan is exceptionally high in sectors where Canada has a strong comparative advantage, such as agri-food, chemicals and related products, and petroleum products

TABLE 6

ESTIMATED TARIFF EQUIVALENTS OF NON-TARIFF BARRIERS IN CANADA AND JAPAN (PERCENT)

TRADED GOODS	CANADA (%)	JAPAN (%)
Agriculture, forestry and fish products	6.5	79.7
Food, beverage and tobacco	25.6	58.1
Textiles	39.0	9.7
Clothing	39.0	4.1
Leather products	0.0	0.0
Footwear	12.8	17.9
Wood products	0.0	0.0
Furniture and fixtures	0.0	0.0
Paper and paper products	0.0	0.0
Printing and publishing	0.0	0.0
Chemicals	0.0	17.9
Petroleum and related products	0.0	17.0
Rubber products	0.0	0.0
Non-metallic mineral products	0.0	7.5
Glass and glass products	0.0	0.0
Iron and steel	0.0	0.0
Non-ferrous metals	0.0	0.0
Metal products	0.0	0.0
Non-electrical machinery	0.0	0.0
Electrical machinery	14.4	0.0
Transportation equipment	11.4*	0.0
Miscellaneous manufacturing	0.0	10.6
Total traded goods	**7.1**	**21.8**

* Includes the effect of the Canada-U.S. trade under the Auto Pact.
Source: Rao (1992).

(Table 6). Similar patterns of non-tariff protection are prevalent in other Asian countries (Rao, 1992).

As in the case of tariffs, Canadian non-tariff protection is substantially higher in industries where the Asia Pacific countries have a strong comparative advantage (labour-intensive manufactured exports, electrical machinery, and transportation equipment). Quantitative restrictions under the Multi-Fibre Agreement account for the very high non-tariff protection (39 percent) in the textiles and clothing industries, for example. Voluntary export restraints on autos and antidumping duties on electrical products, particularly against the Asia Pacific countries, are the main components of Canada's high non-tariff protection in the transportation-equipment and electrical-machinery industries.

Technical and Other Barriers to Trade

Rigid technical barriers to trade regarding standards, testing, labelling, and certification procedures in many Asia Pacific countries often act as severe obstacles for Canadian exports to the region (Conference Board, 1994). Canadian exporters who are forced to introduce product formulation changes to conform to national standards may not be able to produce in sufficient volume and may thus face a competitive disadvantage.

In addition, many current regulations still have no scientific basis and deviate substantially from international practice in both substance and implementation. These regulations serve mainly to keep imported products out of the domestic market. Moreover, the efforts of companies to export to countries such as China and South Korea are often hampered by requirements for higher quality standards than those applied to domestic products (U.S. Trade Representative, 1993).

The lack of protection for intellectual property (e.g., inadequate patent, copyright, and trademark regimes) is pervasive across the Asia Pacific region. For Canadian firms, this issue poses a serious obstacle to exports and FDI, particularly for goods and investments that are knowledge- and technology-intensive. According to the respondents to the Conference Board's 1994 survey, the problem appears to be a combination of the absence of rules and the lack of an effective enforcement mechanism and credible jurisprudence. Most violations are now being dealt with through business deals rather than trade rules. Although governments in the region have taken some steps in recent years to reduce violations of intellectual-property protection, more concrete measures need to be implemented in order to achieve real and lasting progress in that area.

Government procurement policies in a number of Asia Pacific countries often discriminate against foreign imports. For example, Japan has yet to develop a fully transparent procurement regime that recognizes the growing importance of the provision of services in government procurement. At sub-central government levels, non-transparent procurement practices – including a lack of uniform procedures, limited access to information on upcoming projects, and the inability of foreign firms to participate in the early development of systems – act as major obstacles to foreign access. In Taiwan, for example, all public enterprises and administrative agencies must procure locally if the goods and services can be manufactured locally, if acceptable substitutes are available locally, or if the price of local products is not more than 5 percent higher than c.i.f. import prices plus tariffs and harbour prices.

The freer flow of goods and services between Canada and the Asia Pacific countries would strengthen their commercial linkages and improve their respective economic performances. A gradual removal of Canadian and Asian tariff and non-tariff protection, either under a multilateral WTO framework or through bilateral agreements or regional arrangements, would therefore result in a considerable increase in two-way trade flows because of the greater opportunities for consolidating and broadening their comparative-advantage positions. Such trade deals would create durable export opportunities for Canadian firms, especially SME, by securing and expanding access to the Asia Pacific markets.

Investment Barriers

IN THE ASIA PACIFIC COUNTRIES, various forms of FDI barriers exist that either formally or informally impede the entry of foreign firms as well as restrict the operations of firms already in the region.

Formal FDI barriers are controls on foreign investment activity that are introduced explicitly through legislation. These rules tend to impede market access to foreign firms by denying them the right of establishment and national treatment, either generally or in particular sectors (Canada, Industry Canada, 1994a).

Informal investment barriers restrict FDI activity in the form of restrictions on mergers, acquisitions, strategic alliances, joint ventures, and "greenfield" investments. These restrictions arise primarily because of differences in market structure and corporate governance practices, unpublished government or corporate policies, non-transparent administrative procedures, and the informal actions of government and private-sector agents in host countries.

Among Asian countries, FDI barriers are particularly significant in Japan, where certain features of the business and regulatory environment act as a severe restraint on foreign investment activity. FDI flows to Japan remain extremely limited, both relative to the size of the Japanese economy and in comparison with Japanese investment abroad. Despite some significant liberalization of the regulatory climate for foreign investment review, investment remains restricted in a number of sectors (Canada, Industry Canada, 1994a).

Acquisitions, a common form of market access for foreign firms in other industrialized countries, are extremely difficult in Japan. Ties between government and industry, the reluctance of Japanese firms to break long-term personal and supplier relationships, cross-shareholding among allied companies, the low percentage of publicly traded common stock relative to total capital in many companies, and the unwillingness of the *keiretsu* (industrial grouping) to allow one of its members to come under foreign control – all tend to either rule out or considerably raise the cost of acquisitions. There are some indications of improvement in this area, as evidenced by increased merger and acquisition activity by foreign firms in Japan, albeit from an extremely low base (U.S. Trade Representative, 1993).

Exclusionary business practices and Japan's multilayered and, in many areas, antiquated distribution system remain impediments to foreign firms because of the extremely high costs they occasion. In some cases, it is virtually impossible for foreign firms to establish their own distribution channels and extremely difficult to use existing channels.

Although formal FDI barriers in many other Asia Pacific countries have been progressively liberalized and existing laws have been made more transparent, in a number of them (China, Indonesia, South Korea, Malaysia, and the Philippines) foreign equity ownership is often broadly restricted, or restricted in sensitive sectors, especially state-owned enterprises. For example, foreign firms are subject to mandatory local-content and trade-balance requirements in China. Effective access to markets continues to be subject to legal and regulatory restrictions, and to administrative "guidance" and practices in South Korea. On the other hand, Hong Kong, Singapore, and Taiwan have relatively few investment barriers.

Exchange-rate restrictions (especially in China) and the lack of infrastructure are some of the other barriers to undertaking FDI in the Asia Pacific countries. Despite recent improvements in foreign-exchange controls (e.g., the scrapping of the dual exchange-rate system in January 1994), access to hard currency remains a problem for many foreign investors in China. Lack of hard currency adversely affects the operations of foreign firms because of the unfavourable impact on imports of intermediary inputs, the remuneration of non-Chinese employees, and the repatriation of profits (Conference Board of Canada, 1994).

The three pillars of international commerce – foreign direct investment, trade, and technology – are closely interrelated via the growing activities of transnational firms. Indeed, they complement each other. Therefore, a gradual removal of various types of barriers to trade and investment, discussed above, either bilaterally or multilaterally, would deepen the economic linkages between Canada and the Asian countries. Such an outcome would stimulate trade, investment, and the economic performance of Canadian firms, especially SME. In this context, a more concerted effort for the gradual removal of the informal barriers to FDI in the region would further economic integration between Canada and the Asia Pacific region.

Micro-Level Problems

The general impediments to doing business in the Asia Pacific region affect SME more or less the same way as large and very large firms. In addition to these broad obstacles, however, all SME, especially transnational firms, face many other specific constraints in doing business in foreign markets. These include firm-level factors such as intra-firm disagreements regarding risks, average rates of return, and the amount of time and resources required to be successful in foreign markets – a category that encompasses the lack of information, inadequate financing, limited managerial resources, and limited foreign experience (UNCTAD, 1993). SME are likely to be more sensitive than large firms to undertaking risks associated with international business because of their very limited financial resources and the lack of adequate experience in doing business in foreign markets.

Availability of Capital

The lack of financial resources is widely considered to be a major obstacle for SME undertaking FDI (Figures 10 and 11). These firms often lack the leverage necessary to obtain access to credit for undertaking acquisitions, business expansions, or new investments in overseas markets. In addition, SME have a disadvantage vis-à-vis large MNE in acquiring information about the possible sources of finance in host countries or how best to approach international credit institutions. Financial constraints often force the abandonment of projects that might otherwise be commercially sound. Limited financial resources will often force SME to look for less attractive modes of market entry such as international joint ventures and other forms of non-equity participation.

Lack of Market Intelligence

In the Conference Board's 1994 survey, Canadian companies singled out the lack of local market information as the most important impediment to doing business in Asia (Figure 10) – a more significant problem for SME than for larger firms. The information-gathering capabilities of SME are often constrained by limited financial resources. And conversely, the high marginal cost of obtaining information can potentially deter Canadian SME from exploring trade and investment opportunities in the region.

The survey indicates that the lack of local market information was viewed as a more serious problem by companies that had been in the region less than 20 years than by those that had been there longer. Information barriers can be alleviated to a certain extent through government assistance in the form of trade missions, export financing, trade shows, fairs, etc.

Limited Managerial Resources and International Expertise

Management makes all the important decisions concerning the firm's domestic and foreign strategies and activities, and is often involved in its day-to-day operations. Unlike large and very large firms, SME are typically managed by a single individual or a small group of individuals. The limited management capabilities of SME tend to severely constrain long-term strategic decisions

Figure 10

Major Impediments to Doing Business in the Asia Pacific Region[1]

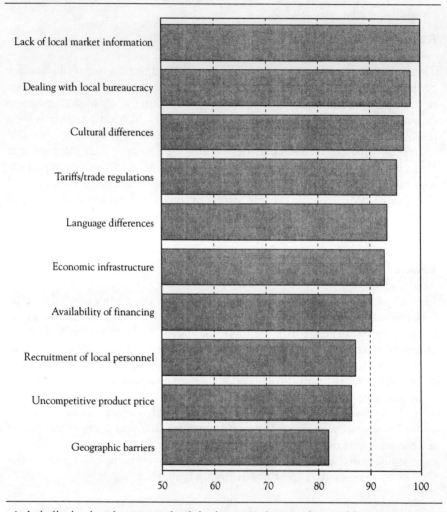

1 Lack of local market information was identified as the most significant impediment and then normalized to 100 for ranking purposes.
Source: Conference Board of Canada (1994).

regarding investment and outward orientation. The UNCTAD surveys indicate that insufficient management capacity is a very strong obstacle for SME, in particular, the smallest or micro-firms, in undertaking FDI (Figure 11).

SME with no previous experience in international business may lack the essential expertise and knowledge to succeed in new markets. For example,

FIGURE 11

IMPORTANT BARRIERS TO FDI BY SME IN DEVELOPING COUNTRIES[1]

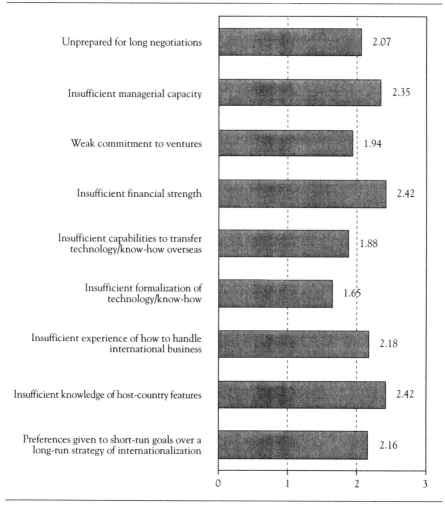

[1] Based on responses to 15-17 public institutes that provide assistance to SME in their FDI. The responses are quantified according to 3 for "frequent," 2 for "sometimes," and 1 for "rare/none."
Source: UNCTAD (1993).

insufficient knowledge of host-country markets can adversely affect the investment decisions of SME with respect to producing locally or importing inputs, selecting a host-country firm as partner, the type and variety of products to manufacture in the host country, etc. In fact, insufficient knowledge of host-country markets is cited as a problem as frequently as the lack of financial resources.

Limited international expertise could also impact on the profitability of SME that are not familiar with the procedures for minimizing tax burdens when operating in more than one country or for moving assets across international boundaries, or that cannot capitalize on their stock of knowledge in negotiating international deals. Similarly, limited exposure to international markets also places SME at a disadvantage in complex negotiations with host-country public officials.

Conclusions

IN ANALYZING FUTURE OPPORTUNITIES AND CHALLENGES of the Asia Pacific region for Canadian business, especially SME, we have attempted to answer a number of important research and policy questions, using three types of information – macro-level data on trade and investment, and on barriers to commerce; company data on financial variables and market orientation; and company surveys. In this section we will summarize the main findings of the study and examine their broad implications for public policy.

Findings

Canada's Commercial Linkages with the Asia Pacific

The Asia Pacific region offers tremendous export and investment opportunities for Canadian business. The Asian countries are expected to continue to post strong economic growth over the next 25 years, thus further increasing the region's importance in the world economy. In addition to the vast export and investment opportunities offered by these developments, Canada will continue to benefit from the high-quality and competitively priced goods sold by the dynamic Asia Pacific countries.

Canada's trade and investment linkages with the Asia Pacific countries have strengthened considerably over the last 25 years, and the Asia Pacific region surpassed the European Union as our second largest trading partner. However, Canada's commercial relations with the region continue to be relatively weak. The Asia Pacific region still accounts for only about 10 percent of Canada's total trade flows (exports plus imports) and FDI stock (inward plus outward). More importantly, Canada's market share in these dynamic markets has declined considerably.

To take full advantage of the vast trade and investment opportunities in the Asia Pacific region, Canada needs to overcome four important challenges: to improve the overall competitiveness (costs, variety, and quality) of its products; to broaden its export base by shifting towards knowledge- and technology-intensive products; to increase the participation of SME in the Asia Pacific region; and to strengthen its investment linkages with the region.

SME and Outward Orientation

The vast majority of business enterprises in Canada are SME. They make a significant contribution to output and employment, and they are the dominant source of job creation. In addition, they are an important source of dynamism and innovation.

SME make a small positive contribution to Canada's exports and direct investment abroad. But only a very small proportion of SME (less than 10 percent) are involved in exporting. Similarly, only a tiny fraction (0.2 percent) of Canadian business enterprises invest abroad. However, Canadian SME are increasingly becoming outward-oriented. Their future commercial activities in the Asia Pacific region could therefore play an important role in determining Canada's success in penetrating the dynamic Asian markets.

The motivations for outward orientation in general are similar for SME and for large firms. Increased growth of sales and assets; improved profitability; access to natural resources, technology, and skilled labour; government incentives; broadening of the comparative advantage position; lower labour and production costs; and formal and informal barriers to exports and investment are some of the important reasons behind the growing globalization of business in Canada and elsewhere.

In our Canadian sample, 56 percent of firms doing business in the Asia Pacific region and elsewhere are SME, compared to just 12 percent in the U.S. sample. However, the sales of Canadian SME represent less than 3 percent of the total sales of all firms doing business in the Asia Pacific and elsewhere.

More importantly, the outward orientation of SME is very low, compared with the large and very large firms in the two samples. Only 21 percent of Canadian SME participate in the Asia Pacific region, compared with 25 percent in the U.S. sample. Both the simple tabulations and the regression results show that the probability of SME participation in the Asia Pacific markets is considerably lower than that of large and very large firms in the two countries.

Notwithstanding these findings, the significant positive relationship between size and outward orientation does not automatically imply that SME are inherently disadvantaged in going global. Firm size may be simply capturing the positive influence of other omitted variables on internationalization. For instance, recent research strongly suggests that firm size is positively correlated with firm age, past experience in international markets, and managerial ability and vision.

The industrial structure of firms (in terms of the number of firms and their respective sales) doing business in the Asia Pacific region and elsewhere is similar to the overall industrial structure of the sample firms in the two countries. In the Canadian sample, technology- and resource-intensive manufacturing firms, along with the finance, insurance, and real estate industry, account for almost 80 percent of the total sales of all firms doing business in the Asia Pacific and elsewhere, compared with about 70 percent in the U.S. sample.

Outward Orientation and Economic Performance

Canadian SME outperformed their large and very large Canadian counterparts with respect to four of the five performance indicators considered in our paper: sales growth, asset growth, capital productivity (sales/assets), and the rate of return on assets. On the other hand, the labour productivity (sales/employment) of Canadian SME is significantly lower than that of large and very large firms. The regression results, however, suggest that part of the productivity gap could be attributed to the lower capital/labour ratio. Nevertheless, the results strongly indicate that Canadian SME do not benefit from scale and scope economies to the same extent as large and very large Canadian firms. More importantly, Canadian SME outperformed their American counterparts by a substantial margin with respect to all five performance indicators.

In general, economic performance and outward orientation are positively correlated in the Canadian sample. In contrast, there appears to be no systematic positive relationship between the two variables in the U.S. sample. These divergent results seem to suggest that outward orientation matters a great deal for the performance of firms from a small and slowly growing domestic market.

Impediments to Participation in Asia Pacific Markets

Both tariff and non-tariff protection in the Asian countries are relatively high in industries in which Canada has a strong comparative advantage – namely, resource and resource-based manufactured products. By the same token, trade protection in Canada is substantially higher in industries in which the Asian countries have a strong comparative advantage – labour-intensive and light manufactured products, electrical machinery, and transportation equipment.

In addition, there are a large number of other formal and informal barriers to trade in the Asia Pacific countries. These include technical barriers to trade (standards, testing, labelling, and certification procedures); lack of adequate intellectual-property protection; discriminatory government procurement policies against foreign firms; and foreign exchange restrictions.

Furthermore, it is extremely difficult to undertake FDI in most of the Asia Pacific countries because of the presence of various formal and informal investment barriers in the region. However, there are relatively few formal and informal restrictions on FDI in Hong Kong, Singapore and Taiwan.

These trade and investment barriers significantly constrain the activities of Canadian SME as well as other Canadian firms in the Asia Pacific region. In addition to these general obstacles to doing business in Asia, SME encounter a significant disadvantage in going international vis-à-vis large and very large firms in four important areas – availability of capital at a reasonable cost; market intelligence; international experience; and managerial depth and dynamism.

Policy Implications

OUR RESEARCH FINDINGS INDICATE THAT Canadian SME in general would be able to successfully manage the challenges of growing competition from Asian firms at home and abroad, and increase their participation in the dynamic Asian markets. Increased outward orientation of Canadian firms, especially SME, would significantly improve the trade, investment, and growth performance of Canadian firms as well as the Canadian economy. An increased participation of Canadian firms in the Asia Pacific could, in turn, help to alleviate the chronic problems of high unemployment, stagnant real incomes, a narrow export base, and the very high dependency on the U.S. market for our exports.

The participation of Canadian SME in the region could be facilitated and considerably strengthened by effective government policies in three key areas – liberalization of trade and investment flows; market intelligence and market awareness; and incentives for exporting and investing.

Our findings suggest that a gradual removal of the various types of formal and informal barriers to trade and investment flows between Canada and the Asia Pacific countries, either bilaterally and/or multilaterally (via WTO or APEC), would considerably strengthen Canada's commercial linkages with the Asian economies.

As noted above, the participation of SME in international markets is severely constrained by lack of adequate market intelligence on export and investment opportunities, competitors, market risk, and government incentives and regulations. The marginal cost of obtaining market intelligence is often very high for SME – a factor that can deter them from fully exploiting trade and investment opportunities in the region. In this context, governments could play a very useful and proactive role by providing timely and up-to-date comprehensive information to SME on the dynamic Asian markets and, more generally, by raising the awareness of SME about the considerable business opportunities in the region.

A better coordination of federal and provincial government trade initiatives would significantly improve their effectiveness in assisting Canadian firms with their internationalization efforts. The successful Team Canada visits to China and South Asia strongly support the notion of a coordinated approach to trade and investment initiatives in the region by federal and provincial governments and business. These initiatives could result in a virtuous cycle of increased outward orientation and improved economic performance of SME.

In addition, other types of government assistance to SME in going global could also be very helpful in increasing their participation in the Asia Pacific markets. These actions could include: increased access to financial resources; a reduction of the risk premium (partial refund of the insurance cost); export financing; greater assistance in matching Asian firms with Canadian firms for joint ventures, strategic alliances, and other equity as well as non-equity ventures in the region; assistance with the adoption and diffusion of new products and process technologies, and so on.

APPENDIX 1

MAIN CHARACTERISTICS OF THE CANADIAN AND AMERICAN SAMPLE FIRMS

TO ANALYZE THE CHARACTERISTICS, outward orientation, strategies, and performance of Canadian SME, and compare these results with those of large firms in Canada and their SME counterparts in the United States, we developed a micro-level database comprising 1,297 Canadian and 2,526 American firms.

DATA SOURCES

THE DATA WERE COMPILED FROM the following sources: the Business Opportunities Sourcing System (BOSS) data bank; the *Canadian Trade Index 1994*; the *American Export Register 1992*; the *Directory of Canadian Business in Japan, 1994*; the *Directory of Canadian Business in Hong Kong 1990-91*; *Who Owns Whom* (Dunn & Bradstreet); Compact Disclosure Canada; the U.S. Securities and Exchange Commission (SEC); and the Worldscope.

BOSS, the *American Export Register*, and *Who Owns Whom* are the primary sources for identifying the outward orientation (exports and/or FDI) and the geographic destination of exports and FDI of Canadian and American firms. The *Canadian Trade Index 1994*, the *Directory of Canadian Business in Japan 1994*, and the *Directory of Canadian Business in Hong Kong 1990-91* were used to identify additional Canadian exporters.

Note, however, that we could not obtain from these sources any data on the exports and FDI of outwardly oriented firms in the two countries. This is a major weakness, because we cannot say anything about the export and FDI intensities of these firms.

Compact Disclosure Canada, SEC, and Worldscope were used to obtain company financial information on outwardly and domestically oriented firms in the two countries. The data on sales, assets, employment, and labour and capital productivity refer to 1994 for most of the enterprises in the two samples. On the other hand, the data on growth of sales and assets, the rate of return on assets, and the debt-to-equity ratio (for American firms only) are annual averages for the past five years.

For analytical purposes, Canadian and American outwardly oriented firms are classified into two broad categories, based on the geographic destination of their exports and FDI: firms doing business in Asia and elsewhere; and firms not doing business in Asia. As discussed later on, a vast majority of outwardly oriented firms in the two countries do business in Asia and elsewhere – i.e., in other international markets as well as in the domestic market. Therefore, these firms could be viewed as "global" firms.

In addition, all the firms in the two countries are grouped into three size classes, based on the dollar value of their total sales: firms with sales between US$1 million and less than US$100 million are classified as SME;[12] firms with sales between US$100 million and less than US$500 million are categorized as large firms; and those with sales in excess of US$500 million are considered as very large.

The sample firms are grouped into the following 12 major industry groups, based on their primary business activity: agriculture; construction; finance, insurance, and real estate; labour-intensive manufacturing; resource-intensive manufacturing; technology-intensive manufacturing; mining; retail trade; wholesale trade; transportation and public utilities; public administration; and other services.

In short, the sample firms in our database are disaggregated by outward orientation, by size, and by industry.

Importance of Sample Firms

THE SAMPLE FIRMS REPRESENT a significant share of the two economies in terms of sales, assets and employment. In 1994, the combined sales and assets of the Canadian firms were US$318.3 and US$1,022.1 billion, respectively. They employed 1.72 million people – about 13 percent of total employment in Canada (Figure A1-1).

Similarly, the American firms command a large amount of resources of the U.S. economy. Their total sales and assets were US$4,627.6 and US$9,945.3 billion, respectively. Their combined employment of 26.1 million represented over 20 percent of U.S. civilian employment in 1994.

Figure A1-1

Main Characteristics of Canadian and American Sample Firms

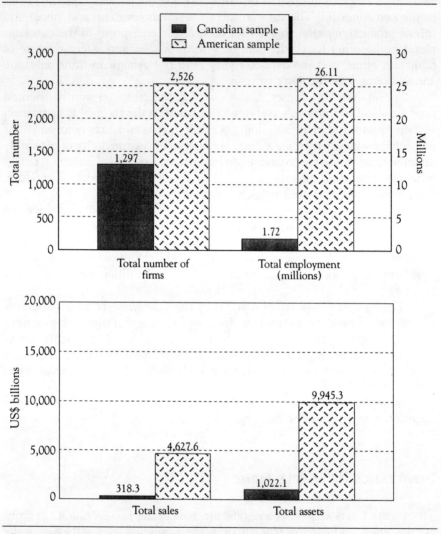

Source: Based on data compiled by the authors from various sources.

Size Distribution

A VERY LARGE PROPORTION (almost three-quarters) of the Canadian sample are SME – firms with sales of less than US$100 million. Firms with sales between US$100 and US$500 million (large firms) represent about 16 percent of the sample. Very large firms (firms with sales over US$500 million) account for the rest of the sample (around 10 percent).

Although SME account for almost 75 percent of the firms in the Canadian sample, they represent less than 10 percent of total sales and assets and 11 percent of total employment of the sample firms (Table A1-1).

In sharp contrast, SME represent only 19 percent of all firms in the American sample, while large (37 percent) and very large (44 percent) firms account for over 80 percent of the sample. Similarly, SME contribute less than 1 percent of total sales, assets, and employment of the American firms.

Industry Distribution

Manufacturing firms represent about 36 percent of all firms in the Canadian sample. Within manufacturing, technology-intensive (19 percent) and resource-intensive (12 percent) firms account for the bulk of manufacturing firms. Finance, insurance, and real estate (17 percent) and wholesale and retail trade (22 percent) make up another 40 percent of the Canadian firms. The remaining firms are mostly in mining, transportation and public utilities, other services, and construction. The industrial distribution of total sales of the Canadian firms is more or less similar to that of the number of firms (Table A1-2).

As in the Canadian sample, firms in technology- and resource-intensive manufacturing, and in the finance, insurance, and real estate group represent over 45 percent of the American firms and about 60 percent of total U.S. sales. But technology-intensive manufacturing firms play a more important role in the U.S. sample than in the Canadian sample in terms of both the number of firms and sales. On the other hand, finance, insurance, and real estate and resource-intensive manufacturing firms contribute significantly more in the Canadian sample than in the American sample. In short, the industrial distribution of firms and sales of the two samples is generally consistent with the comparative-advantage position of the two countries (Canada, Industry Canada, 1994*b*; Eden, 1991).

Outward Orientation

The outward orientation pattern of the Canadian and American firms is similar. Nearly 30 percent of all the Canadian firms do business in Asia and other international markets. More importantly, these firms account for about two thirds of the sales of all Canadian firms, implying that a large proportion of these firms are either large or very large, according to the definitions given above. The proportion of global firms (firms doing business in Asia and elsewhere) in the U.S. sample is significantly higher than in the Canadian sample – 39 percent compared with 28 percent. In the two samples, fewer than 20 percent of all firms have no business in Asia but either export to or/and invest in

Table A1-1
Size Distribution of Canadian and U.S. Sample Firms, by Number of Firms, Sales, Assets, and Number of Employees (Percent)

	NUMBER OF FIRMS			SALES			ASSETS			NUMBER OF EMPLOYEES		
	SME	LARGE	VERY LARGE	SME	LARGE	VERY LARGE	SME	LARGE	VERY LARGE	SME	LARGE	VERY LARGE
Canada	74.0	16.3	9.7	7.0	14.8	78.2	3.7	8.2	88.1	11.0	18.8	70.2
United States	19.1	36.9	44.0	0.6	5.0	94.4	0.8	5.4	93.8	0.9	7.5	91.6

Source: Compiled by the authors using data from various sources.

TABLE A1-2
INDUSTRIAL DISTRIBUTION OF CANADIAN AND U.S. SAMPLES, BY NUMBER OF FIRMS AND SALES (PERCENT)

	AGRICULTURE, FORESTRY, AND FISHING	CONSTRUCTION	FINANCE, INSURANCE, AND REAL ESTATE	LABOUR-INTENSIVE MANUFACTURING	MINING	PUBLIC ADMINISTRATION	RESOURCE-INTENSIVE MANUFACTURING	RETAIL TRADE	SERVICES	TECHNOLOGY-INTENSIVE MANUFACTURING	TRANSPORTATION AND PUBLIC UTILITIES	WHOLESALE TRADE	TOTAL
Canada													
Number of firms	0.3	2.9	16.9	4.9	5.2	0.6	12.0	7.3	8.4	18.9	7.9	14.7	100.0
Sales	0.2	0.4	29.0	2.2	7.6	0.4	15.6	4.7	1.9	21.2	9.6	7.4	100.0
United States													
Number of firms	0.4	1.5	16.9	6.2	3.7	0.0	9.5	8.4	18.4	28.0	11.4	3.6	100.0
Sales	0.2	1.0	13.6	2.6	6.7	0.0	12.9	13.0	4.1	29.5	13.3	3.1	100.0

Source: Compiled by the authors using data from various sources.

Table A1-3

Distribution of Canadian and U.S. Sample Firms, by Market Orientation

Market Orientation	CANADA		UNITED STATES	
	Number of Firms	Total Sales	Number of Firms	Total Sales
Firms doing business in the Asia Pacific region	28.1	65.1	39.4	68.5
Firms doing business in international markets but not the Asia Pacific region	15.1	14.6	18.4	12.6
Firms doing business only in domestic markets	56.8	20.3	42.2	18.9
Total	100.0	100.0	100.0	100.0

Source: Compiled by the authors using data from various sources.

other international markets. These firms presumably do business mostly in North America (Table A1-3).

At the other end of the spectrum, almost 57 percent of the Canadian firms do business exclusively in the domestic market, compared with only 42 percent in the United States. However, the domestically oriented firms account for only about 20 percent of total sales in each sample.

Average Firm Size

Table A1-4 summarizes the average size of Canadian and American sample firms in terms of sales and employment, for all firms as well as for those doing business in Asia and elsewhere. The American firms, in general, are considerably larger (between 8 and 10 times) than their Canadian counterparts. The average sales and employment of the Canadian firms are US$250 million and 1,324, respectively, compared with US$1,830 million and 10,336 for the American sample.

Similarly, the average size of SME in Canada is less than half that of American SME in terms of both sales and employment. For example, the average employment in Canadian SME is 197, compared with 491 in the American sample. At the same time, very large American firms, on average, are more than twice the size of very large firms in Canada – 21,508 employees per firm in the United States versus only 9,565 in Canada. On the other hand, the average size of large Canadian firms (with sales between US$100 and US$500 million) is only slightly smaller than that of their American counterparts

TABLE A1.4
AVERAGE SIZE OF CANADIAN AND U.S. SAMPLE FIRMS

	CANADA				UNITED STATES			
	ALL FIRMS		FIRMS DOING BUSINESS IN THE ASIA PACIFIC REGION		ALL FIRMS		FIRMS DOING BUSINESS IN THE ASIA PACIFIC REGION	
Firm size	Sales (US$ millions)	Employment	Sales (US$ millions)	Employment	Sales (US$ millions)	Employment	Sales (US$ millions)	Employment
Very large	1,970	197	2,270	11,176	3,900	21,508	5,600	30,061
Large	220	1,532	220	1,590	250	2,096	250	2,244
SME	23	197	30	217	51	491	52	542
All firms	250	1,324	570	2,950	1,830	10,336	3,182	17,368

Source: Compiled by the authors using data from various sources.

(Table A1-4). In the two country samples, a similar structure is observed with respect to the average size of the firms doing business in Asia and elsewhere. However, the average sales and employment of these firms are more than double the average for all firms in each country. This result arises mainly because of a significantly higher proportion of very large firms doing business in the Asia Pacific region than in the total sample. In addition, in both Canada and the United States the average size of very large firms is significantly bigger for the sample firms doing business in the Asia Pacific than for those firms which are absent from the region.

APPENDIX 2
SME Performance Indicators

Table A2-1

Share of Employment Accounted for by SME in Canada, by Industry, 1992

	Industry Employment (Thousands)	Fewer than 50 ALUs[1] (%)	Fewer than 100 ALUs[1] (%)	Fewer than 500 ALUs[1] (%)
Agriculture and related services	108.9	90.4	95.9	100.0
Fishing and trapping	9.1	93.4	96.7	100.0
Logging and forestry	41.2	73.8	73.8	81.6
Mining, quarrying and oil wells	144.1	16.7	21.6	37.7
Manufacturing industries	1,830.5	21.6	31.7	53.7
Construction industries	503.1	74.4	82.7	95.0
Transportation and storage	396.0	31.9	39.9	56.3
Communication and other utilities	365.3	6.8	8.7	14.8
Wholesale trade	622.7	51.6	64.2	83.4
Retail trade	1,553.3	46.0	53.7	63.7
Finance and insurance	687.2	14.1	17.5	24.9
Real estate and insurance	231.1	55.5	66.6	83.2
Business services	660.8	45.1	53.5	72.5
Educational services	888.5	4.7	7.4	21.0
Health and social services	1,106.1	24.9	32.1	53.4
Accommodation, food and beverage services	806.1	54.7	66.1	80.8
Other service industries	721.9	57.2	65.3	80.0
To be classified[2]	60.9	77.8	77.8	85.2
Total	**10,727.7**	**36.0**	**43.8**	**59.2**

1 ALU = average labour units employed by a firm. Generally speaking, the ALU is equivalent to the number of employees.
2 Includes firms whose classification has not yet been determined.
Source: Estimates prepared by Entrepreneurship and Small Business Office, Industry Canada, based on data supplied by Statistics Canada.

Figure A2-1

Economic Performance of Canadian SME Doing Business in International Markets but not the Asia Pacific Region, by Firm Size

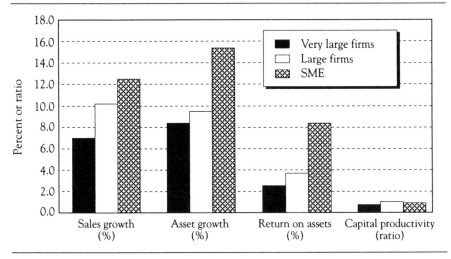

Source: Based on data compiled by the authors from various sources.

Figure A2-2

Labour Productivity of Canadian Firms, by Size

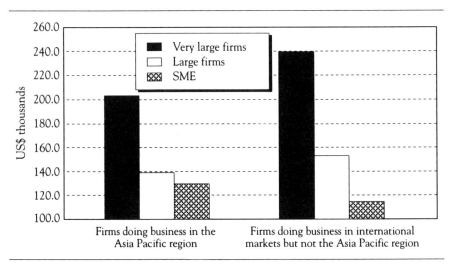

Source: Based on data compiled by the authors from various sources.

FIGURE A2-3

ECONOMIC PERFORMANCE OF CANADIAN AND AMERICAN SME DOING BUSINESS IN INTERNATIONAL MARKETS BUT NOT THE ASIA PACIFIC REGION

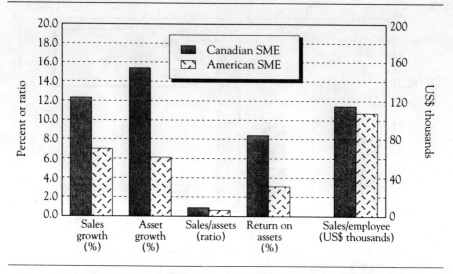

Source: Based on data compiled by the authors from various sources.

APPENDIX 3

REGRESSION RESULTS

IN ORDER TO DETERMINE EMPIRICALLY THE RELATIONSHIP between outward orientation (the geographic destination of exports and/or FDI) and firm size and industry (and other explanatory variables), we estimated two linear probability models (LPM) for the Canadian and American sample firms.

$$Y1_{out} = f(DSIZE, DMI, DEQ) \quad (1)$$
$$Y2_{out} = f(DSIZE, DMI, DEQ) \quad (2)$$

where,

$Y1_{out}$ and $Y2_{out}$ are dichotomous independent variables capturing the outward orientation and geographic thrust of the firms;
DSIZE = size class dummy variable;

DMI = major industry group dummy variable;
DEQ = five-year debt-equity ratio (for regression of U.S. sample only).

Thirteen dummy variables are included in the regression equations to control for the effects of firm size (sales) and type of industry on the outward orientation of the firms. The 12 major industry groups are represented by 11 dummy variables. The three size classes (SME, large, and very large) are represented by two dummy variables. As noted, DEQ is used as an additional explanatory variable in the U.S. regression equation.

The structure of the two dichotomous response (dependent) variables in the equation is as follows:

$Y1_{out}$ = 1 if firms are doing business in the Asia Pacific region and elsewhere or if they are doing business in other international markets excluding the Asia Pacific region;
= 0 if firms are doing business in the domestic market only;

$Y2_{out}$ = 1 if firms are doing business in the Asia Pacific region and elsewhere;
= 0 if firms are doing business in other international markets excluding the Asia Pacific region.

The dummy variables in the two regression equations were assigned the following values:

DSIZE = 1 if firms are SME (DSME) or large firms (DLARGE);
= 0 if firms are very large firms (control variable);

DMI = 1 if firms are in either one of the following industries:
Wholesale trade (DWHTR);
Agriculture, forestry and fishing (DAGR);
Public administration (DPADM);
Construction (DCONST);
Labour-intensive manufacturing (DLIM);
Retail trade (DRET);
Transportation and public utilities (DTPUB);
Other services (DSERV);
Resource-intensive manufacturing (DRESIM);
Finance, insurance and real estate (DFIRE);
Technology-intensive manufacturing (DTECH);

DMI = 0 if firms belong in mining (the control or base category).

The regression results of equations (1) and (2) are shown in Tables A3-1 and A3-2, respectively.

Table A3-1

Regression Results for Canada and the United States: Outward Orientation

Explanatory Variables	CANADA		UNITED STATES	
	Coefficient	t-statistics	Coefficient	t-statistics
Constant	0.798248	14.544*	0.803434	18.133*
DSME	−0.447665	−12.232*	−0.309414	−13.406*
DLARGE	−0.244384	−5.720*	−0.157952	−8.543*
DWHTR	−0.219725	−4.107*	−0.100546	−1.648***
DAGR	0.236681	1.217	−0.202413	−1.410
DPADM	−0.181950	−1.288	—	—
DCONST	−0.199408	−2.573*	−0.298489	−3.762*
DLIM	0.300047	4.532*	0.027684	0.512
DRET	−0.330244	−5.486*	−0.443265	−8.588*
DTPUB	−0.238027	−4.027*	−0.400752	−8.103*
DSERV	−0.056345	−0.960	−0.026248	−0.529
DRESIM	0.337392	6.134*	0.033875	0.672
DFIRE	−0.248948	−4.747*	−0.431563	−9.174*
DTECH	0.413076	7.960*	0.201487	4.443*
DEQ (U.S. only)			−0.00001387	−1.507
R^2	0.42461		0.30966	
F-statistic	72.83115		86.67526	
Total number of cases	1,297		2,526	

— Not enough observations.
* Significant at the 1 percent level.
** Significant at the 5 percent level.
*** Significant at the 10 per cent level.

First, in both equations (1) and (2), the coefficient of the two size dummy variables, DSME and DLARGE, are negative and highly significant. Moreover, the coefficients of DSME in both equations and for both country samples have a larger negative value than those of large firms (DLARGE) in the two regression equations. These results strongly suggest that firm size is a very strong determinant of outward orientation. For instance, the slope value of −0.447 for the Canadian SME coefficient in equation (1) implies that, holding all other factors constant, the probability of Canadian SME participating in the Asia Pacific region, on average, is lower by about 45 percent compared with the very large Canadian firms (the base category). Similarly, the probability of large Canadian firms doing business in the Asia Pacific region is lower, on average, by about 25 percent compared with that of the very large firms.

The coefficient of the dummy variable for technology-intensive manufacturing, DTECH, among other industry dummy variables, is positive and sta-

TABLE A3-2

REGRESSION RESULTS FOR CANADA AND THE UNITED STATES:
GEOGRAPHIC DESTINATION

	CANADA		UNITED STATES	
EXPLANATORY VARIABLES	COEFFICIENT	T-STATISTICS	COEFFICIENT	t-STATISTICS
Constant	0.6824	7.511*	0.6786	11.495*
DSME	−0.2213	−3.865*	−0.2207	−6.298*
DLARGE	−0.1912	−2.994*	−0.1563	−6.129*
DWHTR	0.1680	1.457	−0.1470	−1.792***
DAGR	0.4551	1.616	0.1490	0.664
DPADM	0.5389	1.580	—	—
DCONST	−0.1328	−0.635	−0.0940	−0.746
DLIM	0.0147	0.132	0.0976	1.385
DRET	−0.5449	−2.623*	−0.2623	−3.266*
DTPUB	0.0629	0.496	−0.2306	−3.170*
DSERV	0.1799	1.540	0.0468	0.708
DRESIM	0.1689	1.800***	0.1551	2.356**
DFIRE	0.1351	1.240	−0.1312	−1.829***
DTECH	0.1952	2.132**	0.2326	3.914*
DEQ (U.S. only)			−0.0000005	0.346
R^2	0.0700		0.14088	
F-statistic	3.1608		18.2395	
Total number of cases	560		1,460	

— Not enough observations.
* Significant at the 1 percent level.
** Significant at the 5 percent level.
*** Significant at the 10 per cent level.

tistically significant in both equations (1) and (2) of the Canadian and U.S. regression results. For example, from Table A3-1, the results show that, holding other things constant, the probability that Canadian and American firms in high-technology industries are doing business in Asia Pacific is higher by 40 and 20 percent, respectively, than that for their counterparts in the mining industry. Similarly, Canadian firms in resource-intensive manufacturing and in the finance, insurance, and real estate group have a higher probability (30 percent or more) of participating in the Asia Pacific region than do those in the mining industry.

To determine empirically the relationship between labour productivity (sales/employment) and firm size, industry, and other explanatory variables, we estimated the following regression equation for Canadian and American sample firms:

$$\text{(sales/employment)} = f\,(\text{DSIZE, DMI, DORIENTATION, K\_LRATIO, DEQ}) \quad (3)$$

where,
- (sales/employment) = a measure of labour productivity;
- DSIZE = size class dummy variable;
- DMI = major industry group dummy variable;
- DORIENTATION = outward orientation dummy variable;
- K_LRATIO = the capital-to-labour ratio (assets/employment) in each firm;
- DEQ = five-year debt-equity ratio (for U.S. sample only).

The two dummy variables for size (DSME, DLARGE) and the 11 dummy variables for industry (DWHTR, DAGR, DPADM, DCONST, DLIM, DRET, DTPUB, DSERV, DRESIM, DFIRE, DTECH) in the regression equation assume the same values as in the previous regression equation. As before, DEQ is used as an explanatory variable in the U.S. regression equation. The dummy variable for outward orientation, DORIENTATION, was assigned the following values:

DORIENTATION = 1 if the firms are doing business in the Asia Pacific region and elsewhere (DASIA) or if they are doing business in other international markets excluding the Asia Pacific region (DNOTASIA);
= 0 if firms are doing business in the domestic market only (control group).

The results of regression equation (3) are shown in Table A3-3.

The coefficient of the capital labour ratio (K_LRATIO) in the labour productivity equation, after controlling for size, industry, and outward-orientation variables, is positive and highly significant in both the Canadian and U.S. regression results. Given that SME are relatively more labour-intensive than other firms, the significance of the capital/labour ratio might explain part of the labour productivity gap (levels) between Canadian SME and large and very large firms. However, the coefficients of the two size dummy variables – DSME and DLARGE – are both negative and significant, suggesting that labour productivity is significantly positively correlated with the size variable in the Canadian sample. Thus a large part of the labour productivity gap can be attributed to the size disadvantage of the SME.

TABLE A3-3

REGRESSION RESULTS FOR CANADA AND THE UNITED STATES:
LABOUR PRODUCTIVITY (SALES/EMPLOYMENT)

EXPLANATORY VARIABLES	CANADA		UNITED STATES	
	COEFFICIENT	T-STATISTICS	COEFFICIENT	t-STATISTICS
Constant	1,201.5040	2.787*	327.2736	5.668*
DSME	-1,285.7116	-4.523*	-42.8627	-1.447
DLARGE	-763.5869	-2.410**	-22.0038	-0.950
DWHTR	621.6415	1.591	111.1715	1.487
DAGR	-145.8988	-0.103	-129.7256	-0.738
DPADM	10,220.0523	9.970*	—	—
DCONST	262.0703	0.465	27.7525	0.285
DLIM	198.3282	0.410	-176.8386	-2.671*
DRET	170.9771	0.387	-187.4723	-2.922*
DTPUB	83.0019	0.192	-68.6291	-1.118
DSERV	191.7042	0.450	-156.3434	-2.575*
DRESIM	124.8701	0.308	-105.8391	-1.711***
DFIRE	244.5157	0.637	-170.2159	-2.901*
DTECH	223.0307	0.577	-147.7340	-2.636*
DEQ (U.S. only)			-0.0053	-0.468
K_LRATIO	0.2230	402.149*	0.0856	160.571*
DASIA	-193.0182	-0.856	-18.1430	-0.652
DNOTASIA	-68.5096	-0.273	23.5655	0.801
R^2	0.9923		0.91337	
F-statistic	10,282.0193		1,653.3767	
Total number of cases	1,297		2,526	

— Not enough observations.
* Significant at the 1 percent level.
** Significant at the 5 percent level.
*** Significant at the 10 per cent level.

ENDNOTES

1 We are grateful to Denis Gauthier for supporting our project. Our thanks go to Richard Harris and James McRae for their many useful comments and suggestions on an early draft of the paper. We also wish to thank Francine L'Espérance for her helpful advice and Clifton Lee-Sing for his assitance throughout the project. We acknowledge the assistance of Rick Zimmerling in compiling data from the Business Opportunity Sourcing System database at Industry Canada.

2 The Asian NIE (also referred to as the "Four Tigers") include Hong Kong, Singapore, South Korea, and Taiwan. The members of ASEAN

are Brunei, Indonesia, Malaysia, Papua New Guinea, Philippines, Singapore, and Thailand.
3. We grouped the sample firms into the three size classes on the basis of the dollar value of sales rather than employment, for two main reasons: 1) the sales definition provides a more standard definition of an SME across industries and countries than does the employment definition; and 2) the quality of employment data is poor, compared with that of sales data. See Appendix 1 for a detailed description of the database.
4. The analysis in this section is based largely on APEC Economic Committee (1995) and Rao (1992).
5. APEC has 18 members. The Asian group is comprised of Brunei, China, Hong Kong, Indonesia, Japan, Malaysia, Papua New Guinea, the Philippines, Singapore, South Korea, Taiwan, and Thailand. The non-Asian members of APEC are Australia, Canada, Chile, Mexico, New Zealand, and the United States.
6. Asian APEC countries accounted for about 11 percent, and the EU for just over 8 percent, of total Canadian trade (exports plus imports) in 1992, compared with 8.4 and 10.8 percent, respectively, in 1980.
7. In 1992, the United States accounted for 63.7 percent of the inward FDI stock in Canada, down from almost 78 percent in 1980. Conversely, 57.5 percent of Canada's outward FDI stock in 1992 was located in the United States, compared with 62.2 percent in 1980.
8. SME in general enjoy greater administrative and management flexibility than large firms. In addition, the increase in competition brought about by the reduction in trade barriers and the emergence of additional countries producing manufactures for exports is cited as an important factor contributing to the resurgence of SME in the global economy. Shifts in the international trade environment, in particular the pressures to introduce new products and improve product quality in order to avoid head-on competition with more efficient producers in low-wage countries, have forced many firms to rationalize some or all of their operations. Finally, technological developments, such as the substitution of electronic processes, numerically controlled machine tools, and computer-aided design systems for mechanical processes, have reduced the importance of plant and product scale economies (Pratten, 1991, 115-23).
9. The figures on SME activity in Canada are largely drawn from Industry Canada (1994d).
10. According to a recent Atlantic Canada Opportunities Agency study (1992), small firms in Atlantic Canada – defined as those employing fewer than 100 people – created 90 percent of the jobs during the 1980-89 period. This compares with an 86.6 percent share for all of Canada. The smallest firms (employing fewer than five people) made a significantly greater contribution to new employment in that region than did firms of comparable size in the whole of Canada.

11 In a regional context, the Small and Medium Enterprises (SME) Action Program contributes to the formulation of the APEC Action Agenda by creating a program that pursues APEC objectives of relevance to SME. At the first APEC SME ministerial meeting, held in Osaka in October 1994, it was agreed that APEC's role in support of SME should focus first on addressing the areas where they face the greatest difficulties. The SME Action Program thus addresses the common difficulties faced by SME in all APEC economies in five priority areas – human resources development; information access; technology and technology sharing; financing; and market access.

12 The threshold of US$1 million in sales was chosen because the quality of the financial data for firms with less than US$1 million of sales was generally very poor. For purposes of analyzing possible differences in the characteristics, strategies and performance among SME of different size, a forthcoming Industry Canada working paper will subdivide SME firms into three broad size groups: firms with sales between US$1 and US$10 million; firms with sales between US$10 million and less than US$50 million; and firms with sales between US$50 and US$100 million.

BIBLIOGRAPHY

APEC Economic Committee. "Foreign Direct Investment and APEC Economic Integration." Singapore: APEC Secretariat, 1995.

Atlantic Canada Opportunities Agency. "The State of Small Business and Entrepreneurship in Atlantic Canada, 1992." New Brunswick, ACOA, 1992.

Baldwin, J.R. "Innovation: The Key to Success in Small Firms." Analytical Studies Branch, Research paper 76, Ottawa: Statistics Canada, 1995.

Baldwin, J.R. and J. Johnson. "Business Strategies in Innovative and Non-innovative Firms in Canada." Analytical Studies Branch, Research paper 73, Ottawa: Statistics Canada, 1995.

Bonaccorsi, Andrea. "On the Relationship Between Firm Size and Export Intensity." *Journal of International Business Studies*, 23, 4 (1992):605-35.

Calof, Jonathan L. "The Relationship Between Firm Size and Export Behaviour Revisited." *Journal of International Business Studies*, 25, 2 (1994):367-87.

Canada, Industry Canada. "Formal and Informal FDI Barriers in the G7 Countries: The Country Chapters." Occasional papers 1, vol. 1. Ottawa: Industry Canada, 1994a.

_____. "Economic Integration in North America: Trends in Foreign Direct Investment and the Top 1000 Firms." Working paper 1. Ottawa: Industry Canada, 1994b.

_____. "Canadian-Based Multinationals: An Analysis of Activities and Performance." Working paper 2. Ottawa: Industry Canada, 1994c.

_____. "Small Business in Canada: A Statistical Overview." Entrepreneurship and Small Business Office. Ottawa: Industry Canada, 1994d.

Caves, R. *Multinational Enterprises and Economic Analysis*. Cambridge, U.K.: Cambridge University Press, 1982.

Conference Board of Canada (The). *Canadian Trade Policy Options in the Asia Pacific Region: A Business View.* Report 122-94. Ottawa: the Board, 1994.

―――. *Canada's Business Investment Intentions in China.* Aeril Project 4. Ottawa: the Board, 1995.

Dunning, J.H. *Multinational Enterprise, Economic Structure and International Competitiveness.* Chichester: John Wiley & Sons, 1985.

Eden, L. "Multinational Responses to Trade and Technology Changes," in *Foreign Direct Investment, Technology and Economic Growth.* Edited by Donald G. McFetridge. Investment Canada Research Series. Calgary: University of Calgary Press, 1991.

Hall, C. "APEC and SME Policy Suggestions for an Action Agenda." Unpublished manuscript. Australian APEC Studies Centre, 1994.

Ogbuehi, A.O. and T.A. Longfellow. "Perceptions in U.S. Manufacturing SME Concerning Exporting: A Comparison Based on Export Experiences." *Journal of Small Business Management*, 32, 4 (October 1994):37-47.

Picot, G., J.R. Baldwin and R. Dupuy. "Have Small Firms Created a Disproportionate Share of New Jobs in Canada? A Reassessment of the Facts." Business and Labour Market Analysis Group, Ottawa: Statistics Canada, 1994.

Pratten, Cliff. "The Competitiveness of Small Firms." Cambridge, U.K.: Cambridge University Press, 1991.

Rao, P.S. "The Asia Pacific Rim: Opportunities and Challenges to Canada." Working Paper 37. Ottawa: Economic Council of Canada, 1992.

Rao, P.S. and T. Lemprière. "An Analysis of the Linkages Between Canadian Trade Flows, Productivity and Costs." Working Paper 46. Ottawa: Economic Council of Canada, 1992.

Reuber, A.R. and E. Fischer. "The Role of Management's International Experience in the Internationalization of Smaller Firms." Unpublished manuscript. Toronto, 1995.

Rugman, A.M. *Outward Bound: Canadian Direct Investment in the United States.* Toronto: McGraw-Hill Ryerson, 1987.

Sullivan, Daniel. "Measuring the Degree of Internationalization of a Firm." *Journal of International Business Studies*, 25, 2 (1994):325-42.

United Nations Conference on Trade and Development (UNCTAD). *Small and Medium-Sized Transnational Corporations: Role, Impact and Policy Implications.* New York: United Nations, 1993.

United States Trade Representative. *The 1993 National Trade Estimate Report on Foreign Trade Barriers.* Washington: Office of the U.S. Trade Representative, 1993.

Comment

James McRae
School of Public Administration
University of Victoria

MY INITIAL REACTION AFTER LEARNING the topic to be addressed in this paper – the performance of Canadian small- and medium-sized enterprises

(SME) in commercial relations with Asia Pacific economies – was one of modest surprise. Given the attention that SME have received in the domestic context, investigating the international contribution they make to Canadian employment and output growth would, on the surface, seem to be a natural extension of a research line popularized in the mid-1980s.

My surprise originated from two sources. First, Canadian commercial linkages to the countries of the Asia Pacific region, while important, are not huge. Merchandise exports from all Canadian firms – large, medium-sized, and small – to the region made up 7 percent of Canadian exports in 1971 and 8 percent in 1994. While this level of merchandise exports exceeds the share of exports going to countries of the European Union, it remains relatively small in aggregate terms. Merchandise imports from the region are larger, and have grown more rapidly, from 7 percent in 1971 to 13 percent in 1994. In terms of investment linkages, the Asia Pacific share of total foreign direct investment (FDI) in Canada rose from 1 percent in 1980 to 6 percent in 1994. The share of total Canadian FDI stock going to countries of the Asia Pacific region increased from 3 percent in 1980 to 7 percent in 1994. Thus, in terms of both trade and investment linkages, the Asia Pacific region represents an area of increasing importance, especially for the western Canadian provinces, but for Canada as a whole it remains of secondary importance, well behind the United States.

The second source of my surprise originated in the observation that the output of many SME tends to be non-traded goods and services. While changes in transportation and communications technology, and reductions in regulatory barriers, have made more SME output tradeable or potentially tradeable, the fact remains that in 1992 only 9 percent of the value of Canadian exports originated from SME. Exporting remains an activity dominated by large firms.

With the twin observations that SME tend to be relatively unimportant in terms of the value of Canadian exports and that the Asia Pacific region is of secondary importance in terms of the destination of Canadian exports, an analysis of SME exports to the Asia Pacific region seemed to be a surprising inclusion.

Fortunately, my initial surprise was soon overtaken by a second reaction, that of admiration for the detailed and back-breaking effort put forth by the authors in terms of assembling the database necessary to shed light on the performance of SME in the Asia Pacific region. The authors produce a sample of 1,297 Canadian and 2,526 U.S. firms, and they are able to classify them with respect to export orientation (domestic; international markets excluding the Asia Pacific region, and Asia Pacific and elsewhere); primary business activity (12 major industry groups); size, based on total sales (SME, large firms, and very large firms); total assets; employment; and labour and capital productivity.

The two samples parallel one another reasonably closely in terms of industrial distribution, with the notable exception that the U.S. sample contains a much larger number of technology-intensive manufacturing firms. However, the two samples differ significantly in terms of firm size: SME make up 74 percent of the sample of Canadian firms but only 19 percent of firms in the U.S. sample. American sample firms are two to three times larger, on average, than their Canadian counterparts in two size categories (SME and very large firms), and approximately the same for firms in the large category. Across the entire sample, and measured in terms of sales revenue, the average sample U.S. firm is seven times larger than the average Canadian firm. Measured in terms of employment, the average U.S. sample firm is nearly eight times larger. This result for the overall sample reflects the disproportionate number of SME in the Canadian sample.

The major empirical observation that emerges from "benchmarking" the export orientation of Canadian firms against their U.S. counterparts is that Canadian firms appear to be less outwardly oriented. Just over 28 percent of the Canadian firms are doing business in the "Asia Pacific and elsewhere" market, compared with 39 percent of the U.S. firms. However, this result may be largely explained by the fact that U.S. firms are much larger than their Canadian counterparts, and larger firms have a much higher probability of participating in the Asia Pacific region than smaller ones. The relationship between firm size and participation in the Asia Pacific region is demonstrated by the authors, using both U.S. and Canadian sample data.

Results more relevant to the title of the paper are obtained when the overall sample of firms is segmented into the three size categories – SME, large, and very large firms. The authors report that 21 percent of Canadian SME do business in the "Asia Pacific and elsewhere" market, compared with 25 percent of the U.S. sample firms. Large-firm results for the two countries are similar, but among the very large firms, the proportion of Canadian firms doing business in the "Asia Pacific and elsewhere" region is greater (64 percent) than that of their U.S. counterparts (49 percent). The authors do not discuss the significance of this observation for very large firms. One possible explanation is that, given the smaller size of the domestic Canadian market, very large Canadian firms, despite their size disadvantage vis-à-vis comparable U.S. firms, are forced to seek customers in the "Asia Pacific and elsewhere" region more aggressively than their American counterparts.

The cross-tab results reported in benchmarking the outward orientation of Canadian firms relative to U.S. sample firms suffer from the significant difference in average firm size between the two samples. For both the overall sample and the SME subsample, it remains unclear whether the lower level of Canadian firm activity in the "Asia Pacific and elsewhere" market is due to the size disadvantage of Canadian firms or to some competitiveness factor. Many answers relevant to the firm-size issue are available or potentially available in the econometric work displayed in the Appendix. Given the importance of firm size in understanding whether or not Canadian SME are fully participating in the "Asia Pacific and elsewhere" market, it is a great shame that more effort is not devoted to integrating the econometric and cross-tab results.

Apparently the "problem" of lower Canadian SME firm participation in the "Asia Pacific and elsewhere" market is not due to poorer competitive performance by Canadian firms. The authors look at five indicators of performance – percentage growth in sales, percentage growth in asset value, capital productivity (sales/assets), labour productivity (sales/employee), and return on assets – and conclude that Canadian SME doing business in the "Asia Pacific and elsewhere" market outperform their U.S. counterparts by a significant margin. Thus it may be concluded that Canadian SME are fit, willing, and able to participate in greater numbers in the "Asia Pacific and elsewhere" market.

In summary, I have the impression that the authors devoted considerable effort assembling their large and detailed database, but had only limited time to determine what questions can and should be answered by the information. Their most important observation – and one that is critical to determining whether Canadian SME participation in the "Asia Pacific and elsewhere" market is "right" when benchmarked against U.S. penetration – is relegated to an appendix. Many interesting and important observations are available or potentially available in this database, but they are not being given the focus they deserve because the dazzling array of cross-tab information is crowding them out. I would concentrate on the firm-size and outward-orientation relationship, and use the database to its fullest by discussing the results for all three firm-size classes, not just SME.

Wendy Dobson
Centre for International Business
University of Toronto

Pacific Triangles: U.S. Economic Relationships with Japan and China[1]

JAPAN AND THE UNITED STATES HAVE BEEN on a trajectory of escalating economic conflict. Both sides agree on the mutual importance of bilateral trade and investment. Yet the relationship has evolved in the postwar period from one of friendship to conflict over two main issues: Japanese export penetration of the U.S. market and, more recently, access by foreign firms to Japanese markets.

Each side has particular goals for the economic dimension of the relationship. Recent U.S. administrations have responded to growing political perceptions, reflecting the anger of U.S. interest groups that U.S. producers do not have the same access to Japanese consumers as Japanese producers do. Yet Japanese producers have relatively free access to U.S. consumers, albeit at the risk (also faced by other foreigners) of opposition by special interests who complain about dumping or subsidies.

The current U.S. administration has been of the view that the only way to go is to apply escalating pressure until Japan acts. Equally troubling is the increasing number of objective observers in the United States who also believe it, at a time when Japan (and indeed others) considers such tactics to be illegitimate. U.S. demands for specific market shares are the antithesis of free trade, and Japan is right to resist.

At the same time, Japan is changing. Market forces are bringing more openness and transparency to the Japanese economy, business practices, and business-government relationships. Forced by the relentless market pressures of a strong yen, production is being internationalized; in order to reduce costs to consumers and provide more choice, domestic markets are being deregulated and cheaper foreign imports are being admitted. Japan is also more willing to reconsider the structure of its society and the functioning of its markets than it was when pressures to do so were first brought to bear during the Strategic Impediment Initiative (SII) talks that were initiated in 1989. But until 1996, the changes were not occurring rapidly enough to fend off political pressures in the United States.

Harmony and discord in the conduct of U.S. trade and investment relations with Japan are more than circumscribed bilateral issues. Growing economic interdependence, which is one cause (and possibly part of the solution), means that bilateral tensions and conflict spill over to affect each country's major trading partners. The intensity of the disputes between Japan and the United States and the means that have been used to resolve them have resulted in the disputes extending beyond bilateral relations. U.S. policies toward Japan, initially based on the assumption of Japan's uniqueness, are likely to be extended to other successful Asian exporters. Indeed, U.S.-Chinese relations are now an area of increasing conflict because of growing external imbalances and intellectual property issues. These conflicts and their bilateral solutions could damage Canada's interests and undermine the international trading system.

The purpose of this paper is to examine the economic relationships in this Pacific "triangle" of U.S. trade with Japan and China and evaluate their implications for Canada. Bilateral relations are driven by an aggressive postcold-war U.S. trade policy. The macroeconomic dimension of this relationship is based on the assumption that, if market forces are allowed to function freely, the international adjustment process will correct the external imbalances. This paper focuses on the political economy of these relationships as well. It assumes that the international adjustment process works, but with significant lags. These lags create "windows of crisis" in which politics dominate economics. In the next section, I evaluate U.S. and Japanese trade policy. In the third section, I assess the economic dimensions of interdependence, both macroeconomic (in the savings-investment imbalance and real exchange rates) and microeconomic (expressed as mutual concerns and demands in the SII and subsequent Economic Framework talks). In the fourth section, I focus on the growing economic presence of Japan and the United States in East Asia and the implications for their relationships with China and the future of the Asia Pacific Economic Cooperation (APEC) forum. A fresh approach to U.S.-Japan disputes is put forward – an approach that should be seriously considered during the relative calm in the bilateral relationship in 1996. In this section, I focus on the implications for Canada of U.S. trade with Japan and China.

POLICY FRAMEWORKS

ONE OF THE GREAT DIVIDES IN POSTWAR JAPANESE-U.S. RELATIONS occurred as world politics and economics were transformed by the ending of the cold war. By the time "the common enemy" of communism had disappeared, Japan and Germany had risen to economic parity with the United States. The United States, however, was preoccupied by domestic problems of declining competitiveness and macroeconomic imbalances, which were eroding its traditional economic buoyancy. Domestic support for its role as democracy's military protector were declining. Other concerns such as fair trade with its major trading partners diluted traditional support for the rules-based multilateral trading system.

Japan's anxiety about the future of U.S. security commitments in Asia increased. This anxiety fed into the nascent debate in Japan about its role in the world now that it had achieved its century-old goal of catching up economically and technologically with the West. Indeed, Japan had become one of the world's technological leaders and largest net creditors. One option was to step up its cooperation with its dynamic East Asian neighbors, both to increase its leverage with the United States and as a hedge against widening and intensifying bilateral trade conflicts. Nationalists even advocated redefining Japan's role as *Asia's* leader.

THE UNITED STATES

THE INCREASING OPENNESS OF THE U.S. ECONOMY compounded the shock of declining U.S. industrial competitiveness. Economic reliance on foreign suppliers and foreign direct investment increased, and foreign competition at home and abroad intensified. The U.S. response was to shift the emphasis in its trade policy from the traditional goal of protecting the U.S. market toward increasing access to foreign markets. Robert Lawrence (1994) coined a metaphor that captures the situation well: "As the water level of protection was lowered, the rocks of divergent domestic policies and practices which separated nations became apparent."

As the United States increasingly focused on the domestic policies and practices of its trading partners, particularly those of Japan, four identifiable thrusts emerged in U.S. trade policy: a unilateral thrust using Section 301 of the 1988 *U.S. Trade Act* to target the policies of trading partners considered unfair by U.S. interests, a bilateral thrust involving negotiations such as the SII that targeted trading partners' domestic practices, which U.S. interests believe obstructed market access, a regional thrust whose goal was to reduce the remaining trade barriers with its Canadian and Mexican neighbors and create a dispute settlement mechanism through NAFTA, and traditional multilateral negotiations in the Uruguay Round that were concluded in 1993 (Lawrence, 1994).

A debate emerged between the views and assumptions of realists and internationalists about its relationship with Japan. Realists view the bilateral relationship in geo-strategic terms; they see the world in terms of "us-versus-them." To the realists, growing interdependence is a function of political power and choice rather than market forces. The importance of economic and technological factors in national security has increased in the post-cold-war era, and they believe the nation's traditional commitment to free markets should be re-examined. Thus the potential advantages of interdependence can be offset by international competition for market share. Under such assumptions, one nation's economic advantage becomes the other's vulnerability and international cooperation is difficult to achieve. According to this view, Japan is a mercantilist threat and a growing hegemonic power in the Asia Pacific region.

Internationalists believe that interdependence is an inevitable outcome of market forces as well as political choice, and that it can be influenced not only by states, but also by actors within nations such as bureaucrats, firms, industry associations, and consumers. Growing interdependence implies that distinctions between domestic and foreign economies are more difficult to draw, and new methods of economic management are required.

De facto, the realist view seems to be dominant in the U.S. administration's use of quantitative sectoral market-opening targets. The precedent set in the 1986 Semiconductor Trade Agreement (STA), which was renewed in 1991 and is up for further renewal in 1996, is considered in the United States to be a model to follow.

Japan

JAPANESE POLICY, IN CONTRAST, has been heavily influenced until recently by the outcome of the Second World War. Its foreign policy has been largely passive and reactive, following the United States' lead. The legacy of Japan's prewar militarism and colonialism has forced the country to preoccupy itself with economic development. Successful as this policy has been, it also contains the seeds of the current problems.

Differences in U.S. and Japanese political institutions and processes have reinforced differences in foreign policy styles. The United States has led, and Japan has followed. Although Japan has sought to accommodate U.S. demands, it has often been too little, too late. *Gaiatsu*, or the exertion of foreign pressure, has almost always been required to bring about economic change. Conflicting domestic interest groups and the requirement for consensus have prevented Japan from reforming itself from within.

However, in accommodating U.S. demands, Japan had expected the United States to respond in kind by reciprocating with sensitivity to Japan's interests. The United States' failure to do so has evoked increasing Japanese resentment. Meanwhile, U.S. resentment has grown at Japan's inability to change without the United States playing the "bad guy" (Sato, 1996).

Detractors of Japanese institutions criticize them as being mercantilist and antithetical to free trade and competition in the modern capitalist economy (Johnson, 1995; van Wolfren, 1990; Fallows, 1989; and Prestowitz, 1988). Even objective observers have come to believe that trade liberalization by itself will not solve the problem, and second best solutions will have to be considered. These might include Japanese interests buying off foreign pressure by allowing foreign firms into domestic oligopolies, or using voluntary import expansion (VIE) measures to encourage the movement toward trade liberalization for foreign suppliers of intermediates and capital goods, who are competitive but hampered by *keiretsu* contracting relationships (Bergsten and Noland, 1993).

More recently, Japan's trade and foreign policy is being "Asianized" for several reasons. First, U.S. and Japanese security roles in the region are seen to require Japan to be a more proactive partner. Second, the economic dynamism of East Asia offers a more attractive market than the United States. Third, the Clinton administration's results-oriented trade policies and the Republicans' aggressive trade liberalism are causing Japanese opposition. Finally, many intellectuals, Japanese and Asian, are calling for a redefinition of Japan's identity in regional terms – as an Asian leader (Mochizuki, 1995). This issue was especially sensitive in 1995 because as the APEC chair before the Osaka summit, Japan was faced with reconciling differing approaches to trade and investment liberalization. The Asian participants favour an evolutionary, voluntary approach (known as concerted unilateralism), while the Americans champion a more transparent, negotiated approach with clear commitments and deadlines.

Implications

Primarily because of savings-investment imbalances, and exacerbated by impatience and even skepticism about slow adjustment, there are short-term political pressures for aggressive, results-oriented unilateralism. In the United States, this approach is generally considered to be effective. In Japan, however, there is widespread resolve to never again accept targets such as those contained in the STA. Disputes increasingly focus on domestic policy and institutions. At the bargaining table, the issues increasingly heighten awareness of national differences. This is where the present danger lies. The collision of market and politics is evident in the latest high-level dispute, which occurred in 1995 in the automotive sector. The outcome, when stripped of its win-lose rhetoric, has become just another chapter in the long-running ritual of U.S. assertiveness and Japanese resistance. The agreement committed Japan to import more and to further internationalize its own production. Politically, both sides claimed victory. The United States claimed the agreement will force open Japanese markets, and that the United States will monitor its own performance targets for this goal. The Japanese argued they had successfully resisted U.S. demands for quantitative targets. Yet in reality the market was already forcing Japanese manufacturers to do what they committed to do in the agreement.

The ironic conclusion is that a similar outcome could have been accomplished as much as six months earlier without the high-level rhetoric and tension.[2] In addition, whatever benefits were sought by the escalation must be weighed against the costs: clearly the June 1995 settlement is not the last round in the dispute, which cost both sides credibility. The dispute has aroused strong antagonism among ordinary people on both sides, which suggests there is a danger that popular support for the relationship is being eroded.

There must be better ways to address these differences in a timely fashion. Before examining such options, however, I turn to an analysis of the economic dimensions of interdependence.

Economic Dimensions of Interdependence

By conventional measures, the Japanese and U.S. economies are becoming increasingly enmeshed and interdependent. What trends have bared the "rocks" of domestic differences, and how will these differences evolve?

Current U.S. policy focuses on large merchandise trade deficits, seen to be both macroeconomic and microeconomic in origin. Few, however, would attribute the volatility in the bilateral merchandise trade deficit to changes in trade policy. Most would argue that it is due to differences in macroeconomic conditions in the two countries. The two countries' global merchandise trade balances have followed opposite paths since 1980. Japan's surplus has soared, while the United States has had an increasing deficit (Table 1). The bilateral merchandise trade balance grew in the early 1980s, but the overall imbalance changed very little between 1986 and 1992 (Table 2).

Mainstream economic theory predicts that flexible real exchange rates will adjust to ensure equilibrium in the overall balance of payments. By accounting identity, the external position of a country, expressed as the current account balance (CA), is the balance of trade, or net exports (NX), less net interest payments. This relationship can be shown as ($CA = S - I + T - G$), that is, aggregate savings (S) minus investment (I) and the government deficit (T-G). In other words, the current account balance is the difference between what a country produces and what it consumes. When a country consumes more than it produces, it must import from abroad. Similarly, when a country invests more than it saves, it must borrow and import goods from abroad. Short-term fluctuations in the current account typically reflect short-term fluctuations in the trade balance, since net interest payments tend to change only gradually. Exchange rate adjustments will reduce external imbalances, but with a lag because of the time required for importers to change sources and exporters to change markets. Thus, if the currency depreciates because of a trade deficit, the value of imports increases faster than the value of exports, causing an initial adverse impact on the trade balance (known as the J-curve).

Of course, neither the savings-investment discrepancy nor the current account imbalance are independent of the government deficit. Fiscal policy will influence all three variables (S, I, and CA) through its influence on interest

TABLE 1

GLOBAL MERCHANDISE TRADE AND CURRENT ACCOUNT BALANCES IN JAPAN, THE UNITED STATES AND CHINA, 1980-94 (US$ BILLIONS)

	TRADE BALANCE			CURRENT ACCOUNT BALANCE		
	JAPAN	UNITED STATES	CHINA	JAPAN	UNITED STATES	CHINA
1980	2.13	−25.50	n.a.	−10.75	1.84	n.a.
1981	19.96	−27.97	n.a.	4.77	6.87	n.a.
1982	18.08	−36.45	4.25	6.85	−8.64	5.82
1983	31.46	−67.08	1.99	20.80	−46.29	4.49
1984	44.26	−112.51	0.01	35.00	−107.14	2.51
1985	55.99	−122.15	−13.12	49.17	−115.16	−11.42
1986	92.82	−144.54	−9.14	85.83	−138.84	−7.03
1987	96.46	−160.28	−1.66	87.00	−153.95	0.30
1988	95.00	−126.96	−5.32	79.61	−127.36	−4.09
1989	76.89	−115.14	−5.62	56.99	−103.97	−4.48
1990	63.58	−109.03	9.17	35.87	−92.91	11.89
1991	103.09	−74.07	8.74	72.91	−7.72	13.02
1992	132.40	−96.10	5.18	117.65	−62.00	5.81
1993	141.57	−131.38	−10.65	131.54	−99.69	−12.40
1994	145.93	−164.33	n.a.	129.24	−150.92	n.a.

n.a. = not available.
Source: IMF, *International Financial Statistics*, various issues.

rates, the exchange rate, and the level of economic activity. Expansionary fiscal policy raises demand and leads to an expansion of output and adjustments through several channels. As output rises, demand for money increases and interest rates rise. Private investment is crowded out, and foreign capital flows in. Exchange rate adjustment occurs in the form of currency appreciation, which brings about a drop in net exports at any given level of output.

While changes are brought about by real exchange rate fluctuations, it is underlying values of domestic savings, investment, and the government deficit that determine a country's external position. Empirical studies of trade elasticities for the United States have found a higher income elasticity of import than of export demand. This means that if output in the United States and the rest of the world were growing at similar rates, a steady depreciation of the U.S. dollar would be required to prevent a widening trade deficit (Krugman, 1991).

The persistence of the bilateral savings-investment imbalance is reflected in recent trends in each country's global current account balance as a share of GDP (Table 3). Japanese households are high savers; U.S. household are not. Public sector dissaving, until recently, has been much higher in the United States. Macroeconomic adjustments have also affected real effective exchange rates such that the Japanese yen has appreciated dramatically since

TABLE 2

BILATERAL MERCHANDISE TRADE BALANCES IN JAPAN, THE UNITED STATES AND CHINA, 1980-93 (US$ BILLIONS)

	BILATERAL TRADE BALANCE		
	UNITED STATES-JAPAN	JAPAN-CHINA	UNITED STATES-CHINA
1980	−12.40	0.76	2.59
1981	−18.52	−0.19	1.54
1982	−19.54	−1.84	0.40
1983	−22.33	−0.18	−0.32
1984	−37.69	1.26	−0.39
1985	−50.80	6.00	−0.42
1986	−62.80	4.22	−2.16
1987	−61.17	0.86	−3.45
1988	−57.16	−0.39	−4.34
1989	−54.44	−2.63	−7.07
1990	−46.96	−5.92	−11.49
1991	−48.95	−5.63	−14.05
1992	−54.52	−5.00	−20.13
1993	−64.45	−3.28	−25.07

Source: OECD Statistics Directory, *Foreign Trade by Commodity*, various issues.

1992 while the U.S. dollar has depreciated in real effective terms. As is well known, the U.S. dollar experienced a real appreciation in the early 1980s with the movement toward expansionary fiscal and tight monetary policy. Monetary policy was tightened in 1980 to rein in accelerating inflation following the 1978 oil price shock and remained tight during the subsequent recession and recovery. At the same time, President Reagan moved to reduce the size of government while increasing defence spending and reducing taxes. The net result was soaring fiscal indebtedness and an appreciating exchange rate. Subsequent macroeconomic policy adjustments in Europe, Japan, and the United States, assisted by coordinated monetary policies in 1985, resulted in international adjustment that reduced the U.S. trade deficit in the late 1980s.

Persistent public dissaving in the United States continued to be an obstacle to adjustment, however. By 1995, fiscal consolidation had become a bipartisan goal, but there were significant political differences over what mix of expenditure and tax changes was required to achieve such consolidation. Nevertheless, by late 1995, prospects were good for a budget agreement that includes expenditure cuts and revenue gains of a magnitude sufficient to exceed tax cuts and eliminate annual deficits.

In contrast, for much of the past two decades Japan has had a virtuous circle of high private savings rates, significant capital accumulation, robust rates of economic growth, low inflation, prudent macroeconomic policies including

TABLE 3

MERCHANDISE TRADE AND CURRENT ACCOUNT BALANCES IN JAPAN, THE UNITED STATES AND CHINA, 1980-94 (PERCENT OF GDP)[1]

	TRADE BALANCE			CURRENT ACCOUNT BALANCE		
	JAPAN	UNITED STATES	CHINA	JAPAN	UNITED STATES	CHINA
1980	0.18	−0.99	n.a.	−0.93	0.07	n.a.
1981	1.71	−0.93	n.a.	0.41	0.23	n.a.
1982	1.56	−1.17	1.62	0.60	−0.28	2.22
1983	2.61	−2.00	0.70	1.72	−1.38	1.58
1984	3.73	−3.02	0.01	2.95	−2.88	1.04
1985	3.55	−3.08	−5.04	3.12	−2.90	−4.39
1986	4.47	−3.45	−3.59	4.14	−3.31	−2.76
1987	3.47	−3.57	−0.55	3.13	−3.43	0.10
1988	3.22	−2.38	−1.41	2.70	−2.38	−1.08
1989	2.78	−2.10	−1.66	2.06	−1.90	−1.32
1990	2.01	−1.97	2.71	1.14	−1.68	3.51
1991	2.86	−1.36	2.35	2.02	−0.14	3.50
1992	3.57	−1.70	1.24	3.17	−1.10	1.39
1993	3.40	−2.26	−1.96	3.16	−1.71	−2.28
1994	3.10	−2.72	n.a.	2.75	−2.49	n.a.

n.a. = not available.
1 Data for China are percent of GNP. Japan and China's GNP data are given in billions of yen and billions of yuan, respectively. The figures were transformed into US$ billions using the IMF end-of-year exchange rates.
Source: IMF, *International Financial Statistics*, various issues.

a budget deficit incurred mainly for capital spending, and stable nominal interest and exchange rates. Japan is the world's largest supplier of net capital flows, providing 53 percent of the total in the 1989-93 period. The United States was the largest net user, acquiring 27 percent of total flows during this period (IMF, 1995a).

Even so, Japan experienced an asset bubble in the late 1980s in which land and stock prices rose rapidly and then fell dramatically in the early 1990s. These events may be linked to the effects of monetary policy ease in the late 1980s and particularly to the strong impact of yen appreciation during that period. Adjustment occurred in two ways: through a wealth effect in the traded goods sector (ability to command more international resources), and a substitution effect, which made non-tradeables relatively attractive (since the rate of return in the traded-goods sector had declined because of yen appreciation). These effects were felt in land speculation. Land was then used as collateral for loans, which contributed to the stock market boom. When monetary policy was subsequently tightened to wring out the speculative excesses, the result was not only a recession in domestic demand as households stopped spending and began saving, but as stock and land prices fell, severe structural problems appeared in the banking industry, which had participated heavily in the domestic property market (Bergsten and Noland, 1993).

Adjustments to real effective yen appreciation in the 1980s and 1990s have occurred as outflows of FDI by exporters seeking to take advantage of more favorable relative input prices offshore. Another adjustment that might be expected is a decline in import prices and a shift by consumers to relatively cheaper imported goods. This has not occurred. The Ministry of International Trade and Industry (MITI) has estimated that between January 1992 and November 1994, consumer prices rose 3 percent overall. During that period, the yen rose 26 percent, import prices dropped 14 percent, but wholesale prices dropped only 3.5 percent. Most agree that the main problem is the vast array of regulations. Even MITI acknowledges that if market mechanisms were working properly, consumer prices would have been allowed to fall (Nakamoto, 1995).

Partly as a result of exchange rate fluctuations and partly due to trade barriers, between 1984 and 1994 Japan was a large net investor in the United States in real estate and acquisition of assets at the leading edge of multimedia strategies. Calculations of the difference between the increase in Japan's net external assets and the cumulative current account surplus during this period yield estimates of losses of as much as Y55.1 trillion in the domestic value of these investments (Wolf, 1995). Twenty percent of these losses is attributed to bad investments and 27 percent to unanticipated yen appreciation. The net result was a huge implicit resource transfer from Japan to the United States.

Bilateral trade imbalances are expected to persist because of problems of access by foreign producers to Japanese markets. Patterns of foreign direct investment (FDI) show marked asymmetry (see Table 4) by any measure.[3] By Japanese measures, the stock of Japanese FDI in the United States outweighs U.S. investment in Japan by a factor of 13:1. By U.S. measures the ratio is nearly 3:1; the U.S. stock in Japan is only three times its investments in Hong Kong and China. This asymmetry is significant for two reasons. First, trade and investment are complements; investors (especially U.S. investors) tend to import equipment from and export output to the home country. Thus more investment in Japan should induce more imports into Japan from the United States (as well as contribute to exports). Second, the investment asymmetry is a proxy for an asymmetry in the number of stakeholders pursuing harmonious bilateral relationships. Many fewer American companies have interests in Japan than the reverse. Since companies tend to participate in bilateral sectoral disputes both as initiators and as mediators seeking to preserve harmony in the business environment, when the numbers of U.S. firms are thin on the Japanese side, their moderating influence is missing in Washington. Cross-national interest groups, which might contribute to the moderation of disputes, are less likely to be formed (Sato, 1996).

In summary, while the adjustment process in the two economies is moving toward reducing the underlying savings-investment imbalances, more adjustment

TABLE 4

STOCK OF FOREIGN DIRECT INVESTMENT,
UNITED STATES, JAPAN (1994) AND CHINA (1993)

	US$ BILLIONS
Japan[1]	
Outward Foreign Direct Investment Stock:	
All Countries	463.6
In United States	194.4
In China and Hong Kong	22.6
Inward Foreign Direct Investment Stock:	
All Countries	34.1
From United States	13.7
From China and Hong Kong	0.8
United States[2]	
Outward Foreign Direct Investment Stock:	
All Countries	612.1
In Japan	37.0
In China and Hong Kong	13.7
Inward Foreign Direct Investment Stock:	
All Countries	504.4
From Japan	103.1
From China and Hong Kong	1.7
China[3]	
Outward Foreign Direct Investment Stock:	
All Countries	7.4
In United States	1.0
In Japan	n.a.
In Hong Kong[4]	4.5
Inward Foreign Direct Investment Stock:	
All Countries	57.1
From United States	5.2
From Japan	5.2
From Hong Kong[4]	45.0

1 Data for Japan are for 1994, based on notifications of approved but not necessarily undertaken projects. Therefore, actual stocks of FDI are overstated. Data also exclude disinvestment. They are on a fiscal year basis (April to March).
2 Data for the United States are on an historical-cost basis at year-end.
3 Data for China are for 1993.
4 Author's estimates.
Source: Japan Ministry of Finance; U.S. Survey of Current Business, June 1995; UNCTAD, World Investment Report. 1994, 1995.

is required. Japan faces a continuing crisis. In 1995 the gap between actual and potential output was still 6 percent (IMF, 1995b, p. 21), despite a dramatic easing of fiscal and monetary policy. Indeed, output in Japan may fall further before picking up again because of flagging consumer confidence, barriers to price adjustments, and the inability of financial and non-financial corporations to produce healthy balance sheets. While Japan's measured unemployment rate is slightly more than 3 percent, corporations continue to employ large quantities of redundant labour, which they must shed to restore profitability. Consumers, expecting higher unemployment, lack confidence. Land prices remain elevated and underpin the balance sheets of financial institutions.

Two policy instruments that have been inadequately employed to facilitate adjustment are fiscal policy and deregulation. The reluctance of Ministry of Finance bureaucrats to run deficits is legendary. Yet Japan's savings position could support a much larger fiscal stimulus (provided it were accompanied by continued monetary ease) than has been the case. Deregulation is slow: major laws still on the books include the *Exchange Control Law* (1937), the *Staple Food Control Law* (1942), and the *Land and House Leasing Law* (1941). There is still a broad regulatory framework; only 1,091 items have been deregulated (Miyoshi, 1995).

The United States has achieved a soft landing from growth rates in 1994 that were unsustainable. Progress on fiscal consolidation is underway. Yet the depth of Japanese-U.S. interdependence was demonstrated rather clearly in the summer of 1995 at the end of a particularly fractious bilateral dispute over autos. Lacking clear signs that the appreciating yen would peak, Japanese institutional investors were increasingly reluctant to continue to hold U.S. bonds, and supplies of foreign savings were in jeopardy. The U.S. Federal Reserve Board successfully led several rounds of concerted intervention in foreign exchange markets aimed at nudging existing market trends toward a weaker yen. A nominal rate of 100 yen per dollar was restored, but whether it will be sustainable depends on further policy adjustment.

The Emergence of China and East Asian Interdependence

The emergence of China since 1979 and of the East Asian economies as the new global economic centre of gravity and dynamism are some of the remarkable events of the late 20th century. Although the domestic policies of governments in the region have been largely responsible for this transformation, the three giants have each played a role, both before and as the cold war ended. The United States supplied a security framework, significant investment and a major export market; Japan was a major supplier of capital and intermediate goods, development aid, and private investment; and China has been an emerging importer and exporter since 1979.

Japan and the United States are the largest trading and investment partners for many APEC countries, as they are for Canada. The largest stocks of U.S. foreign direct investment in APEC countries after Canada, Japan, and

Mexico are in Australia, Hong Kong, and Singapore. Each is also a significant inward U.S. investor (U. S. Department of Commerce, 1995). Australia, Indonesia, Hong Kong, Singapore, Thailand, mainland China, Malaysia and Korea are among the largest holders of accumulated Japanese FDI (Japan, Ministry of Finance, 1995). None of the APEC countries, however, is yet a major investor in Japan.

Asians are fond of pointing to growing intra-regional shares of total trade as evidence that East Asia's dynamism is making it self-sufficient and will enable it to "decouple" from the rest of the world. The basic issue, however, is whether the shares of intra-regional trade have increased more rapidly than would be predicted within a systematic framework measuring a normal level of trade. Gravity models can shed light on such questions, using gravity variables of distance, economic size, and growth rates to explain bilateral flows among countries in a group. If there were nothing to the notion of trade blocs, these variables would soak up most of the explanatory power. Frankel (1993) finds some evidence of a special regional effect, implying that intra-regional trade goes beyond what can be explained by proximity alone, with a drop in the bias in the first half of the 1980s and a slight increase in the second half.[4] Historical gravity models show a steady decline in the regional bias of East Asian trade in all but the last five years (Petri, 1993). This is true for most individual East Asian economies as well.[5]

Much as East Asians like the idea of decoupling from industrial countries, their dependence on these markets will continue for several reasons. First, the leading Asian exporters (Korea, Taiwan, and Singapore) need demanding and sophisticated customers who will cooperate with them in upgrading their technological capabilities. These customers are located mainly in industrialized markets. Second, while Japan is one of these customers, it is a long way from opening its markets to East Asia as an alternative to North America and Europe (see Table 5).

In the economic reforms that started in 1979, China first freed up market forces in agriculture. Then Special Economic Zones were permitted in the coastal regions to encourage international direct investment into export-oriented production, mainly light manufacturing. Chinese policy is now intent on upgrading its industrial structure in ways that are largely consistent with a social market economy. The largest entrepreneurial units are owned by state-owned enterprises (SOE), and the most plentiful are township and village enterprises (TVE) owned by municipal authorities.

As its growth rate has accelerated, China has become more integrated into the world economy through trade and capital markets. It has attracted international investors with economic incentives and the potential size of the local market. By some estimates, in 1994 China was the world's top destination for FDI. Ethnic Chinese abroad have dominated this investment and facilitated the skill transfers necessary for international business. Hong Kong, not Japan or the United States, is the largest investor in China.[6] China is also

TABLE 5

JAPAN'S TRADE BALANCE: AGGREGATE AND WITH MAJOR TRADING PARTNERS, 1990-94 (US$ BILLIONS)

	1990	1991	1992	1993	1994
United States	37.95	38.22	43.56	50.17	54.90
Canada	−1.67	−0.45	−0.60	−1.78	−3.02
Western Europe	20.72	29.73	34.13	28.11	22.45
Australia	−5.47	−6.52	−5.40	−4.52	−4.91
China	−5.92	−5.62	−5.00	−3.29	−8.88
Asia-8	28.48	38.03	46.94	57.29	71.08
Hong Kong	10.90	14.25	18.70	20.70	23.59
South Korea	5.75	7.73	6.19	7.44	10.85
Singapore	7.14	8.80	9.88	13.00	14.96
Taiwan	6.93	8.76	11.70	12.40	13.04
Malaysia	0.11	1.16	1.54	2.01	4.13
Indonesia	−7.68	−7.16	−6.67	−6.46	−5.25
Philippines	0.35	0.31	1.18	2.44	3.24
Thailand	4.98	4.18	4.42	5.76	6.52
Total	52.15	77.79	106.63	120.24	120.86

Source: Japan Ministry of Finance.

already a significant outward investor, partly because Chinese investors have sought to secure asset values from the ravages of high inflation.

China has sucked in imports, particularly from Asian countries, and its export performance has propelled it as a world exporter from 32nd place in 1978 to tenth place in 1993. In that year, its share of global exports was 2.5 percent, accounting for nearly 20 percent of its output (Falkenheim, 1995).

Barring political setbacks or environmental crises, China is expected to grow into the world's largest economy within a generation. Purchasing power parity estimates of its current size, which abstract from exchange market distortions, already rank China as the world's third largest economy after Japan and the United States (Bell et al., 1993).

China's growth is propelled both by microeconomic and macroeconomic reforms. The latter are still incomplete. Monetary policy has been characterized by stop-go policies based on credit allocation in the absence of mature financial institutions, particularly commercial banks, to play a disintermediating role. Continued reliance on large, inefficient SOE in heavy industry utilities and the like has complicated the conduct of both monetary and fiscal policy. Until recently, SOE depended on credits from the fiscal authorities for working capital or to cover their losses. The fiscal authorities in turn met such requirements by issuing bonds to the central bank, in effect by printing money. Reforms are gradually being introduced to make the central bank more independent, to modernize financial institutions, and to require SOE to meet a bottom line. The threat of anticipated negative employment and the political consequences of credit stringency or allowing bankruptcies have been used by political pressures to justify double-digit inflation as the price of sustaining double-digit real growth rates.

Regional disparities are another key internal issue. In 1991, China's coastal regions accounted for 23 times more FDI stocks than the interior regions. GNP and gross domestic capital formation per capita in the coastal regions were nearly twice those in the interior (Dobson and Chen, 1994). Little that has happened since then would suggest the gap has narrowed.

China's formal trade policy is shaped in measure by its desire to become a member of the World Trade Organization (WTO). Since 1991, it has committed itself to the principle of bilaterally balanced trade (this objective was achieved with the United States and Japan until the late 1980s [Table 2]); conforming with international practices with respect to subsidies, tariffs, customs regulation; and increasing transparency of its trade system and import approval process (Bell et al., 1993). In early 1994, it took major steps toward currency convertibility.

Informal practices, however, are another issue. Intellectual property issues (e.g., Chinese reluctance to pursue the makers and distributors of counterfeit software) and trans-shipping to obscure bilateral trade balances are problematic. The bilateral trade balance with the United States is an increasingly contentious issue. By U.S. measures, its trade deficit with China was US$30 billion in 1994 (and therefore second only to its deficit with Japan). These measures are disputed by Chinese authorities who complain of "arbitrary practices" in determining the origin of goods from the three Chinas (Holberton, 1995).

China's large size and unique view of its place in the world make its continued integration problematic. Delays in reforming its macroeconomic policy could have repercussions for its neighbours. For example, stringent measures to rein in inflation will now have negative spillovers through the trade account, whereas in the late 1980s similar stringency was unnoticed abroad. Trade issues create mistrust on both sides. The U.S. policy is still to attempt to change China rather than adjust to it. Bilateral policies are confused on both sides: U.S. policy toward China is not coordinated as each government agency carries out its own activities (Walker, 1995), and China's policy is affected by the uncertainties of the power transition.

In summary, East Asian interdependence is significant for the United States' evolving relationship with Japan and China for a number of reasons. The area's economic clout is important, but there are also institutional factors such as East Asia's values that favor relationships, informality, and trust more than rules, and mediation more than confrontation, in handling disputes. China's traditional view of its central place in the world will be moderated by its growing interdependence and by more open economic relations. Its size and unique approach to international relations mean that the world, and particularly the United States, will have to adjust to China. Realists see Japanese strategy in East Asia as extending its influence with its neighbours, diverting trade to itself, and deflecting U.S. pressure to reduce its trade deficits with its neighbours, but it is not so simple. Japan faces conflicting pressures – from the United States to conform to western assumptions and institutions, and from Asia to provide more leadership in the region.

TRIANGULAR RELATIONSHIP: THE CANADIAN-U.S.-ASIAN TRIANGLE

IN THIS RAPIDLY-EVOLVING TRIANGLE, Canada is affected by any discord. Yet it is invisible commercially and politically. Asian markets are distant and unfamiliar to Canadian businesses. Most of Canada's political relationships are bilateral. Although it has a history of bilateral foreign policies, as recently as 1992 Canadian exports to the region were declining (Ries and Head, 1995). Part of this puzzle can be explained by the fact that NAFTA is Canada's *de facto* international business policy. Canada is prominent only in the context of its regional trade agreements with the United States and Mexico.

In Asia, Canada sends mixed signals. Governments have long focused on gaining access to Asian markets for Canadian exports and attracting Asian investment to Canada, rather than on reducing the barriers to Canadian investment in Asia from which future exports would flow (Ries and Head, 1995).

Canada shares with Japan the distinction of being heavily dependent on the United States, its single largest trading partner. Canada's second most significant partner lags far behind. The United States accounted for 81 percent of Canada's exports in 1993, followed by Japan, which accounted for a mere 4.6 percent (see Table 6). The United States accounted for 30 percent of Japan's exports, followed by Hong Kong and Taiwan, each accounting for only 6 percent. No other industrialized economy has such a large disparity between its number one and number two trading partners than does Canada.[7]

Where Canada and Japan differ is in their shares of U.S. imports and investment. Canada is a large importer and major destination for U.S. investment, and Japan is not. Two-way Japan-U.S. trade was US$200 billion in 1993. Two-way Canada-U.S. trade in the same year was nearly US$215 billion. That between Japan and Canada was US$14 billion, less than 7 percent of the total between Canada and the United States (see Table 6).

Canada's exports to Japan nearly doubled in nominal terms in the 1980-90 period, but Japan's exports to Canada have dropped. They are heavily weighted toward food, minerals and raw materials, which accounted for 78.5 percent of the total in 1992 (Ursacki and Vertinsky, 1995).

Japan's exports have more than doubled, so that its trade was nearly balanced before the recession of the early 1990s. Canada's share of Japan's total

TABLE 6

CANADA'S TRADE WITH THE UNITED STATES AND JAPAN,
1965, 1980 AND 1993

	CANADA-U.S. TRADE				
	CANADIAN EXPORTS TO UNITED STATES (US$ BILLIONS)	PERCENT OF TOTAL CANADIAN EXPORTS	U.S. EXPORTS TO CANADA (US$ BILLIONS)	PERCENT OF TOTAL U.S. EXPORTS	MERCHANDISE TRADE BALANCE (US$ BILLIONS)
1965	5.05	60.8	5.64	23.1	−0.82
1980	39.82	63.1	33.81	15.6	7.08
1993	115.19	80.9	99.37	21.8	15.82
	CANADA-JAPAN TRADE				
	CANADIAN EXPORTS TO JAPAN (US$ BILLIONS)	PERCENT OF TOTAL CANADIAN EXPORTS	JAPANESE EXPORTS TO CANADA (US$ BILLIONS)	PERCENT OF TOTAL JAPANESE EXPORTS	MERCHANDISE TRADE BALANCE (US$ BILLIONS)
1965	0.29	3.6	0.21	2.5	0.08
1980	3.73	5.9	2.44	1.9	1.29
1993	6.56	4.6	6.33	1.8	0.23

Source: United Nations, *UN International Trade Statistics Year Book*, 1993.

exports increased little, from a small base. The commodity composition of these exports, as is well known, is complementary to that of Canadian exports: 56 percent of the total is vehicles and machinery, less than 20 percent is electrical products, and the rest is distributed among other manufactured products (Ursacki and Vertinsky, 1995).

Flows of FDI between Japan and Canada are small (Table 7). In 1990, the stock of Japanese investment in Canada was US$5.7 billion: 44 percent in manufacturing and more than 47 percent in trading and financial activities. Only two Canadian firms are listed in the 30 most profitable companies in Japan – Toppan Moore and Nippon Light Metal, an Alcan affiliate (*Tokyo Business*, 1994).

Studies of Canada's trade and investment relations with Japan point out a number of well understood reasons for the disappointing performance: Japan's barriers to its market, inadequate effort by Canadians, the concentration of Canadian products in Japan's low-growth sectors, and the importance of direct investment by Japanese firms as a determinant of subsequent trade flows.

Two trends may create even more dilemmas in the future. First, Japan's population is aging, and its economy is becoming labour-scarce, highly-skilled information-based. Japan's traditional demand for Canadian raw materials is in

TABLE 7

CANADA'S INVESTMENT POSITION WITH THE UNITED STATES, 1960, 1980 AND 1990; AND JAPAN, 1980 AND 1990

	CANADA-U.S. INVESTMENT			
	CANADIAN FDI IN THE UNITED STATES (US$ BILLIONS)	PERCENT OF TOTAL CANADIAN FDI ABROAD	U.S. FDI IN CANADA (US$ BILLIONS)	PERCENT OF TOTAL U.S. FDI ABROAD
1960	1.67	64.0	11.56	36.2
1980	14.36	62.2	43.09	20.0
1990	45.24	60.1	69.35	16.4
	CANADA-JAPAN INVESTMENT			
	CANADIAN FDI IN JAPAN (US$ BILLIONS)	PERCENT OF TOTAL CANADIAN FDI ABROAD	JAPANESE FDI IN CANADA (US$ BILLIONS)	PERCENT OF TOTAL JAPANESE FDI ABROAD
1980	0.09	0.40	0.92	2.52
1990	0.66	0.88	5.66	1.82

Source: Statistics Canada, *Canada's International Investment Position*, 1992.

secular decline. Japan's increasing FDI in Asia and integration into the Asian economies will result in its finding alternative suppliers. Japanese multinationals no longer scour the world for raw materials, which are the traditional Canadian industries. Now they seek technologies. Second, the solution of Japan's trade disputes with the United States will divert trade by creating preferences for U.S. products over possibly more efficient Canadian suppliers.

Bilateral trade and investment between Canada and Japan is dominated by the relationship of each with the United States. There are few bilateral irritants, so few mechanisms have evolved for dealing with them. (Only as recently as 1995 did the Forum 2000 Follow-up Report make a bilaterally-supported recommendation for a bilateral problem resolution mechanism [Canada-Japan Forum 2000, 1995]). Official contacts tend to be regular, unremarkable (in that they are largely delegated to middle-level and senior officials), and ritualistic. Each country considers the other to be excessively preoccupied with the U. S. relationship, to the detriment of the bilateral relationship.

There are three implications of all this for Canada. First, Canada can learn from the United States to be more proactive in identifying and advancing its interests in the relationship. In technology-intensive industries such as software and biotechnology, in which Canada has world-recognized expertise and products, it can help Japanese firms compete more effectively. To achieve this kind of cooperation, however, it needs to develop more proactive strategies to capture the attention of Japanese companies (Rapp, 1995).

Second, Canadian diplomacy can play a valuable role as an interested-but-objective third party, a member of the G-7, and a close partner of both Japan and the United States (provided they both take it more seriously [Kitamura, 1991]).

Third, the Canadian-U.S. relationship holds lessons for Japan (which Canadians should point out to the Japanese). In important ways, the Americans view Japan very differently than they do Canada. While the Japanese and American economies are increasingly interdependent, it is a kind of "competitive interdependence" in which Japan's competitive challenges in trade and technology are viewed as threats. Where Americans view Canada as "like us" (however much Canadians might disagree), they view Japan as "unique." More effective management of trade and investment tensions, given the increased politicization of these issues in the U.S. Congress, requires more resort to the rule of law through multilateral and bilateral mechanisms.

Despite the differences in the way the two bilateral relationships are conducted, as their interdependence increases, the probabilities of both conflict and cooperation increase. Yet the history of the Canada-U.S. relationship has shown that when the volume of transactions is large and reasonably symmetrical, so is the number of stakeholders who contribute to, and benefit from, harmonious relations. The difference between the volume of trade and FDI between Canada and the United States and Japan and the United States bears repeating. Both Canada and Japan run merchandise trade surpluses with the United States. However, in 1993, Canada's surplus was about 27 percent of Japan's (Table 8, line 3), yet its economy is about 15 percent the size of Japan's.

Why is Japan's surplus so much more political? A major contributing factor is asymmetry in trade and investment. The size of two-way trade between Canada and the United States is immense: export flows between Canada and the United States are roughly equal. Japan's exports to the United States, in contrast, outweighed those of the United States to Japan by more than 100 percent in 1993 (Table 8, line 4). Some of the bilateral export asymmetry between Japan and the United States can be explained by the striking asymmetry in bilateral investment relations.

Foreign direct investment is not promoted here as an end in itself, but as a major vehicle for increasing the number of transnational private-sector interest groups who, while working to promote their own interests, can be expected to create cross-border ties, thereby promoting harmony between nations. China provides an example of an APEC member whose inward investors have played a significant role in moderating political positions at home. Both the U.S. and Japanese governments have threatened China: the United States over human rights abuses (in 1993 in the debate on the most favoured nation [MFN] renewal) and Japan over nuclear testing (in 1995). In

TABLE 8

BILATERAL TRADE PATTERNS IN CANADA, JAPAN AND THE UNITED STATES, 1993

CANADA-UNITED STATES		JAPAN-UNITED STATES	
1 Canadian-U.S. exports (US$ billions)	115.20	1 Japan-U.S. exports (US$ billions)	106.35
Share of total Canadian exports (%)	81.0	Share of total Japanese exports (%)	29.5
2 U.S.-Canada exports (US$ billions)	99.40	2 U.S.-Japan exports (US$ billions)	47.87
Share of total Canadian exports (%)	21.8	Share of total Japanese exports (%)	10.5
3 Merchandise trade balance (Canada-United States)	15.82	3 Merchandise trade balance (Japan-United States)	58.48
4 (1)/(2)	1.15	4 (1)/(2)	2.22

Source: U.S. Department of Commerce.

each case, multinational investors in China pressured their home government to moderate its actions in order to minimize commercial damage. Since 1993, U.S. exports to China and its rapidly-growing business ties became part of the political calculus in Washington over renewal of China's MFN status. The range of direct investment had expanded beyond production ventures into services including computer software, accounting, insurance, consulting, and marketing. It was estimated in 1993 that removing the MFN status might jeopardize 171,000 U.S. jobs if China were to retaliate (Dunne, 1993).

CANADA AND CHINA

CHINA HAS LONG BEEN A PRIORITY MARKET for Canadian exports, particularly of natural resources. As recently as 1988, non-cereal exports accounted for 35 percent of total exports to China and manufacturing exports for only 3.5 percent. By 1994, these shares had risen to 71 and 32 percent, respectively. Yet these exports accounted for less than half the China's average imports from the rest of the world in 1985 (Falkenheim, 1995). Canadian FDI in China in 1993 was less than 1 percent of China's total stock. Direct investment in a target market is becoming increasingly important to promote trade, but the awareness of this among Canadian policymakers is very recent. In the case of China, it is especially important because of the turbulence of the Chinese business environment. There is competition among different levels of government, particularly among industrial ministries, most of which are directly involved in running businesses. The domestic market is still protected. To do business there, one must invest. Canadians are latecomers and their business presence is weak. There are several reasons for this, most of them shortcomings on Canada's part, such as inadequate financing facilities, lack of aggressiveness by the private sector, and the predominance of small- and medium-sized enterprises (SME) among Canadian companies doing business abroad (many subsidiaries of multinational enterprises (MNE) are prohibited from acquiring an international presence).

In summary, Canada's economic ties with China and Japan continue to be dominated by traditional natural-resource-based products. While all three giants of the Asia Pacific – the United States, Japan, and China – are key to Canada's objectives, Canada is seen as insignificant to theirs. This is reinforced by Canada's adoption of a bilateral policy focus, instead of a regional one in which synergies can be exploited. Changes are needed that proactively build a Canadian policy commitment and business presence that is perceived to be comparable to that provided by NAFTA.

IMPLICATIONS AND CONCLUSIONS

FOR AT LEAST THE NEXT DECADE, the United States' economic relations with Japan and China will be important determinants of international political and economic stability. Ideally, the international adjustment process will open Japanese markets in response to market forces, and structural policies will bring about single prices in domestic and international markets. Opening the Japanese market to greater import penetration will create alternative markets for Asia's exporters, particularly China. Deficit reduction in the United States will reduce its demands on world savings and its attractiveness as an export destination. Conflicts will be taken to the World Trade Organization (WTO) and settled expeditiously.

In the future, four issues will be important: how bilateral relations are conducted within the Pacific triangle of the United States, Japan, and China; these countries' roles in APEC and how they view its potential; their policies toward the rest of Asia; and managing China's emergence.

THE FUTURE OF JAPANESE-U.S. TRADE POLICY

U.S. TRADE POLICY IS LIKELY TO CONTINUE TO BE DRIVEN by realist assumptions about the world and preoccupations with the employment consequences of structural and technological change, and to be conducted in an aggressive, confrontationist style. This is a trajectory toward conflict, a recipe – if not for disaster – for collisions, policy errors, and misunderstandings that will spill over into the rest of the world.

Like all large nations have historically done, the members of the Pacific triangle will insist on dealing bilaterally with each other, especially on issues that are beyond the WTO's jurisdiction or require rapid attention. APEC may act as a buffer because its leaders meet regularly and its officials are working hard to develop an institutional framework. But its reach is limited by its determinedly lean institutional structure.

There is a need for a fresh approach to the U.S.-Japan bilateral trade relationship. Both sides were discredited in the 1995 auto dispute. Japanese officials and politicians vowed to resist renewal of the 1991 Semiconductor Trade Agreement in 1996. There are signs that the United States' actions toward China are becoming a ritual – identify the irritant in the fall and threaten sanctions in the spring (human rights annually, intellectual property (IP) in 1994 and IP and the trade balance in 1995).

Thus the stage is set for continued tensions. There must be a better way. Greenwald (1996) demonstrates that sectoral and systemic differences can be moderated on the basis of objective appraisals and the rule of law rather than by purely political means. A dispute settlement mechanism (DSM) should be set up by both governments to provide expert panels like those used in the Canada-U.S. Free Trade Agreement (FTA) to address bilateral disputes. Panel membership should be limited to three, one each drawn from rosters of independent experts by each of the two countries and the third (the chairman) an expert from a third country.

In an ideal world, bilateral disputes would be addressed just like other disputes. They would be referred to the WTO. But increasingly, U.S.-Japan differences are over structural issues, anti-competitive behavior and restrictive business practices, which WTO provisions do not cover. These issues could, however, be addressed by expert dispute panels of the kind now used in the FTA and NAFTA. Greenwald identifies several areas where objective panels could play a role in settling disputes in the absence of a free trade agreement. He outlines a wide range of issues that could have been addressed by DSMs, but that were dealt with by political means, thereby contributing to tension and rancor.

In the absence of a free trade agreement, the key issue is the terms of reference for such panels. Three ways of developing terms of reference can be envisaged. First, as Greenwald points out, since there are more than 30 sectoral agreements, the terms of those agreements can be subjected to adjudication by panels. Greenwald identifies contentious sectoral issues such as negotiating provisions for the Nippon Telephone and Telegragh (NTT) procurement as well as the commitments and status of the so-called "secret" side letter in the Semiconductor Trade Agreement. These issues could have been delegated to a DSM. With sufficient expertise, such a panel might have produced creative reports that would either settle a legal issue or provide an objective basis for a political resolution, much as happened in a Canada-U.S. dispute over salmon and herring.

The second way is to evaluate whether domestic laws and regulations are being effectively enforced, as was done in NAFTA. Using the examples of such contentious issues as flat glass and paper products, Greenwald demonstrates how the positions of the two governments on enforcement could be subjected to expert review that might even lead to cooperation between the two governments in effective and non-discriminatory enforcement of anti-monopoly laws.

The third way is in areas such as anti-competitive behavior, foreign investment, and distribution where governments could reach agreement on rules or principles where common laws or regulations do not yet exist. Greenwald's examples include a DSM to promote agreement on strengthening Japan's antitrust law and its enforcement. Since informal barriers inhibit foreign investment, a DSM might also accelerate the implementation of plurilateral initiatives such as the OECD standards of liberalization and investment protection. A U.S.-Japan dispute oversight body should be established to ensure that the panel system is properly constituted and administered. Such a body would parallel the oversight structure in the Canada-U.S. Free Trade Agreement.

Why should the United States be interested? It is important to recognize that the United States is likely to view such a proposal as yet another way for Japanese bureaucrats to slow down the market-liberalization process. We believe, however, that this risk is offset by the potential benefits. First, the Japan-U.S. relationship is in danger of eroding under unremitting pressure for quantitative targets, because there is widespread public support in Japan to say "no." Ways must be found to depoliticize the handling of disputes. Expert third-party opinion and judgment is itself a potentially effective and very valuable substitute or even complement. As Goldstein (1994) suggests, the Canada-U.S. FTA dispute settlement mechanism has helped to restore presidential flexibility vis-à-vis Congress and within the executive branch itself, in the direction of safeguarding free trade and foreign policy considerations.

Why should Japan be interested? The legal systems in the two countries are very different. The proposed DSM would function in a depoliticized and non-litigious manner, which is more congruent with Japanese traditions. Other Asian countries also strongly prefer to settle disputes in a non-litigious manner. The Japanese political culture is such that people are generally reluctant to file legal suits to avoid conflict and maintain harmony. By contrast, in the United States and Canada, economic disputes have tended to be channelled into the courts, to be resolved by common law, civil codes, and statutory principles of business law. Both countries have tended to make greater use of the GATT dispute settlement mechanisms.

APEC

THE MITI WAS ONE OF THE ORIGINAL CHAMPIONS OF APEC. As the cold war ended and worries mounted about possible failure of the Uruguay Round, a pan-Pacific institution was seen to be desirable as a focus for U.S. interests in Asia. Today, APEC has the potential to be a mechanism within which Japan can link its regional thrust to build an Asian community and its desire to help strengthen global economic governance systems (Funabashi, 1995). Yet in the wake of the Osaka leaders' meeting, at which Asian voluntarism clashed once again with a U.S.-led push for commitments and deadlines to create a free trade area, Japan appears to be destined to play the role of APEC middleman rather than leader.

One of the major fears among Asian leaders is that U.S. unilateralism toward Japan will become a general policy toward other successful exporters in the rest of the region, and particularly toward China. APEC members should be invited and encouraged to observe the deliberations of the proposed Japan-U.S. DSM. As noted earlier, multinational firms that invested in China have acted at times to moderate U.S. bilateral policies and are active in alliances in the Asian region to influence national policies through their local affiliates (Richardson, 1995).

APEC members have found the idea of a dispute mediation service attractive enough to accept recommendations in 1995 from the Eminent Persons Group (EPG) for such a mechanism to be set up as soon as possible to help deal with trade conflicts in the region. The EPG makes two arguments for such a service. First, it recognizes that a number of trade problems relating to competition, investment, government procurement and environmental policies lie outside the scope of WTO rules. Second, it recognizes the incongruence to Asians of the WTO's highly legalistic procedures and the "win-lose" confrontations inherent in arbitration. The proposal offers the intermediate route of mediation. All interested parties in a dispute can then come together voluntarily with a mediator as catalyst for a settlement (APEC, 1995, p. 11). This proposal for mediation is an important adaptation of western legalistic procedures in recognition of the importance of "saving face" in Asian relationships.

If mediation fails to settle a dispute, the dispute could be presented to a special review panel organized by the service. As Dobson and Sato (1996) propose, the review panel could make an expert objective assessment based upon the covered bilateral or multilateral agreement. The disputants would then follow the expert assessment on a voluntary basis.

The attraction of these ideas is based on the assumption that growing economic interdependence through webs of trade and financial flows is an irreversible phenomenon and that economic conflicts will damage the interests of both protagonists and innocent bystanders. By contrast, in realists' views of the world, national interests and power relationships play a larger role.

Policies toward Asia

JAPAN-U.S. TENSIONS ARE LIKELY TO EXTEND to the East Asian economies. Up to now, Japanese policy toward Asia has reflected U.S. priorities. Today, however, Japan is pursuing with some success a more independent, though complementary, thrust. But the Japanese economic presence in the Asian economies, through aid flows and large and growing flows of direct and portfolio investment, is increasing in several key manufacturing industries such as electrical and electronics and autos.[8] Although U.S. investment flows, which declined in the 1980s, are picking up again, and although U.S. auto interests largely ignored East Asian markets outside Japan until very recently (Dobson, 1993), realist views promote the idea of rivalry and use of market opening measures beyond Japan.

Another factor is Japan's asymmetry with the Asian economies. Trade balances are in surplus although Japanese data show imports picking up, and FDI flows into Japan are minimal. While Asians stand to benefit from the U.S. opening the market on an MFN basis, they are also ambivalent, fearing they may become targets in future trade conflicts.

China

CHINA'S GROWING ECONOMIC CLOUT, different economic institutions, unique view of its place in the world, and historical predisposition to use force in settling disputes contribute to uncertainty in the international community. Yet its growing external deficits with the United States are putting the two countries on a collision course. Diverse, often conflicting, U.S. policy initiatives are contributing to the problems.

What is the best policy stance to adopt toward China? Should it be contained? Should it be integrated? The answer is clear to the economist. China should become integrated into the world economy by means of nurturing trade and investment ties, encouraging the freeing up of market forces, and reinforcing in positive ways its sense of international responsibility. As China's economic integration proceeds, military conflict with its neighbours would amount to shooting itself in the foot. Japan is more integrated into the international system, but China is likely to share the suspicions of some of its neighbours about the United States' motives and interest in APEC and to be more defiant in its bilateral dealings with the United States. Clearly, China will be a force to which the international system will have to adjust. Adjustment on the part of the United States is unlikely to be smooth, however.

Implications for Canada

CANADA NEEDS TO BECOME MORE PROACTIVE in identifying its own policy agenda for the region. Such an agenda begins with addressing the question of whether we are a Pacific nation, and, presuming we are, following through with a greater policy focus on bilateral relations in a regional framework.

Canada is invisible. It is not taken seriously. Yet Australia and New Zealand, which also have been perceived to be frittering away their magnificent birthrights, are gaining notice and respect from their Asian neighbors because they have redefined their trade and foreign policy thrusts to accelerate their integration into the region. Canada should do the same.

A clearer policy focus should begin with measures that assist Canadian business to penetrate Asian markets as latecomers. Attention should be paid to reducing information barriers and the fixed costs of entry. This can be done through alliances among large and small firms and Canada's large ethnic population. Efforts should focus on Canadian industries that are internationally competitive. Better data are required to monitor progress, especially in the service sector. The competitiveness of Canadian firms in these dynamic but turbulent economies, once established, can then become the basis for Canadian initiatives to reduce trade barriers.

Finally, Canada should take more seriously the facts that Japan and the United States are its two largest trading partners and U.S.-China conflicts are an important and growing threat. Conflicts on the Big Three triangle (U.S.-Japan and U.S.-China) will affect Canada's trade and other economic interests in the Asian region. Canada's external economic policies should build more systematically on its links with all three neighbors in order to buffer and depoliticize the inevitable bilateral disputes that lie ahead.

Endnotes

1   Research assistance by Birgitta Weitz is gratefully acknowledged.
2   This point was made in a July 1995 speech in Tokyo by Commerce Under Secretary Jeffrey Garten. He was subsequently rebuked.
3   Japanese FDI is reported as accumulated FDI based on notifications, which overstate the size of the stock because such investments are not necessarily undertaken. These data also exclude disinvestment.
4   None of these results was statistically significant, though.
5   The exceptions are China where the decline has been rapid and the Philippines where there has been an increase. Both these changes are due to political reasons.
6   Estimates of inward FDI must be tempered by the knowledge that possibly significant amounts of FDI are domestic resources channelled through Hong Kong to take advantage of incentives available only to foreign capital.
7   Pointed out by T.J. Pempel.

8   FDI stocks from both sources are larger in the non-manufacturing, however.

BIBLIOGRAPHY

APEC. *EPG Report: Implementing the APEC Vision*. Singapore: APEC, 1995.
Bell, Michael W. et al. "China at the Threshold of a Market Economy." Occasional Paper No. 107. Washington, DC: International Monetary Fund, 1993.
Bergsten, C. Fred and Marcus Noland. *Reconcilable Differences?* Washington, DC: Institute for International Economics, 1993.
Canada-Japan Forum 2000. "Partnership Across the Pacific." Uunpublished manuscript, 1995.
Dobson, Wendy and D. Chen. "Foreign Investment in China." University of Toronto (unpublished manuscript).
_____. *Japanese Trading and Investment Strategies in East Asia*. Singapore: ISEAS, 1993.
Dobson, Wendy and Hideo Sato (eds). *Managing US-Japanese Trade Disputes: Are There Better Ways?* Ottawa: Centre for Trade Policy and Law, 1996.
Dunne, Nancy. "Clinton's $7 Billion Dilemma on China." *Financial Times of London*, 20 May 1993.
Falkenheim, Victor C. "The China Market: Dancing With A Giant," in *Benchmarking the Canadian Business Presence in East Asia*. Edited by A.E. Safarian and Wendy Dobson. Toronto: University of Toronto, 1995.
Fallows, James. *More like Us*. New York: The Wendy Weil Agency, 1989.
Frankel, Jeffrey A. "Is Japan Creating a Yen Bloc in East Asia and the Pacific?" in *Regionalism and Rivalry*. Edited by Jeffrey A. Frankel and Miles Kahler. Chicago: The University of Chicago Press, 1993.
Funabashi, Yoichi. *Asia Pacific Fusion: Japan's Role in APEC*. Washington, DC: Institute for International Economics, 1995.
Goldstein, Judith. "International Law and Domestic Institutions: Reconciling US-Canadian 'Unfair' Trade Laws." Unpublished manuscript. Stanford University, 1994.
Greenwald, Joseph A. "Binational Dispute Settlement Mechanisms," in *Managing US-Japanese Trade Disputes: Are There Better Ways?* Edited by Wendy Dobson and Hideo Sato. Ottawa: Centre for Trade Policy and Law, 1996.
Holberton, Simon. "Beijing Protests over US Claims of $30 billion Trade Deficit." *Financial Times*. (October 9, 1995).
International Monetary Fund. *World Economic Outlook*. Washington, DC: IMF. (May 1995a).
_____. *World Economic Outlook*. Washington, DC: IMF. (October 1995b).
Japan, Ministry of Finance. "Zaisi Kinyu Tokei Geppo." June 1995.
Johnson, Chalmers. "Atarashii Shihonshugi no Hakken" ("A Discovery of New Capitalism"). *Leviathan*, 1,1 (1995).
Kitamura, Hiroshi. "The Position of Canada in Japan's Diplomatic Framework." *Pacific Affairs*, 64 (1991):226-34.
Krugman, Paul R. "Has the Adjustment Process Worked?" *Policy Analyses in International Economics*. Washington, DC: Institute for International Economics, 1991.

Lawrence, Robert Z. "US Trade and Investment Priorities: Asia's Place," in *Pacific Trade and Investment: Options for the '90s*. Edited by Wendy Dobson and Frank Flatters. Kingston: Queen's University, 1994.

Miyoshi, Masaya. "Deregulation in Japan: The Road Ahead." *Nikkei Weekly*. (June 26, 1995).

Mochizuki, Mike. "Japan Sees New Opportunities in Asia." *JEI Report*. 23A (June 23, 1995). Washington: Japan Economic Institute.

Nakamoto, Michiyo. "Red Tape Stops Benefit of Strong Yen Trickling Down." *Financial Times*. (July 1, 1995).

Petri, Peter. "The East Asian Trading Bloc: An Analytical History," in *Regionalism and Rivalry*. Edited by Jeffrey A. Frankel and Miles Kahler. Chicago: The University of Chicago Press, 1993.

Prestowitz, Clyde V., Jr. *Trading Places: How We Allowed Japan To Take the Lead*. New York: Basic Books, 1988.

Rapp, William V. "Capturing Japan's Attention: Canada's Evolving Economic Relationship with Japan," in *Benchmarking the Canadian Business Presence in East Asia*. Edited by A.E. Safarian and Wendy Dobson. Toronto: University of Toronto, 1995.

Richardson, Neil R. "Trade Blocs and the U.S.-Japan Relationship." Unpublished manuscript. University of Wisconsin-Madison, 1995.

Ries, John and Keith Head. "Canada's Business Presence in East Asia," in *Benchmarking the Canadian Business Presence in East Asia*. Edited by A.E. Safarian and Wendy Dobson. Toronto: University of Toronto, 1995.

Safarian, A.E. and Wendy Dobson (eds). *Benchmarking the Canadian Business Presence in East Asia*. Toronto: University of Toronto, 1995.

Sato, Hideo. "Introduction and Overview," in *Managing US-Japanese Trade Disputes: Are There Better Ways?* Edited by Wendy Dobson and Hideo Sato. Ottawa: Centre for Trade Policy and Law, 1996.

Tokyo Business. "Gaishi 300." 1994.

United States, Department of Commerce. *Survey of Current Business*. Washington, June 1995.

Ursacki, Terry and Ilan Vertinski. "Canada-Japan Trade in an Asia-Pacific Context." *Pacific Affairs*, 1995.

van Wolfren, Karel. "The Japan Problem Revisited." *Foreign Affairs*, 69,4 (1990).

Walker, Tony. "The Doubt Behind the Angry Mask." *Financial Times*. (August 25, 1995).

Wolf, Martin. "High Noon in the Pacific." *Financial Times*. (June 26, 1995).

Comment

Masao Nakamura
Faculty of Commerce and Business Administration
University of British Columbia

PROFESSOR DOBSON PRESENTS A COMPREHENSIVE VIEW of the current issues surrounding U.S. bilateral relationships with Japan and China and their implications for Canada. She also suggests a role that Canada might be able to play to improve U.S. relationships with Asian countries.

Dobson emphasizes the politico-economic aspect of these issues. This is quite appropriate, since most of the bilateral economic disputes discussed in her paper involve domestic stakeholders. Certainly in the bilateral disputes between the United States and Japan the economic issues are intermingled with political ones. Today, unlike during the cold war, mutual security concerns do not ensure that economic disputes are held in check and do not blow up into major domestic political confrontations.

Some of the issues Dobson considers important from a Canadian perspective are the following:

- Because of growing economic interdependence, bilateral tensions and conflict between the United States and Japan may spill over to affect their major trading partners, including Canada.

- Policies towards Japan may be extended to other successful Asian exporters, particularly China. Canada's interests could suffer as a result.[1]

Dobson does not specify the ways Canada might suffer. However, there are several actual situations where Canadian or other non-U.S. firms were disadvantaged in comparison with U.S. firms in securing Japanese firms' procurement. In each of these cases, Japanese firms got more credit with the Japanese government for buying from U.S. firms than from non-U.S. firms.

Dobson goes on to analyze the political economy of a number of bilateral issues between Japan and the United States. The first is Japan's trade imbalance with the United States at the macroeconomic level, and the second is the market access problem faced by non-Japanese firms. This market access problem is thought to be caused by invisible barriers attributed to certain Japanese business practices. She correctly points out that these two issues are not connected logically.

She believes the high-level rhetoric and tension of the U.S.-Japan negotiations is unnecessary. She argues (as do many others) that market realities were already forcing Japanese manufacturers to do what they committed to do in their recent agreement with the United States – namely, to import more from the United States and further internationalize their own production. Arguing that the costs associated with these negotiations are high relative to their benefits, she sees a role for Canada in helping the two countries address their differences in a more timely and less costly fashion.

She suggests that a dispute settlement mechanism of the kind used in the Canada-U.S. Free Trade Agreement be set up to address bilateral Japan-United States disputes, and that Canada help with its establishment. I think this is a good idea. However, it would require mutual understanding of the differences between the countries' economies and the legal and political systems of the countries. These differences are matters of degree, and not of kind, since Japan and the United States share essential democratic values.

It is true that Japan could learn from Canada's experiences in dealing with the United States. Relatively little is known in Japan about the mechanisms through which Canada and the United States resolve their disputes, which range from lumber, automobile, and farm product trade issues to the management of ocean salmon.

Nevertheless, Canada should be cautious in presenting itself as a completely impartial party in bilateral disputes involving the United States and Asian countries. The fact is that Canada's economic and political interests are often aligned quite closely with those of the United States.[2] But they do have their differences, and Canada has demonstrated skills in dealing with the United States.

I am not sure that Canada is invisible in its trade relationships with Japan or with China, as Dobson claims in her paper. It seems more likely that in Asia Canada is perceived, perhaps incorrectly, to be an extension of the United States. In many cases this perception has worked in the favour of Canadian firms that want to do business there.

Finally, Dobson emphasizes the potential implications for Canada of Asian regional trade blocks. Trade among Asian countries has been growing much faster than that among Western developed countries. Nevertheless, the magnitude of the former is still considerably smaller than the latter.

Japan has had a significant impact on the thinking and development of free trade zones in Asia. First, the fact that Japan has been having difficulties with the United States has alerted other Asian countries to the possibility that they might face the same kinds of difficulties if they become overly dependent on U.S. markets. China, Singapore, Malaysia, Taiwan, and South Korea, in particular, are fully aware of the nature of Japan's experiences in its trade negotiations with the United States.

Japanese firms have been playing an important role in integrating Asian production activities. In this regard, it would be interesting to compare the roles of different geographical areas as markets for the products of Japanese firms. An analysis such as the one that follows would give us a deeper understanding of the structural basis for geographical zoning in economic activities.

North America and Western Europe are large and sophisticated markets for Japanese products. Most Japanese FDI in these regions, except for resource-based FDI, is set up to serve local markets. Less than 5 percent of the output from Japanese firms' North American operations is exported back to Japan or to third countries. In contrast, more than 40 percent of the output from Japanese firms' operations in Asia is exported back to Japan or to third countries.[3]

There is also a considerable flow of technology from Japan to other Asian countries. In 1993, Japan's technology exports to other Asian countries amounted to US$1.9 billion. In contrast, it paid about US$2.6 billion to the

United States for licensing and other fees and received about US$1.2 billion from the United States for technology exports, a technology trade deficit of US$1.4 billion. Similarly, it paid about US$1 billion to Europe for technologies in 1993 and received about US$0.7 billion for technology exports, a technology trade deficit of US$0.3 billion. However, these technology trade deficits for Japan were more than made up by Japan's technology exports to other Asian countries.

A significant portion of Japan's technology exports to Asia were licensed to Japanese-affiliated operations. There are several possible reasons for the considerably higher profitability of Japanese operations in Asia than in North America and Europe.[4] One is the successful implementation of proven production methods and technologies. In addition, geographical proximity and the abundance of trainable labour provide Japanese firms with the flexibility for deploying their global strategies.

It is not clear to what extent cultural and other non-economic factors are responsible for Japanese firms' success in Asia. A number of factors may have facilitated implementation of some, if not all, Japanese business practices there. For example, the absence of serious labour movements and the lack of established interfirm relationships in most newly developed and developing economies may make it easier, or cheaper, for Japanese firms to implement the management practices best suited for Japanese production methods. In contrast, in North America successful Japanese manufacturing operations tend to be limited to those which have adopted the greenfield approach.[5] Moreover, even on a greenfield, Japanese methods are generally expensive to implement in North America.

So, where does Canada fit in in Asia? Most Canadian firms seem to find it advantageous to concentrate their business activities in Canada and the United States, which enables firms to generate scale economies. The potential for high growth opportunities in Asia has not changed many Canadian firms' calculations. It is not likely that Canadian firms face higher risks or more problems in Asia than do firms from other countries. We should also keep in mind that the North American market is risky for Japanese firms. In recent years, about 70 percent of Japanese-affiliated manufacturing firms that were set up in the United States and Canada did not survive their first 15 years of operation. (Survival rates for Japanese affiliates are considerably better in Asia, about 50 percent.[6]) Perhaps the connections Canadian firms have established in Canada with firms from Japan and the United States could be carried over to the latters' Asian locations.

Canadian firms could also take advantage of incentives offered by foreign governments. Even Japan, which traditionally discouraged foreign firms from entering its market, is now in the global race for foreign direct investment! Japan now offers tax, financial, and other incentives to foreign companies' operations in Japan.[7] Canadian firms might also implement new production methods and technologies that allow them to achieve scale economies while producing many differentiated but related products in smaller quantities. These production methods, sometimes referred to as the "Toyota production system," are well suited for Asian markets, where products that are specifically tailored to the market segments are required.

ENDNOTES

1. It is likely that as early as 1996 or 1997 China will take over from Japan as the country with which the United States registers the largest bilateral trade deficit. This will undoubtedly attract the attention of U.S. trade policymakers. So far the United States has had little impact on Chinese behaviour towards international trade. For example, despite formal, government-level agreements between the United States and China, Chinese infringements of U.S. intellectual property rights seem to continue.
2. Historically, Canada has often followed the United States' lead where trade restrictions against Japan are concerned. These restrictions include the grain export ban under the Nixon administration and voluntary export restraint agreements for Japanese autos.
3. See, for example, M. Nakamura and I. Vertinsky, *Japanese Economic Policies and Growth: Implications for Businesses in Canada and North America*, University of Alberta Press, 1994.
4. See Nakamura and Vertinsky (1994) and also M. Nakamura, "Japanese Direct Investment in Asia-Pacific and Other Regions: Empirical Analysis Using MITI Survey Data," *International Journal of Production Economics*, 25 (1991): 219-29.
5. For analysis of issues related to Japanese management practices, see M. Nakamura, "Japanese Industrial Relations in an International Business Environment," *North American Journal of Economics and Finance*, 4 (1993): 225-51.
6. Survival rates for Japanese foreign direct investments and United States-Japan joint ventures located in Japan are discussed, respectively, in *Japanese Overseas Investment '93*, Tokyo: Toyo Keizai Shimposha, 1993, and M. Nakamura, M. Shaver, and B. Yeung, "An Empirical Investigation of Joint Venture Dynamics: Evidence from United States-Japan Joint Ventures," *International Journal of Industrial Organization*, forthcoming.
7. The Japanese government established the Japan Investment Council in 1994 to promote foreign direct investment in Japan. For example, the Japan Development Bank offers low interest loan programs to attract foreign corporations to Japan. For a detailed description of this Council and incentive measures for FDI, see the JETRO section of the Japanese government's page on the World Wide Web.

Murray G. Smith
Norman Paterson School of International Affairs
Carleton University

Canadian Trade and Investment Policies and the Asia Pacific Region: Confronting Ambivalence?

13

THIS PAPER EXAMINES THE BARRIERS TO TRADE and investment that affect Canada's economic links with the Asia Pacific region, and explores the options for expanding these economic linkages. In section one I first review the post-NAFTA, post-Uruguay-Round agenda for trade and investment policies, including the implementation and evolution of the World Trade Organization and the Uruguay Round agreements. Second, I examine the competitive regional liberalization agendas, including the overlapping initiatives in the Asia Pacific region and Western Hemisphere after the Bogor and Miami Summits, the NAFTA expansion process, the enlargement of the European Union, the negotiation on investment at the OECD, and the transatlantic initiatives.[1]

In the second section, I examine the Canadian trade barriers and investment policies as they apply to trade and investment links with the Asia Pacific region. In particular, the special regimes for the agricultural and textiles and apparel trades and the application of anti-dumping duties to Asia Pacific trade will be examined. This section draws on all the sources that are available to measure the non-tariff barriers affecting Canadian trade, but there are gaps in the data on the incidence of these measures on Canadian imports from non-NAFTA countries.

In the third section, I examine the trade and investment barriers and impediments in the Asia Pacific region that affect Canada's trade and investment links with the region. This section relies upon published sources including the GATT-WTO reports under the Trade Policy Review Mechanism and the PECC/APEC study of trade and investment impediments.

In the fourth section, I examine a range of policy options that may enable Canada to expand trade and investment links with the region. The interaction between APEC initiatives, other regional initiatives and multilateral initiatives are explicitly examined. The options for Canada are assessed critically in light of some of the obstacles to APEC as evidenced in the objections to making the APEC investment declaration binding. Is *open regionalism* a basis for liberalizing dynamic and diverse economies, or is it a veil for trade and investment regimes that lack transparency?

A more basic question is, what are Canada's economic interests in the region, apart from the trade and economic relationship with the United States? Since the implementation of the Canada-U.S. Free Trade Agreement and the North American Free Trade Agreement (NAFTA), the share of Canadian exports going to the United States has increased to about 80 percent, and the most rapid growth has been in non-automobile, manufactured goods. Does it follow from this fact that Canada should ignore economic and policy developments in the rest of the Asia Pacific region and focus on NAFTA?

The answer is no; one should be careful of fallacies in analyzing bilateral trade and economic issues. First, Canada may be competing with other countries in the Asia Pacific region for the U.S. market or for investment. Second, Canada trades with the Asia Pacific region through the United States. Canadian products may be transshipped through the port of Seattle or the Los Angeles airport, for example. More significantly, Canadian components may be sourced by U.S. multinational enterprises (MNE). Undoubtedly the FTA and NAFTA have intensified Canadian participation in U.S.-based MNE production networks, thus increasing Canada's indirect economic interaction with other countries in the Western Hemisphere and the Asia Pacific region. Thus what appear to be bilateral trade flows and investment links can be deceptive in an increasingly globalized world economy. Canada has significant economic interests in the economic and policy developments in the Asia Pacific region. Futhermore, the integration of the Canadian economy into U.S. MNE production networks – which was one of the goals of the FTA and the NAFTA – and the substantial increase in indirect interaction with other Asia Pacific economies, raises particular policy issues for Canada in its effort to promote integration in the Asia Pacific region.

GLOBALIZATION AND THE ASIA PACIFIC

THE INTERPENETRATION OF NATIONAL ECONOMIES by trade and direct investment, or "globalization," has been a consistent trend since the Second World War and has accelerated during the 1980s and the 1990s. The globalization that took place in the 1980s, as popularized by Kenichi Ohmae, Michael Porter and Robert Reich, was conceived as a Triad phenomenon.[2] In a widening range of industries, largely those based in OECD countries, multinational enterprises needed to operate in the U.S., European and Japanese markets in order to remain competitive. The OECD was an increasingly (although not completely) borderless economy. The GATT trading system focused trade liberalization among the members of the OECD club, and the OECD did have some nonbinding codes for investment.[3]

A detail missing from this broad-brush portrait of globalization is that Japan was not integrated with respect to investment to the same degree that North America and Europe were, as characterized by large and rapidly expanding two-way direct investment links. Japanese foreign direct investment abroad was modest until the 1980s, and the level of foreign direct investment in Japan is still low compared to other OECD economies.[4]

Most "developing" countries were little touched by globalization, at least until very recently. Consider the case of India, which was a founding member of the GATT, but which used its diplomatic skill and knowledge of the GATT to maintain a very restrictive, even autarchic trade, investment, and payments policy. Of course this diplomatic victory – an exercise of sovereignty – was realized at great cost to the Indian economy and to the living standards of Indian citizens. India's share of world trade has stagnated, and the share of trade in GDP is among the lowest in the world. India was a *de jure* member of the trading system, but it was not a *de facto* member.[5]

Over the last two or three decades, outside the OECD region, the countries that did not conform to the norm of restrictive trade and payments regimes and relative economic decline were the Asian Newly Industrializing Economies (NIE). Although there were significant differences among their policies – Hong Kong, for example, pursued a *laissez-faire* nineteenth century model while other dynamic Asian tigers were more interventionist – in general they pursued outward-oriented strategies.[6] Their selective interventions were geared to promoting the industries that were most successful in export markets, not in backing declining industries.[7] All had pro-export biases with stable macroeconomic regimes, relatively low tax shares of GDP, public expenditures focused on education and infrastructure, and high savings rates. All have opened their economies to international trade and investment.

The combination of the success of the Asian NIE and the failure of import substitution policies that was reflected in the debt crisis had a persuasive effect on many of the developing countries. As the 1980s turned into the 1990s, they and former socialist economies worldwide engaged in competitive liberalization. Starting with Mexico in the mid-1980s, the rejection of the *dependencia* model and the swing to neoliberalism spread through much of Latin America. As an indication that policy in these areas matters, in 1980 Venezuela had greater absolute exports than Mexico, but by 1989 Mexican exports were three times Venezuelan exports.

Perhaps the most dramatic development of the late 20th century is the reintegration of the centrally planned economies into the world economy. This process started very gradually in the late 1970s with China's open-door policy, which has been incremental but without major reversals ever since. The fall of the Berlin Wall and the collapse of the Soviet Union were very dramatic political events with significant long-run economic consequences.

Presently, China is striding while the former Soviet Union is stumbling back into the world economy.

There is a sharp contrast between China and the former Soviet Union's reintegration into the global economy. In both cases there was much untapped potential in the form of underutilization of human and natural resources, forced underconsumption and grave misallocation of savings and capital investment in their centrally planned systems. But in China, the opening of the economy has unleashed a remarkable period of sustained economic growth, whereas in Russia and the other former Soviet republics it has unleashed a severe economic decline and the implosion of the state. Undoubtedly, this contrast between the two regions' economic performances in the transition will be debated. In the short run, Russia has emulated the statist hyperinflation of Latin American economies of previous decades rather than the Asian NIE. China is not a *de jure* member of the GATT-based trading system, however, it is *de facto* more integrated into the global economy than India is.[8]

Why has the GATT system fostered integration/globalization among the OECD countries' economies, but not the developing countries' economies? Why is India less integrated into the world economy than China? The historical answer is that, starting in the 1950s, at the height of the cold war and anticipating the wave of decolonization, the key players in the GATT accepted the rhetorical plea from developing countries for special and differential treatment.

The Asia Pacific countries were the first to join the OECD countries in reducing barriers to trade and investment, but they have reduced barriers on a unilateral basis rather than through reciprocal negotiations, which have been the norm for the OECD countries.[9] The unilateral and competitive liberalization of trade and investment policies has spread to Latin America and even the former centrally planned economies. What does this mean for the Asia Pacific region? Will the Asia-Pacific economies continue to lead the globalization of globalization? Were the Asia Pacific countries, including the developing ones, not strong proponents of the multilateral system and supporters of the successful conclusion of the Uruguay Round? Before attempting to answer these questions, let us review briefly the outcome of the Uruguay Round and explore the linkages between it and the globalization process.

An Overview of the Uruguay Round and the World Trade Organization

THE RESULTS OF THE URUGUAY ROUND ARE EMBODIED in a single agreement under the auspices of the World Trade Organization (WTO). This single undertaking encompasses the GATT 1994, which replaces the GATT 1947 and 12 separate agreements pertaining to trade in goods as well as the General

Agreement on Trade in Services (GATS) and the new rules for intellectual property rights protection. All these agreements are subject to a single, integrated dispute settlement mechanism.

Membership in the WTO requires negotiation of market access schedules for manufactured goods, agricultural products, and services. The prohibition of quantitative restrictions is strengthened and existing quantitative restrictions, such as those on agricultural products, are to be replaced by tariff equivalents, which are to be gradually reduced over time.

What does this mean for the conduct of trade or investment decisions? This new legal structure represents a complete overhaul of the multilateral trading system. For example, the concept of special and differential treatment, which has been a key element of the GATT since the 1950s, will be largely phased out in five to seven years. This is a significant change in the global trading system. Until now, the major obligations for binding tariffs with respect to market access, national treatment for imported products, the prohibition on quantitative restrictions, and all the other elements of the multilateral trading rules have only applied to the small group of developed countries, primarily the members of the OECD. Many developing countries have been formal participants in the GATT system, but in practice the developed countries have had obligations to them, while developing countries have had few effective obligations to the developed countries.

Although there are still some elements of special and differential treatment, when the WTO is implemented and the obligations are phased in over the next decade, the trading rules will be truly global and apply to many developing economies.

China and the republics of the former Soviet Union are now renewing their participation in the GATT-WTO system. In one way or another, all are seeking membership in the World Trade Organization. As this process unfolds, in the context of the World Trade Organization there appears to be a substantial consolidation and reinforcement of this very significant process of competitive liberalization.

Undoubtedly the exemplary effects and competitive challenges from the Asian NIE helped stimulate competitive liberalization in every region of the world. Yet the liberalization of the NIE was largely unilateral (bilateral disputes with the United States may have accelerated the process in a number of Asian countries). Among the Asian economies, only Japan has participated fully in reciprocal trade negotiations in the GATT, before the Uruguay Round. (The liberalization of the New Zealand and Australian economies during the 1980s was also largely unilateral.)

As a result, there was considerable ambivalence and dissonance in the policy positions of most of the East Asian countries during the Uruguay Round. At the rhetorical level they strongly supported a successful conclusion to the Uruguay Round, but they were reluctant to bind further liberalization of their industrial sectors and to make extensive offers for liberalization of barriers to trade in sectors such as financial services ((including investment). Of course, Canada joined Japan and Korea in seeking to keep special quotas for import-competing agricultural producers.

The countries of the Asia Pacific region and even within the APEC are a very diverse group economically. This diversity in their levels of and their approaches to economic development is both a stimulus and a challenge for economic integration, both at the regional and global levels. Moreover, the East Asian approach to economic integration is different from the more formal European and North American approaches.

The Asia Pacific countries, and to some extent the APEC process itself, contributed substantially to the successful conclusion of the Uruguay Round. All the APEC countries, including the East Asian developing countries, are taking on substantial obligations as the agreements under the WTO are phased in.[10] Yet reciprocal bargaining and accepting contractual international obligations about their trade and investment regimes is a novel experience for the East Asian developing countries. Thus the implementation of the WTO will both consolidate the liberalization of the APEC economies and transform the conduct of international economic relations among the East Asian developing countries.

Although the promise of the WTO and the Uruguay Round Agreements is not yet fully realized, the process of competitive liberalization is reflected in various regional and multilateral negotiations. The APEC Eminent Persons Group (EPG) has played an important role in articulating a vision of liberalization in the Asia Pacific region. In November 1994 at the Bogor Summit, the leaders announced a vision of free trade among the developed APEC members by 2010 and among developing members by 2020. In addition, they adopted the Non-Binding Investment Principles (NBIP).

Not to be outdone, the Miami Summit of the Americas, which followed within a few weeks of the Bogor Summit, announced that a Free Trade Area of the Americas (FTAA) will be achieved by 2005. In addition, the three NAFTA members announced that negotiations with Chile for NAFTA accession would be launched. There were ripple effects of these declarations: at the same time that FTAA was announced and NAFTA accession proposed for Chile, the four countries of Mercosur, Brazil, Argentina, Uruguay and Paraguay achieved significant progress toward a customs union.

At the same time developments were underway in Europe. Some of the remaining members of EFTA (already largely replaced by the EEA), Austria, Finland, and Sweden became members of the European Union on January 1, 1995. The European Union also was involved in negotiating bilateral agreements with countries in Eastern Europe and a customs union with Turkey, as well as proposing a regional trading arrangement with the Maghreb countries.

Transatlantic bridges also were proposed. Canada proposed a NAFTA-EU linkage. The Mercosur and the EU discussed some kind of linkage agreement. A transatlantic dialogue was established involving the European Union and the United States and resulted in a business summit in Seville in November 1995 and the EU-U.S. Madrid Summit in December 1995.

Some critical questions could be raised about these initiatives. Will they be compatible with each other and with the multilateral system? How serious are they? Will the regional initiatives achieve deeper integration by dealing with issues beyond the multilateral system? Or will they be essentially broadening agreements dealing with the elimination of tariffs and other border measures through FTAs or customs unions?

A different challenge is posed by the efforts to deepen integration, which is focused on the OECD countries. As a result of U.S. frustrations with the inadequacies of the APEC investment code, discussions are now underway at the OECD for a multilateral arrangement on investment. Negotiations over telecommunications issues are also underway under the General Agreement on Trade in Services (GATS), and there are various proposals on the table to address trade and competition policy and trade and environment issues.

Canadian Barriers to Trade and Investment with the Asia Pacific Region

ALTHOUGH CANADA HAS PARTICIPATED (more or less) in every round of the GATT negotiations, after the Tokyo Round Agreements were fully implemented in 1987, Canadian trade barriers were still relatively high compared to those of most OECD members.[11] In much of the manufacturing sector, the Canadian tariff was about 9 percent, but in certain sectors, such as textiles, clothing and footwear, there were tariffs in the 15-35 percent range as well as quantitative restrictions. The structure of the Canadian barriers to trade and to investment were transformed by the FTA, NAFTA and the Uruguay Round, but those changes are still being implemented and adjusted.

Impact of the FTA and NAFTA

THE OBVIOUS IMPACT OF THE FTA WAS TO ELIMINATE TARIFFS for most trade between Canada and the United States on products that met the FTA rules of origin. Except for the most sensitive industries, which remain on a 10-year phase-out schedule, tariffs are largely eliminated from bilateral Canada-U.S. trade. Most other non-tariff barriers were also removed, with the notable exception of those on agricultural products. With the implementation of the NAFTA, trade barriers between Canada and Mexico will be phased out by 2005, again with exceptions in agricultural products.

Concern has been expressed by non-NAFTA member countries that reducing trade barriers among NAFTA members would divert international trade away from non-NAFTA to NAFTA countries. Mexico's economic and export structure is perceived to be similar to those of other developing countries, especially those of the countries in East Asia.

Sunder Magun has examined the potential for trade diversion between China and Mexico with the existence of NAFTA.[12] The size of the trade diversion depends, in part, on the degree of similarity between Mexican and Chinese exports. It will be significant only if there is a substantial overlap between Mexico and China on a product-by-product basis in the North American market. He analyzed individual commodity categories to identify which product categories of China's and Mexico's exports currently have a high degree of overlap in the U.S. markets. In these product categories, China would probably face the highest competition from Mexico in the U.S. markets. China will face greater competition from Mexico in natural-resource-based products such as petroleum products; products requiring low-skilled, intensive labour such as men's outer garments, children's toys, games and sporting goods, and footwear; and light manufactured goods like motor vehicle parts and accessories, furniture and parts, electrical/mechanical domestic appliances, calculating machines, cash registers, lighting fixtures and fittings, taps, cocks, valves, and radio broadcasting receivers.

Magun also conducted a similar analysis of individual commodity categories in order to identify the export product categories of China and Mexico with a high degree of overlap in Canadian markets. China will face greater competition in such products as television receivers, motor vehicle parts and accessories, carpets and rugs, children's toys, electrical/mechanical domestic appliances, glassware, telecommunication products and in the manufacture of base metals.

Although the overlap in production and trade by industry has an impact on the potential for trade diversion, the height of the external barriers to trade imposed by the NAFTA countries is a critical factor determining whether trade is actually diverted. As the discussion in the next section indicates, there are significant barriers to trade between Canada and countries in the Asia Pacific region, which need to be addressed if trade diversion is to be reduced or avoided.

CANADIAN TARIFFS

CANADIAN TARIFFS ON TRADE with non-FTA and NAFTA partners are being reduced significantly as a result of the Uruguay Round. In some sectors, such as textiles and apparel, they remain high, but in most others they have been cut to relatively low levels. In addition, Canada has conducted a review of the General Preferential Tariff (GPT) duties, since the value of the preferences has been reduced in light of the Uruguay Round reductions. However, sensitive products such as textiles and apparel are exempt from GPT.

Although Canadian tariffs since the Uruguay Round are relatively low, at least compared to those of developing countries, in parts of the manufacturing sector they remain higher than U.S. tariffs. This would create a competitive disadvantage for Canadian manufacturers, who must pay higher tariffs for offshore imports, except that duty drawback permits manufacturers and processors to obtain inputs at world prices in order to export. Under the FTA, duty drawback was to be stopped as of January 1, 1994, but was extended until January 1, 1996 in the NAFTA, in part because of the delays in completing the Uruguay Round.

The looming elimination of duty drawback has put pressure on the Canadian external trade barriers. Thus Canada is not only implementing the Uruguay Round, but unilaterally reducing tariffs either through the GPT – a tariff review conducted internally in the Department of Finance – or the special reference to the Canadian International Trade Tribunal on textile tariffs. Of course, reducing these external trade barriers against non-NAFTA members will reduce the scope for trade diversion.

Canadian Non-Tariff Barriers

SOME NON-TARIFF BARRIERS REMAIN IMPORTANT in Canada/Asia Pacific trade and economic relations, namely Multifibre Arrangement (MFA) quota restrictions on imports of textiles and clothing from low-cost suppliers, agricultural trade barriers, and Canadian anti-dumping law and practice.

MFA Quota Restrictions on Imports of Textiles and Clothing

Canada has regularly negotiated quota restrictions on exports of textiles and clothing under the MFA with Asia Pacific countries. The four largest suppliers of apparel to Canada that are subject to MFA restrictions are Hong Kong, South Korea, Taiwan and the People's Republic of China. India, Thailand and Indonesia are also significant suppliers. However, while the quotas of major suppliers such as Hong Kong, Taiwan and South Korea were cut to annual growth rates of 1 percent or less in the last several years, China's was allowed to expand at a rate of 5 percent annually. The United States and the European Union are substantial suppliers of clothing and the major suppliers of textiles to Canada, but neither is subject to MFA restrictions, and textiles and clothing from the United States enter Canada at preferred, zero or low duty rates under the FTA and the NAFTA. Mexico has not been a major supplier of apparel to the Canadian market and has not been subject to MFA quota restrictions.

Nevertheless, these quota restrictions are equivalent to tariffs of 20-40 percent on top of high MFN tariffs, ranging from 15-30 percent. In 1994 the average MFN tariff for textiles and clothing was 18.1 percent. This will be reduced to 12.7 percent by the year 2000. However nominal they are, clothing tariffs are higher than those for textiles, which implies high effective rates of protection relative to non-NAFTA imports. When the tariff equivalents of the quotas are combined with tariffs, the barriers to imports of clothing and apparel from the Asia Pacific countries range from 40-75 percent.[13]

Under the Uruguay Round Textiles Agreement, textiles and apparel quota restrictions are to be phased out within 10 years. However, under the Textiles Agreement most of the liberalization occurs in the last two years of the Agreement. This is certainly the case with Canada's early implementation of the Textiles Agreement, because Canada has "reintegrated" textile and tariff lines that were not subject to MFA restrictions, such as work gloves. Some of Canada's highest external trade barriers are imposed on these products, which are key exports in some of the East Asian economies.

Agricultural Trade

The external trade barriers on textiles and apparel do not have the distinction of being the highest imposed by Canada. That distinction is reserved for the supply-managed sectors – dairy and poultry products. The conclusion of the Uruguay Round has allowed Canada to replace import prohibitions and quotas on these commodities with tariff-rate quotas and tariff equivalents. For example, in 1995, the tariff equivalents for dairy products included 351.4 percent for butter, 289 percent for cheese, 283.8 percent for milk and 237.2 percent for skim milk powder. By the year 2000, these tariffs will be reduced by 15 percent but they will still be prohibitive.

Canada has moved to restructure and reduce the substantial subsidies granted the agri-food sector. The Western Grain Transportation Assistance program is being eliminated with short-term compensation payments. Thus, in this sector, Canada has moved far in advance of its obligations to reduce export subsidies under the Uruguay Round Agreements.

Canadian Anti-Dumping Law and Practice

Canada is an active user of anti-dumping laws and procedures. Appendix 1 is a list of Canadian anti-dumping cases. The United States is the most frequent target of Canadian anti-dumping complaints, which is not surprising since about two-thirds of Canada's imports come from the United States. Forty-two anti-dumping cases were brought in 1990-95, and the average duties were about 35 percent.

Anti-dumping cases in Canada involving imports from Asia Pacific countries other than the United States tend to be of two types. The first involves the more industrialized countries being subject to a broad anti-dumping case. For example, Australia, Japan, Korea, and New Zealand were subject to some of the large steel cases in 1992-93 that also involved exporters from the United States, Europe and Latin America. The second type is against imports of labour-intensive standard technology products from lower-wage countries. Some examples are ladies footwear from China, Taiwan and Thailand; rubber footwear from China, Hong Kong, Korea and Malaysia; photo albums from China, Indonesia, the Phillipines, and Thailand; and bicycles from China and Taiwan. Anti-dumping duties have also been applied to other labour-intensive products from the Asia Pacific region.

It is notable that China is invariably included in these cases and frequently is subject to the highest duties. Under Canada's *Special Import Measures Act* (SIMA), China continues to be treated as a non-market economy in dumping investigations. Under SIMA, Revenue Canada uses a surrogate approach according to which an export price or a production cost of a surrogate market economy replaces the comparable export price from China or imputed production cost in China, in order to decide the "normal" value of the Chinese domestic price in question. This practice is allowed by the GATT-WTO. However, it does not take into account the progress achieved by China's economic reforms, particularly its price reforms. Even assuming it is still appropriate to apply the surrogate approach in Chinese dumping cases at the present time, the choice of the surrogate countries by Revenue Canada is still questionable, because most surrogate countries Revenue Canada has chosen have much higher levels of economic development than China. Most surrogate countries in the Chinese dumping cases do not appear to have been chosen on the basis of cost comparability. In 11 Canadian cases against Chinese exports, six surrogates are developed countries and the remaining five are developing economies that are all at a higher level of development than China. Therefore, the surrogate approach has led to high normal values for Chinese products and resulted in large dumping margins. Also, because the choice of the surrogate country is not known in advance, Chinese producers and exporters and Canadian importers are unable to predict prices and implement stable pricing policies that comply with Canadian anti-dumping law.

The analysis of the Chinese dumping cases demonstrates that the present Canadian anti-dumping law can constitute a serious trade barrier to Chinese exports. One option might be to amend the Canadian law and to abandon the surrogate approach, or to make the surrogate approach non-applicable to the Chinese cases. It can be argued that the surrogate approach is no longer appropriate to China's exports because of its economic reforms, changes to its pricing system and greater participation in the world trading system. If it is not realistic to alter the Canadian anti-dumping law and practice in the short term, Revenue Canada could modify its surrogate approach to select surrogates that are more comparable to the development level of the Chinese economy.

Canada's Investment Barriers

CANADA HAS BEEN IMPLEMENTING A MORE OPEN INVESTMENT REGIME in recent years. Basically it maintains few significant barriers to new investment from Asia Pacific countries, apart from excluded sectors such as uranium mining and basic telecommunications. However, under the *Investment Canada Act*, Canada reserves the right to review large foreign acquisitions in order to ensure "net benefit to Canada" of the foreign investment. Until recently the threshhold for screening foreign acquisitions was $153 million for NAFTA partners and $5 million for investors from other countries, but this disparity is being removed as part of Canada's implementation of the Uruguay Round.

This review process is applied more stringently to "culturally sensitive sectors" such as newspaper, magazine, periodical and book publishing and distribution; film and video; audio music recording; and music in print or machine-readable form. As a result, foreign investment in these cultural sectors is subject to review regardless of its size, whether it is new, or whether it is through direct or indirect acquisition.

Asia Pacific Trade and Investment Barriers

Japan

SOME ARGUE THAT JAPAN IS A SPECIAL CASE, where informal barriers impede trade and investment. Japan has low nominal tariffs that averaged 3.9 percent in the industrial sector in 1994. This average will be reduced 56 percent on a trade-weighted basis to 1.7 percent in 1999 as a result of the Uruguay Round.[14] However, the tariff structure escalates according to the degree of processing involved, which creates obstacles to further processing of resource products such as converting logs to lumber.

Japanese agriculture is highly protected. According to OECD calculations of nominal assistance coefficients, Japanese agricultural producer prices are on average about three times world levels.[15] Products such as rice, wheat, coarse grains and milk have nominal assistance coefficients for producers in excess of 700 percent, reflecting import quotas and other non-tariff barriers. The Uruguay Round Agriculture Agreement will lead to significant increases in import competition through minimum access commitments for rice and conversion of quotas into tariff equivalents for other products. However, Japan's over-quota tariffs are as high as 306 percent for barley and 413 percent for wheat.

There is debate about whether Japan's industrial structure, combined with the lack of foreign direct investment (FDI), creates private obstacles to trade. Some analysts attribute the relatively low level of FDI in Japan in part to the historic effects of restrictions on investment and capital movements that were imposed by the Japanese government until the late 1960s. These restrictions were progressively removed up to 1980, yet Mason notes that efforts by Cargill in the early 1980s and Toys-R-Us to invest in Japan were impeded by regulatory restrictions.[16] Wakasugi expresses the alternative view that the low level of inward direct investment into Japan can be explained by economic factors.[17]

The combination of low levels of FDI, the high degree of vertical integration in many Japanese industries, and what is perceived as tolerance of restrictive business practices have led to a preoccupation with competition policy. The Structural Impediments Initiative, negotiated by the Bush administration with Japan, called for vigorous enforcement of competition policy by the Japan Fair Trade Practices Commission as well as changes in the regulations for large stores. More recently, Kodak has complained that the structure of the Japanese film industry, which is dominated by Fuji, impedes the sale of Kodak film, leading the U.S. government to complain that market access has been nullified and impaired. More generally, there is concern that vertical integration of Japanese industry and the tolerance of restrictive practices has led to the foreclosure of markets. For example, sources in the Canadian pulp and paper industry have complained that vertical integration between paper mills and newsprint distributors in Japan is a significant impediment to the export of newsprint and may create an advantage for Japanese firms in acquiring pulp mills.

INDONESIA

ALTHOUGH INDONESIA HAS SOME NOTABLE TARIFF PEAKS, notably automobiles at 275 percent, most products are subject to a 40 percent tariff ceiling. The average Indonesian tariff was 20 percent in 1994 compared with 22 percent in 1990.[18] Import licensing has declined from about 40 percent of imports in 1985 to under 10 percent in 1994, but it is estimated that 30 percent of manufacturing production and 35 percent of agricultural production are covered by import licenses.[19] The industries subject to import licensing are subject to local content requirements, thus the non-tariff protection for these industries far exceeds the 40 percent tariff ceiling. Indonesia imposes a joint venture requirement on foreign investors, and the required minimum investment is $1 million.[20] More than 30 industries are closed to foreign investment. Indonesia does have investment incentives, including a two-year exemption from duties on imported capital goods as well as special export processing zones in which imported inputs are duty free and investment rules are more liberal but domestic sales are limited to 25 percent of production.

KOREA

DURING THE 1970S, KOREA INTRODUCED a heavy and chemical industries drive, which included preferential extension of credit and partial reversal of import liberalization in the machinery and transport equipment sectors. However, this program has been gradually dismantled in the last decade. The Republic of Korea has reduced its average unweighted tariff from about 24 percent in 1982 to about 8 percent in 1994.[21] In the mid-1980s, about 40 percent of Korea's tariff lines were subject to restrictive import licensing, but this had been reduced to less than 2 percent of tariff lines by 1995. Korean agricultural producers, like their Japanese counterparts, are highly protected, with producer subsidy equivalents of more than 300 percent for rice, barley, sesame, and dairy products in the early 1990s.[22] As a result of complaints to the GATT by Australia, New Zealand and the United States, Korea has greatly liberalized imports of beef.

The Philippines

The Philippines has also engaged in unilateral liberalization. In 1981, average nominal tariffs were 41 percent. This had declined to about 25 percent by 1992.[23] The range of products subject to import licensing has been reduced. Automobiles are an exception: in order to encourage the import of kits for assembly, passenger motor vehicles are still subject to import licensing, and the tariff is 34.3 percent. Efforts are underway to rationalize the production of, and trade in, automotive products within the ASEAN.

China

China is an interesting and important case. Even though there are some questions about the data, China has been a magnet for foreign direct investment and has rapidly expanded its trade. Yet its trade and investment regime remains highly restrictive, at least on paper. Tariffs range up to 200 percent, and a subtantial proportion of trade is covered by import licensing in sectors dominated by state enterprises. This apparent paradox can be explained by the discretionary dispensation of access to imports, often on a duty-free basis, for joint ventures. These are subject to re-export requirements and other performance requirements. Although the complicated, multipart exchange-rate system was dismantled and replaced by a unified exchange rate, not all enterprises have access to foreign exchange. In recent years, there has been investment in joint ventures in China, such as in the manufacture of automobiles aimed at the domestic market. These ventures are highly protected, but subject to extensive performance requirements.

Summary

While there has been much liberalization of trade and investment in the Asia Pacific countries in the last 15 years, which has contributed to their strong economic performance and growth, there are still significant barriers to trade and investment in the region. In particular there are critical issues about the transparency of their trade and investment regimes. In at least some cases, the discretionary and voluntary nature of market and investment access generates potential for corruption.

Canadian trade and economic interests could be affected by the non-transparent trade and investment regimes. Business people may need to be tutored on local business practices and the regulatory process. There could also be significant indirect effects on Canada's trade with the United States. Since the FTA and NAFTA, Canada-U.S. trade has been considerably restructured, a significant portion of it within multinational production networks. The position of Canadian-based firms and multinational subsidiaries in the multinational production network could be significantly affected by Trade Related Investment Measures (TRIM) and performance requirements in the Asia Pacific countries.

CANADA-ASIA PACIFIC TRADE AND INVESTMENT: AN ASSESSMENT OF THE OPTIONS

THIS SECTION EXAMINES A RANGE OF POLICY OPTIONS that may enable Canada to expand trade and investment links with the region. These include unilateral reductions in trade barriers against non-NAFTA partners associated with elimination of duty drawback within NAFTA; bilateral building block agreements such as Foreign Investment Protection Agreements (FIPA); bilateral framework agreements that would facilitate trade and investment; bilateral free trade agreements analogous to what is expected to emerge from the current Canada-Israel negotiations; accession to NAFTA for Chile and other potential trading partners, and Asia Pacific Economic Co-operation (APEC) initiatives to accelerate the timetable for WTO obligations and to engage in future liberalization beyond WTO obligations either on an open or preferential basis. What is the interaction between APEC initiatives, other regional initiatives such as Western Hemisphere and FTAA, and multilateral initiatives at the WTO and the OECD? This section assesses the options for Canada critically in light of some of the obstacles to APEC, as evidenced in the objections to making the APEC investment declaration binding. Is *open regionalism* a basis for liberalization of dynamic economies or is it a facade to veil trade and investment regimes that lack transparency?

This section also addresses some the questions about the impact of Canadian policy on trade with the Asia Pacific, considers the issues posed by China's accession to the WTO, explores the scope for APEC leadership on issues such as subsidies and countervailing measures, comments on bilateral options for Canada, and considers the issue of deeper integration.

Canadian firms face particular challenges in breaking into the expanding markets in Asia Pacific countries outside NAFTA. During the 1970s, when demand for natural resource commodities was expanding rapidly in Japan, Canadian exports across the Pacific expanded with only modest investment links, primarily through Japanese investment in the development of Canadian resources. The 1980s were a time of transition, when demand for Canadian natural resources in the Pacific Rim countries lagged while the countries on the western Pacific built up large trade surpluses with the United States. During the 1990s, trade flows and investment links among the western Pacific countries are likely to continue to expand. For example, manufactured products constitute an expanding share of Japanese imports, but the main sources of those products are other Asian countries. At the same time, Japan's exports to, and investments in, ASEAN countries now exceed those of the United States.

The implementation of the Uruguay Round Agreements and the establishment of the WTO will lead to significant internationalization of MNE activity in Canada. Many MNE subsidiaries in Canada, Canadian enterprises participating in MNE production networks, and Canadian-based MNE have restructured to take advantage of opportunities under the Canada-U.S. Free Trade Agreement and NAFTA, by seeking out niches in the North American market.

Policies already in train will affect Canada's economic relations with Asia Pacific countries, which are not members of NAFTA. The reduction of trade barriers under the Uruguay Round Agreements will have three main effects on Canadian trade and investment patterns. First, it will reduce by roughly 40 percent the tariff preferences on Canadian exports to the United States and U.S. exports to Canada under the FTA and NAFTA. Similarly, preferences in the Mexican market under NAFTA will be reduced, but this is a prospective effect because the phase-in schedule for the Uruguay Round reduction in trade barriers follows closely behind the reductions under NAFTA. Second, Canadian access to offshore markets will significantly improve, including access for processed resource products in the zero-for-zero sectors. In addition, there are significant potential gains for manufactured goods, including high-technology products, in export markets. Third, the reductions in trade barriers under the tariff schedules that Canada has offered in Geneva will lead to a reduction in input costs for many Canadian industries. All these effects will make Canadian-based exports more competitive in the global marketplace.

Because of increased competition in the resource sectors, the need to reduce Canada's overall current account deficit, and the implementation of the Uruguay Round under the WTO, there is pressure to expand and diversify exports. This diversification pressure will be market-driven, but the strong role of multinational enterprises in the Canadian economy presents some challenges. Participation in multinational production networks may facilitate globalization of Canadian business, but foreign-based multinational firms may rely upon other trading powers for advocacy of their business interests. Sometimes this works to Canada's advantage, while at other times it does not. For example, a Canadian supplier to a multinational may suspect, or the Canadian subsidiary of a multinational may know, that its interests are being damaged by discriminatory practices in foreign markets, yet it may be unable or unwilling to document the complaint. More insidious, MNE may negotiate performance requirements to gain investment access to China or other Asian economies that are prejudicial to Canadian interests.

Trade development and policy are being transformed by globalization. Trade development is becoming international business development, as the export of goods and services frequently requires market presence in the form of direct investment, joint ventures, or technology licensing. Indeed, both inward and outward direct investment can facilitate trade and access to technology through participation in multinational production networks. Trade policy is encompassing an ever broader array of non-tariff barriers, and new issues such as the rules for investment and trade in services and intellectual property rights are being brought within the ambit of trade policy.

The resources for international business development are coming under pressure as part of the application of fiscal restraint policies by Canadian governments. Yet fiscal restraint implies that growth must be export-led, which increases pressure on international business development programs. At the same time, the pressures for fiscal restraint are resulting in the retrenchment of some of the international business development activities that were undertaken by the federal and provincial governments in the 1980s.

With the exception of Quebec, there is much more interest in a "team Canada" approach to international business development. This may involve more private sector participation, either through contracting out or through fee-for-service activities. In searching for the appropriate model or vehicles for international business development, it is important to recognize the strategic role of government in advocating Canadian commercial interests. Clearly, however, Canadian governments have less capacity to engage in competitive export financing initiatives.

Key issues still need to be addressed in Canadian trade and investment policies toward non-NAFTA economies. Some of the highest remaining barriers in Canada's trade regime are imposed on imports from the Asia Pacific region. These restrictions include post-MFA restrictions on imports of textiles and apparel, high agricultural tariffs, anti-dumping duties and procedures and relatively high tariffs on some products from the Asia Pacific. Thus there are questions for Canadian policy as well as questions about how the overlapping agenda for regional and multilateral initiatives will evolve.

WHAT CAN BE DONE IN THE ASIA PACIFIC?

AMONG THE ISSUES TO BE ADDRESSED BY CANADA in the Asia Pacific region are normalizing U.S.-China relations and bringing China into the WTO (properly), identifying common issues for the APEC agenda, which may also serve as preparation for the Singapore Ministerial for the WTO in late 1996, pressing ahead with implementing the Uruguay Round commitments in the Asia Pacific region, bilateral initiatives that Canada could pursue, and identifying issues for deeper integration in the Asia Pacific. There is also the broader question of the potential global impact of efforts to achieve the Bogor vision of free trade in 2020.

China's Participation in the World Trade Organization

China has sought to renew its formal ties with the GATT since the early 1980s as part of its economic reforms and open-door policy. Strong economic growth and the progressive opening of the trade and investment regime in the past few years have created the basis for China's full participation in the trading system. However, further significant policy changes by China and greater transparency in its economic policies and legal regimes are required for it to become a member of the new World Trade Organization.

Membership in the WTO requires negotiation of market access schedules for goods, for agricultural products, and for services. The prohibition of quantitative restrictions is to be strengthened, and existing quantitative restrictions, such as those on agricultural products, are to be replaced by tariff equivalents. These tariff equivalents are to be gradually reduced over time. In addition, WTO members must provide adequate and effective protection of intellectual property.

China attached high priority to its attempt to become an original member of the World Trade Organization. However, the attempt was unsuccessful because not all the outstanding issues could be resolved and it sought special and differential treatment as a developing country. There have been very significant economic reforms in China since the open-door policy began in 1978, but important sectors of the economy are still dominated by state-controlled enterprises, and many questions are raised by trading partners about the transparency and openness of its trade and payments system.

The incremental liberalization that has occurred in China's trade and investment policies under the open-door policy has stimulated a dramatic period of sustained dynamic economic growth. This may give policymakers in China more confidence to continue the reform that will be required to resolve the many outstanding issues if China is to be a full participant in the WTO and the trading system.

The new WTO and the single undertaking ups the ante for the integration of China and other countries with transition economies into the global system. Stricter rules, the elimination of quantitative restrictions for imports and exports, new standards for intellectual property protection under the trade-related aspects of intellectual property rights agreement (TRIPs agreement), and the provisions for the elimination of export subsidies or subsidies contingent on domestic sourcing requirements under the Subsidies Agreement all raise specific challenges for the transition countries.

Some of the provisions of the WTO agreement include special transition arrangements for the transition economies. Others, however, only have transition arrangements for developing countries. China claims developing country status and, like other developing countries with large populations and relatively low per capita incomes (as measured by the World Bank in 1988), will likely claim special Annex 7 status under the Subsidies Agreement.[24]

Canada was the first of the Quadrilateral Group of trading nations to enter into bilateral negotiations with China under the GATT auspices. Other countries such as the United States refused to enter into such bilateral negotiations, because they wished to retain the right to refuse to apply the GATT to China. However, new arrangements under the WTO have permitted the United States to enter into bilateral negotiations over the GATT accession/WTO membership without giving up this right, if it considers the negotiations to be unsatisfactory.

Canada has clearly signalled its support for China's participation in the GATT-WTO, but has sought clarification of many aspects of its trade regime. Canada has significant interest in obtaining improved and more predictable and transparent access to the Chinese market. Thus it wishes to ensure that China has a well-defined program and schedule of commitments to bring China's trade and legal regimes into conformity with the obligations of the WTO and commitments to increase its market access for manufactured goods, agricultural products, and services.

Since the MFA will be phased out in 10 years after the WTO comes into force, there is some pressure for China to resolve its GATT-WTO membership issue as soon as possible. Otherwise, now that the WTO has come into force, China is the only major country that will be left under the MFA quota arrangements. Thus Canada and China will have significant bilateral issues to negotiate with respect to trade in textiles and clothing.

In general, the growing economic linkages between China and the global economy should serve to sustain the process of economic reform in China and facilitate China's eventual participation in the WTO, notwithstanding the policy uncertainties associated with dynastic change in China. However, as more firms make substantial investments in China aimed at the domestic markets, such as the joint ventures in automotive production, and agree to performance requirements, they will acquire vested interests in the complex Chinese trade and investment regime.

SUBSIDIES AND COUNTERVAILING MEASURES

THERE IS A GREATER COMMONALITY OF INTERESTS and perspectives over the prospect of limiting subsidies in the Asia Pacific region than there is over the broader multilateral system. The most obvious avenue for Canada to take over Asia Pacific cooperation on subsidies is to work together with the Asia Pacific countries to make certain that there is effective multilateral cooperation and that the ambitious subsidy notification and surveillance mechanisms under the World Trade Organization work effectively.

Making the Subsidy Agreement operational and effective will in itself be a significant accomplishment. Although its implementation will not generate much enthusiasm at the level of heads of government, it is a very practical and significant objective for increasing cooperation in the Asia Pacific region to pursue multilateral goals.

The potential also exists for the Asia Pacific region to provide international leadership in building upon the subsidies rules negotiated in the Uruguay Round. This could occur either through multilateral negotiations or through regional initiatives.

Certainly, the Asia Pacific nations have an interest in multilateral negotiations on subsidies, but with the Uruguay Round just concluded and the implementation of the World Trade Organization about to begin, multilateral negotiations seem remote. However, the Agriculture Agreement will have to be renegotiated within six years, and that renegotiation will be linked to subsidies negotiations, because some of the special provisions for agricultural subsidies expire after six years and the general rules under the Subsidies Agreement will apply to agricultural trade. Thus the preparation for these negotiations will need to start in the next couple of years, and the Asia Pacific nations should explore the development of common perspectives.

Regional initiatives to limit the use of subsidies and countervailing duties could be pursued in conjunction with multilateral initiatives, either as a complement to the multilateral process or as a fallback if the multilateral initiatives to strengthen subsidies rules encounter resistance. One area of potential regional cooperation is to limit or even eliminate agricultural export subsidies within the Asia Pacific region. Since such an initiative would require going further and faster than the WTO Agriculture Agreement would require, this raises the question of how to deal with countries from outside the region, for example, the EU, who are unwilling to limit the use of export subsidies any more than required under the Agreement. This has been a problem within the NAFTA, where both Canada and the United States maintain subsidies on agricultural exports into the Mexican market. Any plurilateral arrangement to limit agricultural export subsidies in the Asia Pacific region would also require some agreement to pursue countervailing procedures against subsidized exports from outside the region. If importing countries were unwilling to take countermeasures against extra-regional subsidized imports, there would be considerable reluctance in the United States to limit or to dismantle agricultural export subsidies. However, Canada might go further in limiting agricultural export subsidies, even on a unilateral basis, largely because of its fiscal pressures.

Similar problems arise in efforts to limit subsidies in the industrial sector, especially in high-technology industries. The shift in the U.S. position by the Clinton administration toward less discipline over subsidies for research and development during the Uruguay Round negotiations may or may not be long term. Even if U.S. policy were to shift away from the use of research and development subsidies toward a more market-oriented approach, the United States would likely wish to retain the flexibility to respond to developments in European industrial policies.

Although there are obstacles to taking regional initiatives to limit subsidies beyond the requirements of the WTO Subsidies and Agriculture Agreements, the opportunity of exploring the possibility in the Asia Pacific region should not be ignored. The fiscal pressure on the United States and Canada to reduce their subsidies, combined with the more limited use of subsidies in most Asian countries, does offer the basis for regional initiatives to limit the use of subsidies and countervailing measures within the region. The clear common interest of the Asia Pacific region lies in making sure the World Trade Organization, and especially the Subsidies Agreement, function effectively, and that the multilateral rules are strengthened in the near future, but perhaps the APEC group can go further faster.

Bilateral Options for Canada

CANADA SHOULD EXPLORE BILATERAL OPTIONS in the Asia Pacific, but it must be recognized that the potential is limited. Certainly Canada should expand the coverage of bilateral FIPA and double taxation agreements to more countries, because these are important elements of the framework for commercial activity.

Beyond these building block agreements, formal bilateral framework agreements with key countries can help manage bilateral relations and facilitate the implementation of Uruguay Round obligations, but ministerial time and attention are not infinite. These agreements seldom contain substantive obligations, but they can be useful vehicles for dealing with bilateral concerns about the implementation of multilateral agreements.

A more aggressive Canadian approach to implementation of international trade and economic agreements now seems warranted for all the countries in the Asia Pacific. Certainly, the U.S. government regards vigorous advocacy as a key element in trade promotion. The report to the U.S. Congress of the Trade Promotion Coordinating Committee states:

> the limitations of the current advocacy process adversely affect the ability of U.S. companies to compete ... The goal of improved advocacy is consistent with U.S. interests and objectives, to provide appropriate U.S. government support to large as well as small and medium-sized U.S. firms as they face foreign competitors who are aggressively backed by their governments.[25]

This view on policy in the United States is reflected in the establishment in the office of the U.S. Trade Representative of a special unit on enforcement of international trade agreements.

The Canadian approach to representation of interests in the Asia Pacific may not necessarily involve a "carpet-bombing" litigation strategy (except perhaps for dealing with the United States), but Canadian representatives must be vigilant about derogations from international agreements and be aggressive in advancing Canadian interests. Perhaps U.S. advocacy on issues such as the protection of intellectual property will serve to benefit Canadian high-technology companies, but the United States may not be aggressive in its advocacy about performance requirements and other TRIM when U.S. companies have acquired a vested interest in these measures.

Bilateral FTAs

There may be possibilities for a bilateral free trade agreement with Chile, now that its NAFTA accession seems gridlocked in Washington. Canada would benefit from a bilateral FTA with Chile, both through expanding the small volume of trade and through deepening and consolidating the investment relationship. This agreement should be fairly straightforward and should parallel NAFTA, except that the rules of origin should be less cumbersome and much more liberal.

The problem is that aside from Chile, there are few plausible partners for a bilateral FTA. A bilateral free trade agreement with Singapore may also be a possibility if ASEAN-NAFTA solidarity concerns do not block it.

Deepening Integration

The APEC group should also address issues of deeper integration. The third EPG group has identified trade and competition policies as an agenda item and drafted terms of reference for a study. This is a priority, but the frame of reference should be very broad. It should encompass everything from restrictive business practices and corruption to competition policy issues such as vertical and horizontal restraints, anti-dumping and investment. The frame of reference should be shifted from the opening of markets to the contestability of markets. This broad agenda may block progress in the short term, but deepening integration involves interlinked issues. It is also being discussed in the transatlantic dialogue and undoubtedly will come up on the WTO agenda. Indeed, competition policy already has become an issue in the telecommunications negotiations.

Technical Barriers to Trade and Environment

As a result of the Uruguay Round, the Agreement on Technical Barriers to Trade (TBT) was adopted with the aim of ensuring that "technical regulations and standards, including packaging, marking and labelling requirements, and procedures for assessment of conformity with technical regulations and standards do not create unnecessary obstacles to international trade."[26] However, there are contentious issues regarding detailed product standards within the TBT and eco-labelling, which some argue is not covered by the TBT agreement. Eco-labelling, which is a trademark certification of environmental friendliness, is not in itself exceptionally controversial, but it does raise the issues of what the criteria are for environmental production processes and whether they can be applied on an extraterritorial basis.

There is also a broader, but related issue: the use of trade measures, including restrictions and sanctions, to implement or enforce international environmental agreements. It is on the multilateral agenda because the United States feels that it can act unilaterally on an extraterritorial basis to enforce environmental process standards. The European Union has made a proposal to renegotiate the provisions for exceptions to the GATT 1994 in the case of multilateral environmental agreements.

This issue may acquire some urgency as a result of the proposal under the Basel Convention on the Control of Transboundary Movements of Hazardous Wastes and their Disposal to ban all exports of recyclable materials from the OECD countries to developing countries by 1997. The ban on exports of recyclables was supported by the European Union and a coalition of G77 countries, with political impetus coming from the African countries. If enacted, this ban would pose particular problems for some of the Asia Pacific countries, notably Korea, the Philippines, and Thailand, which depend heavily on imported scrap metals to supply their processing industries. The United States is not a party to the Basel Convention, and both Australia and Canada have expressed reservations about the proposed ban. Thus there is scope for policy dialogue within APEC about the appropriateness of this particular measure, but there also needs to be a broader dialogue about whether this is a sensible policy response to the problems of hazardous materials.

IMPACT ON THE GLOBAL TRADING SYSTEM

THERE IS A DEBATE BETWEEN THOSE WHO TAKE THE VIEW that there is a constructive and creative tension between regional economic integration and multilateral liberalization, and those who take the opposite view that regional integration efforts can be corrosive and even cancerous to the multilateral system. Economists who broadly share a consensus about the benefits of trade liberalization have divergent views on the issue. In the 1940s and 1950s, there was considerable debate about the fundamental concepts of trade creation and trade diversion in economics. The basic conclusion reached was that if there is sufficient trade creation, then removing barriers among a group of countries can be beneficial, not only to the partners concerned but to other countries as well. However, the economics literature established conditions that ensure that trade creation effects dominate the trade diversion effects.[27]

At this point, discussion of the trade creation or trade diversion effects of an Asia Pacific free trade area is premature. Commentators and analysts have emphasized the role of unilateral liberalization and complementary integration initiatives such as infrastructure investments.[28] One such commentator counsels against an East Asia economic group and the expansion of NAFTA, in favour of the open regionalism or super-regionalism of a strengthened APEC.[29] This is the approach supported by the Eminent Persons Group chaired by Fred Bergsten in its successive reports, which have been adopted by APEC.

The Osaka meetings in 1995 were focused on cautious consolidation after the bold declarations of Seattle and Bogor. However, countries tabled the first steps toward implementation of the liberalization program. Most countries accelerated commitments made in the Uruguay Round. Thus the measures likely are trade-promoting, because they are on a most-favoured-nation basis.

At the same time, the process of NAFTA expansion, which is a source of anxiety among Asian nations, is blocked, not because of a high policy of multilateralism, but because of the low politics of divided government in the United States. The Republican Congress and Democratic administration have been unable to agree on the terms to extend fast-track negotiating authority. Thus de facto, if not de jure, open regionalism is the operative policy among APEC countries.

There is still considerable debate within APEC about whether a strictly MFN approach should be followed or whether achieving the final stages of integration in the Asia Pacific region will require reciprocal liberalization on a preferential basis, such as in a free trade area. This is going to remain unresolved for some years because of a strategic dimension to the issue. Many in APEC would prefer to continue the process of multilateral liberalization, but if APEC is unequivocally commited to MFN liberalization, that may undermine the prospects for progress in future multilateral negotiations. Another uncertainty is whether the East Asian economies will continue to open their trade and investment regimes on a unilateral basis.

Clearly APEC's strategy will be influenced by developments elsewhere in the world. There is a risk that the European Union will become mired in the internal politics of the intergovernmental conference, the challenges to creating a European currency, and the problems of including countries in Central Europe. Africa and South Asia have not participated in the process of competitive liberalization except to a very modest degree. If they begin to pursue more outward-oriented growth strategies, that will improve the prospects for progress on multilateral negotiations.

Conclusion

Despite the siren song of the globalization gurus, the Asia Pacific is not a borderless economy. Indeed, there has been much unilateral liberalization of trade and investment in East Asia and the Antipodes. However, much remains to be done if the APEC vision of free trade in 2020 is to be achieved. Given the predilection to oxymorons in Pacific discourse – concerted unilateralism, open regionalism – it is easy to be cynical about the prospects for APEC.

Canada has its own ambivalence toward trade with the Asia Pacific. The Canadian economy continues to become more open to trade and investment, but Canada imposes high trade barriers on some key Asia Pacific exports. It should continue to press for acceleration of WTO obligations and ensure that the implementation of the Uruguay Round Agreements is effectively monitored. Canada must become more aggressive in representing its interests, as multinational firms may agree to performance requirements that have adverse consequences for Canada's trade with NAFTA partners.

The ability of Canadian enterprises to compete effectively in the global marketplace depends on whether the federal government, in conjunction with the private sector, can make the new international trade agreements work for Canadian interests. This will require a reorientation of its trade commissioner service, which is already subject to fiscal pressures. It will also require support from Ottawa to commit resources to documenting impediments to Canadian business and building cases that can be used for advocacy, either through bilateral lobbying or formal dispute settlement channels. Although domestic framework policies influence the overall competitiveness of the Canadian economy, it will be the vigilance of the federal government in monitoring the implementation of international trade agreements and its effectiveness in advocating Canadian interests that will make or break competitive opportunities for many Canadian enterprises in the global marketplace. It is very important to recognize the potential significance of performance requirements imposed on U.S. MNE by Asian economies for Canada's trade with NAFTA partners.

For the near future, there seems little likelihood of APEC conflicting with FTAA and the transatlantic dialogue, but they all address issues such as trade and competition policy and trade and environment interactions, which are on the various regional agendas as well as the multilateral agenda. Competition among forums can be healthy.

The immediate challenge for the APEC countries is to contribute to the implementation of the World Trade Organization and the Uruguay Round Agreements (and for China to continue the reform process and become a member of the WTO). The main focus of the APEC agenda will be policy dialogue, and the Singapore Ministerial for the WTO should provide a forum to bridge these overlapping regional dialogues. The Singapore Ministerial may not yield much in substantive results, but it will be important for setting the agenda for negotiations. At the same time, initiatives could be taken at the Singapore Ministerial to help ensure the compatibility of these regional initiatives with the multilateral system.

In the longer term, APEC will have to address the question of whether to maintain a strict MFN approach, a plurilateral approach such as an Asia Pacific subsidies code, or to pursue a free trade area in the region. How quickly this issue must be addressed depends upon the prospects for, and the pace of, NAFTA expansion and the prognosis for the FTAA initiative, as well as the success of multilateral efforts to negotiate issues such as investment, both in the OECD and the WTO.

Unfortunately, there are few plausible candidates for bilateral FTAs, apart from Chile, and even this is a steppingstone to NAFTA accession. Nevertheless, Canada has clear economic interests in participation in APEC because of its bilateral trade and investment links, and because of its substantial indirect economic interaction through its intensive links with the U.S. economy. Canada must be prepared to confront its own ambivalence on trade policy, if it hopes to achieve greater integration into the Asia Pacific region. Through unilateral reductions in barriers, or in negotiations in regional and multilateral fora, Canada must be prepared to reduce its own trade and investment barriers.

APPENDIX 1

KEY TO ABBREVIATIONS

- A/D – Dumping investigation.
- C/V – Countervailing investigation.
- DD – Definitive duties applied as injury was found by the Canadian International Trade Tribunal (CITT). Injury finding still in place.
- DDR – Definitive duties applied as injury was found by the CITT. Injury finding since rescinded.
- FD – Final determination followed by no injury finding by CITT.
- T – Terminated prior to Final Determination.
- U – Undertaking accepted and investigation suspended. Undertaking still in effect.
- UX – Undertaking accepted but has since expired.
- I – Investigation underway, no final decision.
- SIMA – Special Import Measures Act.

TABLE A1-1
LIST OF CANADIAN ANTI-DUMPING AND COUNTERVAILING CASES, 1985-95

CASE NUMBER	COMMODITY/PRODUCT	A/D AND/OR C/V	COUNTRY AFFECTED	HS CODE	MARGIN OF DUMPING (PRELIMINARY) %	MARGIN OF DUMPING (FINAL) %	INITIATION DATE	DISPOSITION
1	Charcoal briquets	A/D	United States	2702.20.00 2704.00.00 4402.00.10 4402.00.90	20.0-40.0	60.5	85.01.18	DDR

TABLE A1-1 (CONT'D)

Case Number	Commodity/Product	A/D and/or C/V	Country Affected	HS Code	Margin of Dumping (Preliminary) %	Margin of Dumping (Final) %	Initiation Date	Disposition
2	Rail car and locomotive axles	A/D	Japan United States United Kingdom	8607.19.10	7.35 10.2 25.6	1.37 10.2 28.5	85.01.24	T FD DDR
3	Polyphase induction motors, 1 to 200 horsepower inclusive	A/D and C/V A/D	Brazil Japan Mexico Poland (UK) Taiwan United Kingdom	8501.51.90.00 8501.52.90.10 8501.53.19.00 8051.52.90.20 8051.52.90.30	6.6 20.3 38.2 52.7 24.6 24.9	6.6 13.7 38.2 49.7 15.9 14.2	85.02.07	DDR DDR
4	Modular automated plants	A/D	United States	n.a.	–	–	85.02.14	T
5	Frozen pot pies and compartment dinners	A/D	United States	1602.31.10 1602.39.10 1602.50.10 1602.49.10	...	...	85.04.24	U
6	Barbed wire	A/D	Argentina Brazil Poland (UK, France) Korea	7313.00.10 7217.33.00	46.21 32.69 54.71 4.26	49.86 31.50 47.30 5.56	85.05.01	DDR

TABLE A1-1 (CONT'D)

CASE NUMBER	COMMODITY/PRODUCT	A/D AND/OR C/V	COUNTRY AFFECTED	HS CODE	MARGIN OF DUMPING (PRELIMINARY) %	MARGIN OF DUMPING (FINAL) %	INITIATION DATE	DISPOSITION
7	Surgical adhesive tapes and plasters	A/D	Japan	3005.10.10 3005.10.91 3005.10.99 3005.90.10 3005.90.20 3005.90.30 3005.90.91 3005.90.92 3005.90.99	57.6	57.6	85.05.08	DDR
8	Polyphase induction motors, 1 to 200 horsepower inclusive	A/D	Romania (UK)	n.a.	35.0	30.1	85.07.05	FD
9	Hot-rolled carbon steel plate	A/D	German Democratic Republic (Federal Republic of Germany)	7208.11.00 7208.12.00 7208.21.00 7208.22.00 7208.31.00 7208.32.00 7208.33.00 7208.41.00 7208.42.00 7208.43.00	...	11.9	85.07.11	UX
10	Pentaerythritol	A/D	Chile	n.a.	24.0-38.0	40.99	85.08.19	FD

Table A1-1 (cont'd)

Case Number	Commodity/Product	A/D and/or C/V	Country Affected	HS Code	Margin of Dumping (Preliminary) %	Margin of Dumping (Final) %	Initiation Date	Disposition
11	Rubber hockey pucks	A/D	Czechoslovakia German Democratic Republic	5607.49.10 5607.49.20 5607.50.10 5607.50.10	66.50 58.92-59.05	63.32 55.15	85.08.21	DDR
12	12-gauge shotshells	A/D	Belgium France Italy United Kingdom	n.a.	37.0 10-21 13-28 14-25	12.52 35.77 14.31 5.97	85.09.12	DDF
13	Photo albums with self-adhesive leaves (imported together or separately)	A/D	China	4820.50.90.00 4820.90.90.20 4820.90.90.90	34-43	68.49	85.09.20	DD
14	Colour televisions	A/D	Korea	n.a.	8.22	4.54	85.09.03	FD
15	Oil and gas well casing	A/D	United States Korea Federal Republic of Germany Argentina Austria	7304.20.90 7305.20.00 7306.20.00	0.3-14.06 12.04 0.67 46/0-52.9 0.6-8.7	0.3-14.06 13.03 0.67 35.0-52.5 0	85.09.20	DD DD DDR DDR FD
16	Boneless manufacturing beef	C/V	EEC	0202.30.00	–	–	85.10.18	DD
17	Whole potatoes	A/D	United States	0701.90.00	30.7	32.4	85.10.18	DD

CANADIAN TRADE AND INVESTMENT POLICIES

TABLE A1-1 (CONT'D)

CASE NUMBER	COMMODITY/PRODUCT	A/D AND/OR C/V	COUNTRY AFFECTED	HS CODE	MARGIN OF DUMPING (PRELIMINARY) %	MARGIN OF DUMPING (FINAL) %	INITIATION DATE	DISPOSITION
18	Single-use hypodermic needles and syringes	A/D	Japan United States	n.a.	22-62 25-74	—	85.11.08	T
19	Drywall screws	A/D	Taiwan	7318.15.00.32	3.7-31.2	32.35	85.12.20	DDR
20	Spandex filament yarn	A/D	Korea	n.a.	10.1-22.6	1.17	86.02.13	T
21	ABS resin	A/D	Korea	3903.30.10 3903.30.90	27.0	17-27	86.03.19	DDR
22	Artificial graphite electrodes	A/D	Belgium Japan Sweden United States	8545.11.12 8545.11.22 8545.90.92	13.62 27.28 9.02 17.08	3.86 23.99-48.31 9.02 14.86-29.10	86.04.30	DDR
23	Pressure cleaners	A/D	United States	n.a.	35.2-46.5	11.71	86.06.16	T
24	Dry pasta	C/V	EEC	n.a.	—	—	86.07.02	FD
25	Grain corn	C/V	United States	1005.90.10 1005.90.90 2039.90.91 2309.90.92 2309.90.99	1.04799/ per bushel	0.849/ per bushel	86.07.02	DDR
26	Drywall screws	A/D	Korea	7318.15.00.32	31.12	14.88	86.08.01	DDR

TABLE A1-1 (CONT'D)

Case Number	Commodity/Product	A/D and/or C/V	Country Affected	HS Code	Margin of Dumping (Preliminary) %	Margin of Dumping (Final) %	Initiation Date	Disposition
27	Carbon steel seamless pipe	C/V	Brazil	n.a.	—	—	86.08.13	FD
28	Oil and gas well casing	A/D	Federal Republic of Germany	7304.20.90	0–8.0	34.12	86.08.20	U
			Japan	7305.20.00 7306.20.00	10.1–22.5	45.0		
29	Tile backer board	A/D	United States	6810.19.00	...	...	86.10.08	UX
30	Yellow onions	A/D	United States	0703.10.91 0703.10.99	35–56	42.58	86.10.14	DD
31	Gasoline-powered chain saws	A/D	Federal Republic of Germany	8467.81.00	30.71	26.44	86.10.24	DDR
			Sweden		20.32	16.55		
			United States		33.35	18.42		
32	Absorbent clay	A/D	United States	n.a.	44–75	16.0	86.11.14	T
33	Fertilizer blending equipment	A/D	United States	n.a.	7.97	9.41	87.03.05	FD
34	Printing plates	A/D	Japan	8442.50.90	45.71	46.18	87.04.01	FD
			United Kingdom		28.6	31.88		DDR
35	High-voltage porcelain station post insulators	A/D	Federal Republic of Germany	8546.20.00	11.9	—	87.04.08	UX
			Japan		60.3			

TABLE A1-1 (CONT'D)

CASE NUMBER	COMMODITY/PRODUCT	A/D AND/OR C/V	COUNTRY AFFECTED	HS CODE	MARGIN OF DUMPING (PRELIMINARY) %	MARGIN OF DUMPING (FINAL) %	INITIATION DATE	DISPOSITION
36	Photo albums with self-adhesive leaves	A/D	Malaysia Singapore Taiwan	4820.50.90.00 4820.90.90.20 4820.90.90.90	45.1 45.6 53.3	49.3 52.9 60.8	87.04.09	DD
37	Phenol	A/D	Spain	n.a.	40.6	39.16	87.05.22	FD
38	Wide-flange steel shapes	A/D	Spain	7216.10.00.30 7216.33.00.11 7216.33.00.12 7216.33.00.13 7216.33.00.20 7216.33.00.91 7216.33.00.92	6.14	7.29	87.05.22	DDR
39	Hot-rolled carbon steel reinforcing bars	A/D	Mexico United States	n.a.	20.62 20.66	20.62 21.04	87.05.27	FD
40	Solid urea	A/D	German Democratic Republic (Ireland) USSR	n.a. n.a.	66.47 61.35	66.51 57.66	87.05.29	FD
41	Drywall screws	A/D and C/V	France	7318.15.00.32	38.2	38.2	87.06.04	DDR
42	Automobiles	A/D	Korea	n.a.	36.3	26.3	87.07.15	FD

TABLE A1-1 (CONT'D)

Case Number	Commodity/Product	A/D and/or C/V	Country Affected	HS Code	Margin of Dumping (Preliminary) %	Margin of Dumping (Final) %	Initiation Date	Disposition
43	Photo albums with pocket sheets	A/D	Federal Republic of Germany	4820.50.90.00	23.54-57.37	44.0	87.08.28	DD
			Hong Kong	4820.90.90.20	13.16-78.43	64.0		
			Japan	4820.90.90.90	9.47-29.73	24.0		
			Malaysia		39.84-67.34	40.0		
			China		63.13-66.75	63.0		
			Korea		25.60-87.84	59.0		
			Singapore		64.46-67.33	66.0		
			Taiwan		31.85-90.90	59.0		
44	Recreational vehicle doors	A/D	United States	n.a.	14.0-25.4	12.65-17.07	87.08.28	FD
45	Carbon steel welded pipe	A/D	Brazil	7305.31.10	26.1	25.1	87.09.16	U and DD
			Luxembourg	7305.39.10	11.7	11.7		U and FD
			Yugoslavia	7306.10.10	33.9	28.1		U and FD
			Poland (Brazil)	7306.10.21	29.7	10.7		FD
			no UT	7306.30.00				
			(Turkey)	7306.60.10	56.3	51.0		FD
			no UT	7306.90.10				
46	Stainless steel buttweld fittings	A/D	Japan	7307.23.00 7307.29.10	40.0	40.0	88.02.08	DDR
47	Steel wool	A/D	United States	n.a.	13.9	13.9	88.05.26	T

CANADIAN TRADE AND INVESTMENT POLICIES

TABLE A1-1 (CONT'D)

CASE NUMBER	COMMODITY/PRODUCT	A/D AND/OR C/V	COUNTRY AFFECTED	HS CODE	MARGIN OF DUMPING (PRELIMINARY) %	MARGIN OF DUMPING (FINAL) %	INITIATION DATE	DISPOSITION
48	Sour cherries	A/D	United States	0809.20.21.00 0809.20.29.00 0809.20.90.00 0811.90.20.10 0811.90.20.99 0812.10.00.90	39.04	35.36	88.06.21	DDR
49	Apples	A/D	United States	0808.10.10.00	31.9	27.45	88.07.08	DDR
50	Lead acid batteries	A/D	Korea	8507.10.00	4.5-44.6	4.6	88.08.30	T
51	Padded coat hangers	A/D	Taiwan United States	44.21.10.00	19.0 20.0	19.0 20.0	88.09.16	FD
52	Grinding balls	A/D	United States	7325.91.90.90	13.7	13.04	88.09.23	UX
53	Motors over 200 horsepower.	A/D and C/V A/D	Brazil France Japan Sweden Taiwan United Kingdom United States	8501.53	19.5 11.7 40.6 34.6 30.6 11.7 19.3	28.7 28.1 38.0 33.8 17.5 9.6 15.3	88.09.30	FD
54	Mini refrigerators	A/D	Poland (Federal Republic of Germany) (Mexico, Brazil)	8418.22.90.90	75.4	75.6	89.01.19	FD

Table A1-1 (cont'd)

Case Number	Commodity/Product	A/D and/or C/V	Country Affected	HS Code	Margin of Dumping (Preliminary) %	Margin of Dumping (Final) %	Initiation Date	Disposition
55	Key blanks	A/D	Italy	8301.70.00.10 8301.70.00.20 8301.70.00.90	42.0	45.75	89.03.17	DDR
56	Plastisol (dispersion, liquid polyvinyl chloride)	A/D	United States	n.a.	29.0–36.0	5.2	89.07.07	T
57	Landing nets	A/D	United States	9507.90.90.20	6.3–45.3	21.0	89.07.17	UX
58	Certain transit concrete mixers	A/D	United States	8474.31.00.20 8705.40.00.00	…	…	89.08.04	UX
59	Ladies footwear	A/D and C/V A/D	Brazil Poland (Portugal) Romania (Spain) Taiwan Yugoslavia (Spain) China (Thailand)		9.72 34.13 38.0 21.69 23.75 22.0	25.8 38.74 20.0 27.5 26.2 47.3	89.08.25	DDR DD
60	Textured polyester yarn	A/D	Mexico		30.12	12.3	89.10.03	T
61	Refill paper	A/D and C/V	Brazil		32.5	32.5	89.12.08	DD DDR
62	Municipal tractors	A/D	Federal Republic of Germany		31.7	28.7	90.03.21	FD

CANADIAN TRADE AND INVESTMENT POLICIES

TABLE A1-1 (CONT'D)

CASE NUMBER	COMMODITY/PRODUCT	A/D AND/OR C/V	COUNTRY AFFECTED	HS CODE	MARGIN OF DUMPING (PRELIMINARY) %	MARGIN OF DUMPING (FINAL) %	INITIATION DATE	DISPOSITION
63	Dry dog food	A/D	United States	2309.10.00.99 2309.10.00.20	34.0-53.0	44.68	90.03.28	T
64	Certain stainless steel bars	A/D	India		38.5	30.6	90.05.03	FD
65	Photo albums	A/D	Indonesia		76.5	73.6	90.06.08	DD
66	Lint rollers	A/D	United States		25.2	25.89	90.07.06	FD
67	Photo albums	A/D	Philippines Thailand		78.0 55.5	78.0 55.5	90.07.10	DD
68	Wedge clamps	A/D	United States	8205.70.10.00 8205.70.90.00	33.5	6.5	90.11.14	UX
69	Carbon steel welded pipe	A/D	Argentina India Romania (Taiwan) Taiwan Thailand Venezuela	7306.30.00.14 7306.30.00.24	46.5 37.1 18.3 13.9 41.4 33.1	46.5 19.8 17.8 13.6 40.8 33.4	90.11.16	DD
70	Stainless steel welded pipe	A/D	Taiwan	7306.40.00	16.9	18.2	90.12.24	DD

TABLE A1-1 (CONT'D)

Case Number	Commodity/Product	A/D and/or C/V	Country Affected	HS Code	Margin of Dumping (Preliminary) %	Margin of Dumping (Final) %	Initiation Date	Disposition
71	Malt beverages (beer)	A/D	United States	2203.00.00.11 2203.00.00.12 2203.00.00.19 2203.00.00.21 2203.00.00.22 2203.00.00.29 2203.00.00.31 2203.00.00.32 2203.00.00.39	30.0	29.8	91.03.06	DDR
72	Carpeting	A/D	United States	5703.20.10.00 5703.30.10.10 5703.30.10.20	21.05	13.23	91.08.06	DD
73	Toothpicks	A/D	United States	4421.90.90.30 4421.90.90.99	18.5	16.9	91.08.19	DD
74	Graphite electrodes and connecting pins	A/D	Austria France Spain FRG United Kingdom	8545.11.12 8545.11.22 8545.90.92	8.6-69.1 7.6-16.6 14-31.5 8.5-38.6 7.1	33.6 11.2 15.3 25.9 19.8	91.09.20	T
75	Aluminum coil stock	A/D	Sweden	7606.11.20.11 7606.11.20.13 7606.12.29.11 7606.12.29.13 7616.90.90.90	47.0	42.4	91.07.12	DD

CANADIAN TRADE AND INVESTMENT POLICIES

TABLE A1-1 (CONT'D)

CASE NUMBER	COMMODITY/PRODUCT	A/D AND/OR C/V	COUNTRY AFFECTED	HS CODE	MARGIN OF DUMPING (PRELIMINARY) %	MARGIN OF DUMPING (FINAL) %	INITIATION DATE	DISPOSITION
76	Christmas trees	A/D	United States	0604.91.30.00	31.0	14.9	91.11.15	T
77	Roller bearings	A/D	Japan	8482.20.10.00 8482.99.10.21 8482.99.10.29	44.72	36.17	91.12.12	FD
78	Bicycles	A/D	China Taiwan	8714.91.00.00 8712.00.00.10 8712.00.00.50 8712.00.00.90	45.0 25.0	34.0 13.0	92.05.15	DD
79	Lettuce	A/D	United States	0705.11.11.00 0705.11.12.00 0705.11.90.00 2005.90.99.69	31.6	31.0	92.06.08	DD
80	Wedge clamps	A/D	United States	7616.90.90.90	...	...	92.05.08	UX
81	Gypsum board	A/D	United States	6809.11.10.00 6809.11.90.00	28.0	27.3	92.06.24	DD
82	Cauliflower	A/D	United States	0704.10.11.00 0704.10.12.00 0704.10.90.00	46.1	50.0	92.06.30	FD
83	Hanging file folders	A/D	United States	4820.30.00.90	...	...	92.07.22	U

527

TABLE A1-1 (CONT'D)

CASE NUMBER	COMMODITY/PRODUCT	A/D AND/OR C/V	COUNTRY AFFECTED	HS CODE	MARGIN OF DUMPING (PRELIMINARY) %	MARGIN OF DUMPING (FINAL) %	INITIATION DATE	DISPOSITION
84	Waterproof footwear	A/D	Czechoslovakia	6401.91.20.00	36.6	47.2	92.07.09	FD
			Slovakia (US)	6401.92.12.00				
			PRC (Taiwan)	6401.92.92.00	18.7	47.2		
			Korea	6401.92.92.00	47.2	47.2		
			Taiwan	6402.91.00.91	5.5	5.5		
				6404.19.90.20				
85	Hot-rolled carbon steel plate	A/D	United States	7208	10.2	10.2	92.08.24	FD
			Brazil		51.3	43.2		DD
			Czechoslovakia		39.8	53.8		
			Denmark		65.0	53.8		
			Federal Republic of Germany		12.8	12.3		
			Romania (US)		45.0	53.8		
			United Kingdom		40.8	10.2		
			Macedonia (US)		40.9	40.9		
			Belgium		23.7	23.6		
			Slovakia		0.0	0.0		
			Slovenia		0.0	0.0		
86	Tomato paste	A/D	United States	2002.90.00.19	13.1	11.2	92.09.01	FD
				2002.90.00.11				

CANADIAN TRADE AND INVESTMENT POLICIES

TABLE A1-1 (CONT'D)

CASE NUMBER	COMMODITY/PRODUCT	A/D AND/OR C/V	COUNTRY AFFECTED	HS CODE	MARGIN OF DUMPING (PRELIMINARY) %	MARGIN OF DUMPING (FINAL) %	INITIATION DATE	DISPOSITION
87	Hot-rolled carbon steel plate-heat treated (joined with #85 above at PD)	A/D	United States	7208	10.2	10.2	92.09.08	FD
			Brazil		51.3	43.2		
			Czechoslovakia		39.8	53.8		
			Denmark		65.0	53.8		
			Federal Republic of Germany		12.8	12.3		
			Romania		45.0	53.8		
			United Kingdom		40.8	32.9		
			Macedonia		40.9	40.9		
			Slovakia		0.0	0.0		
			Slovenia		0.0	0.0		
88	Hot-rolled carbon steel sheet products	A/D	Federal Republic of Germany	7208	55.4	56.6	92.09.16	FD
			France		17.3	18.9		
			New Zealand		39.2	39.2		
			Italy		55.4	56.5		
			United Kingdom		36.3	40.8		
			United States		10.9	11.6		
89	Cold-rolled carbon steel sheet products	A/D	France	7209.11.00	38.1	38.1	92.11.16	DD
			Germany	7209.12.00	31.2	31.2		
			Italy	7209.13.00	87.3	87.3		
			United Kingdom	7209.14.00	55.7	55.7		
			United States	7209.21.00	9.5	9.5		
				7209.22.00				
				7209.44.00				
				7211.30.00				

529

Table A1-1 (cont'd)

Case Number	Commodity/Product	A/D and/or C/V	Country Affected	HS Code	Margin of Dumping (Preliminary) %	Margin of Dumping (Final) %	Initiation Date	Disposition
90	Fibreglass pipe coverings	A/D	United States	7019.39.00.12 7019.39.00.13	51.0	51.0	93.02.04	DD
91	Copper and brass pipe fittings	A/D	United States	7412	51.6	51.6	93.02.05	DD
92	Synthetic baler twine	A/D A/D and C/V	United States Portugal	5607.41.00.00	17.4 0.0	16.4 0.0	93.07.30	DD T
93	Hot-rolled carbon steel plate	A/D	Italy Spain Ukraine Korea	7208.11.00.10 7208.12.00.10 7208.13.00.10 7208.14.00.10 7208.21.00.11 7208.23.00.00 7208.24.00.00	45.6 27.5 43.2 45.6	44.5 38.2 39.1 44.5	93.10.18	DD
94	Corrosion-resistant steel sheet	A/D	United States Australia Brazil France Germany Spain Sweden United Kingdom Japan Korea New Zealand	7210.31 7210.39 7210.49 7212.21 7212.29 7212.30 7225.90.00.10 7226.99.00.10	24.1 (9.8) 33.0 55.9 34.1 62.0 28.0 24.0 24.3 62.0 12.0 30.0	17.0 (2.2) 32.7 51.4 32.8 60.8 28.4 60.8 23.5 60.8 11.3 32.1	93.11.17	DD

TABLE A1-1 (CONT'D)

CASE NUMBER	COMMODITY/PRODUCT	A/D AND/OR C/V	COUNTRY AFFECTED	HS CODE	MARGIN OF DUMPING (PRELIMINARY) %	MARGIN OF DUMPING (FINAL) %	INITIATION DATE	DISPOSITION
95	Plastic shrinkable bags	A/D	United States	3923.21.00.41 3920.10.00.11 3923.21.00.42	...	...	93.11.19	U
96	12-gauge shotshells	A/D	Hungary Czechoslovakia Poland (Article 19 (b) of SIMA)	9306.21.00	44.0 35.0 0.0	37.0 32.0 0.0	93.11.24	DD T
97	Memorials	A/D and C/V	India	6802.23.00.10 6802.23.00.20 6802.93.00.10 6802.93.00.20	47.0	27.9	93.12.22	DD
98	Apples (red delicious)	A/D	United States	0808.10.10.20 0808.10.10.60	35.0 (red delicious) 11.0 (golden delicious)	24.0-28.0 (red delicious) 18.0 (golden delicious)	94.07.14	DD
99	Residential steel storage building	A/D	United States	9406.00.99.20 9406.00.99.90	19.1	21.9	94.09.13	U

Table A1.1 (cont'd)

Case Number	Commodity/Product	A/D and/or C/V	Country Affected	HS Code	Margin of Dumping (Preliminary) %	Margin of Dumping (Final) %	Initiation Date	Disposition
100	Refined sugar	A/D and	United States	1701.91.00 1701.99.00	71.0-85.0		95.03.17	I
		C/V	EU	1702.90.31 1702.90.32 1702.90.33	EU 0-59.0			
		C/V	Denmark Germany	1702.90.34 1702.90.35				
		A/D	Netherlands United Kingdom Korea	1702.90.36 1702.90.37 1702.90.38 1702.90.40 1702.90.50	179.0			
101	Jars and jar caps	A/D	United States	8309.90.00.91 7010.90.90.51 7010.90.90.52 7010.90.90.53 7010.90.90.54 7010.90.90.55	36.0-39.0	9.0-46.0	95.03.24	DD
102	Pasta	A/D C/V	Italy	1902.10.10.30 1902.19.91.30 1902.19.12.30 1902.19.92.30	37.0		95.08.30	I

TABLE A1-1 (CONT'D)

CASE NUMBER	COMMODITY/PRODUCT	A/D AND/OR C/V	COUNTRY AFFECTED	HS CODE	MARGIN OF DUMPING (PRELIMINARY) %	MARGIN OF DUMPING (FINAL) %	INITIATION DATE	DISPOSITION
103	Portable file cases	A/D	China (United Kingdom as surrogate)	4820.30.00.90	62.0		95.09.21	1
104	Culture media	A/D	United States United Kingdom	3821.00.00.00 3822.00.00.10 3822.00.00.20	38.7-80.8		95.09.29	1

TABLE A1-2

CASES INITIATED UNDER THE PREVIOUS ANTI-DUMPING ACT, BEFORE DECEMBER 1, 1984. THEY REMAIN IN EFFECT, I.E., SIMA DUTIES ARE APPLICABLE

PRODUCT	COUNTRIES
Canned ham	Denmark Netherlands
Canned luncheon meat	European Community
Carbon welded pipe	Korea
Paint brushes	China
Photo albums	Hong Kong Korea
Potatoes (russet skins)	United States
Rubber footwear	Bosnia-Hercegovina China Croatia Czec Republic Hong Kong Korea Malaysia Poland Slovenia Taiwan Yugoslavia
Synthetic rope	Korea
Tillage tools	Brazil

APPENDIX 2

KEY TO ABBREVIATIONS
A/D – Dumping investigation
C/V – Countervailing investigation
SIMA – Special Import Measures Act

TABLE A2-1
LIST OF MEXICAN ANTI-DUMPING AND COUNTERVAILING CASES, 1986-95

CASE NUMBER	COMMODITY/ PRODUCT	A/D AND/OR C/V	HS CODE	COUNTRY AFFECTED	EXPORTING COMPANY	MARGIN OF DUMPING (PRELIMINARY) %	MARGIN OF DUMPING (FINAL) %	INITIATION DATE	DISPOSITION
1	Caustic soda	A/D		United States				87.01.29	Negative
2	Trietilamina	A/D		Federal Republic of Germany				87.03.09	Negative
3	Trietilamina	A/D		United States				87.03.09	Negative
4	Monoisopropilamina	A/D		United States				87.07.10	Negative
5	Vatihormetros	A/D		United States				87.07.10	Negative
6	Revista Orbit	A/D		United States				87.07.10	Negative

Table A2-1 (cont'd)

Case Number	Commodity/Product	A/D and/or C/V	HS Code	Country Affected	Exporting Company	Margin of Dumping (Preliminary) %	Margin of Dumping (Final) %	Initiation Date	Disposition
7	Carton Kraft	A/D		United States				87.08.24	Negative
8	Corindon Artificial Cafe	A/D		Brazil				87.09.21	Negative
9	Acetato de Eter Monoetico	A/D		United States				87.09.22	Negative
10	Carburo de Silicio	A/D		Brazil				87.09.22	Negative
11	Potassium carbonate	A/D		United States				87.09.23	Negative
12	Potassium carbonate	A/D		Belgium				87.09.23	Negative
13	Potassium carbonate	A/D		Federal Republic of Germany				87.09.23	Negative
14	Potassium hydroxide	A/D		United States				87.09.23	Negative
15	Potassium hydroxide	A/D		Belgium				87.09.23	Negative
16	Potassium hydroxide	A/D		Federal Republic of Germany				87.09.23	Negative
17	Monoetilmina	A/D		United States				87.12.04	Negative
18	Graphite electrodes	A/D		Spain				87.12.21	Negative
19	Microcomputers ATT	A/D		United States				88.01.07	Negative

CANADIAN TRADE AND INVESTMENT POLICIES

TABLE A2-1 (CONT'D)

CASE NUMBER	COMMODITY/ PRODUCT	A/D AND/OR C/V	HS CODE	COUNTRY AFFECTED	EXPORTING COMPANY	MARGIN OF DUMPING (PRELIMINARY) %	MARGIN OF DUMPING (FINAL) %	INITIATION DATE	DISPOSITION
20	Tiles	A/D		United States				88.01.11	Compromise on price (exporters)
21	Tiles	A/D		Federal Republic of Germany				88.01.11	Compromise on price (exporters)
22	Urethra probes	A/D		Malaysia				88.02.18	Negative
23	*Toluen Disocianato*	A/D		United States				88.08.10	Negative
24	Steel bands	A/D	7211.30.01 7211.41.01 7211.49.01	Brazil	(1) Brasmetal Waelholz, SA (2) Maxitrade SA and/or Sao Bernardo SA (3) Other companies		(1) 8 (2) 38 (3) 38	88.09.13	Positive
25	Steel rods	A/D		Brazil				88.09.15	Negative
26	Steel products	A/D		European Community				88.09.21	Negative

537

Table A2-1 (cont'd)

Case Number	Commodity/Product	A/D and/or C/V	HS Code	Country Affected	Exporting Company	Margin of Dumping (Preliminary) %	Margin of Dumping (Final) %	Initiation Date	Disposition
27	Rodamientos de Bolus y Tomillados	A/D		Japan				88.12.02	Negative
28	Cellophane film	A/D		United States				88.12.06	Negative
29	Kitchenware of iron and/or died steel and enamel and porcelain (pewter)	A/D	7323.94.03	Taiwan	10 specific companies and all other exporters		17	88.12.08	Positive
30	Sorbitol	A/D	2905.44.01	France	All other exporters		52	89.03.13	Positive
31	Tiles	A/D		Brazil				89.04.12	Negative
32	Filas Alcalinas	A/D		United States				89.04.12	Negative
33	Ceramic wall tiles	A/D	6908.90.01	Brazil	22 specific companies and all other exporters		46	89.07.29	Positive
34	Amortigua-Dores			Brazil				89.09.04	Negative
35	Acrylic fibre	A/D	5501.30.01 5503.30.01	United States	Cynamid International		3	89.09.25	Positive
36	Hilo de Cauclło	A/D		Spain				89.10.09	Negative

TABLE A2-1 (CONT'D)

CASE NUMBER	COMMODITY/ PRODUCT	A/D AND/OR C/V	HS CODE	COUNTRY AFFECTED	EXPORTING COMPANY	MARGIN OF DUMPING (PRELIMINARY) %	MARGIN OF DUMPING (FINAL) %	INITIATION DATE	DISPOSITION
37	Polyvinyl chloride (PVC)	A/D	39.04.10.01	United States	Shinthec Inc.; Vista Chemical Co.; Occidental Chemical Co.; and all other exporters		46	90.02.06	Positive
38 39	Tweed cloth	A/D	5209.42.01	Hong Kong United States	All the exporting companies, with the exception of NanYang Cotton Mill; The Quicken Textiles Ltd.; Alwick Textile Export Co.; Yuki Textiles (H.K.)Ltd.; Moo Fung Weaving Factory Ltd.; Fung Fu Dyeing and Weaving Co., Ltd.; Yee Luen Cloth Co., Ltd.; Kes Dhari Ltd.; Tak Ming Textiles Ltd.; Sun Ping Weaving and Dyeing FTY. Ltd., Esraymo Co., Ltd.; Darvik Enterprises Ltd.; Far East Network (HK) Ltd.; Seaphone Textile Ltd.; Tarrant Co., Ltd.; Everben Co., Ltd.; and Amertex International Ltd.		47	90.05.31	Positive

TABLE A2-1 (CONT'D)

CASE NUMBER	COMMODITY/ PRODUCT	A/D AND/OR C/V	HS CODE	COUNTRY AFFECTED	EXPORTING COMPANY	MARGIN OF DUMPING (PRELIMINARY) %	MARGIN OF DUMPING (FINAL) %	INITIATION DATE	DISPOSITION
40	*2 Etil Hexanol*	A/D		United States				90.09.05	Negative
41	*Carton Kraft*	A/D		United States				90.09.11	Negative
42	*Diyodohidroxiquinoleina*	A/D		India				90.10.11	Negative
43	Aluminum ingots	C/V		Venezuela				90.11.08	Negative
43	Rods and bars of iron or non-alloy steel	A/D	7214.20.02	United States	Florida Steel Company		13	90.11.28	Positive
44	Rods and bars of iron or non-alloy steel	A/D	7214.20.02	Venezuela	(1) Siderurgica Del Turbio, CA (2) Acerotec, CA and all other exporters with exception of Siderurgica Del Orinoco CA (SIDOR)		(1) 31 (2) 36	90.11.28	Positive
45	Steel plates	A/D		United States				90.11.30	Negative
46	Hot-rolled carbon sheet	A/D		United States				90.11.30	Negative
47	Cold-rolled steel sheet	A/D		United States				91.11.30	Negative

TABLE A2-1 (CONT'D)

Case Number	Commodity/Product	A/D and/or C/V	HS Code	Country Affected	Exporting Company	Margin of Dumping (Preliminary) %	Margin of Dumping (Final) %	Initiation Date	Disposition
48	Rolled vinyl	A/D	3918.10.99 3918.90.01 4815.00.01 4815.00.99 5904.10.01	United States	(1) Tarkett, Inc. (2) 4 specific companies and all other exporters with the exception of Armstrong World Industries and Congoleum Corporation		(1) 16 (2) 179	91.03.21	Positive
49	Artificial guts (sausage casings) of regenerated cellulose	A/D	3917.39.01	United States	(1) Viskase Corp. (2) Teepak Inc.		(1) 21 (2) 18	91.05.23	Positive
50	Artificial guts (sausage casings) of regenerated cellulose			Spain	Ind Navarra de Envolturas Celulosicas		28	91.05.23	Positive
51	Table service	A/D	6911.10.01 6912.00.01	China	All the table service originating in China, except that imported through a third country		26 for the porcelain table service 23 for the ceramic table service	91.05.23	Positive
52	Almidón Catiónico	A/D	35.05.10.01	Low Countries	Avebe BA		16	91.07.29	Positive

Table A2-1 (cont'd)

Case Number	Commodity/ Product	A/D and/or C/V	HS Code	Country Affected	Exporting Company	Margin of Dumping (Preliminary) %	Margin of Dumping (Final) %	Initiation Date	Disposition
53	Plastic syringes	A/D	9018.31.01	United States	Terumo Corp and all other exporters		Duty is different depending on type of toy; average is 28	91.07.29	Positive
54	*Alambron*	A/D		United States				91.09.20	Negative
55	Telephone connectors	A/D	8536.90.23	United States	3M Company		20	91.10.28	Positive
56	Electric power transformers	A/D	8504.23.01 8504.22.01	Brazil	(1) Trafo Equipamientos Electricos, SA (2) Construcciones Electromecanicas, SA		(1) 29 (2) 35	91.11.15	Positive
57	*Cierres de Cremallera*	A/D		Venezuela				92.02.10	Negative
58	Sodium triple-phosphate	A/D	2835.31.01	Spain	Rhone Poulenc Quimica		36	92.02.18	Positive
59 60	*Pelicula de Polipropileno*	A/D		Colombia Brazil				92.03.20	Negative

TABLE A2-1 (CONT'D)

CASE NUMBER	COMMODITY/ PRODUCT	A/D AND/OR C/V	HS CODE	COUNTRY AFFECTED	EXPORTING COMPANY	MARGIN OF DUMPING (PRELIMINARY) %	MARGIN OF DUMPING (FINAL) %	INITIATION DATE	DISPOSITION
61	Tube or pipe fittings	A/D	7307.19.02 7307.19.03 7307.19.99 7307.99.99	China	Tianjin Malleable Iron Works; Tianjin Malleable Cast Iron Factory; Louis Delius Gingh and Co.; and all other exporters		33.34	92.04.01	Positive
62	Synthetic staple fibres, not carded, combed or otherwise processed for the spinning of polyesters	A/D	5503.20.01 5503.20.02 5503.20.03 5503.20.99	Korea	a) Samyang Co. b) Daewoo Co. c) Cheil Synthetics d) Other companies		a) 3.74 b) 14.81 c) 4.49 d) 32	92.05.07	Positive
63	Textiles of cotton and synthetic fibre	A/D		China				92.05.22	Negative
64	Textiles of cotton and synthetic fibre	A/D		China				92.05.22	Negative
65				Taiwan					
66				South Korea					
67				Hong Kong					
68				Pakistan					
69				Argentina					
70				Brazil					
71				Colombia					

TABLE A2-1 (CONT'D)

CASE NUMBER	COMMODITY/ PRODUCT	A/D AND/OR C/V	HS CODE	COUNTRY AFFECTED	EXPORTING COMPANY	MARGIN OF DUMPING (PRELIMINARY) %	MARGIN OF DUMPING (FINAL) %	INITIATION DATE	DISPOSITION
72	Bandas de Huile	A/D		Korea				92.05.25	Negative
73	Rolled carbon steel plate	A/D	7208.12.01 7208.22.01	United States	a) National Steel Corp. b) Geneva Steel Corp. c) Bethlehem Steel Co. d) USX Corp. e) LTV Steel Co.		a) 37.05 b) 4.18-32.84 c) 16.62 d) 11.19 e) 39.92	92.05.29	Positive
74	Hot-rolled carbon steel sheet	A/D	7208.13.01 7208.14.01 7208.23.01 7208.24.01	United States	a) Bethlehem Steel Co. b) Geneva Steel Co. c) LTV Steel Co. d) National Steel Co. e) USX Corp. f) Weirton Steel Corp. g) Samsung America h) Hubbell Int'l i) Hemas Empresas			92.05.29	

TABLE A2-1 (CONT'D)

CASE NUMBER	COMMODITY/ PRODUCT	A/D AND/OR C/V	HS CODE	COUNTRY AFFECTED	EXPORTING COMPANY	MARGIN OF DUMPING (PRELIMINARY) %	MARGIN OF DUMPING (FINAL) %	INITIATION DATE	DISPOSITION
75	Cold-rolled carbon steel sheet	A/D	7209.12.01 7209.13.01 7209.22.01 7209.23.01	United States	a) Bethlehem Steel Co. b) Geneva Steel Co. c) LTV Steel Co. d) National Steel Co. e) USX Corp. f) Hubbell Int'l		a) 02.73 b) 12.88 c) 12.88 d) 12.88 e) 06.88 f) 07.82	92.05.29	Positive resolution excludes regime or management of compensating quotas
76	Lighting sets for Christmas trees	A/D	9405.30.01	China		1.97 Dls. 6.15 Dls. 7.05 Dls. 9.44 Dls.	00.0	92.08.19	Negative
77	Hydrogen peroxide	A/D	2847.00.01	United States	Degussa Corporation		34.5	92.10.08	Positive, appeal denied
78	Fluorspar: containing, by weight, more than 97% calcium fluoride	A/D	2529.22.01	China	Minmetals Shangai Corp.; Metalurgica Imp.; and Exp. China Cosmetals Inc.		16	92.11.26 (without compensation quota)	Positive

TABLE A2-1 (CONT'D)

Case Number	Commodity/ Product	A/D and/or C/V	HS Code	Country Affected	Exporting Company	Margin of Dumping (Preliminary) %	Margin of Dumping (Final) %	Initiation Date	Disposition
79	Candles, tapers, and the like	A/D	3406.00.01	China	Universal Candle Co.; Samuelson and Co.; Fedmet Trading Co.		103	92.12.24 (provisional resolution)	Positive
80	Flat-rolled products of iron, plated or coated	A/D	7210.31.01 7210.31.99 7210.39.99 7210.41.01 7210.41.99 7210.49.01 7210.49.99 7210.70.01 7210.70.99	United States	a) Bethlehem Steel Co. b) US Steel Int'l c) New Process Steel d) All the rest of the companies	a) 05.84 b) 09.74 c) 29.0 d) 29.0	38.21 (all the companies)	92.12.24	Positive
81	Carbon steel plate	A/D	7208.32.01 7208.33.01 7208.42.01 7208.43.01	United States	a) US Steel Corp. b) National Steel Co. c) Geneva Steel Co. d) Gulf States Steel e) Steel Incorporation	a) 28.73 b) 28.25 c) 38.13 d) 28.73 e) 28.73	c) 03.86 7208.4201 7208.4301 c) 78.46 7208.3201 7208.3301	92.12.24	Positive

TABLE A2-1 (CONT'D)

CASE NUMBER	COMMODITY/ PRODUCT	A/D AND/OR C/V	HS CODE	COUNTRY AFFECTED	EXPORTING COMPANY	MARGIN OF DUMPING (PRELIMINARY) %	MARGIN OF DUMPING (FINAL) %	INITIATION DATE	DISPOSITION
Cont'd.									
81	Carbon steel plate	A/D			f) Inland Steel Co.	f) 28.73			
					g) Sharon Steel Co.	g) 28.73			
					h) LTV Steel Co.	h) 28.73			
					i) Hemas Empresas	i) 28.73			
					j) Samsung America	j) 23.05			
					k) Bethlehem Steel Co.	k) 28.73	k) 46.18		
					l) Lukens Steel Co.	l) 28.73	l) 44.90		
					m) USX Corp.	m) 17.66	m) 76 all the others: 78.46		
					n) Weirton Steel Corp.	n) 22.56			
					o) Hubbell Int'l	o) 28.73			
82	Padlocks made of brass	A/D	8301.10.01	China	*Master Lock Company	484	181 *0.0	92.12.24	Positive
83	Polymers of styrene	A/D	3903.19.02 3903.19.99	United States	a) Ashlan Chemical	44.32	a) 44.32	93.03.03	Positive
					b) Chemister US Inc.	11.82	b) 11.62		

TABLE A2-1 (CONT'D)

CASE NUMBER	COMMODITY/ PRODUCT	A/D AND/OR C/V	HS CODE	COUNTRY AFFECTED	EXPORTING COMPANY	MARGIN OF DUMPING (PRELIMINARY) %	MARGIN OF DUMPING (FINAL) %	INITIATION DATE	DISPOSITION
Cont'd.									
83	Polymers of styrene				c) Dow Chemical Co. d) BASF Corporation e) Mueshlstein Int'l and other exporters		c) 0.0 d) 28.10 e) 44.32		
84	Polymers of styrene	A/D	3903.19.02 3903.19.99	Federal Republic of Germany				93.03.03	Negative
85	Meat of bovine animals, frozen	A/D C/V	0202.10.01 0202.20.99 0202.30.01	European Community	Emborg Foods Aalborg	43.18	45.74	93.03.05	Positive
86	Live swine, meat of swine, fresh, chilled or frozen	A/D	0103.01.01 0103.91.99 0103.92.99 0203.11.01 0203.19.99 0203.21.01 0203.22.01 0203.29.99 0206.30.01	United States	Mitchell's Pigotel; National Farms; The Pig Company; John Morell & Co;; Farmland Foods; Excel Corp.; Sioux-Preme Packing;; Bonanza Meat Co;;	0.0	0.0	93.03.05 Declares untimely the application for safeguards	Negative

CANADIAN TRADE AND INVESTMENT POLICIES

TABLE A2-1 (CONT'D)

CASE NUMBER	COMMODITY/ PRODUCT	A/D AND/OR C/V	HS CODE	COUNTRY AFFECTED	EXPORTING COMPANY	MARGIN OF DUMPING (PRELIMINARY) %	MARGIN OF DUMPING (FINAL) %	INITIATION DATE	DISPOSITION
Cont'd.									
86	Live swine, meat of swine, fresh, chilled or frozen		0206.30.99 0206.41.01 0206.49.01 0206.49.99 0209.00.01 0210.12.01 0210.90.99		El Paso Meat Co.; Exim Commodities Exchange; Penton Meats; E.R. & A. Trading Co.;				
87	Slide fasteners fitted with chain scoops of base metal	A/D	9607.11.01	Colombia	Industrias Yidi, Ltda.; Industrias Eka, Ltda.	00.0 00.0	00.0	93.03.05	Negative
88	Footwear	A/D	a) 6401 b) 6402 (except HS code 6402.11.01)	China	All exporters		a) 165 b) 232	93.04.15	Positive, except annex 7
89	Footwear	A/D	6403 (except HS code 6403.19.99 and 6403.99.99)	China	All exporters		323	93.04.15	Positive, except annex 7

TABLE A2-1 (CONT'D)

CASE NUMBER	COMMODITY/ PRODUCT	A/D AND/OR C/V	HS CODE	COUNTRY AFFECTED	EXPORTING COMPANY	MARGIN OF DUMPING (PRELIMINARY) %	MARGIN OF DUMPING (FINAL) %	INITIATION DATE	DISPOSITION
90	Footwear	A/D	6404 (except HS code 6404.11.99)	China	All exporters		313	93.04.15	Positive, except annex 7
91	Footwear	A/D	6405 (except HS code 6405.10.99)	China	All exporters		1105	93.04.15	Positive, except annex 7
92	Footwear	A/D	6406	China	All exporters		0.0	93.04.15	Positive, except annex 7
93	Bicycles	A/D	40.11	China		279		93.04.15	Positive
94	New pneumatic tires, of rubber	A/D	40.13	China		279		93.04.15	Positive
95	Inner tubes, of rubber	A/D	87.12	China		144		93.04.15	Positive
96	Textiles	A/D	a) 52.01 to 52.12	China	Aldwick Textile Export; Kendall Yantai Medical; Shangai Textiles; Shangai Hua Shen; Zheijang Medicine	54	a) 331	93.04.15	Positive

TABLE A2-1 (CONT'D)

CASE NUMBER	COMMODITY/ PRODUCT	A/D AND/OR C/V	HS CODE	COUNTRY AFFECTED	EXPORTING COMPANY	MARGIN OF DUMPING (PRELIMINARY) %	MARGIN OF DUMPING (FINAL) %	INITIATION DATE	DISPOSITION
97	Textiles	A/D	a) 53.01 to 53.11	China	Aldwick Textile Export; Kendall Yantai Medical; Shangai Textiles; Shangai Hua Shen; Zheijang Medicine	331	a) 331	93.04.15	Positive
98	Textiles	A/D	b) 54.01 to 54.08	China	Aldwick Textile Export; Kendall Yantai Medical; Shangai Textiles; Shangai Hua Shen; Zheijang Medicine	331 501	b) 501	93.04.15	Positive
99	Textiles	A/D	55.01 to 55.16 and 5402.49.05	China	Aldwick Textile Export; Kendall Yantai Medical; Shangai Textiles; Shangai Hua Shen; Zheijang Medicine	501 54	b) 501	93.04.15	Positive

TABLE A2-1 (CONT'D)

CASE NUMBER	COMMODITY/ PRODUCT	A/D AND/OR C/V	HS CODE	COUNTRY AFFECTED	EXPORTING COMPANY	MARGIN OF DUMPING (PRELIMINARY) %	MARGIN OF DUMPING (FINAL) %	INITIATION DATE	DISPOSITION
100	Textiles	A/D	c) 30.05, 58.03, 59.11	China	Aldwick Textile Export; Kendall Yantai Medical; Shangai Textiles; Shangai Hua Shen; Zheijang Medicine	54	a) 54	93.04.15	Positive
101	Clothing (finished garments)	A/D	61.01 to 61.17	China	(1) Algunos Exports (2) Others	533	(1) 0 (2) 533	93.04.15	Positive
102	Clothing (unfinished garments)	A/D	62.01 to 62.17	China	(1) Algunos Exports (2) Others	533	(1) 0 (2) 533	93.04.15	Positive
103	Clothing (other articles of textiles and clothing)	A/D	63.01 to 63.10	China	All	533	379	93.04.15	Positive
104	Chemical organic products	A/D	29.01 to 29.42	China	China National Chemical	208.81	208.81	93.04.15	Positive
105	Hand tools; hand saws; files, rasps, pliers and hand-operated spanners and wrenches	A/D	82.01 82.03 82.04 82.05 82.06	China	Asics Tiger Corp.; US Trade Corp.; Shangai Machinery Imp Export Corp.	312	312	93.04.15	Positive (except annex 1.1)

TABLE A2-1 (CONT'D)

CASE NUMBER	COMMODITY/ PRODUCT	A/D AND/OR C/V	HS CODE	COUNTRY AFFECTED	EXPORTING COMPANY	MARGIN OF DUMPING (PRELIMINARY) %	MARGIN OF DUMPING (FINAL) %	INITIATION DATE	DISPOSITION
106	Electric motors and generators and their accessories	A/D	85.01 to 85.48	China		129	129	93.04.15	Positive (annex 3.1)
107	Recording apparatus for the reproduction of sound and for the reproduction of TV sounds and images.	A/D	Chapter 85	China	(1) Black & Decker (2) Others		(1) 51.4 (2) 129	93.04.15	Positive
108	Iron and non-alloy steel	A/D	7207.10.01 7207.19.02 7214.20.01 7214.40.01 7214.50.01 7214.60.01 7224.10.04 7228.30.01 7228.30.99 7228.40.01 7228.50.02 7228.50.99	Brazil	Acos Ipanema, SA; Acos Villares, SA; Villares Ind. de Base; Acos Especiais Itabira	00.0	(*)	93.04.15	Positive (except annex 9)

TABLE A2-1 (CONT'D)

Case Number	Commodity/ Product	A/D and/or C/V	HS Code	Country Affected	Exporting Company	Margin of Dumping (Preliminary) %	Margin of Dumping (Final) %	Initiation Date	Disposition
109	Sulphuric acid	A/D	2807.00.01	Japan	Dowa Mining Co.; Hibi Kyodo Mitsubishi Metal Mining Co.; Nippon Mining Co.; Sumitomo Metal Mining Co.; IMCI, SA; Interacid Trading Inc.	166.67		94.04.15	Positive (provisional)
110	Toys	A/D	9501.00.02 9503.60.01 9503.90.01 9504.40.01 9505.90.99 9506.62.01	China	MB Sales Ace Novelty Mattel Inc. Hasbro Inc. Tayco International Woolworth Overseas	351	351	93.04.15	Positive (except annex 2.1)
111	Pencils	A/D	9609.10.01	China	China 2. Pencil Fac.; Guang Dong Stationary & Sporting; Shangdong Light Prod.; Sichuan Light Ind.; Tianjin Stationery Imp.	451	451	94.04.20	Positive
112	Homopolimeros	A/D		United States				93.04.23	Negative

TABLE A2-1 (CONT'D)

CASE NUMBER	COMMODITY/ PRODUCT	A/D AND/OR C/V	HS CODE	COUNTRY AFFECTED	EXPORTING COMPANY	MARGIN OF DUMPING (PRELIMINARY) %	MARGIN OF DUMPING (FINAL) %	INITIATION DATE	DISPOSITION
113	Refrigerator-freezer, fitted with separate external doors	A/D	8418.10.01 8418.21.01	Korea	Daewoo Electronics Co.; Goldstar Co. Ltd.; Samsung Electronics	0.0		93.04.23	Negative
114	Flat-rolled products of iron, plated or coated	C/V	7210.31.01 7210.31.99 7210.39.01 7210.39.99 7210.41.01 7210.41.99 7210.49.01 7210.49.99 7210.60.01 7210.70.01 7210.70.99	United States	a) Armco Steel Co. b) Bethlehem Steel Co. c) Block Steel Co. d) California Steel Co. e) Cyclops/Armco Steel Incorporation f) Gregory Galvanizing g) Gulf Steel Inc. h) I/N KOTE (Inland) i) Inland Steel Co. j) Metaltech Co. k) National Steel Co. l) Pinole Point Steel m) Rouge Steel Co. n) Sharon Steel Co. o) Triumph Industries p) USS Corp. (USX) q) Warren Consolidated			93.06.30	Negative

Table A2-1 (cont'd)

Case Number	Commodity/ Product	A/D and/or C/V	HS Code	Country Affected	Exporting Company	Margin of Dumping (Preliminary) %	Margin of Dumping (Final) %	Initiation Date	Disposition
Cont'd.									
114	Flat-rolled products of iron, plated or coated				r) Weiton Steel Corp. s) Wheeling Pittsburgh t) Wheeling-Nisshing u) L-S II-Electro-Galv. v) LTV Steel Co. w) USS-Posco Ind. x) World Metals Inc. y) New Process Steel z) Mitsui & Co. Inc. aa) US Steel Corp. bb) Hubbell Int'l				
115	Cold-rolled carbon steel sheet	A/D C/V	7209.12.01 7209.13.01 7209.23.01	a) Canada (A/D) b) United States (C/V)	a) Dofasco Inc.; Samuelson & Co; The Titan Industries b) US Steel Group & US Steel Corp.; Bethlehem Steel Corp.; Inland Steel Company; LTV Steel Co.; Armco Steel Co.	a) 0.0	a) 0.0 b) 0.0	93.08.28	Positive

TABLE A2-1 (CONT'D)

CASE NUMBER	COMMODITY/ PRODUCT	A/D AND/OR C/V	HS CODE	COUNTRY AFFECTED	EXPORTING COMPANY	MARGIN OF DUMPING (PRELIMINARY) %	MARGIN OF DUMPING (FINAL) %	INITIATION DATE	DISPOSITION
Cont'd.									
115	Cold-rolled carbon steel sheet			c) South Korea (A/D)	c) Pohan Iron & Steel Co.; Samsung Company Ltd.; Daewoo Corporation; Keoyang Company		c) 0.0		
				d) Venezuela (A/D, C/V)	d) i) C.V.G. Sidor; ii) Other Exporters		d) i) 0.0 undertaking ii) 55.75		
				e) Federal Republic of Germany (A/D)	e) Krupp AG.; Hoesch-Krupp; Krupp Hoesch Stahl AG.; Ferrostahl aktiengesellschaft; Thyssen Stahlunion GmbH.; Preussag Stahl Aktiengesellschaft.		e) 185.76		
				f) Brazil (A/D, C/V)	f) i) Usinas Sederurgica de Minas Gerais, SA; ii) Cia. Siderurgica Nacional; iii) Cia. Siderurgica Paulista; iv) Other exporters		f) i) 17.36 ii) 31.57 iii) 31.57 iv) 31.57		
				g) Australia (A/D)	g) BHP Steel BHP Trad.		g) 0.0		

Table A2-1 (cont'd)

Case Number	Commodity/ Product	A/D and/or C/V	HS Code	Country Affected	Exporting Company	Margin of Dumping (Preliminary) %	Margin of Dumping (Final) %	Initiation Date	Disposition
116	Hot-rolled carbon steel sheet	A/D C/V	7208.13.01 7208.14.01 7208.23.01 7208.24.01	h) Federal Republic of Germany (A/D)	h) Thyssen Stahl Union Ferrostaal	b) 0.0	h) 0.0	93.08.27	Positive
				i) South Korea (A/D)	i) Daewoo Company Samsung Pacific		i) 0.0		
				j) Canada (A/D)	j) i) Dofasco Inc. ii) Other exporters		j) i) 15.37 ii) 45.86		
				k) Netherlands (A/D)	k) i) Hoogovens Groep ii) Other exporters		k) i) 56.54 ii) 56.54		
				l) Brazil (A/D, C/V)	l) i) Usinas Siderurgica ii) Cia. Siderurgica Paulista iii) Other exporters		l) i) 10.25 ii) 31.41 iii) 31.41		
				m) Venezuela (A/D, C/V)	m) i) C.V.G. Sidor ii) Other exporters		m) i) 0.0 undertaking ii) 80.95		
117	Carbon steel sheet	A/D C/V	7208.32.01 7208.33.01 7208.42.01 7208.43.01	n) Canada (A/D)	n) Algoma Steel Inc. Stelco Inc.	c) 0.0	n) 0.0	93.08.27	Positive
				o) United States (C/V)	o) Klockner Steel Trade All exporters		o) 0.0		
				p) Brazil (A/D)	p) i) Usinas Siderurgica de Gerais, SA ii) Cia. Siderurgica Paulista iii) Other exporters		p) i) 36.28 ii) 37.32 iii 37.32		

CANADIAN TRADE AND INVESTMENT POLICIES

TABLE A2-1 (CONT'D)

CASE NUMBER	COMMODITY/ PRODUCT	A/D AND/OR C/V	HS CODE	COUNTRY AFFECTED	EXPORTING COMPANY	MARGIN OF DUMPING (PRELIMINARY) %	MARGIN OF DUMPING (FINAL) %	INITIATION DATE	DISPOSITION
118	Rolled carbon steel plate	A/D C/V	7208.12.01 7208.22.01	q) United States (C/V)	q) Bethlehem Steel Corp. United States Steel Co. National Steel Co. LTV Steel Co. Geneva Steel Co.	d) 0.0	q) 0.0	93.08.28	Positive
				r) South Korea (A/D)	r) Pohan Iron and Steel Co. Daewoo Company Samsung Pacific Inc.		r) 0.0		
				s) South Africa (A/D)	s) All exporters		s) 0.0		
				t) Canada (A/D)	t) All exporters		t) 31.08		
				u) Brazil (A/D, C/V)	u) i) Usinas Siderurgicas de Minas Gerais, SA ii) Cia. Siderurgica Nacional iii) All exporters		u) i) 14.15 ii) 23.95 iii) 23.95		
				v) Venezuela (A/D, C/V)	v) i) C.V.G. (SIDOR) ii) Other exporters		v) i) 0.0 ii) 84.05		
119	Fluors of fish	A/D	2301.20.01	Chile	w) Pesquera Coloso x) Pesquera Eperva y) Pesquera Iquique z) Pesquera Indo aa) Other exporters	e) 2.45 f) 24.6 g) 17.64 h) 28.42 i) 28.42	0.0	93.10.05	Negative

559

Table A2-1 (cont'd)

Case Number	Commodity/ Product	A/D and/or C/V	HS Code	Country Affected	Exporting Company	Margin of Dumping (Preliminary) %	Margin of Dumping (Final) %	Initiation Date	Disposition
120	Paper and paperboard	A/D	4802.52.01 4802.52.99	United States	a) Georgia Pacif Corp. b) International Paper Co. c) James River Corp. d) Perez Trading Co. Inc.	a) 14.69 b) 30.67 c) 00.00 d) 00.00		93.10.05	Negative
121	Valves made of iron or steel	A/D	8481.20.01 8481.20.04 8481.20.99 8481.30.04 8481.30.99 8481.80.01 8481.80.03 8481.80.09	China	a) Newmans Inc. b) Corp. Suzhou Valve Factory Sufa Sufa Valves, C-MEC Intersev Int'l Serv. American Energy Serv. Newco Zidell Valve Company Crane Valves	105	a) 4.0 b) 105.0 All the other companies	93.11.01	Positive
122	Caustic soda	A/D		United States				93.11.06	Negative
123 124 125 126 127 128 129	Urea	A/D	3102.10.01	(*) Russia (*) Belarus (*) Uzbekistan (*) Tadzhikistan (*) Lithuania (*) Estonia (**) Ukraine	a) Mitsui & Co. b) Ferico Limited	a) 57.23 b) 41.56 all the other companies: 57.23	(*) 0.0 (**) (annex 10)	93.11.08	(*) Negative (**) Positive (annex 10)

TABLE A2-1 (CONT'D)

CASE NUMBER	COMMODITY/ PRODUCT	A/D AND/OR C/V	HS CODE	COUNTRY AFFECTED	EXPORTING COMPANY	MARGIN OF DUMPING (PRELIMINARY) %	MARGIN OF DUMPING (FINAL) %	INITIATION DATE	DISPOSITION
130	Additives	A/D		United States				93.11.08	Negative
131	Sodium hydroxide – caustic soda	A/D	2815.12.01	United States	Trans Marketing Houston Inc. Occidental Chemical Co. Helm US Chemical HCI ChemicalOverseas Int'l Petrochemical Atlantic Chemical Int'l Enichem America Inc. Greenwood Chemicals P.P.G. Industries Vulcan Chemicals Div.	a) 24 All other companies: 116	35.83	93.11.08	Positive (annex 111)
132	Rods and bars of iron or non-alloy steel	A/D	7214.20.02	Brazil Spain	Cia. Siderurgica Pains. Dedini, SA Siderurgica Ferrostaal Aktiengesellschaft Marcial Ucin, SA	0.0	57.69	93.11.19	Positive
133	Pencil sharpener	A/D	8214.10.02	China	Import & Export de Shangai Anhui Light Ind.	145	145	93.11.19	Positive

TABLE A2-1 (CONT'D)

CASE NUMBER	COMMODITY/ PRODUCT	A/D AND/OR C/V	HS CODE	COUNTRY AFFECTED	EXPORTING COMPANY	MARGIN OF DUMPING (PRELIMINARY) %	MARGIN OF DUMPING (FINAL) %	INITIATION DATE	DISPOSITION
Cont'd.									
133	Pencil sharpener				China National Light Industrial Products Guangdong Stationary & Sport Good Branch Tonghua Pencil Sharpener Factory				
134	Maletas and Dolsas	A/D		China				93.11.29	Negative
135	Rods and bars of iron or non-alloy steel			Spain				93.11.29	Negative
136	Padlock and locks	A/D	8301.40.99	China	National Metals Prod. Guangzhou Metals & Minerals Corp. China National Metals Guangxi Metals & Minerals Co.	236	236	94.05.02	Positive
137	Tubes, pipes and hollow profiles, seamless, of iron or steel	A/D	(a) 7304.31.01 7304.31.08 7304.31.99 (b) 7304.31.03	United States	Commercial Steel Inc. Damille Metal Co. Jd. Fields & Co. Marubeni Tubulars Inc. Tubamerica	0.0	a) 82.41 b) 0.0	94.02.10	Positive

TABLE A2-1 (CONT'D)

CASE NUMBER	COMMODITY/ PRODUCT	A/D AND/OR C/V	HS CODE	COUNTRY AFFECTED	EXPORTING COMPANY	MARGIN OF DUMPING (PRELIMINARY) %	MARGIN OF DUMPING (FINAL) %	INITIATION DATE	DISPOSITION
138	New pneumatic tires, of rubber (of the kind used on bicycles)	A/D	4011.50.01	India	Govin Rubber Ltd. Metro Tyres Ltd. Ralson (India) Ltd. Sawhney Rubber Ind. Dewan Rubber Ind. Ltd. Dunlop India Ltd.	116		94.11.16	Positive (provisional)
139	Carbon steel sheet	A/D C/V	7208.32.01 7208.33.01 7208.42.01 7208.43.01	Brazil (A/D, C/V) Canada United States (C/V)	Algoma Steel Inc. Stelco Inc. Klockner Steel Trade Pohan Iron & Steel Co. Ilva USA Inc. Cia. Siderurgica Nal USIMINAS, SA Bethlehem Steel Co. Geneva Steel Co. LTV Steel Co. National Steel Co. Samsung Pacific Inc. USA Steel Co.	0.0		93.10.27	Positive
140 141 142 143 144	Cold-rolled carbon steel sheet	A/D C/V	7209.12.01 7209.13.01 7209.23.01	Brazil (A/D,C/V) Venezuela (A/D,C/V) Germany	Hoesch Export AG Ferrostaal Samuelson & Co. Fedmet Trading Co. Steel Co.	0.0		93.10.28	Positive (provisional)

Table A2-1 (cont'd)

Case Number	Commodity/Product	A/D and/or C/V	HS Code	Country Affected	Exporting Company	Margin of Dumping (Preliminary) %	Margin of Dumping (Final) %	Initiation Date	Disposition
Cont'd.									
145	Cold-rolled carbon steel sheet			Australia	Dofasco				
				Canada	Algoma Steel Inc.				
				United States (C/V)	Co-Steel Inc.				
					The Titan Industries				
					Daewoo Company				
					Keo Yang Company				
					Samsung Company				
					BHP Trading				
					BHP Steel BHP Trad.				
					Cia. Siderurgica Nal				
					Cia. Siderurgica Paulista				
					USIMINAS				
					Dufercco Steel				
					Marubeni America				
					Ferrostal Do Brazil				
					BHP Trading Inc.				
					C.V.G. SIDOR				
					Bethlehem Steel Co.				
					LTV Steel Co.				
					US Steel Corp.				
					National Steel Co.				
					Geneva Steel Co.				
					Samsung Pacific Inc.				
					Sharon Steel Co.				
					Stemcor USA Inc.				

TABLE A2-1 (CONT'D)

CASE NUMBER	COMMODITY/ PRODUCT	A/D AND/OR C/V	HS CODE	COUNTRY AFFECTED	EXPORTING COMPANY	MARGIN OF DUMPING (PRELIMINARY) %	MARGIN OF DUMPING (FINAL) %	INITIATION DATE	DISPOSITION
146	Hot-rolled carbon steel sheet	A/D	7208.13.01	Germany	Thyssen Stahl Union	0.0		93.10.27	Positive (provisional)
147		C/V	7208.14.01	Brazil	Ferrostaal				
148			7208.23.01	(A/D, C/V)	Samuelson Son & Co.				
149			7208.24.01	Canada	Fedmet Trading				
150				Korea	Algoma Steel, Inc.				
151				Low Countries	Steelco, Inc.				
				Venezuela	Co-Steel Inc.				
				(A/D, C/V)	Daewoo Company				
					Samsung Pacific				
					Cia. Siderurgica Nal.				
					COSIPA				
					USIMINAS				
					Duferco Steel Inc.				
					Ferrostal Do Brazil				
					Marubeni America				
					BHP Trading Inc.				
					C.V.G. SIDOR				
					Bethlehem Steel				
					LTV Steel Co.				
					USX Corp.				
					ARMCO				
					Weirton Steel Corp.				
					Inland Steel Corp.				
					Geneva Steel				

TABLE A2-1 (CONT'D)

CASE NUMBER	COMMODITY/ PRODUCT	A/D AND/OR C/V	HS CODE	COUNTRY AFFECTED	EXPORTING COMPANY	MARGIN OF DUMPING (PRELIMINARY) %	MARGIN OF DUMPING (FINAL) %	INITIATION DATE	DISPOSITION
Cont'd.									
151	Hot-rolled carbon steel sheet				Lukens Steel Corp. Nucor Steel Corp. Sharon Steel Corp. Stemcor USA Inc. Gulf States Steel Inc				
152	Rolled carbon steel plate	A/D	7208.12.01	Brazil	Fedmet Trading Co.	0.0		93.10.28	Positive (provisional)
153		C/V	7208.22.01	(A/D, C/V)	The Titan Ind. Corp.				
154				Canada	Daewoo Company				
155				Korea	Coutinho Caro & Co.				
156				United States (C/V)	Maurice Pincoffs Co. Cia. Siderurgica Nal. Cia. Siderurgica Paulista				
				South Africa	USIMINAS, SA				
				Venezuela (A/D, C/V)	Duferco Steel Inc. Marubeni America Ferrostaal BHP Trading Inc. Inc. C.V.G. (SIDOR) USA Steel Co. National Steel LTV Steel Co. Samsung Pacific Inc. Bethlehem Steel Geneva Steel				

CANADIAN TRADE AND INVESTMENT POLICIES

TABLE A2-1 (CONT'D)

CASE NUMBER	COMMODITY/ PRODUCT	A/D AND/OR C/V	HS CODE	COUNTRY AFFECTED	EXPORTING COMPANY	MARGIN OF DUMPING (PRELIMINARY) %	MARGIN OF DUMPING (FINAL) %	INITIATION DATE	DISPOSITION
157	Diammonium Phosphate	A/D	3105.30.01	United States	a) Cargill Fertilizer b) Mobil Mining & Minerals Co. c) J.R. Simplot Co. Phoschem, Inc.	a) 2.07 b) 12.66 c) 16.43 and all other exporters		94.06.23	Positive (provisional)
158	Rolled carbon steel plate	A/D	7208.12.01 7208.22.01	Armenia	Magnitogorskiy Metallurgischeski Kombinat.	29.30		94.11.22	Positive (provisional)
159				Azerbaijan	Novolipetskiy Met. Zavod.				
160				Belarus	Cheropovets Iron & Steel Works				
161				Estonia					
162				Georgia					
163				Kazakhstan					
164				Kirghizistan					
165				Latvia					
166				Lithuania					
167				Moldova					
168				Tadzhikistan					
169				Turkmenistan					
170				Ukraine					
171				Uzbekistan					
172				Russia					
173	*Dolvinos*	A/D		United States				94.06.03	Provisional stage

567

Table A2-1 (cont'd)

Case Number	Commodity/ Product	A/D and/or C/V	HS Code	Country Affected	Exporting Company	Margin of Dumping (Preliminary) %	Margin of Dumping (Final) %	Initiation Date	Disposition
174	Styrene-butadiene rubber	A/D	4002.19.01 4002.19.02 4002.19.99	Brazil	Petroflex, Industria e Comercio, SA	0.0		94.10.27	Positive (provisional)
175	Meat of swine	C/V	a) 0203.21.01 b) 0203.22.01 c) 0203.29.99	Denmark	Towers Thompson Danish Crown Danish Swedish Meat Dat-Schaub. Consuma A/S Peter Holm A/S Slagteriregion Syd. Sorensen Food HP A/S A-S Jens Christiansen, Kroller Hans & Co. Lyka Snacks A/S Findane A/S Emborg Foods-Aaborg Ess-Food	a) 42.83 b) 23.79 c) 21.81		94.11.22	Positive (provisional)
176	Tube or pipe fittings: couplings, elbows, sleeves; of iron	A/D		Brazil				95.04.11	
177	Ball bearings	A/D		Taiwan				95.10.11	

TABLE A2-1 (CONT'D)

CASE NUMBER	COMMODITY/ PRODUCT	A/D AND/OR C/V	HS CODE	COUNTRY AFFECTED	EXPORTING COMPANY	MARGIN OF DUMPING (PRELIMINARY) %	MARGIN OF DUMPING (FINAL) %	INITIATION DATE	DISPOSITION
178	Durum wheat	C/V	1001.10.01	Canada	Cargill Ltd. (Canada)			94.04.04	
179			1001.90.99	United States	Harvest States Corp. Farmland Industries Cargil Inc. Continental Grain Co. Archer Daniels Midland Co. The Scoular Company Bunge Corporation Louis Dreyfuss Corp. Bartlett Company De Bruce Grain Inc. Toepfer International				
180	Polymers of propylene	A/D	3902.10.01	United States	a) Huntsman Polypropylene b) Hulmont USA Inc. c) Phillips Petroleum d) Exxon Chemical Trading Inc. e) Amoco Chemical Co. f) Aristech Chemical	a) 0.0 b) 2.45 c) 4.36 d) 25.69 e) 5.11 f) 2.33	0.0	93.04.23	Negative

TABLE A2-1 (CONT'D)

CASE NUMBER	COMMODITY/ PRODUCT	A/D AND/OR C/V	HS CODE	COUNTRY AFFECTED	EXPORTING COMPANY	MARGIN OF DUMPING (PRELIMINARY) %	MARGIN OF DUMPING (FINAL) %	INITIATION DATE	DISPOSITION
Cont'd.									
180	Polymers of propylene				g) Fina Oil Chem. Co. h) Lyondell Polymers i) Solvay Polymers Inc. Eastman Chemical USA Inc., and all other exporters	g) 11.42 h) 0.0 i) 25.69			
181	Chemical wood pulp soda of sulphates	A/D	4703.21.01	EE.UU.		0.0		94.05.02	Negative
182	Sodium bicarbonate	A/D	2836.20.01	EE.UU.		0.0	0.0	93.08.19	Negative
183	Meat of bovine animals, frozen, chilled and edible offal	A/D	0201.30.01 0202.30.01 0206.21.01 0206.22.01 0504.00.01	United States	Excel Corporation Monfort Inc. H & H Foods Co. Alpha Star Int'l Inc. IBP, Inc. MB Int'l Corp. AJC Int'l Inc. Local & Western Texas Inc. Sun Land Beef, Co. Sigma Alimentos Int'l	0.0		94.06.03	Positive (provisional)

ENDNOTES

1 The November 1994 meeting of leaders in the Asia Pacific Economic Cooperation group of countries was held in Bogor, Indonesia and the December 1994 meeting of the Summit of the Americas was held in Miami, Florida.

2 Kenichi Ohmae et al. popularized the term Triad to refer to the United States, Europe and Japan. See Kenichi Ohmae et al. *The Triad Power: The Coming Shape of Global Competition*. New York: Macmillan, 1985.

3 The rounds of the GATT negotiations involved reduction of tariffs and some non-tariff barriers through reciprocal negotiations on a most-favoured-nation (MFN) basis. However, because of the concept of special and differential treatment, only the OECD countries actually reduced their trade barriers. For a thorough discussion of the concepts of reciprocity and MFN, see William Diebold. "The History and The Issues," in *Bilateralism, Multilateralism and Canada in U.S. Trade Policy*. Edited by William Diebold, Cambridge, Mass.: Ballinger, 1988, pp. 1-36.

4 See Edward M. Graham. "Global Corporations and National Governments." Washington: Institute for International Economics, 1996.

5 For a discussion of the legal and historical aspects of special and differential treatment, see Robert Hudec. *Developing Countries in the GATT System*. Thames Essay no. 50, Aldershot, U.K.: Gower for the Trade Policy Research Centre, 1987.

6 The World Bank. *The East Asian Miracle: Economic Growth and Public Policy*. New York: Oxford University Press, 1993.

7 Not surprisingly, this scope for discretionary selective intervention created opportunities for corruption, as recent revelations about South Korea have indicated. Perhaps the interesting question for political economy is why was the corruption and rent-seeking behaviour apparently efficiency and growth promoting? One of the debilitating results of the import-substitution regimes seems to be the insidious effects of rent-seeking on the allocation of entrepreneurial capital and the focus of enterprise management.

8 There is no perfect measure of integration, but the following indicators are suggestive. In the mid-1970s both China and India had a ratio of exports (or imports) to GDP of about 5 percent. By 1991 China's had risen to almost 20 percent, while India's had risen to only 6 percent. Also, India's share of world trade has remained stagnant at around 0.5 percent while China's has risen sharply to more than 2 percent. See A. Alexandroff et al. *Canada-China Trade and Economic Relations: Building a Strategic Partnership*. Ottawa: Centre for Trade Policy and Law, Carleton University, 1995, Chart 2, p. 5.

9 The Asian NIE also had other complementary policies to promote growth and high savings, but these are the not the focus of discussion here.

10 Except for the People's Republic of China (and the Republic of China), which is (are) not yet a WTO member(s).
11 A few OECD economies such as Australia's had even higher barriers in the industrial sector, but this was the exception.
12 This analysis draws on research conducted by Sunder Magun for the Canada-China trade project at the Centre for Trade Policy and Law.
13 This estimate is based on a detailed study by M.G. Smith and J.F. Bence. *Tariff Equivalents for Bilateral Export Restraints on Canada's Textile and Apparel Trade.* Ottawa: Institute for Research on Public Policy, 1989, prepared for the Canadian International Trade Tribunal. The methodology used in this study was to use the quota premia in Hong Kong as a measure of the incremental impact of the MFA quota restrictions. This was the methodology used by G.P. Jenkins. *Cost and Consequences of the New Protectionism: The Case of Canada's Clothing Sector.* Ottawa: The North-South Institute, 1985. This method is increasingly less useful, simply because Hong Kong has become a relatively high-cost supplier of apparel.
14 GATT Secretariat. *Trade Policy Review: Japan, 1994.* Geneva: GATT Secretariat, 1995, p. 15.
15 Ibid., p. 105.
16 Mark Mason. "Japan's Low Levels of Inward Direct Investment: Causes, Consequences and Remedies," in *Corporate Links and Foreign Direct Investment in Asia and the Pacific.* Edited by E. Chen and P. Drysdale, Pymble: Harper Collins, 1995, pp. 129-51.
17 Ryuhei Wakasugi. "On the Causes of Low Levels of FDI in Japan" in *Corporate Links and Foreign Direct Investment in Asia and the Pacific.* Edited by E. Chen and P. Drysdale. Pymble: Harper Collins, 1995, pp. 112-27.
18 GATT-WTO Secretariat. *Trade Policy Review: Indonesia.* Volume 1. Geneva: GATT-WTO Secretariat, 1995, p. 50.
19 Ibid., p. 60.
20 APEC Secretariat. *Guide to the Investment Regimes of the Fifteeen APEC Member Economies.* Singapore: APEC Secretariat, 1995, p. 151.
21 GATT Secretariat. *Trade Policy Review: Republic of Korea.* Volume 1. Geneva: GATT Secretariat, 1992, pp. 6, 64.
22 Ibid., p. 146.
23 GATT Secretariat. *Trade Policy Review: The Philippines.* Volume 1. Geneva, GATT Secretariat, 1993, pp. 4, 67.
24 Annex 7 of the Agreement on Subsidies and Countervailing Measures, concluded in the Uruguay Round Agreements under the World Trade Organization, has a special list of countries, such as Egypt and Nigeria, that can claim equivalent treatment as least developed countries as long as their per capita income is below US$1,000.
25 Trade Promotion Coordinating Committee. *Toward a National Export Strategy.* Report to the United States Congress, 30 September 1993.

26 See GATT Secretariat. "Final Act Embodying the Results of the Uruguay Round of Multilateral Trade Negotiations, Marrakesh, 15 April 1994." Geneva: GATT Secretariat, p. 117.

27 See Richard Lipsey's classic article: "The Theory of Customs Unions: A General Survey." *The Economic Journal*, 70 (September 1960), which emphasizes scale and growth effects as well as the static efficiency gains and losses.

28 See Peter Drysdale and Ross Garnaut. "The Pacific: A General Theory of Economic Integration," in *Pacific Dynamism and the International Economic System*. Edited by C.F. Bergsten and M. Noland. Washington: Institute for International Economics, 1993; and Soogil Young. "East Asia as a Force for Globalism," in *Regional Integration and the Global Trading System*. Edited by K. Anderson and R. Blackhurst. New York: Harvester Wheatsheaf, 1993.

29 Hadi Soesastro. "Implications of the Post-Cold War Politico-Security Environment on the Pacific Economy," in *Pacific Dynamism and the International Economic System*. Edited by C.F. Bergsten and M. Noland. Washington: Institute for International Economics, 1993, pp. 365-87.

Comment

Edward M. Graham
Institute for International Economics
Washington, D.C.

MURRAY SMITH'S ARTICLE IS A RATHER ECLECTIC ESSAY dealing with the integration or "globalization" of the world economy. Smith maintains that this integration was largely limited to the OECD nations, but that the creation of the WTO, as well as certain regional arrangements, implies that other nations will soon be forced to become part of the "global" economy. In this context, Smith discusses Canadian trade and investment barriers with respect to the Asia Pacific (APEC) nations, the need to integrate China into the global economy (largely via WTO accession), and policy issues facing Canada.

One of Smith's major themes is that developing countries, including those of the Asia Pacific region, have until very recently been little affected by globalization. Interestingly, this has been, in part at least, the result of the GATT Part IV exemptions for developing countries from the obligations to which GATT contracting parties (member countries) putatively all subscribed. The obligations guarantee a certain openness of these nations' economies. The exemptions were meant to allow developing countries to "catch up" with the advanced nations, with the implication that to accept the GATT obligations would doom these countries to perpetual underdevelopment. But with the benefit of hindsight it is clear that the effect of the exemptions has been to retard rather than stimulate development.

Smith uses China to illustrate this by way of reverse example: China's recent remarkable growth has been fostered by a slow but steady (and thus far irreversible) opening of its economy. China, ironically, was not a GATT contracting party, nor is it yet a WTO member. The opening has been unilateral and voluntary on China's part. India, by contrast, is a founding member of the GATT but has largely been untouched by the GATT obligations. India's economy has been among the most closed and autarkic in the world. Also, relative to China's, India's economy has been stagnant.

All of this, Smith duly notes, is about to change. Over the next 10 years, virtually all of the developing nations' exemptions from WTO obligations will be phased out; indeed, many already have been. The question is, will it make a difference? Will economic performance in the developing world as a whole look less like that recently in India and more like that in China? The answer is, of course, that we do not know. Experience seems to suggest that openness of a developing nation's economy enhances dynamic performance, but there are a lot of factors that determine such performance, of which relative openness is but one. Smith seems to think that the answer is "yes," and I hope that he is right. But, in the end, only time will tell.

Smith spends some time discussing Canadian trade and investment barriers. Being neither a Canadian nor an expert on Canada per se, I have little to say about this. He notes that on the important issue of whether NAFTA will create trade diversion that will affect the Asia Pacific economies (Smith specifically singles out China), most of the diversion will take place with respect to the U.S. market, where in some commodity categories Mexican products might displace Chinese ones. With respect to Chinese-Canadian trade, the main problems (from the point of view of Chinese exports) seem to lie in the textile/apparel sector and with respect to Canadian anti-dumping practices.

Smith returns to China in the concluding section of his paper, indicating that a high priority for Canadian policy should be accession of China to the WTO. I second this. However, it must be noted that the main force blocking it is the United States. Exactly what Canada can do to affect U.S. policy is an interesting question. The main controversy centres around whether China should be subject to the transition arrangements for "annex 7" developing nations. On the basis of per capita income, as measured by the World Bank (a measurement which, in the view of a number of Chinese experts, is too low), China would qualify for these arrangements. As a practical matter, however, the United States is probably right in insisting that China is too big, growing too rapidly, and is in some sectors too advanced to be accorded these preferences.

Beyond this, however, the U.S. approach to Chinese accession is not constructive. It consists mostly of a whole series of bilateral demands that the U.S. government insists must be met before the United States will approve Chinese accession. More constructive would be the approach suggested by WTO Director General Renato Ruggiero, whereby a special deal might be cut for China that would put it neither in the "developing" ("annex seven") nor the "developed" category. To some, a special deal might seem onerous, but it is probably the only practical solution to the dilemma of Chinese accession. What such a deal would imply is that, item by item, issue by issue, special terms must be worked out for China that take into account both the backwardness of many sectors and regions of the Chinese economy and the size and dynamism of others. Canada could do much worse than to become an ardent backer of this approach.

Richard G. Lipsey
Canadian Institute for
Advanced Research
Economic Growth and Policy Program
Simon Fraser University

& Russel M. Wills
Cognetics International Research Inc.
Vancouver

14

Science and Technology Policies in Asia Pacific Countries: Challenges and Opportunities for Canada

IN THIS PAPER WE FIRST REVIEW recent policies in South Korea and Singapore with respect to science and technology and industrial development. We then suggest some of the opportunities and challenges that are created by these policies for small open industrialized countries such as Canada. Finally, we suggest an agenda for future research.

Approach

THERE IS A GROWING DEBATE IN ACADEMIC CIRCLES over the causes of the "East Asian Miracle" (see, for example, Young, 1994). Whatever the causes, the success of the Asian newly industrialized countries (NIC) – Singapore, South Korea, Taiwan and Hong Kong – is undeniable. They have experienced massive growth that the other Asian developing nations have latterly attempted to emulate with more or less success. All of the NIC, apart from Singapore, in their unbridled push for industrial growth, have created major environmental problems in many of their larger cities. There has been considerable loss of life when industrial premises and apartment complexes have collapsed as a result of unregulated building practices. But whatever the unfortunate side effects, the NIC have grown on a scale that seemed impossible in the two decades after World War II.

Pack and Westphal (1986) destroyed the then-common myth that the NIC's success was based on *laissez-faire* market policies. Although all four NIC employed outward-looking, market-oriented policies, all but Hong Kong created what Wade (1990) calls "governed market economies" and one of the present authors has called "massaged market economies" (Lipsey, 1994). We accept the importance of a market orientation, expressed in what John Williamson

(1990) calls "the Washington consensus." But what we are concerned with here is the "governed" or "massaged" part of these economies, the roles that governments undoubtedly played, using very specific and detailed instruments.

There is, in fact an ongoing debate over the role of government policies in the NIC's success. The policies are often considered generically, in terms of tax relief for investment or assistance to R&D, for example. Sometimes such policy analysis involves the construction of various indices of patent rates and "technological importance" for East Asian countries of the type that common sense might anticipate, as is the case with the recent work by the group at the National Science Foundation (Rausch, 1995) analyzing Asia's new high-tech competitors.

Getting beyond these generalities requires laying out the actual instruments in more detail than we have found in any single current source. Although the detail may seem tedious at times, the paradigm of the managed market economy is embedded in these very details, along with the institutional capacity to run the policies effectively.[2] We try not to pose choices between extremes that are seldom the objects of real policy, and instead to consider policies that are typically found in those countries. For example, the top-down, bureaucratic picking of winners sometimes found in Europe and North America is not what East Asian policy is all about. Instead, it is about finding a middle ground of cooperation between the government and the private sector to help identify future winners, with each sector using its particular expertise.

With respect to the policies that we consider in this paper, the important issue in most countries is not *laissez faire* versus heavy-handed interventionism, but finding an effective balance between private- and public-sector involvement that suits each country's needs and capacities.[3] To deal with this issue, it would be necessary to examine the successes and failures of specific policies and their possible applicability to other industrialized countries – a task that would take us well beyond what we can accomplish in this paper. Since most economic policy papers analysing science and technology (S&T) support measures in the NIC involve only the most general discussions, in the following we present the actual tax, fiscal, human resource and industrial incentives and programs used by two successfull NIC – Korea and Singapore – to integrate contemporary S&T into private sector economic advancement. While we take Canada as our example because we are concerned with Canadian policy, our analysis applies to other industrialized countries.

East Asian Background

It can be argued that the most successful science and technology policies and strategies have been articulated and implemented not in Europe but in Japan and the Asian NIC. With their eastern ethic, which stresses the claims of the community over the individual, countries such as Singapore and Korea have also become strong in cooperative policy making that involves both the government and the private sector. Also, several of the NIC have integrated industrial and science policies and instruments. Their institutional capacity to make and carry out such policies is impressive, and any other country seeking to copy these procedures and policies must determine whether it has institutional capacity to do so effectively.

In Japan, Singapore, South Korea, and Malaysia, S&T policies and strategies have guided such issues as:

- What industries should government emphasize during the next 5 to 10 years?

- What are the priority products in these industries that should be emphasized and supported via the tax and incentive structure?

- What specific technologies are necessary to produce such priority products efficiently?

- Which of these technologies should the government help the private sector to develop, and which should be imported from abroad through arrangements with multinationals?

The World Bank (1993) has argued that rapid economic growth in the NIC is partly a result of well-articulated technology policies and strategies embedded in industrial support measures. Singapore, for example, chose to emphasize software and computer/financial services 16 years ago at a time when computer software was embedded in machines and given away freely. At that time, there was almost no distinct software industry anywhere in the world. How did they make this visionary prognosis, given the alleged inability of government officials to pick industrial winners? Singapore has also effectively tied its public research effort to industrial needs, and has now gone on to develop and apply sophisticated tax/fiscal incentives that support certain sectors of a biotechnology industry, mobile personal communications, multimedia, and other sectors. (In contrast, the Canadian biotechnology industry is shrinking.)

New Developments

THE NIC AND THEIR MORE SUCCESSFUL FOLLOWERS provide striking examples of the substitution of knowledge for a natural resource base. The governments of these countries believe that industrial competition in the next century will be over innovation of sophisticated technologies, and in one of the most significant policy developments in Asia over the last few decades, the most successful of the Asian industrializing countries are increasingly emphasizing innovation rather than adopting and adapting existing technologies.

As a result, important new developments are taking place in regional S&T integration both in East Asia and internationally. Given the growing technical competence of the NIC, Japanese firms are being compelled to transfer state-of-the-art manufacturing technology to countries such as Taiwan and Korea, something they have until recently been reluctant to do. One of the striking characteristics of the state of science and technology in the Asian Pacific countries is the considerable intra-regional cooperation. Japanese chip multinationals have contributed to regional technology diffusion and transborder technical cooperation. They have done this through their considerable efforts to set up advanced computer training institutes in South-East Asian countries and a large number of manufacturing, assembly, and R&D networks. Examples are NEC in Hong Kong and China, and SONY in South-East Asia. Hitachi recently concluded an agreement with the Korean firm, Goldstar, to manufacture four megabit Dynamic Random Access Memories (DRAMS) in Korea, and Matsushita has instigated several design and production facilities in Taipei and Malaysia for consumer electronics. Indeed, by the end of 1994, the majority of big Japanese computer firms, including NEC and Fujitsu, had set up pre-competitive research agreements or research joint ventures with NIC universities and public research firms to produce applications software.

Cooperation is also growing between regions. It is well known that many American computer and chip companies have set up R&D facilities in Japan during the past half-decade to tap into Japan's growing technological expertise in areas such as advanced computing and to help forge strategic alliances with Japanese partners. Less well known is the growing technological expertise of countries such as Korea, Singapore, and Malaysia. Indeed, U.S. chip manufacturers, who were initially in countries such as Malaysia and Taiwan for the low-cost semiconductor assembly labour, are now tapping into this expertise. For example, the Texas Instruments-Acer (Taipei) alliance designs and produces integrated circuits and semiconductors in Taiwan; the venture between Taiwan Semiconductor and NEC produces application-specific integrated circuit (ASIC); and Hewlett-Packard and Samsung of Korea have concluded an agreement to produce work stations in Korea in competition with those of Sun Microsystems.

There has been a "globalization" of both R&D and manufacturing facilities. National innovation support systems have been internationalized, and related parts of both research and production are frequently done in several countries. Cooperative international R&D projects have facilitated the international diffusion of technologies at increasingly early points in their product life cycles. During the 1970s and 1980s, technologies were generally transferred to Asia only after they had "matured" enough to be appropriate in countries with low labour costs.[4] In the 1990s, multinationals and smaller firms alike have become more willing to make know-how available internationally.

For example, the Japanese have shown an increased willingness to transfer microelectronics technology to Korean firms by making both research and production alliances in that country. One reason for this recent openness is the high cost of undertaking R&D projects for individual enterprises, which has led to "the formation of a multiplicity of network-like organizations whereby R&D, manufacturing and even distribution activities are carried out of the basis of varied relationships – some of which are equity-based and others which occur on a non-equity, contractual basis" (Pacific Economic Cooperation Council [PECC], 1994).

In another recent change in international technology markets, the traditional licensing, equipment transfer and foreign investment channels have been augmented by new ones such as government-instigated, international, pre-competitive research projects; the activities of international consulting and technology search firms; and technical service agreements.

This paper focuses on two countries that have greatly benefitted from these changes, Korea and Singapore. The S&T policies and programs of these countries create challenges and opportunities for small, open industrialized countries such as Canada. For example, the growing technological sophistication of the NIC should, a decade from now, make these countries potential sources of collaborative R&D and mature markets for Canadian firms. Canadian policymakers should be able to learn valuable lessons from the successes and failures of these policies. Canadian governments might be able to learn from the cooperative nature of public/private sector relations in Asia, which contrast starkly with the often deeply adversarial relation between the government and private sectors in Canada (in spite of some appearances to the contrary) and the limited involvement of the latter in policy making.

S&T Policy in Korea

IN THE EARLY SIXTIES, Korea was a poor, war-devastated country with a per capita GNP of US$87. But between 1962 to 1987 the Korean economy experienced an average annual growth rate of around 9 percent, with exports rising to $61.6 billion and per capita GNP to $4,936 by 1989. Korea started out exporting labour-intensive products such as textiles and plywood, but within a decade their ship and steel producers were challenging the world. By 1983-84 the country had diversified into cars, VCRs, PCs, and memory chips.[5] By 1989, the R&D/GNP ratio had surpassed those of some industrialized countries, as had scientists and engineers as a proportion of the work force.

Korea's science and technology policies and their supportive tax/fiscal incentives and programs appear to have contributed substantially to this stunning success. In a remarkably short period, Korea was able to assimilate and adapt modern technologies in a range of products and manufacturing processes. Important in this success were a Confucian ethic, which stresses the claims of the community over those of the individual, and the judicious application of a reverse engineering and imitative capability (which was achieved before Korea had any formal R&D facilities whatsoever).

Both the government and the private sector made major educational expenditures early on, with percent total government budgetary expenditures devoted to education rising to 26 percent by 1991. Furthermore, this was only one-third of total educational expenditures, the remainder came from Confucian parents with a high commitment to educating their children and from the private sector.

Another factor was the established tradition of government support for overseas training. A recent example is the program for overseas postgraduate training operating through the Korean Science and Engineering Foundation (KOSEF). It plans to send over 10,000 science and engineering PhDs overseas for post-graduate training by the year 2000 (Kim, 1993).

The Confucian and Korean devotion to hard work also contributed. For example, Kim (1988) has found that Korean manufacturing workers work, on average, 53.8 hours weekly, compared to 44 to 48 hours in Japan and the other Asian NIC.

Perhaps most important was the fact that Korea relied on imported foreign technology and at the same time implemented a policy of import substitution and export promotion.

In this section we analyze these factors for two periods, from the beginning of industrialization in 1960 up to 1979, and from 1979 to the present.

1960-79

BEFORE 1979, THE KOREAN GOVERNMENT, in its development policies, consistently selected mature industries and mature manufacturing technologies. These technologies were comparatively easy to reverse-engineer, and acquiring them did not require foreign investment or licensing from offshore companies. As a result, direct foreign investment as a percentage of total foreign borrowing was lower in Korea (6.1 percent) than in the other Asian NIC; for example, it was 91.9 percent in Singapore and 44.6 percent in Taiwan (Korea Exchange Bank, 1987). Thus, from the very beginning of its industrialization program, Korea tried to act independently of MNE control and selected mature sectors in which they could obtain manufacturing technologies through the purchase of turnkey factories, such as steel, cement, and paper. The government supported both import substitution and export industries from the outset; in supporting both policies it was unique among developing countries.

Korean policy was biased against the development of an indigenous capital goods industry (Kim and Kim, 1985), and of all the NICs, Korea has the highest proportion of payments for foreign technology in the form of capital goods (as compared to foreign direct investment, licensing, for example) (Westphal et al.,

1985). Korea turned this apparent disadvantage into an advantage by ripping apart and reverse engineering the barrage of capital goods that entered Korea during this period.

Another factor contributing to Korea's success was a strategic approach to technological development in which the technologies selected were matched to each specific stage of industrialization (Hyung Sup Choi, 1983). The Koreans assumed that a policy or a technology that is appropriate at one stage of industrialization may be inappropriate at another, and tried in selecting them to have economically realistic goals.

The Koreans selected strategic industries and technologies during each period according to the following criteria (Hyung Sup Choi, 1983):

- its historical background and environmental circumstances;

- the international marketing prospects;

- the appropriateness of a specific industry in terms of its accumulated S&T capabilities; and

- a realistic assessment of the potential for further development of these capabilities.

Another factor in the development of a rapid and advanced technological capability was procurement contracts with U.S. military sources. Westphal et al. (1985) note that U.S. military procurement with Korean firms gave sectors such as plywood and tire manufacturers the experience of "assisted learning through doing," which enabled them to comply with precise product specifications. Some of these industries subsequently became major exporters. Contractors in construction learned American product specification, modern techniques for managing construction projects, and how to prepare bid documents from the U.S. Engineering Corp. All of these skills enabled local firms to innovate incrementally and continually in construction techniques during this period, and this industry too became a major export. Another influence of the procurement process may have been to help Korean producers of consumer electronics to convert to the industrial electronics necessary for the large turnkey factories, which Korea would eventually export to much of Asia. In all of these areas, the informal technical assistance given by buyers to make sure that product specifications were met was a big boost to Korea's acquisition of technological capacity.

During each five-year period, the government designated a set of "strategic industries" for development through import substitution and export. This designation process was aided by an army of hired Western consultants. The strategic industries targeted for the second five-year period included machinery, ship building, and construction. Having selected the industries, groups consisting of representatives from the government and the private sector decided on the products in these industries that should be emphasized and supported via the tax and incentive structure. Then studies were undertaken to determine what specific technologies were needed to produce the priority products cheaply and efficiently. The strategic industries were then supported through a variety of policy measures such as risk-sharing loans, protection from foreign competition, favourable financing, and a variety of tax/fiscal incentives. To achieve economies of scale in the exploitation of mature technologies, the government created the *chaebols* – large firms that were the chosen instruments in the export drive.

Other NIC also created large firms to act in the national interest, but only Korea exercised the prerogative of "penalizing poor performers and rewarding only good ones" (Kim, 1988). Successful firms received licenses in other industries, thus leading to *chaebol* diversification. The Korean government also refused to bail out poorly managed companies in healthy sectors, letting the *chaebols* take them over instead (Amsden, 1989). As a result, Korean industry is highly concentrated, and in 1990, the *chaebols* accounted for over 50 percent of all industrial production.

During this period of industrialization based on mature technologies, the indigenous S&T infrastructure was a small factor in industrial growth. With no research capacity in the universities during the early part of this period, the government created the Korea Institute of Science and Technology (KIST), a type of Battelle Institute to meet industry's technical needs. But until the late 1970s KIST and its spinoffs were staffed with advanced Korean researchers trained overseas who had little manufacturing knowledge (Kim, 1989). Until the early 1980s, the scientific and technological infrastructure created by the government had only marginal effects on the development of the mature technology industries, and formal R&D only played a small role during this stage of reverse engineering mature technologies (Westphal et al., 1985).

Another factor in Korea's industrial success during this period is the rapid increase in investment demand in the early 1960s. Rodik (1995), for example, argues that in the early 1960s the government engineered a significant increase in the private return to capital by eliminating a number of investment impediments and establishing a sound investment climate, and more importantly, by alleviating the failures in coordination that had been blocking an economic take-off.

1980-95

BY THE EARLY 1980S, Korea had lost its low-wage advantage, since real wages had grown almost 8 percent annually during the 1970s. Furthermore, the United States and other countries forced Korea to impose and tighten patent, copyright, and industrial property legislation, thus eliminating much of the imitative, reverse engineering of manufacturing technologies that formed the basis of the industrial and technological strategies of the previous period. The response of the government was to liberalize market mechanisms, reduce *chaebol* support, and control some of the *chaebol* abuses through antitrust legislation such as the *Fair Trade Act* (even so, the *chaebols* continued to thrive, and by 1984 their sales were still more than half the annual GDP [Korea Industrial Technology Association, 1993]).

Market Liberalization

Part of the new approach has been to reduce tariff rates and permit significantly more imports. Technology transfer policy has been liberalized, since Korea needed more complex manufacturing technologies to sustain its growth than it did in 1960-79. By 1990, 95 percent of industrial sectors had been opened to foreign direct investment. Unlike in 1960-79, when offshore firms invested in Korea because of low labour costs, companies have begun investing to reap collaborative benefits in producing technology-intensive products.

During this period, the government has also given strong support to small and medium-sized technology-intensive firms, establishing 205 "business territories" in which large firms cannot operate. The growth of these territories has been supported through the Compulsory Lending Ratio Program, according to which all Korean commercial banks must give at least 35 percent of their loans to firms in these territories. Through the *Small and Medium Enterprise Formation Act* of 1986 and other measures, the government has played a central role in the creation of a venture-capital industry and in directing venture funds toward small, technology-intensive firms.

The government has abolished virtually all of the special tax and customs incentives as well as the industry-specific promotion acts and incentives given to the "strategic" industries during the previous period. These have all been replaced by a single industrial promotion act, designating new, high-tech industries for promotion and tying all industrial incentives together with activities such as R&D promotion and developing S&T human resources.

Kim notes that Korean firms became so adept at reverse engineering, which requires the same skills as R&D, that it was comparatively easy for them to make the shift to generating new product and process innovations.

> Reverse engineering involved activities that sensed the potential needs in markets, activities that located knowledge or products that would meet the market needs, and activities that would fuse the two activities into a new project. Reverse engineering also involved purposive search of relevant information, effective interactions among technical members within a project group and with marketing and production departments within a firm ... and with suppliers, customers, local research institutes ... skills and activities required in these processes are in fact the same in ... R&D (Kim, 1993).

R&D Growth

Between 1971 and 1987, the R&D/GNP ratio rose from 0.32 to 1.93 percent, as the economy diversified away from industries based on mature technologies toward those based on technological innovation. Public sector R&D decreased from 68 percent of total research expenditures in 1971 to 18 percent in 1990, with the private sector gradually taking over the research function. The state-controlled banks now provide preferential financing for corporate R&D. Recently, however, new attempts have been made to bolster public-sector R&D expenditures. For example, in 1994 the budget of the Korean Ministry of Science and Technology was increased by 21 percent from the previous year.

Fiscal Incentives and Program Support

The government has instigated a variety of mechanisms to support industrial R&D. In the mid-1980s, it introduced two programs for direct research subsidies: National R&D Projects, which offers direct research subsidies in new technology areas, and the Industrial Base Technology Development Project. These programs concentrate on new materials development, biotechnology firms, basic research in the universities, design of semiconductors, energy conservation technology, nuclear energy, and minicomputer development.

The government has also introduced a series of tax incentives to support industrial innovation. These include reducing or eliminating tariffs on imported R&D equipment and supplies, allowing non-capital and human resource expenditures associated with research to be deducted from taxable income, allowing accelerated depreciation on R&D facilities, and exempting research facilities from property tax. It also set up the Technology Development Reserve Fund, which allowed firms to set aside 30 percent of their before-tax profits to use for research purposes for a four-year period.

During this period, the establishment of private research laboratories has been promoted through a series of tax and financial incentives, and the number of private research laboratories increased from 126 in 1984 to 680 by 1989. It is widely acknowledged, however, that during the past decade the government underinvested in the research aspects of higher education, emphasizing undergraduate teaching rather than research. As a result, it is often claimed that Korea currently lacks the base of scientists and engineers with advanced training to maintain its international competitiveness. But recently the government has moved to correct this defect, (as discussed below). Until this can be done, Korea's public R&D institutes carry out the advanced, industrially relevant research.

In spite of government support, cooperative projects between both universities and industry and public R&D institutes and industry appear to have met with little practical success during this period. Nonetheless, the *chaebols* have created successful research ties with multinationals. A variety of firms, including Hewlett-Packard, Honeywell, ATT, IBM, Monsanto, and Hitachi have set up high-tech joint ventures with *chaebols*, which have also developed working relations with the public R&D institutes and strengthened their own in-house research capabilities.

Technology Deepening and Diversifying[6]

In addition to its R&D effort, Korea continues to look outward to reduce its technological reliance on Japan and the United States. Since Korea is now a real competitor, Japan still sometimes refuses to transfer technology in key areas such as components, software, capital goods, and machinery. To surmount this situation, the government has two technology strategies to restructure the industrial sector: technology deepening (indigenous efforts to improve the industrial base) and technology diversification (finding sources of technology outside the United States and Japan).

These strategies have involved the government, industry, and academia in new R&D areas. The *chaebols* have significantly increased their R&D expenditures, as has the government through the Ministries of Science and Technology and Trade and Industry. For the first time, there was a major attempt to strengthen the university base in S&T when 16 science and engineering research centres were set up in 1989-91. One hundred such centres are planned by the year 2000, and most of the government research institutes are now in the process of moving away from industry-related R&D toward more generic, advanced research. As large firms have recently been setting up their own research labs, the government R&D units have been searching for a new role (Hobday, 1991).

During the 1980s, Korea developed strengths in several high technology and capital-intensive industries, based mainly on imported technologies. The field of electronics components, for example, has been strong in mass-produced standard semiconductors, but weak in application-specific integrated circuits, computer-aided design systems, factory automation, capital goods, and material inputs. In telecommunications, Korea is strong in telephones, handsets, small rural exchanges, private branch exchanges, facsimile machines, optical transmission, and digital microwave. Korean producers are weak in areas such as large-scale public switches, ASICS for public exchanges, and laser technology.

Korea has also been successful in consumer electronics, shipbuilding, iron and steel, passenger vehicles, and fine chemicals, and it also has an emerging aerospace industry (Hobday, 1991). Developing each of these areas further will require research and advanced training. Until the early 1990s the universities had made little contribution to industrial technologies. The current strategy, however, includes an effort to improve industrially relevant R&D. During the early 1990s, 16 centres were set up to do science and engineering research in topology and geometry, theoretical physics, semiconductor physics, organic chemistry, molecular microbiology, plant molecular biology and gene manipulation, advanced fluid engineering, artificial intelligence, sensor technology, spacecraft research, thin film fabrication and crystal growing for advanced materials, biotechnology process engineering, and animal resources.

One hundred centres are planned by the year 2000, with a mandate to develop technology in collaboration with firms, government research institutes, and universities, and these centres have been encouraged to form links and cooperative research with leading international research groups. For instance, some have formed a cooperative venture with the CNRS in France. There has also been a rapid growth of private research facilities, which increased from 52 in 1980 to 831 in 1992, with a total of 37,481 researchers.

Finally, with its Highly Advanced Nation (HAN) project, Korea is attempting to develop its domestic scientific and industrial base to achieve technological parity with the G-7 nations by the year 2000. This project involves 11 nationally defined technology and product areas, including high-definition TV, integrated services and data networks, and DRAM memory chips. The total funding of US$4.6 billion for 1992-2001 for the HAN project is being provided by government, the private sector, and public corporations such as Korea Telecom and Korea Electric.

In summary, Korea is trying to reduce its technological dependence on Japan and the United States. One way of doing this is to accelerate its domestic R&D. Another is to look outward to countries other than Japan and the United States. Canada could become an important source of new production technologies for Korea.

Challenges and Opportunities for Canada

THE GOVERNMENT-FUNDED RESEARCH INSTITUTES such as the Korea Advanced Institute for Science and Technology (KAIST) and KIST and Korea's new research centres are possible candidates to host cooperative research projects between Korean and Canadian concerns, as are Canadian universities and the National Research Council (NRC) network. (The NRC has already signed a cooperative research memorandum of understanding with the Korea Research Institute of Chemical Technology in the field of membrane separation technology and is attempting to initiate research projects with the Korea Institute of Industry and Technology Information and the Korea Academy of Industrial Technology.) Given Canada's comparatively poor R&D record, such collaboration is a means of gaining access to Korean research results 5 to 10 years from now.

Barriers to Collaboration

There are, however, real barriers to scientific collaboration that must be addressed before useful, long-term involvement in S&T can be established between Korea and Canada. The first is the protection of intellectual property rights (IPR). Transitional arrangements in favour of the United States seriously disadvantage Canadian interests in sectors requiring patent or copyright protection such as pharmaceuticals, cosmetics, agrochemicals, semiconductors, printed materials, computer software, and sound recordings.

The second is restrictions on foreign investment. Some industrial categories are still restricted with respect to foreign investment, and Canadian investors face real problems with respect to caps on the transfer of equity capital and local loan facilities. Finally, there is a variety of problems with respect to market access, such as some high import duties, tax discrimination against imported products, quantitative restrictions, and protectionist safety standards. If these problems could be overcome, there would be multiple benefits for both countries to engage in complementary collaboration in scientific and technological R&D.

Potential Benefits of Long-Term Complementary and Collaborative S&T Research

We have seen that Korea is greatly expanding its S&T base and is actively seeking new research partners and new sources of offshore technology. If the market can be liberalized and foreign investors are able to get better access to the Korean market, then Canadian-Korean cooperation in S&T may benefit both parties through the principle of complementary collaboration. Korea now has significant strengths in manufacturing process and product technology but is weak in basic and applied research. To give an example of complementary collaboration, a Korean producer could, in exchange for advanced process or standards technology, transfer leading-edge manufacturing technology to a Canadian firm. This could take place in a joint R&D venture, a strategic partnership alliance, or a jointly-owned company (Hobday, 1991).

Strategic research projects in Korea could also help Canadian firms to access this growing market more efficiently than direct foreign investment. Limited versions of such research collaboration have existed with some European firms since the early 1990s. For example, the British firm AMSTRAD manufactures personal computers in Korea for sale mainly in the European markets but also for the Korean domestic market, and it also does collaborative software research. Hewlett-Packard has a strategic partnership with Samsung in computing technology.

Research collaboration could also benefit academic research in Canada. At present, many Canadian universities suffer from financial constraints, and in some S&T fields it is hard for experienced, senior researchers to recruit talented, young postdoctoral researchers, many of whom have migrated to the United States. Korea experiences exactly the opposite situation. It has very few world-class research labs with senior project leaders, but it has a large population of young, inexperienced researchers. One possible model of complementary research collaboration, to which Korean government authorities are open, is for groups of Korean researchers to engage in research work in Canadian universities, funded from Korea. Such an arrangement might facilitate links that would benefit both parties in the long run.

In short, Korea could be a window of opportunity for Canada in the Pacific Basin, providing Canadian firms with access to East Asian markets, low-cost labour, and low-cost components. S&T collaboration at the academic level could also benefit Canadian universities with the infusion of funds and young researchers. Pre-competitive S&T collaboration at the academic, industry, and government research laboratory levels would improve relations and set up the personal contacts that are vital for future cooperation. In addition, it would ensure long-term Canadian access to the fruits of Korean R&D. (At the end of 1995 five Canadian companies had some form of strategic alliance with the Korean private sector, which is about one-tenth the average of other developed countries.)

Why would Korea be interested in such arrangements? As already noted, Korea is diversifying its international sources of technology, and Canada is seen as an excellent source outside Japan and the United States. The fostering of such collaboration is also part of the current government's technology strategy. In the past three years, the government has made an effort to strengthen international technical cooperative agreements (such agreements are often empty political shells in many countries) and has tried to use them to access international technological developments. In addition to the usual exchange of researchers and participation in international technical cooperative agreements, the government is encouraging pre-competitive, joint R&D projects among Korean small and medium-sized firms and is soliciting the active participation of foreign companies in these ventures.

Hobday (1991) has suggested the following government-led initiatives to help overcome some of the evident difficulties in establishing collaborative research between the two countries: ministerial-level agreements or statements of S&T intent; Korean-Canadian working parties from industry, academia, and government labs to identify specific projects for collaboration; government supporting mechanisms in both countries; and schemes to assist collaboration among firms and universities in the two countries.

An initial step would be to identify S&T areas of research collaboration that would provide unequivocal benefits to Canadian and Korean companies and S&T agencies. Korean officials have suggested the following areas for collaborative research:

- Energy, including nuclear power, nuclear waste management, fuel cycle technology
- Energy conservation technology, including alternative energy sources
- Offshore oil development, including gas combustion technology

- Natural gas liquification
- Clean coal technology
- Biotechnology
- Ocean sciences
- Environment, pollution (air and water)
- Medical research
- Housing and construction
- Information technology (mainly software)
- Policy research in all of these areas

If there is progress in resolving the problems, it may be possible to establish a national scheme within Korea to agree to terms and conditions of collaboration with Canada. An agreed Canadian-Korean S&T procedure would facilitate collaboration between firms, research laboratories, and universities in the two countries.

The Korean and Canadian industry ministers recently signed an Arrangement on Industrial and Technological Cooperation (October 20, 1995) to promote strategic alliances and partnerships between Canadian and Korean companies, especially in the priority sectors of telecommunications, environment, energy, biotechnology, manufacturing technologies, chemicals, and new materials. This is a first step. To reap real benefits, sustained coordinated effort is needed to see that more specific arrangements are actually made and that they really achieve their goals. This requires substantial institutional capabilities in Canadian governments and firms.

S&T Policy in Singapore

SINGAPORE, A SMALL ISLAND NATION at the tip of the Malay Peninsula, has few natural resources save the ingenuity of her three million or so people. It has thus traditionally been almost entirely dependent on commerce and entrepôt trade. Since independence in 1965, Singapore has established itself as a base for foreign multinationals, which were initially attracted by the low-cost, well-educated, and hard-working labour, and an efficient telecommunications, road, and port infrastructure (Porter, 1990). One of the four Asian NIC, its real GDP grew at an average rate of around 9 percent annually between 1960 and 1990. Its per capita income level was US$13,240 (S$22,900) in 1990.

In the last 30 years, Singapore has transformed itself from a labour-surplus, entrepôt-trade-based, urban mercantile centre with a rural hinterland, into a high-tech island, which has several times diagnosed industrial trends and emerging technologies 5 to 10 years before anyone else – for example, software in 1978 and multimedia and wireless communications in 1990. Within the constraints of its limited human resources, Singapore conceived and implemented an S&T strategy in the early 1980s that involved attracting high-value-added technology and knowledge-based industries such as software, computer service industries, financial services, and medical consultancy services. This strategy consisted of three sets of interlinked measures: development support measures directed at these new industries, export support, and human resources training. We examine these measures in two periods, 1980-90, and 1991-95.[7]

1980-90

DURING THE FIRST PERIOD, government policy concentrated mainly on attracting multinationals without placing much emphasis on fostering Singaporean companies.

Training Programs

One of the main thrusts of Singapore's technology strategy during this period was training workers and managers to meet the demands of the knowledge-based industries they were attempting to attract. The government operated special training centres in collaboration with the private sector and other governments. The partners were the Japan-Singapore Institute of Software Technology, the German-Singapore Institute of Production Technology, and the French-Singapore Institute of Electrotechnology. By 1982, 2,298 apprentices and technicians had completed two-year training programs in these institutes and had been placed in 215 companies. In total, these training institutes covered specialized areas such as computer-based tool and die design, computer control of production processes, robotics, microprocessor applications, and software writing.

Before 1980, Singapore's software professionals came from the in-house training programs of multinational computer manufacturers and firms that used computers widely, such as the national banks and insurance organizations. The government first instituted financial and export incentives to upgrade and expand the quality of these programs and then established the computer-related training institutes. At the same time computer programs throughout the school system were upgraded and expanded, the Institute of System Science was created (with IBM) together with the three training institutes mentioned above.

In 1980, the government set up a ministerial Committee on National Computerization to draw up a strategic plan with the aim of "establishing Singapore as a centre of computer services and software." Chaired by the Minister for Trade and Industry, the committee was asked to create a master plan to build up microelectronics expertise and to create an export-oriented software industry. One of their recommendations involved establishing the National Computer Board (NCB), which would implement the plan for the development of computer software. The NCB then set out a series of innovative programs and incentives to induce firms to set up software development, electronics, semiconductor, and financial services operations in Singapore.

In its first report in 1980, the Committee on National Computerization identified the absence of trained software and computer professionals as the main barrier to further developing Singapore as a centre for computer services, and between 1980 and 1990 the government managed to augment the software writing pool by a factor of six. In addition, during 1983-84 Singapore set up three specialized training units: the Computervision CAD/CAM training unit in partnership with Computervision Corp. of the United States, the Asza/Economic Development Board Robotics Training Unit, and a cooperative training group with Japax, Ikegai, and Hamai of Japan.

Industrial and Export Incentives

The government also used fiscal, tax, and export incentives during this period to move industry toward more technology-intensive activities. Tax incentives were given to firms involved in technology upgrading, automation, and software training. The following are the major incentives that were offered by the Economic Development Board to industry during this period.

Export Incentives: There was a 90 percent tax exemption on all profits above a certain level resulting from export sales. It was granted for three or five years; the longer period was granted to companies that had not received a Pioneer Status Incentive.

Pioneer Status: This allowed for total tax exemptions of 40 percent of a corporation's income tax for 5 to 10 years. It was granted, among others, to firms that engaged in software development and export, and was the most extensively utilized incentive by software firms.

Investment Allowance: This was an incentive for both manufacturers and technical service companies. It was utilized as an alternative to the above two incentives, mainly to promote automation. Firms were granted tax exemptions on a specified amount of profits equal to the approved investment allowance, which was a percentage of the fixed investment in plants, machinery, and factory buildings. Such allowances were granted as a bonus over the normal capital allowances given to firms, and any firm could claim a three-year accelerated depreciation on plants and machinery and not lose any benefits of the investment allowance.

International Consultancy Services Incentive: This was instigated to increase exports among software consultancy and engineering firms in Singapore. It was a five-year, 20 percent tax write-off on export profits that were above a specified level. Export services eligible for this incentive included software, engineering design, machinery production, data processing, and all technical advisory services. Only companies with a gross export income of at least

US$1 million were eligible, which disqualified most Singapore software firms. There was also difficulty with consulting firms in determining what percent of exports was due to consulting and what percent was due to software production.

The Capital Assistance Scheme (CAS) and Related Incentives: The CAS promoted projects of strategic value to industrial development. Under it, projects could obtain fixed interest loans at very low interest rates for long terms. The Small Industry Finance Scheme (SIFS) attempted to encourage small local industries with fixed assets of less than S$2 million at the time of application. Realizing that small companies such as software companies may provide vital support to larger firms, the SIFS provided fixed-cost funds to participating banks to lend to small local manufacturing or high-tech service enterprises. The Skills Development Fund (SDF) focused on equipping employees with new skills and knowledge. It provided financial assistance to firms to train their employees with the skills that were critical to Singapore's restructuring effort and was financed through a levy on employers of 4 percent of the salaries of all employees who earned less than S$750 per month. There were two types of financial incentives offered under the SDF. First, the Skills Training Scheme provided direct grants to employers for existing training programs to undertake skills upgrading. This scheme covered from 30 to 90 percent of allowable training costs, and priority was given to technical skills, software skills, product design and research skills. The second was the Interest Grant for Mechanization.

Interest Grant for Mechanization: This scheme, introduced in December 1980, encouraged firms to invest in new automated machinery and equipment to eliminate labour-intensive methods. Companies renting or purchasing machinery and equipment could apply for a grant to reduce the interest costs. The interest grant comprised half of the actual interest from financing, to a maximum of 9 percent yearly. (In 1983, the scope of this scheme was widened to allow assistance in financial leasing for automated machinery and to provide full equipment subsidies for priority support industries.)

Tax Incentives for R&D

Singapore early on introduced several R&D incentives: a double deduction of 200 percent on any R&D expenditures, accelerated depreciation over three years for machinery and plants used for R&D, investment allowances of a total of 50 percent of R&D capital investment, and capitalization of lump-sum payments for manufacturing licenses for five years.

Between 1978 and 1990, reported R&D expenditures increased 26 percent annually, from 0.2 percent of GDP to 1.0 percent by the end of the decade. The number of research scientists and engineers rose from 818 to 4,276 during the same time period, or 28 per 10,000 workforce. The training grants and many of the incentives were an immediate success, and though wages gradually increased during this period, Singapore successfully discouraged labour-intensive factories and graduated from an assembler of consumer products to an integrated manufacturer of advanced consumer electronics. It also branched out into high-grade components for industrial electronic equipment. But in spite of these generous incentives, little real R&D was done during this period in Singapore, nor were there any significant Singaporean spin-offs from the training or research units. Rather than limiting the activities of foreign-controlled multinationals, the Singaporeans embraced them and postponed supporting indigenous industry until the 1990s.

1990-95

DURING THIS PERIOD, Singaporean policymakers have been operating under the assumption that to maintain their international competitiveness they must move to more innovation and promote activities with more design content. To do this, they must increase their capacity to undertake original, commercially useful research and product development.

As is the case with Korea, in the 1990s Singapore has been pushing a number of basic objectives: support basic and applied research in selected strategic industries; rapidly expand the pool of trained R&D personnel; develop technology belts in the country as areas for diffusion; and improve the capacity to commercialize research results and market internationally.

R&D Incentives and Programs

By 1990, R&D expenditures in Singapore amounted to around $638 million per year, or approximately 1.0 percent of GDP. The foreign multinational private sector accounted for 59 percent of this total, with electrical/electronics industries leading the way. The number of private organizations conducting research increased from 135 in 1980 to 264 in 1990. This rise was accompanied by a corresponding increase in the total number of R&D personnel, which rose from 1,193 in 1980 to almost 4,500. Almost one-fifth of this pool consisted of foreign nationals (Singapore Ministry of Labour, 1993).

Singaporeans are now concentrating on process R&D. Although industrially relevant research must eventually result in marketable products or services, given the increasingly short product life cycles, product research often generates only short-period advantages for companies, while process development can help a company to maintain its advantage over a longer period and larger product group.

An apparently reasonable division of research labour has now been established. The universities are concentrating on basic research, the research centres and government institutes on applied research, and industry on product and process development. Since industry-driven R&D seeks to produce proprietary products that can be sold for a profit, government policy accepts that the allocation of resources for this R&D should be governed by the willingness of companies to perform it. The government, by establishing the National Science and Technology Board (NSTB), is now playing a more proactive coordinating and facilitating role in promoting industrial R&D and transferring useful technology to the private sector. Again, as in Korea, the government research institutes support and complement industry efforts by emphasizing pre-competitive research and process development, usually in niches relevant to their strengths. They also undertake specific competitive stage R&D for industry clients on a fee basis.

The strategy of the NSTB, the leading facilitator and promoter of research, is to encourage companies to undertake more research activities through a variety of financial and fiscal incentives; identify the resources required for industry to build a longer-term sustainable advantage in such research, particularly with respect to process technology; and fund centres and institutes that can train the people or provide the technical support to enable companies to undertake their R&D.

The main initiatives to support this strategy are the following:

- A commitment of $2 billion in public sector support for industry-driven R&D between 1990 and 1995, with the government matching up to half of the country's total R&D expenditures as a proportion of GDP.

- A human resources development plan to provide the research personnel to support industry-driven R&D that addresses the reward and status system involved in a research career in Singapore, and also concentrates on developing more specialized capabilities in a few selective key technologies.

- Enhanced fiscal incentives for industrial research.

- Support for research commercialization.

- The development of a physical infrastructure in selected locations to support research activities, in particular, a technology corridor along the southwestern portion of Singapore, which includes the National University of Singapore, the Science Park, and tertiary institutes (NSTB, 1991).

Key Technologies and Strategic Policy Thrusts

Given Singapore's small base, efforts were concentrated in developing core competencies in technologies relevant to its economy. With the aid of dozens of R&D practitioners from both the private and public sectors, the NSTB identified nine technology areas that are most relevant to Singapore's future economic growth: information technology (IT), microelectronics, electronics systems, advanced manufacturing technology, materials technology, energy, water and environment, biotechnology, food and agriculture, and medical sciences.

It is instructive to examine briefly the ways in which the Singaporeans determined key product areas in one domain, information technology. IT was identified as a key industry for economic development in the early 1980s. Due to many of the measures taken in 1980-90, the IT industry has been growing at 28 percent annually and its output is expected to reach $5 billion by 2000. The total number of IT professionals grew from 1,400 in 1984 to over 10,000 in 1990. Singapore has been repeatedly ranked number one in the world by the *World Competitiveness Report* with respect to effective use of IT in business. The software support measures have proven effective, with many multinationals setting up software development centres in Singapore. These include Primefield, Microsoft, CSA Research, Creative Technology, and Apple. Singapore is now well tied into the Internet, has effectively computerized its civil service, and modern telephony is available nationwide.

Then in 1989-90, dozens of Singaporeans, Western consulting firms, industries and researchers were asked to place technologies into three groups, according to how immediately and intensely they could be applied in Singapore industry. Group 1 technologies were those that were mature and ready for immediate exploitation, included CAD/CAM, Asian Language Computing, Expert Systems, EDI Networks/VANS, Videotex, and LANs. Group 2 technologies were those considered reasonably mature and expected to be exploited by industry within two years, included multimedia, image process and handwriting recognition, voice processing, computer graphics and distributed processing, natural language processing and robotics, and wireless communication. Since it was anticipated that these technologies would have considerable impact in industrial applications in the immediate future, the relevant research institutes initiated applied research in these areas immediately. Group 3 technologies were those considered still to be in the experimental stage but to have potential industrial applications in five years or more, included parallel processing, neural networks, fuzzy logic, and broadband ISDN.

This group also identified critical IT technologies necessary to sustain the software industry. These included open systems computing, new software development methodologies, object-oriented technologies, and CASE (computer-aided software engineering) technologies. Since changes in these areas could cause basic changes in the ways in which software was developed, Singaporean software developers were encouraged to adopt appropriate state-of-the-art software technologies and maintain rigorous software engineering discipline to maintain high productivity and quality control.

Having identified (in confidence) priority products that Singapore might produce in many of these technology areas, the NSTB set forth policy recommendations under four distinct strategic thrusts.

Strategic Thrust 1, Applications-Driven R&D – It was recommended that the NSTB should adopt an industry-driven funding strategy for all public R&D institutes to ensure that they assume an industry perspective in the selection of projects; set up focused centres to develop strategic applications in specific industries, for example, a multimedia applications centre; follow the example of Taiwan and encourage public R&D institutes to generate spinoffs, starting with the development of expert systems; establish a nationwide program on software engineering that will eventually lead to certification by the international ISO 9000 standard; set up a research centre for wireless technologies, which are expected to have a major impact on service-oriented industries; and regularly identify strategic technologies and subproducts, as the National Computer Board did.

Strategic Thrust 2, Globalization – The main recommendations here were that the NCB attempt to develop Singapore as the Asian Language Competency Centre for the Asia Pacific region with an emphasis on translating software into Asian language versions; set up technology intelligence offices in the United States, Europe and Japan that would scan industry and technology developments and identify high-tech business opportunities for Singapore; and strengthen the incentives to induce multinationals to set up software development centres in Singapore.

Strategic Thrust 3, Supportive Business Infrastructure – The three main recommendations were that the public R&D institutes be encouraged to take up equity positions in local software development start-ups through either cash investment or the provision of human resources; local software companies be assisted in developing their international marketing channels; and that all government assistance schemes and incentives be extended to support the entire life cycle of IT R&D from feasibility study through to commercialization.

Strategic Thrust 4, Human Resources – In 1990 there were around 7,000 people (or approximately 28 per 10,000 work force) engaged in R&D, 4,298 research scientists and engineers and the remainder technicians and support staff. It was recommended that the output of university graduates be increased, and also that Singapore undertake an aggressive foreign recruitment program, concentrating on China, Hong Kong, and the developed economies that were experiencing economic slowdowns. Just as they had opened the doors to offshore investors in manufacturing and services and benefitted from the inflow of technology, expanded markets and business practices, they should now open their doors to research personnel to get the skills that would take Singapore years to develop on its own.

The government has also been attempting to nurture a research culture, realizing that research is perceived by many Singaporeans as "anonymous, invisible, and having little impact on the real world" (NSTB, 1991). Arguing that such nurturing should start as early in life as possible, funds were offered to encourage science, mathematics and engineering among secondary students to reinforce their interest in science careers. Young Inventors Awards were set up and highly publicized, and the NSTB conducted extensive career counselling among undergraduates and put them in contact with potential employers. In addition, a Research Assistant Scheme was established to allow graduates to work as research assistants on NSTB-funded programs at the various research institutes and centres, where they could work for two years under supervision before advancing to postgraduate courses. Through the Industry Research Education Program, researchers from the private sector can earn credits for a master's degree doing private enterprise research, with NSTB paying a training allowance for approved candidates (in addition to the company salary). This scheme encouraged companies to sponsor research training for their staff.

During this period, the NSTB has narrowly focused on the type of postgraduate training that is thought to be required for Singaporean industry's R&D activities, and has sponsored only scholarships that produce students employable by industry (Wong, 1993).

Although we are not analyzing specific Canadian S&T policy measures in this paper, it must be noted here that Canada – through the National Centres of Excellence Program, the National Biotechnology Strategy, the Strategic Technology Program, and other measures – has conducted a similar exercise of identifying strategic products and technologies, but with apparently less success than countries such as Singapore. Why did Singapore succeed when we did not? At least part of the answer is provided by Chowdhury and Islam (1994) in their description of the East Asian style of government.

1. An elite bureaucracy (such as the Economic Planning Board in South Korea).

2. A political system that has allowed ... technocratic insulation. In other words key bureaucratic agencies have been given sufficient scope to take policy initiatives.

3. Relatively close government-big business interactions.

Another part of the answer may be that in these "emerging product exercises," Industry Canada's research results were not supported through tax/fiscal incentives and programs by other branches of government, nor were they seriously followed up on.

Some Canadian measures such as the reorientation toward information technologies are too recent to discern specific effects. In some parts of Canada's information technology strategy – such as that concerned with education – the public sector has taken the innovative lead with programs such as CANARIE and School-Net. These matters are discussed in the concluding section.

Fiscal Incentives

The Research and Development Assistance Scheme was directed mainly toward small and medium-sized firms with limited resources. It has been extended to cover expenses related to R&D feasibility studies, the development of prototypes, and all expenses incurred with patent applications. Pioneer Status had previously been granted to companies that undertook manufacturing activities of high value-added and skill content, allowing them total exemption from profit taxes up to 10 years. During this period, it has been granted for an additional two years to companies that undertook significant R&D operations in Singapore.

Double deduction of R&D expenses has been extended to include activities relating to adaptation, modification and design, and quality improvement of a product or process. The Operational Headquarters Incentive provides a favourable tax regime for multinationals to site their Asian headquarters in Singapore. To qualify, a company must have substantial operations in Singapore to support technical support services, marketing and sales, funds management, business planning and administration, product development, and R&D.

Previously, capital expenditures incurred in getting approved know-how or patent rights qualified for a five-year writing down allowance at a 20 percent rate. To encourage greater technological transfer, the writing down allowance has been granted over a shorter time period at the rate of 33.3 percent. The 50 percent investment allowance incentive has been extended to all fixed capital expenditures incurred for approved R&D purposes, thus stimulating capital investments for R&D in industries and services. Companies are allowed to set aside 20 percent of taxable income over a three-year period and place it in a pool to fund future R&D work. It is still too early to evaluate the effectiveness of these extended incentives.

Commercialization of Research Results

Singapore has traditionally been weak in its ability to commercialize research results. For example, it averaged around 0.2 patents per 10,000 residents during 1986-89. During the same period, the research institutes commercialized only around 20 new products, and there have been no spinoffs. One of the reasons for this lack of commercialization is that since the country has such a short history of industrialization, it does not have a strong base for applied research. Also, the multinationals on which the economy relied brought with them their own technology and markets and did their research in their headquarter countries. It is also said that Singaporeans, like Canadians, are risk-adverse and prefer to work at salaried positions.

Successful exploitation of R&D results requires complementary assets such as finance, marketing, and competitive manufacturing capabilities, including the ability to produce prototypes cheaply. Although Singapore is an international financial centre, entrepreneurs still face major difficulties obtaining project financing, and Singaporean venture capital funds still prefer mature start-ups. With respect to marketing, Singapore has a very small domestic market to drive technical research or serve as a product test bed, and its past experience has been mainly in manufacturing rather than in marketing or distribution.

To overcome some of these commercialization problems, the research institutes have been directed to seek industry funding for their projects. They have also been encouraged to enter joint ventures with multinationals in technology areas central to Singaporean competitiveness and to do consulting work for enterprises. The Singaporeans are also fostering the growth of venture capital firms and have been developing indigenous expertise to manage such funds, while large multinationals have also been encouraged to base a portion of their design and research work in Singapore. The research institutes have been encouraged through fiscal incentives to spin off commercial companies involving entire project teams or divisions. It is too early to determine the effectiveness of these measures.

In summary, during the 1990-95 period, Singapore has been intensifying its efforts at knowledge-intensive industrialization and attempting to achieve developed economy status within the next 20 years and position itself as a global city in the Asia Pacific region. It is continuing to emphasize attracting knowledge-intensive industries and enhancing its human resources to work in them. Through a variety of measures, Singapore has also now begun to promote high value-added innovative activities and the internationalization of local firms as a way of overcoming small domestic market constraints. As part of this internationalization strategy, Singapore is participating in the development of a "growth triangle" with southern peninsular Malaysia and the Riau Province of Indonesia, and in 1994-95 Singapore-based firms, for the first time, began to relocate to these areas.

CHALLENGES AND OPPORTUNITIES FOR CANADA

SINGAPORE EMPLOYS, AS DOES KOREA, a large array of policies for the transfer and, more recently, the creation of technologies. Many of the NIC's policies, Singapore's in particular, sound like overkill. But from a growth point of view, this may be a more productive way to spend government money non-optimally than the massive array of transfers and non-industry-directed spending that one finds in the advanced industrialized countries of Europe and North America. Furthermore, both Singapore and Korea have developed the institutional capacity to manage their battery of policies relatively effectively.

People in Canada and the United States who are convinced that government can do little good by attempting to manage the market economy, in

Wade's sense of the term, will be little worried by all of this. Those who believe that institutionally capable governments can manage a market-oriented economy and who believe that it is important for a country to be in the forefront of the commercialization of some important technologies will be concerned by the contrasts between the newly-industrialized countries and North America in these areas.

Some, although by no means all, of the tax, fiscal, and industrial incentives and programs initiated by Singapore to support S&T in the last five years may be applicable to Canada. Determining such applicability would require considerable research beyond the scope of this paper. Thus it may be more useful to concentrate here not on specific incentives but on a successful procedure initiated by the Singaporean government – the effort to support the private sector in identifying new strategic technologies and subproducts where comparative advantages could be developed (before other countries have identified them), and then promoting their development through a variety of highly focused incentives and programs. We shall propose a similar Canadian exercise in a restricted domain of potential comparative advantage to Canada: non-timber forest products and some of their production technologies.

It has become a cliché that civil servants in North America are not adept at "picking the winners," while organizations created by Asian governments, ranging from Singapore's NSTB to Japan's MITI, have often succeeded in doing so. Why is this? The deep tradition of cooperation between the private and public sectors in these countries making for less adversarial relations, and the authoritarian prerogatives of their governments, may have some explanatory power. There is, however, a more specific reason. Korea and Singapore have periodically helped the private sector in identifying industries, products, and new production technologies that might profitably be developed. Often, by the time contemporary and future production technologies have been identified in government and academic studies, the technologies are three to six years old. Yet in 1979 the Singaporeans managed to identify software as a future growth industry in which they could excel when it was still generally given away, embedded in computers or written on an individual application basis. They also identified other growth industries such as multimedia products and wireless personal communications in the mid-1980s before there was much development anywhere. Their accurate and profitable prognoses were arrived at by spending considerable sums of money and time consulting a variety of knowledgeable sources such as industry participants; international consulting firms such as Arthur D. Little, Stanford Research Institute, and the Boston Consulting Group; and active research firms such as Xerox Parc. Singapore probably spent more than US$1 million in 1979-80 just to identify computer software as a future growth industry.

Of course, the uncertainties surrounding technological change make predicting fundamental breakthroughs hazardous. For example, the confident predictions made in the 1950s about the early onset of the age of nuclear power have largely been dashed, and the predicted arrival of practical energy transmission through superconductivity continues to be postponed. Nonetheless, Singapore's experience shows that a cooperative initiative between the private and public sectors can do very well *what firms must attempt in any case*: predict profitable lines for the next generation of product and process technologies in which firms need to invest now.

This discussion suggests that it is a potentially useful role for government, *in cooperation with the private sector*, to help identify and develop new areas of comparative advantage, often based on existing resources, skills, and experience. One of Canada's potential new comparative advantages lies in the diversity of new products and services that are emerging from its temperate forests. The British Columbia forestry industry has been experiencing a variety of growth and supply problems, and for several years forestry companies have been attempting to diversify away from the export of raw logs or timber. The thinking about diversification has been mainly limited to other wood products such as plywoods, and yet, through "bioprospecting," the temperate forests of Canada in general and British Columbia in particular can yield a potential variety of new non-toxic pesticides and crop protectors, agrochemicals, cosmetic bases, biodetoxifiers, and other forest-based products and services.

Bioprospecting involves the search for fauna and flora whose underlying genetic and chemical makeup may provide the basis for pesticides, foodstuffs, medicines (the economic benefits of which have been exaggerated), and other new products and services. Bioprospecting is now part of the industrial strategy of all of the Asian developing countries with forests, such as Malaysia, Indonesia, Thailand, and Lao PDR. Emerging product categories include biocides, herbicides, and fungicides; medicinal and dental products such as spermicides and non-toxic contraceptives; birth facilitators; high-end oils, dyes, and dammars; preventative health care products; detoxifying agents for both industrial processes and human tissue; and crop protectors.

To give one example, during the summer of 1994 one Clayoquot Sound organization identified a centipede exudate which is non-toxic, evaporates in about a year, and acts as a crop protector by repelling birds and some insects. The world market for nontoxic crop protectors exceeds US$70 million.

The new technologies facilitating these industries involve revolutionary advances in sophisticated screening of natural products – computer-automated enzyme and receptor screening. Since these tests are becoming more efficient, cheaper, and faster, large-scale testing of a variety of insects, plants, fungi, bacteria and marine organisms is now possible. There are currently natural products research programs at Merck & Company, Beecham, Smith Kline, Monsanto, Glaxo, and other companies. These new screens use low-cost assays that mimic desirable pharmacological (or other) properties as filters with which to eliminate the vast majority of samples, so as to allow researchers to home in on the tiny minority that have real prospects of commercial success in a particular application (Wills et al., 1994).

Given these developments, if the provincial and federal governments are serious about helping the forestry diversification effort and supporting potential future growth industries, they might, following Singapore's example in other fields, undertake economic and industrial research on non-timber forest products and services directed to the following items.

- The identification of new products and services of comparative advantage based on wildlands substances, which forestry industry firms and new entrants could produce and export.

- The review and analysis of the world markets for these potential products and services.

- The formulation of a technology strategy, including tax and fiscal incentives, human resources training, export support and specialized technology support measures, to help develop these sectors of the economy.

This would require more than throwing money and rhetoric at the problem; intense and extended cooperation between the private and public sectors is necessary.[8] Also, the governments must have the institutional capacity to do the job. Careful, sustained cooperation with the private sector, and the ability to select a battery of competent analysts and advisors, is required of the governmental institutions. Do Canadian governments have such institutional competence? If not, could they develop it? If not, how much will this affect the ability of the private sector to keep up with cutting-edge developments in new areas of potential high-value-added production?

Conclusions and Agenda for Future Research

SEVERAL ASIA PACIFIC REGION COUNTRIES, such as Korea, Singapore, and Malaysia have had extraordinary growth rates during the past decade. This growth is in no small part a result of their consistently funding programs and incentives that support technology advancement in the private sector and integrating these measures into their industrial policy. In contrast, Canada has recently eliminated many of its technology support programs for the private sector. No doubt many of these programs and incentives were not competitive with those of the NIC in the first place and were not evaluated usefully. For example, when we reviewed the success stories of the Department of Foreign Affairs' Technology Inflow Program (TIP), six companies presented as successful users in 1993 had gone bankrupt by 1995. However, the exigencies of deficit elimination raises the real risk that valuable and effective programs will be eliminated along with the deadwood. We are not confident, for example, that the Canadian government's Manufacturing Assessment Services, Manufacturer's Visitors Program, and Technology Outreach Program were eliminated after any careful assessment of their costs and benefits. We suspect, instead, that the demands of welfare transfers and debt interest payments squeezed these programs out of the already small battery of Canadian technology support policies.

There is a general belief in the Asian countries that the industrial competition of the next century will be based on sophisticated technological production. Consequently, for more than a decade, these countries have set forth, evaluated, and revised an entire gamut of policies, strategies, and incentives that support technology advancement in production. They have also provided sophisticated market intelligence work and industrial analysis to help the private sector identify new product areas. Given the current trends, it is very likely that Taiwan, South Korea, and Singapore will all increase their R&D/GDP ratios to 2 percent or greater by the end of the current decade, with their private sectors assuming around half of this total, thus surpassing Canada.

It would therefore seem prudent, at the very least as a form of insurance, for the federal government to internationalize Canada's R&D effort by supporting a variety of specific cooperative research projects between Canadian and South Korean concerns. Given the fact that general S&T cooperative agreements and pre-competitive research projects between countries are often empty shells with little emerging from them, it is important that the projects be made as specific and commercial as possible. We also argue that it is a legitimate role of government, following procedures pioneered by the NIC, to assist the private sector in providing sophisticated market intelligence and industrial analysis of areas of potential new comparative advantage such as the new non-timber forest products that could be harvested from Canada's temperate forests, and then to assist in the pre-commercial stages of their development.

With respect to future research, several major thrusts would be useful. First, a detailed comparative analysis and critique of Canada's federal and provincial S&T policies, strategies, and tax/fiscal incentives in the context of comparable recent efforts in the Asia Pacific countries such as Singapore, South Korea, and Malaysia should be undertaken. Such issues as which Canadian policies seem to be working and failing and which of the NIC's policies might be transferable to Canada – taking into consideration our different social and economic structures, institutional capacities and potential sources of new comparative advantages – should be examined. This would be particularly timely given the recent budgetary changes and elimination of technology support programs in Canada, and it could lead to recommendations for a specific set of policies, programs and tax/fiscal incentives, which Canada might adopt to be competitive with the NIC's technology support effort during the next 20 years. This review should 1) analyze the economic effects of specific incentives and programs discussed above for South Korea and Singapore, plus those involved in the S&T strategies of Taiwan and Malaysia; 2) identify and analyze equivalent Canadian incentives; 3) determine the reasons for the differences in economic effects between these countries and Canada (means of implementation, scope, etc.); and 4) determine what new incentives and programs Canada should adopt to be competitive with these countries.

Second, in major new emerging product areas such as the non-timber forest products, major surveys and research should be regularly undertaken to identify emerging products at as detailed a level of specificity as possible, determine forms of government tax/incentive support, and to show how these could be developed by private-public sector cooperative efforts.

Third, given the extraordinary achievements of Korea and Singapore, beyond those of their S&T policy measures that we might emulate, we should also examine the complementarities that exist between Canada and the Asian NIC and capitalize on them. For example, Singapore, built mainly on multinational assembly plants, now needs telecommunications software expertise and environmental engineering – in which Canada excels. Australia has regularly sold the type of educational services widely available in Canada to Singapore and throughout Asia. We should also review previous complementarity studies, such as the Japan-Canada complementarity work of the late 1980s; the effects of the resulting Japan S&T Fund; and the lessons to be learned from the Singaporean, Korean and Malaysian contexts. Also, did the programs that were designed to link Canada in different ways with the NIC achieve business advances or greater linkages?

In other words, we still need quantitative studies on the benefits of previous federal S&T programs and policies, particularly as they relate to support of Canadian businesses in the Asia Pacific region. None of the above tasks will be easy. As was the case in Singapore and Korea, major work would be needed from teams of leading industrial experts and others. But if these tasks are not accomplished, Canada risks both loosing out in many areas in which others will develop new leading-edge products, and having a seriously flawed policy mix in terms of continuing counterproductive policies while failing to adopt potentially productive ones.

ENDNOTES

1   For funding, we are indebted to Industry Canada and the Canadian Institute for Advanced Research. For their many comments and readings, we are grateful to Cliff Bekar, Ken Carlaw, Colin Fiddler, Adam Holbrook, and Simon McInnes.

2   This is no trivial condition. We are currently studying cases in which policies that worked well at the one level of government ran into diffi-

culties when they were transferred to a higher level because of apparently small differences in how they were administered.
3   Lipsey and Carlaw (1996) use case studies to compare and contrast the European and the Asian approaches.
4   One of the stylised facts about technologies outlined in Lipsey and Bekar (1995) and Lipsey and Carlaw (1996) is that they begin in crude forms that require much skilled labour to create and operate. Then, as they are improved and extended in application, in an evolution that often stretches over decades, less skilled labour is required.
5   We do not suggest that these policies were successful everywhere or were without serious problems. See, for example, Westphal (1990).
6   All technical details in this section are based on Hobday (1991) and two trips made by one of the authors to South Korea in 1993-94.
7   Whether other policies might have resulted in an even faster rate of growth of per capita income is debated (see, for example, Young, 1994). Be that as it may, these policies certainly achieved their industrial objectives and per capita income did grow faster than in many other Asian developing nations (and environmental quality was maintained, which was in stark contrast with all of the other NIC).
8   Another example is "ecotourism," which, although it is the fastest growing component of the world's tourism industry, was left out of the government's most recent tourism strategy (Canada, Supply and Services Canada, 1995).

Bibliography

Amsden, A. H. *Asia's Next Giant: South Korea and Late Industrialization.* New York: Oxford University Press, 1989.

Canada. Supply and Services Canada. "Tourism; Canada's Export Strategy – An Integrated Plan for Trade, Investment and Technology Development." Ottawa: Supply and Services Canada, 1995.

Chowdhury, A. and I. Islam. "Public Policy and International Competitiveness: What We Can Learn From East Asia." Unpublished manuscript. Federal Reserve Bank of San Francisco, 1994.

Hobday, M. "The Needs and Possibilities for Cooperation Between Selected Advanced Developing Countries and the Community in the Field of Science and Technology; Country Report on the Republic of Korea." Report by Sussex Research Associates Ltd. The Commission of the European Communities, 1991.

Hyung Sup Choi. *Bases for Science and Technology Promotion in Developing Countries.* Bangkok: Asian Productivity Organization, 1983.

Kim, Ji-Soo. "Management of Technology: Case of Korea." Presented at the Comparative Pacific Rim Strategies of Developing Technology Managers conference. Institute of Southeast Asian Studies, Singapore, 1993.

Kim, L. "Korea's Acquisition of Technological Capability for Internationalization: Micro and Macro Factors." *Business Review* 22, 1 (1988).

───────── "Science and Technology Policies for Industrialization in Korea," in *Strategies for Industrial Development.* Edited by J. Suh. Kuala Lumpur: Asia and Pacific Development Council, 1989.

Kim, L. and Y. Kim. "Innovation in a Newly Industrializing Country: A Multiple Discriminate Analysis." *Management Science*, 31, 3 (1985).

Korea Advanced Institute for Science and Technology. *Bulletin*, 1993.

Korea Exchange Bank. "Direct Foreign Investment in Korea." *Monthly Review*, October 1987.

Korea Industrial Technology Association. "Major Indicators of Industrial Technology." 1993.

Korea Science and Engineering Foundation. *KOSEF Data Book.* 1992.

Lipsey, R.G. "Markets, Technological Change and Economic Growth." Quaid-I-Azam Invited Lecture. *The Pakistan Development Review*, 33, 4 (winter 1994):327-52.

Lipsey, R.G. and C. Bekar. "A Structuralist View of Technical Change and Economic Growth," in *Technology, Information and Public Policy.* Edited by T.J. Courchene. Kingston Ont: John Deutsch Institute, 1995.

Lipsey, R.G. and Carlaw. "A Structuralist View of Innovation Policy," in *The Implications of Knowledge Based Growth."* Edited by P. Howitt. Calgary: University of Calgary Press, 1996.

National Science and Technology Board. *Science and Technology: Windows of Opportunity – National Technology Plan.* Singapore: NSTB, 1991.

Nelson, R. R., ed. *National Innovation Systems.* New York: Oxford University Press, 1993.

Pacific Economic Cooperation Council (PECC). *Pacific Economic Development Report 1995.* Singapore: Craft Print Pte. Ltd., 1995.

Pack, H. and L. Westphal. "Industrial Strategy and Technological Change: Theory Versus Reality." *Journal of Development Economics*, 22 (1986):8-128.

Porter, M. E. *The Competitive Advantage of Nations.* New York: The Free Press, 1990.

POSTECH. *Bulletin.* (1994)

Rausch, L. *Asia's New High-Tech Competitors.* NSF 95-309. Washington, DC.: National Science Foundation, 1995.

Rodik, D. "Getting Interventions Right: How South Korea and Taiwan Grew Rich." *Economic Policy, A European Forum*, 20 (April 1995).

Singapore Ministry of Labour. *Singapore Yearbook of Labour Statistics.* Singapore, 1993.

Wade, R. *Governing the Market: Economic Theory and The Role of Government in East Asian Industrialization.* New Jersey: Princeton University Press, 1990.

Westphal, L. "Industrial Policy in an Export-Propelled Economy: Lessons from South Korean Experience." *Journal of Economic Perspectives*, (Summer 1990).

Westphal, L.E., L. Kim, and C. Dahlman. "Reflections on the Republic of Korea's Acquisition of Technological Capability," in *International Technology Transfer: Concepts, Measures, and Comparisons.* Edited by N. Rosenberg and C. Frischtak. New York: Praeger, 1985.

Williamson. J., ed. *Latin American Adjustment: How Much Has Happened?* Washington: The Institute for International Economics, 1990.

Wills, R., I. Townsend-Gault and G. Hearns. "Policy and Legislative Options to Develop Non-Timber Forest Products and Services in Vietnam and Lao PDR." Report to Lao PDR PMO and the International Development Research Centre. Centre for Asian Legal Studies. Vancouver: University of British Columbia, 1994,

Wong, P.K. "Singapore's Technological Development Strategy," in *The Emerging Technological Trajectory of the Pacific Rim Region*. Edited by D. F. Simon. New York: M. E. Sharpe, 1993.

World Bank. *The East Asian Miracle: Economic Growth and Public Policy*. New York: Oxford University Press, 1993.

Young, Alwyn. "The Tyranny of Numbers: Confronting the Statistical Realities of East Asian Growth Experience." NBER Working Paper No. 4680. Cambridge, Mass.: National Bureau of Economic Research, 1994.

Comment

Louise Séguin-Dulude
Institut d'économie appliquée
École des Hautes Études Commerciales

LIPSEY AND WILLS' OBJECTIVES IN THIS PAPER are twofold: 1) to provide a detailed, accurate description of the policies implemented by South Korea over the last 35 years and Singapore over the last 15 years in the areas of science, industry, and technology; and 2) to explore the possibility of implementing in Canada one of the science and technology policies put in place by the government of Singapore between 1990 and 1995.

They argue that a debate on the impact of science and technology policies on the growth of newly industrialized Asian countries cannot take place without an in-depth examination of those policies. They note, moreover, that most of the existing studies are restricted to overviews. Their paper is an attempt to remedy the situation by providing a painstakingly detailed account of the various incentive measures and programs launched by the governments of South Korea and Singapore to propel their respective economies into the technological age.

The first part of the paper contains an abundance of detailed and valuable information on various aspects of the policies put in place by these governments to support the transfer, dissemination, adaptation, and creation of technologies. To further their strategy in this area, the South Korean and Singaporean governments have, over time, taken a wide range of measures and initiatives, which include:

- promoting certain industries, sectors, products and technologies

- developing programs designed to help small- and medium-sized high-technology firms receive loans or locate in high-technology parks

- giving these firms access to R&D loans at preferential rates

- introducing tax and financial measures favouring R&D, innovation, and plant modernization

- establishing and funding manpower training centres in conjunction with private enterprise and foreign governments

- strengthening the role of universities through the creation of research and training centres specializing in science and engineering

- awarding scholarships to graduate students in disciplines with industrial applications.

The first part of the paper is very well documented and acquaints us with the various aspects of the science and technology policies developed. However, while it is true that in the period examined by Lipsey and Wills the real growth in the total and per capita GDPs of South Korea and Singapore exceeded those of most the others in the region, the authors fail to provide evidence that science and technology policies played a role in this economic success. Policies vary from one country to the other. They evolve and sometimes undergo extensive transformation. Sometimes they promote the transfer of technology, sometimes its creation. In short, the policies have been transitory, while the growth has been sustained.

The authors admit that they did not set out to assess the costs and benefits of all of the science and technology policies adopted by the governments of South Korea and Singapore. This would have been too ambitious an undertaking. Now and then, however, we learn that some programs have had to be delayed, others abandoned, and still others added in order to compensate for certain discriminatory practices. The following questions arise:

- To what degree was each of these measures necessary, useful, or effective?

- Why were certain measures abandoned, replaced, delayed or, conversely, strengthened?

- To what extent were these measures independent of each other, complementary, or even contradictory?

- What is the cost-benefit ratio of the various aspects of the science and technology strategy implemented by South Korea for the last 35 years and by Singapore for the last 15 years?

Without answers to these questions, it is difficult for us to endorse Lipsey and Wills' second objective, which is to explore the possibility of importing one or more of these policies into Canada. Without knowing the impact of this surfeit of science and technology programs, we are ill-equipped to select a program or an initiative suitable for trial in Canada, for such a choice would seem arbitrary.

Moreover, there is no guarantee, in our view, that the supposed success of South Korea's and Singapore's science and technology policies in their national environments would be replicable to an equal or comparable degree in a Canadian environment.

Marked differences in the culture and ethics of South Korea and Singapore on one hand, and Canada on the other, raise doubts about the possibility of successfully transplanting to Canada the policies of either of these Asian countries, where the rights of the community have primacy over the rights of the individual. Lipsey and Wills touch on this difference, but perhaps minimize its implications.

At various times, and fully in keeping with the culture and ethics of their societies, the governments of South Korea and Singapore implemented programs designed to stimulate technological development by favouring certain industries, sectors, companies, regions or university disciplines. Such decisions were easy to make and had to be made because they benefited society as a whole, even though they discriminated against other industries, companies, regions, or disciplines.

In Canada, where most people believe in the equality of individual rights, it would be difficult to secure the support of society and business for proactive technology-development policies that are based on discrimination. Our economic policies are largely based on the principle of universal eligibility – whether they target business, the work force, or the general public. When the federal or provincial governments adopt "discriminatory" measures, it is generally with a view to assisting the have-nots, not the haves; in other words, it is done to offset inequalities, rarely – and certainly not openly – to create them.

These cultural and ethical differences also explain why our federal and provincial governments play such a small interventionist and regulatory role in the economy compared to the governments of South Korea and Singapore.

Lipsey and Wills conclude by proposing that the next phase of the research on this subject will consist of an in-depth comparative analysis of the science and technology policies of four Asian countries and Canada. This study, they say, should take into account differences in the social and economic structures of the countries, the capabilities of their respective institutions, and the sources of the renewed comparative advantages for the various economies. Lipsey and Wills are thus fully mindful of the cultural, social, and economic differences. They also argue that the effectiveness or ineffectiveness of the science and technology policies of each of the five countries studied must be demonstrated and evaluated before we try to answer the questions before us, which are: What science and technology policies should Canada adopt to make its economy competitive? What lessons can it learn from the Asian experience?

Well before reaching their conclusion, however, Lipsey and Wills decide that Canada can replicate some of the incentive measures and joint government/private-sector initiatives undertaken in Singapore between 1990 and 1995. This decision seems premature. In describing the wide range of measures and initiatives developed by the governments of South Korea and Singapore to promote the technological development of their economies, this paper lays the essential groundwork for a debate. However, the real debate has yet to begin. In my opinion, no single policy should be endorsed at this time.

Richard Pomfret
Department of Economics
University of Adelaide, Australia

15

Australian Experience with Exporting to Asia[1]

AMONG ALL THE EUROPEAN COUNTRIES and countries settled by Europeans, Australia has been the most successful in exporting to the high-performing economies of East Asia. In the early 1950s one-seventh of Australian exports went to that region; today the fraction is around three-fifths. In striking contrast to other member countries of the Organisation for Economic Cooperation and Development (OECD), Australia also has a substantial trade surplus with East Asia. The primary aim of this paper is to explain this export success.

A second aim is to draw lessons from Australia's experience that might apply to other countries. At first sight, it may seem that Canada has much to learn from such lessons because its economy has some similarities to Australia's. However, the differences between them are also substantial.[2] Although both Australia and Canada are perceived as primary product exporters, Australia's exports are much more heavily weighted towards primary products than are Canada's, and the product mix is somewhat different. Canada is a more industrialized country, with comparative advantage in a wider range of manufactures.

The first two sections of the paper set the scene by describing Australia's trade patterns and policies. The exposition in the first section is data-driven, following the Standard International Trade Classification (SITC) categorization of goods. The key policy change in Australia has been the liberalization of imports since the early 1980s, although the government continues to pursue a complex range of restrictive policies with respect to imports of cars, textiles, clothing and footwear, as well as in export promotion.

The remainder of the paper seeks to explain the growth of Australian exports to Asia since the 1950s. To what extent has the geographical reorientation of Australia's trade been driven by exogenous changes, such as the United Kingdom's accession to the European Community, or rapid economic growth in East Asia and a favourable complementarity in the commodity composition of Australian exports? To what extent has it been driven by policy measures taken by the Australian government, including domestic reforms and international economic diplomacy?

The paper analyzes the largest component of Australia's exports – primary products – which reflects the complementarity between labour-abundant East Asia and resource-rich Australia. It also reviews a major debate that has taken place in Australia concerning the performance of manufactured exports. The underlying judgment here is that Australia should be trying to benefit from its human capital rather than its finite (and low value-added) resources, and a similar distinction impinges on discussions of service exports, which have been an undoubted growth area. This judgment underlies many of Australia's interventionist economic policies, which reflect dissatisfaction with the industrial structure created by the market mechanism. The industry policy debate is relevant to the subject of this paper, because lessons from exporting location-specific natural-resource-intensive goods and services are of less interest to other countries than lessons from exporting human-capital-intensive goods and services.

The paper also reviews the role of discriminatory trade policies in Australia's commerce with East Asia. This discussion is very brief because such policies have played a minimal role in Australian export performance in the recent past, but the manoeuvring behind the formation of Asia Pacific Economic Cooperation (APEC) is seen by some observers as a step on the road to East Asian regionalism – in which Australia's place is uncertain.

Finally, the paper draws lessons for Canada from Australia's experience in expanding exports to East Asia and some general conclusions from the study.

Australia's Trade Patterns

The Composition and Pattern of Australia's Foreign Trade

AUSTRALIA'S TRADE HAS ALWAYS CONSISTED OF the exchange of predominantly primary product exports for manufactured imports. Apart from the gold rush eras, most of the exports before the 1960s came from farmers and grazers. In 1950, rural products, of which wool was the most important, accounted for 86 percent of Australia's exports. This percentage has continuously declined since then (Table 1). During the 1950s this decline was partly due to the growing importance of manufacturing and service exports. The big transformation in the composition of exports, however, occurred during the minerals booms of the 1960s – which contributed to the delay in liberalizing trade by boosting prosperity – and the early 1980s. Although manufacturing exports have grown in importance during the early 1990s, their share of exports is not much higher than it was in 1960. Service exports, on the other hand, have grown steadily, to account for one-fifth of Australia's exports in the 1990s.

TABLE 1

COMPOSITION OF AUSTRALIA'S EXPORTS

	RURAL PRODUCTS	FUELS, MINERALS AND METALS	OTHER GOODS	SERVICES
	(Percent)			
1950-51	86	6	3	5
1960-61	66	8	13	13
1970-71	44	28	12	16
1980-81	39	34	11	16
1985-86	32	42	9	17
1992-93	23	41	16	20

Source: Anderson (1995), p. 33.

The composition of imports has been much more stable. Manufactures have always dominated Australia's imports. The share of primary products in total imports declined during the 1960s, from 18 to 11 percent, recovered to 18 percent in 1980 and then declined to 11 percent by the early 1990s. Since 1960, services have always accounted for between one-fifth and one-fourth of imports, primary products for less than one-fifth, and manufactured goods for over three-fifths (Anderson, 1995, p. 34).

Following the floating of the Australian dollar in 1983 and the subsequent financial market liberalization, Australia's trade deficit increased as the country attracted capital inflows. There has been a major debate over whether the government should worry about the size of the current account deficit if the foreign debt is owed by the private sector (which is largely the case). Leaving aside issues of optimality, the debt/GDP ratio has increased substantially since 1983, and net interest payments overseas, which were stable around 2 percent of GDP in the three decades before 1983, have mounted sharply.

The most striking aspect of Australia's foreign trade since 1950 has been the dramatic shift in its geographical pattern. During the first half of the 1950s, 63 percent of exports went to Europe and 60 percent of imports came from Europe (Table 2). By the early 1990s, Europe was buying 16 percent of Australia's exports and supplying 26 percent of Australia's imports. This change in trade patterns was caused in part by an increase in trade with North America, but that shift was more or less complete by the late 1960s. The main element was the rapid growth of Australia's trade with East Asia.

Table 2 reveals two aspects of Australia's trade reorientation in favour of East Asia. First, that realignment has been more pronounced on the export side. East Asia's share of Australian exports grew from 14 percent in 1950-55 to 38 percent in 1968-72, 47 percent in 1980-84, and 56 percent in 1990-92,

TABLE 2

Direction of Australia's Merchandise Trade, 1951-92

	1951-55	1968-72	1980-84	1990-92
	(Percentage shares)			
Exports				
United Kingdom	36	11	4	4
Other Europe	27	16	10	12
North America	10	16	13	12
Japan	8	26	27	26
New Zealand	4	5	5	5
Other East Asia	6	12	20	30
Middle East	1	2	8	3
Other	8	12	13	8
Imports				
United Kingdom	45	21	7	6
Other Europe	15	19	17	20
North America	15	27	25	26
Japan	2	13	21	18
New Zealand	1	2	3	5
Other East Asia	7	7	14	19
Middle East	4	5	9	3
Other	11	6	4	3

Source: Anderson (1995), p. 37.

while the corresponding import shares were 9, 20, 35, and 37 percent. Second, while trade with Japan led the way during the 1950s and 1960s, the rest of East Asia was the source of increased Asian orientation during the 1970s and 1980s.

Australia's Merchandise Trade in 1994

TABLE 3 LISTS AUSTRALIA'S TOP 15 EXPORTS AND IMPORTS (by SITC three-digit categories) in 1994. The ongoing specialization in primary products is clear. Of the top 15 exports, only two come from SITC classes 5-8, and these are rather special. Aluminum is one of the outputs of the integrated aluminum industry, which exports bauxite, alumina, and aluminum (the combined exports of categories 285 and 684 place the integrated industry as Australia's third exporter, after coal and gold). The other manufactured good in the top 15 exports is computers and office machinery parts (759), a category in which Australia is a net importer but which includes a large amount of intra-industry trade.

Table 3

Australia's Major Merchandise Exports in 1994

SITC	Description	Value
		(A$ Millions)
Exports		
321	Coal	6,700
971	Gold, non-monetary	4,751
268	Wool and other animal hair	3,588
011	Bovine meat	2,981
281	Iron ore and concentrates	2,592
041	Wheat and meslin	2,286
285	Bauxite and alumina	2,161
684	Aluminum	2,126
061	Sugar, molasses and honey	1,625
333	Crude petroleum and oils	1,546
334	Refined petroleum and oils	1,267
343	Natural gas	1,081
759	Computers and office machinery, parts	1,071
022	Milk and cream products	855
036	Crustaceans and molluscs	851
Imports		
781	Passenger motor cars	3,830
752	Computers	2,856
333	Crude petroleum and oils	2,139
759	Computers and office machinery, parts	1,961
782	Goods and special purpose vehicles	1,817
764	Telecommunications equipment, n.e.s.[1]	1,813
784	Motor vehicles parts	1,697
792	Aircraft and associated equipment	1,339
641	Paper and paperboard	1,289
874	Measuring and checking instruments	1,210
778	Electricity distributing equipment	1,199
542	Medicaments (including veterinary)	1,094
723	Civil engineering equipment	1,010
898	CDs, tapes, software and musical instruments	964
713	Internal combustion piston engines	936

1 Not elsewhere specified.
Source: Australia. Department of Foreign Affairs and Trade (1995a), pp. 22 and 28.

Australia's exports are concentrated in the outputs of a small number of resource-based activities: coal ($6,700 million); gold ($4,751 million); bauxite, alumina, and aluminum ($4,286 million); crude and refined oil and natural gas ($3,895 million); wool ($3,588 million); beef ($2,981 million); iron ore ($2,592 million); wheat ($2,286 million); and sugar ($1,625 million). These provide almost half of all Australia's merchandise exports ($32 billion out of a $65 billion total).

Adding other resource-based items from the top 75 three-digit SITC export categories brings the total up to $46 billion, or over two-thirds of all Australia's merchandise exports. These categories are listed in Table A-1, and casual observation suggests that they are based on a small number of well-defined natural resources (some of which are closely related to the leading exports listed in the previous paragraph – e.g., iron and steel, much of which is primary or semi-finished, and gold coins). The total resource-based component of Australian exports will be substantially over two-thirds because this count omits heterogeneous top-75 categories such as 098 (edible products n.e.s. [not elsewhere specified], $158 million), 522 (organic chemical elements, $187 million), and 592 (starches and glues, $164 million), as well as smaller but cumulatively non-trivial categories from SITC classes 0-4.

On the import side, the list of leading sectors is dominated by manufacturing activities. Only one of the major imports (SITC 333, crude petroleum and oils) is not from SITC classes 5-8. This dominance by manufactures continues much further down the import list. In the top 50 import categories, only 4 are from SITC classes 0-4: 333 and 334, refined oil (17th); 248, wood (30th); and the catchall group 098, edible products n.e.s. (41st).

Turning to the geographical pattern of trade, Table 4 shows the strong Asian orientation of Australia's foreign trade. Six of the top 10 trading partners are in East Asia. The United States remains Australia's largest import supplier, and the ranks, in Australia's total trade, of New Zealand (3rd) and the United Kingdom (fifth) imply some residual historical effects. Otherwise, the non-East-Asian countries uniformly have a weight in Australian trade below their weight in the world economy. Conversely, South Korea, China, Singapore, Taiwan, Hong Kong, Malaysia, Indonesia, and Thailand all rank higher among Australia's trading partners than they do in world trade.

The most extraordinary aspect of Table 4, however, is the size and pattern of bilateral trade surpluses and deficits. Canada is unique in having almost balanced bilateral trade with Australia. Every other country has a large imbalance. With Australia, the United States runs a trade surplus of over $10 billion on total trade of $19 billion; this is a large imbalance, and the sign is especially striking given the United States' large aggregate trade deficit. Australia's imports from Germany are almost four times its exports to that country; with Sweden, the ratio is 1 to 10. Thus the pattern is that Australia has trade deficits with every member of the European Union (EU) and with the United States and that these deficits are often large relative to the bilateral trade flows.[3]

With Asian trading partners the picture is reversed. Australia runs trade surpluses with all of its significant Asian partners except China, Vietnam, and Papua New Guinea. The China deficit may be a statistical artefact, because many imports are entering China through Hong Kong, so that the exports to China figure is likely to be biased downward; on combined China and Hong Kong trade, Australia runs its "Asian" surplus. Australian success in running

TABLE 4

AUSTRALIA'S TRADE, BY COUNTRY, 1994

	EXPORTS	IMPORTS	NET EXPORTS	RANK[1]
	(A$ Millions)			
Brunei	54	0	+54	65
China	2,821	3,374	–553	6
Hong Kong	2,654	841	+1,813	10
Indonesia	2,014	1,038	+976	12
Japan	16,011	12,100	+3,910	1
South Korea	4,736	1,766	+2,970	4
Malaysia	1,956	1,220	+736	11
Philippines	724	231	+493	23
Singapore	3,454	2,065	+1,389	7
Taiwan	2,847	2,455	+392	8
Thailand	1,368	876	+492	15
Vietnam	164	289	–125	31
India	874	476	+398	20
Iran	419	18	+401	32
New Zealand	4,386	3,384	+1,002	3
Pakistan	201	147	+54	35
Papua New Guinea	967	1,121	–154	17
Canada	1,104	1,103	0	16
United States	4,640	14,840	–10,200	2
European Union				
Austria	30	234	–203	42
BLEU[2]	435	583	–149	22
Denmark	66	323	–258	33
Finland	162	508	–346	28
France	762	1,638	–876	14
Germany	1,043	4,077	–3,033	9
Greece	16	54	–37	58
Ireland	31	332	–301	34
Italy	1,146	1,835	–690	13
Netherlands	737	693	+44	18
Portugal	32	72	–40	52
Spain	285	395	–110	27
Sweden	135	1,224	–1,090	19
United Kingdom	2,365	4,058	–1,693	5
World	64,984	68,103	–3,119	

1 Ranking by value of total trade (columns 1+2).
2 Belgium Luxemburg Economic Union.
Source: Australia. Department of Foreign Affairs and Trade (1995b), pp. 38-42.

large trade surpluses with Japan and South Korea is in stark contrast to North American and European perceptions of these markets as being "hard to crack."

In sum, Australia has succeeded in establishing a triangular trade pattern of importing from Europe and the United States and exporting to East Asia. It has benefited from the rapidly growing Asian markets in order to finance imports from its traditional suppliers. This is quite different from the response to the East Asian high-performing economies by North American and western European countries: they have mainly benefited from importing manufactured goods from Asia and have complained of difficulty in accessing Asian markets.

MERCHANDISE TRADE WITH EAST ASIA[4]

NOT SURPRISINGLY, AUSTRALIA'S MERCHANDISE TRADE with East Asia has the same commodity pattern as its aggregate trade, with primary products being exported and manufactures being imported. Australia's two biggest exports to East Asia are gold and coal, worth about $4.6 billion apiece in 1994, followed by iron ore, wool, aluminum, and beef, each worth around $2 billion. Crude petroleum ($1,260 million) and natural gas ($1,023 million) were the only other exports to Asia valued above $1 billion in 1994.

On the import side, Australia's trade with Asia exhibits a greater concentration within the broad SITC categories 5-8 than do its aggregate imports. Motor vehicles (category 78); computers and office machinery (75); radios, television sets, videocassette recorders (VCRs), etc. (76); clothing (84); textiles (65); and the assorted light industrial products in SITC category 89 (toys, plasticware, CD tapes, etc) – each accounts for imports of well over a billion dollars, even though they include the areas of residual protectionism in Australia.[5]

Japan is Australia's largest trading partner. The initial reorientation of Australia's exports towards Asia primarily involved Japan, but since then Japan's share of Australian exports has reached a plateau. Between 1989 and 1994 Australian exports to Japan grew by 5 percent per year, which is non-negligible but well below the average growth of Australian exports to Asia. This may reflect a high base level, but it is still surprising: especially in the context of the appreciation of the yen, Japan should have become a more attractive market. The relatively sluggish growth (and lack of response to exchange rate changes) may reflect the composition of Australia's exports: 71 percent of exports to Japan are primary products, and manufactured exports have actually been decreasing by 2 percent per year since 1989-90.[6]

The commodity composition of Australia-Japan trade is concentrated. Five commodities accounted for over half of Australia's exports of $16 billion in 1994: coal ($3.2 billion), beef ($1.6 billion), gold ($1.4 billion), iron ore ($1.2 billion), and natural gas ($1.0 billion). Out of Australia's $12 billion imports from Japan in 1994, $5 billion consisted of motor vehicles and their components.

Although *South Korea*, which became Australia's second largest export market in 1994, remains well behind Japan as a trading partner, the Australian trade surplus with that country is even more striking than that with Japan, with exports reaching $4.7 billion and imports, $1.8 billion. Again, Australia's exports are dominated by minerals, although coal and natural gas are not major exports to Korea. Detailed analysis is hampered by the large entries in categories 2* (a "dump" category for unclassified crude materials) and 988 (confidential items), which together account for almost a third of 1994 merchandise exports. Gold is the largest single export, accounting for 16 percent of the total. Apart from aluminum and metals, internal combustion engines are the most important manufactured exports and have expanded rapidly during the 1990s (to $140 million in 1994).

China and *Hong Kong* currently rank seventh and eighth among Australia's export markets; combined, they would come second. A large but unknown amount of trade with Hong Kong comes from or is destined to China. Practically all of Australian imports from Hong Kong are manufactures, led by integrated circuits (IC) and computer parts, many of which are made in South China. More important in the entrepot role of Hong Kong is likely to be the transshipment of Australian exports. Australian exports to Hong Kong ($2,654 million in 1994) are led by gold ($410 million), dried crustaceans ($159 million), and pearls and precious stones ($142 million), but a distinguishing characteristic is the wide range of manufactured exports going to the territory. Most of these export items are likely to be retailed in Hong Kong to traders or consumers from other parts of East Asia as well as to Hong Kong residents.

Unusually among Asian countries, China has a trade surplus with Australia, but this probably reflects reporting biases. Reported exports to China ($2.8 billion in 1994) are difficult to analyze, because over $1 billion are listed as confidential. Well over half of the non-confidential trade consisted of two items – wool ($667 million) and iron ore ($406 million) – and the reported trade consists almost entirely of producer goods. The highly visible imported consumer goods in China's coastal cities are often clandestinely routed through Hong Kong, and this presumably accounts for some of Australia's exports to Hong Kong.

Another special feature of Australian economic relations with China concerns foreign direct investment (FDI). Australia is the only major country for which FDI in China is less than the flow in the opposite direction. This situation is mainly due to two very large Chinese investments in Australia – in iron ore mining and aluminum smelting – that are obviously intended to secure raw material supplies.

Taiwan is Australia's sixth largest export market ($2.8 billion in 1994), with a by now familiar story of a small number of primary exports in return for manufactured imports. The composition of exports is a mixture of the Japanese and China/Hong Kong patterns. Coal ($409 million) and aluminum

($256 million) were the largest items in 1994, followed by gold ($202 million), wool ($182 million), crustaceans ($177 million), iron ore ($139 million), copper ($135 million), and beef ($126 million).

The *Association of South-East Nations* (ASEAN) has been the most dynamic area of Australian export growth over the last decade, with annual growth rates of around 20 percent, from around $2 billion in the mid-1980s to $9.6 billion in 1994. Singapore is the most important ASEAN market, but in the 1990s exports to Singapore have flattened out, and Thailand, Indonesia, and Malaysia have been the fastest-growing export markets. Gold has been the major export in recent years, but sales also include a diversified range that includes minerals, wool, dairy products, and other processed foodstuffs.

Other Asian markets are of little current significance. There has been great interest in the opening up of the Vietnamese economy and Australia has been among the leading sources of FDI, but the value of trade is still small compared to that with other Asian countries. Australia also has a high profile in Laos, with the opening of the first bridge over the Mekong in 1994 (financed with Australian aid), but exports were a mere $39 million in 1994 (of which gold was $31 million). There are some hopes of exploiting Australian expertise in mining, telecommunications, and livestock as a basis for trade with the Russian Far East and other Soviet successor states; an Australian embassy was opened in Almaty, Kazakhstan, in 1995.

EXPORT OF SERVICES

ALTHOUGH AUSTRALIA'S SERVICES EXPORTS have grown faster than merchandise exports over the last half-century, Australia has remained a net importer of services. Since the mid-1980s, however, the size of the net import ratio $[(M-X)/(X+M)]$ has declined. A large component of this turnaround has been in tourism, where Australia switched from being a persistent net importer to a net exporter around 1988. Today the tourism sector is Australia's largest services export income earner.

Export of education services has also increased rapidly. Gross revenue from education exports in 1993 has been estimated at $1.6 billion, larger than exports of, say, iron and steel or aluminum (Findlay, 1995, p. 130). In 1992-93 the Australian aid agency supported 660 students from developing countries, at a cost of $110 million in fees and living costs. The number of private fee-paying students is less precisely known, but it is estimated to be around 70,000 in 1995, with the majority coming from ASEAN countries.

Over two-fifths of Australia's service exports in 1992-93 went to East Asia (Australia, Department of Foreign Affairs, 1995b, p. 31). Japan accounted for 19 percent; ASEAN, for 14 percent; Hong Kong, for 5 percent; and South Korea, Taiwan, and China, for 2 percent each. Other important markets were the European Union, with 18 percent; the United States, 13 percent;

and New Zealand, 6 percent. Canada accounted for 1.3 percent. The only countries with which Australia ran a surplus on services trade in 1992-93 were Japan ($1,770 million), Papua New Guinea ($187 million), Taiwan ($182 million), New Zealand ($106 million), Malaysia ($73 million), and South Korea ($62 million). As with merchandise trade, Australia consistently has a services trade deficit with non-Asian countries.

Tourism and education exports to Asia are growing rapidly. In both areas, proximity is an advantage. The open spaces (with beaches and golf courses) and the unique fauna (koalas and kangaroos) are major tourist attractions, while the climate is a further plus for tourists from Northeast Asia. Education exports benefit from Australia's location as the closest English-speaking country to East Asia and from established secondary and tertiary systems with a good reputation. Proximity appears to be important for parents in South-East Asia (who often make the educational location decision on behalf of their children), as does Australia's reputation for safety (in contrast to the image of the United States).[7]

Both tourism and education exports could expand substantially, as demand is income-elastic and supply, price-elastic.[8] Demand for both could exhibit bandwagon effects as more satisfied tourists and students report home and as Australian suppliers become more attuned to Asian customers. Liberalization of Australia's international air transport policy would also help to increase the competitiveness of Australia as a provider of both services.

Some Reservations

THE REST OF THIS PAPER SEEKS TO DETERMINE how Australia succeeded in shifting and expanding its exports to Asia. Before doing this, let me raise some reservations about the exercise.

Trade theorists generally do not worry too much about bilateral trade patterns in a world of multilateral trading based on non-discriminatory policies (i.e., on the granting of most-favoured-nation treatment, as required by Article I of the General Agreement on Tariffs and Trade, or GATT). The theoretical perspective has its clearest counterpart in the type of homogeneous primary products that dominate Australia's exports: if coal is paid for in a convertible currency at a world price that is the same across locations, does it matter whether Australian coal is sold to the United States, the European Union, or Japan? This is relevant to the trade flows described above, because part of the bilateral trade deficit with the United States is due to the redirection of Australian petroleum exports from that country to Asia, adding to Australia's surplus on Asian trade; such redirection may result in minor savings on transport costs, but the net economic effect on Australia is trivially small.

A second difficulty in analyzing trade flows is the practical matter of identifying the nationality of a good. This applies to the manufactured goods that Australia imports. Computers and office equipment and their parts figure prominently in Table 3, on both the import and the export side. The intra-industry trade arises because the process of making a computer can easily be broken down into a number of processes that involve different inputs of skilled and unskilled labour and capital equipment, and are therefore suited to differing locations. Identifying a computer's nationality by its final point of assembly (or by the headquarters of the company whose name appears on it) may be a misleading guide to where it was really made. Indeed, the various components and assembly stages may have come from such a variety of locations that the computer may truly be an international good. Such products are a real headache when imports from different trading partners are subject to differing treatment, thus raising the thorny issue of defining rules of origin. Otherwise, the customs official may not care too much about the national origin, even though how it is entered into the trade statistics will have implications for a study like the present one.

A third, and probably less significant, problem has already been alluded to. Once an Australian export leaves the port with the destination marked in the bill of lading, it normally passes beyond Australian surveillance. Apart from the usefulness of statistical tracking for analysis, the Australian authorities are not particularly concerned about the ultimate destination of the good. This is especially a problem with goods shipped to the entrepots of Hong Kong and Singapore, whence they may be transshipped to other Asian destinations. It is also a problem if purchasers of Australian services have dual citizenship, which is not unusual within the economically dynamic overseas Chinese community. These last examples are not relevant to analysis of aggregate trade with Asia, but they will distort reported patterns of trade with individual Asian countries.

I will not belabour the point because trade policy debates invoke bilateral trade balances so often that economists must treat them as a matter of public concern, even though their importance may often be overstated.

Australia's Trade Policies

The Evolution of Australia's Trade Policies

Australia's trade policy during the first half of this century was part of a package intended to redistribute the benefits of the country's abundant natural resources more widely in the economy and to encourage settlement and capital inflow. After Federation in 1901, the debate over tariff policy was won by the protectionists, and tariffs were sharply increased in 1907, 1921, 1926, and in the 1930s. Australia also granted "Imperial preference" to imports from Britain, the dominions, and the British colonies.

Thus, in broad outline, Australian trade policies during the first third of this century were similar to Canada's.[9] Both countries had a comparative advantage in primary products and sheltered their manufacturing sectors behind tariff barriers. Australia had, however, erected higher tariff barriers by the mid-1920s (Table 5) and had a less developed manufacturing sector than Canada. Canada also had different trade patterns than Australia, as the United States had become Canada's major trading partner, while Australia's trade remained heavily oriented towards Great Britain throughout the first half of the twentieth century.

The big divergence between Australian and Canadian trade policies came after 1945. While Canada followed the U.S. lead in dismantling tariffs on a multilateral basis under the aegis of the GATT, Australia chose to sit out the GATT-based trade liberalization of the 1950s and 1960s, identifying with the primary product-exporting developing countries rather than with the industrialized countries. By 1970, Australia, together only with New Zealand among the OECD countries, had an average manufacturing tariff rate of over 20 percent; the corresponding figure for Canada was 14 percent; for Japan, 12 percent; for the United States, 9 percent; and for the European Community, 8 percent (Anderson, 1995, p. 41).

An important policy change from the 1950s to the 1960s was the replacement of import quotas by tariffs. During the 1960s publicity was given to the high dispersion of tariffs, especially when using effective rates of protection, a concept developed by the Australian economist Max Corden. The Tariff Board, previously a pliant justifier of requests for protection, instigated a review of the tariff in 1971 and increasingly became a source of economic expertise as its staff and consultants provided empirical evidence of the costs of protection. The Tariff Board, renamed the Industries Assistance Commission in 1973 with wider terms of reference and now called the Industry Commission, has been an important force for transparency in Australian trade policy-making over the last quarter century.

TABLE 5

AVERAGE TARIFF LEVELS, 1925

	PERCENT
Australia	25
Canada	16
United States	29
United Kingdom	4
France	12
Germany	12
Italy	17
Spain	44

Source: Pomfret (1988), p. 25, based on League of Nations *Tariff Level Indices* (Geneva, 1927), p. 15.

Australia's protectionist policy was finally reversed in 1973 when a 25 percent tariff cut was implemented. Trade liberalization stuttered later in the decade, especially with respect to textiles, clothing and footwear, and motor vehicles, and then resumed in the 1980s. The effective rate of assistance to manufacturing other than these two groups is now close to zero and will be phased out by 2000 (Figure 1). The goal of free trade is widely accepted and current trade policy debates concern the appropriate speed with which protection for automobiles and textiles, clothing and footwear should be dismantled.[10]

Australia has generally pursued a non-discriminatory trade policy during the postwar period. The value of Imperial preferences was quickly eroded and disappeared completely after Britain's accession to the European Community. Australia was the first country to introduce a system of tariff preferences for developing countries, in 1966, but this and other preferential treatment for Pacific islands have had little impact on the country's trade flows.[11] The most significant regional trading arrangement in which Australia is a member is the Closer Economic Relations agreement (CER) with New Zealand, which in 1983 replaced a more limited free trade agreement between the two neighbours (McLean, 1995).

A salient consequence of Australia's trade policies was on the openness of the economy. In the mid-1800s Australia's ratio of trade (exports plus imports) to GDP was over 50 percent, one of the highest in the world. In the early 1900s, when Australia had the world's highest per capita income, the trade-to-GDP ratio was still a relatively high 42 percent. The corresponding figures for Canada in those two periods were 31 and 32 percent. Australia's trade/GDP ratio continued to fall, however, reaching 27 percent in 1984, when it was one of the lowest in the OECD; Canada's trade/GDP ratio that year was 51 percent. The declining trend has been reversed over the past decade, although Australia's trade ratio remains well below Canada's.[12]

A striking way to see the policy-induced closing of the Australian economy is to look at the relationship between the trade/GDP ratio and the level of GDP. Larger economies do more internal trading and tend to have lower trade/GDP ratios. This relationship is illustrated by the regression lines in Figure 2, which is taken from Anderson and Garnaut (1987, pp. 14-15). Around 1870, Australia was an outlier with an exceptionally high trade/GDP ratio for the size of its economy. In 1984 Australia was again an outlier, but in the opposite direction, with an exceptionally low trade/GDP ratio given the size of its economy.

Export Promotion Policies

THE AUSTRALIAN TRADE COMMISSION (Austrade) is the main instrument for export promotion. Austrade's mandate is to support the international expansion of Australian business by promoting Australia as a source of supply for

FIGURE 1

EFFECTIVE RATES OF ASSISTANCE TO AUSTRALIA'S MANUFACTURING AND AGRICULTURAL SECTORS, 1968 TO 1992, WITH PROJECTIONS TO 2000

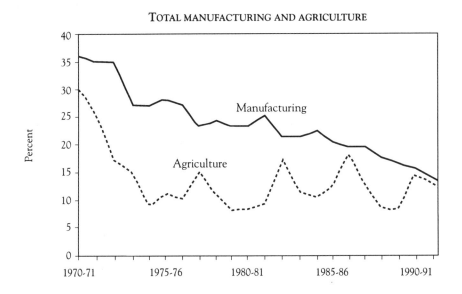
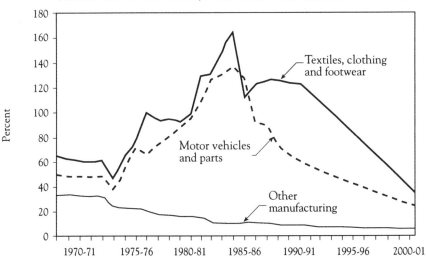

Figure 2

Trade Orientation and the GDP of Various Industrial Countries, Circa 1870 and 1984[1]

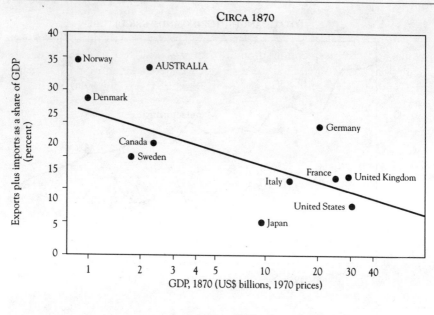

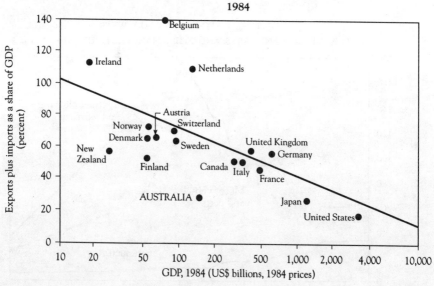

1   The least-squares regression line was estimated excluding Australia.

goods, services, and investment funds, as well as a location for investment. It provides a range of advising and counselling services through a network of state offices within Australia and regional offices in the major export markets. Austrade also manages an extensive trade fair and display program around the world (85 expositions in the 1994-95 fiscal year, of which 41 were in East Asia).

Austrade provides services that Australian exporters value. The extent to which its activities actually promote exports is difficult to assess, and this difficulty has itself contributed to Austrade's indifferent reputation by reducing its accountability. A widely publicized report in the early 1990s criticized Austrade's management and questioned whether the social benefits from Austrade outweighed the costs to Australian taxpayers.

The Export Market Development Grants (EMDG) scheme provides grants of up to $250,000 for reimbursement of export-marketing expenses. The government committed $230 million to the EMDG scheme in 1994-95. A report on the scheme claimed that every grant of $1 generated $15 to $25 in exports for "experienced exporters." The International Trade Enhancement Scheme and the Innovative Agricultural Marketing Program are similar but smaller schemes. The Export Finance and Insurance Corporation provides insurance, financing, bonds, and guarantees to exporters.

Australia's aid program, administered by the Australian Agency for International Development (AusAID, formerly AIDAB), has an explicit mandate to promote Australia's commercial interests. AusAID selects projects in which Australia is a competitive supplier and directly sources a large share of its spending to Australian producers (AusAID estimates that 90 percent of its aid spending is on Australian goods and services). In 1994-95, $130 million (9 percent of the aid budget) was allocated to the Development Import Finance Facility (DIFF), which assists Australian capital-goods producers in competing for contracts against foreign aid-supported competitors. Food aid in 1992-93 was valued at $91 million, much of which involved direct purchases from Australian farmers at market prices, making AusAID one of the largest buyers of wheat (over 200,000 tonnes) and of rice (nearly 40,000 tonnes) that year; aid purchases came to $5,760 per rice grower. In 1992-93, over 6,600 students from developing countries received full assistance from AusAID to study in Australia; another 5,000 received some assistance.

Australian goods and services provided as aid do play a promotional role and may tie users into buying Australian-made parts or replacements, but just what volume of additional exports is generated by AusAID's spending on Australian goods and services for delivery to developing countries is debatable. An AIDAB study undertaken in 1992-93 found that business valued at $575 million had resulted from Australia's cumulative aid to China between 1980 and 1992 of $285 million and that a further $263 million in follow-on business was expected by the end of 1993; thus every $100 of aid had generated $200 in business, with a long-term effect of over $300 being anticipated. Outside observers have, however, been sceptical of such claims, which are not based on a thorough comparison with the no-aid alternative scenario; in other words, some of the claimed follow-on trade might have happened even without the aid. An even stronger leverage claim has been made for AusAID's program in Thailand, where $23 million in inputs to a lignite mining project are said by AusAID to have generated commercial contracts for Australian exports worth over $350 million.[13]

In addition to the programs of the federal government, the individual states also have a variety of export promotion programs. The full array of these measures is best captured in the detailed sectoral studies published by the Bureau of Industry Economics, but some idea is given by the summaries of assistance to six sectors in Sheehan, Pappas and Cheng (1994, pp. 84-103). Total federal government expenditure on export promotion and industry development going to the manufacturing sector in 1993-94 amounted to $1,128 million – just under 1 percent of total budget outlays – but this understates the full cost of industrial policies. In particular, it excludes expenditures by other levels of government and, more importantly, much of the economic cost that is borne by consumers and does not enter into the government budget. The automotive industry illustrates this point.

Australian governments have a record of intervening in the automotive industry ever since the first General Motors Holden (GMH) was produced in 1948. Auto policies initially focused on protecting domestically made cars from competing imports, and then became more involved in trying to restructure the industry, reducing model proliferation in order to realize scale economies. The Export Facilitation Scheme (EFS) is part of this evolution; the EFS grants credits to car and component producers on the basis of Australian value added in exports, and the credits can be used to offset duty on imported cars or components or can be traded. Given the continuing high level of automotive protection, the EFS represents a substantial export subsidy. Australian car exports increased in the early 1990s, led by the convertible model that Ford sold in North America as the Mercury Capri. The major automotive export is engines, and it is clearly to the car makers' benefit to realize scale economies in engine manufacture.

Government economists have praised the EFS on the grounds that when combined with import protection, it generates fewer distortions than would import protection alone.[14] There is no longer a "home market bias" (Corden, 1995, p. 14), whereby producers are encouraged to produce for domestic rather than foreign markets. This is correct, but the welfare analysis must also take into account the fact that the incentives for the favoured industry are increased and are thus likely to expand production, which may be welfare-reducing. More generally, the EFS is a classic second-best measure, and the net welfare effect cannot be established a priori.

Albon (1992) estimated that the social benefits of the EFS narrowly exceeded the costs. Albon also reports that Australian costs in making the

Capri were at least 30 percent above the world price, implying that the expansion of output of this car involved a net welfare loss to Australia. Engine exports may make more economic sense, but it is difficult to demonstrate this, given the distorted structure of Australia's car industry. In any case, the net welfare gain reported by Albon only arises because his comparison was with a no-EFS scenario; the EFS was beneficial because it reduced the costs of other protective measures by easing restrictions on imports. Dismantling the restrictions on automotive trade would make the EFS redundant and raise national welfare.[15]

What has been the impact on exports of government policies? The most important export promotion policy has been the general trade liberalization and other deregulation measures introduced since the early 1980s. These measures have increased the competitiveness of niche manufacturers and service exporters.

The export promotion measures described in this section are hangovers from the previous era, reflecting an ongoing belief among Australian policymakers that government assistance is necessary for national economic success. Indeed, it may well be that particular schemes benefit particular sectors or groups of exporters. Assessment studies have focused on such direct effects, and, as emphasized above, there is a suspicion that some studies have been biased towards ascribing too many direct effects to policies, often by using a before-and-after comparison rather than the more appropriate with-and-without comparison. It is difficult to detect an impact on aggregate exports; the export promotion schemes have tended to focus most on manufacturing exports and on smaller and poorer markets, while Australia's export success over recent years has been mainly in primary products and to the richest markets in Asia (Japan, South Korea, Singapore, Taiwan, and Hong Kong).

The overall package of export promotion policies is complex, which in itself is a bad omen for the policies' effectiveness. Complexity reduces transparency, and hence the accessibility of government assistance to those most likely to benefit from it.[16] Complexity also breeds a rent-seeking mentality, encouraging managers to devote time to gaining government assistance; time has an opportunity cost in terms of its use in more direct applications to exporting. In a June 1995 statement, the Minister of Trade, Bob McMullan, set out his vision of export policy: "The Team Australia co-operative approach to strengthen Australia's export effort is being reinforced by extending the web of partnership and alliances between all levels of government and the private sector. It seeks to increase effectiveness and avoid duplication" (McMullan, 1995, p. 3).

The two sentences are likely to be contradictory in practice, but more striking is the continuing belief in the government's ability to manage an export promotion package that is being made even more complex.

The automotive EFS has been dwelt upon because it illustrates several of these points. The scheme was proposed by General Motors Holden and adopted in 1982 by the government, in return for a GMH commitment to build an engine plant (Capling and Galligan, 1992, p. 204). The initial impact was small, and after 1985 the EFS became more complex: credits were made transferable and a wider range of imports made eligible, but the conditions for use of the transferable credits had to be concisely defined in order to retain the sectoral nature of the scheme. In the early 1990s, the EFS was heralded as a success on the basis of the Capri sales, a visible sign that Australia could match any country in producing complex manufactures. Comeuppance was quick, however. Not only were the Capri's fit and finish well below international standards, leading to a loss of reputation and of North American market share, but production costs were high and initial export sales were heavily subsidized. Nevertheless, the EFS remains in place and engine exports are lauded as a manufactured export success, because the costs (in terms of lost consumer surplus caused by high car prices supported by the trade barriers plus EFS policy) are less visible.

Related Policy Changes

THE RAPID SHIFT TOWARDS A MORE LIBERAL TRADE REGIME has been part of a broader move to deregulate the economy since 1983.[17] In particular, financial and foreign exchange rate markets were liberalized during the 1980s with the adoption of such measures as the floating of the Australian dollar in December 1983, the removal of controls on bank deposits in 1984 and 1985, and the May 1985 invitation to foreign banks to seek licences. A more general program of microeconomic reform is being pursued, but its consequences are sometimes hard to distinguish from the pro-competitive effects of technological change – in the broad area of telecommunications and information technology, for example. A striking exception in the present context is the failure to deregulate domestic and international aviation.[18]

The post-1983 Labour government came to power with a commitment to promote an active industrial strategy. This activism has been a source of tension in the context of the opening up of the economy and the deregulation described so far (Stanford, 1993). Some of the industrial policies have been sectoral – for example, the automobile and textile, clothing and footwear policies, which included delayed trade liberalization, and the steel plan, which was terminated in 1988 when steel producers were outperforming the plan targets. Others have been more broadly targeted at promoting "sunrise industries." Measures to encourage research and development include tax concessions, the Grants for Industrial R&D (GIRD) scheme, the "partnership for development" programs announced in 1987 as an umbrella for deals between large corporations and state or federal governments, and other schemes. A more detailed description of the range of Australia's industrial policies is beyond the scope of this paper, but they remain the stuff of public debate as state and federal politicians retain a predilection for detailed intervention to promote industrial restructuring.[19]

An important measure of the macroeconomic performance of the Australian economy after the change of government in 1983 was the Prices and Incomes Accord between the Labour Party and the Australian Council of Trade Unions. Under the accord, centralized wage fixing was reinstated, providing an instrument for preventing nominal-wage increases from eroding the nominal depreciation of the dollar in the mid-1980s. The cut in real wages contributed to growth in exports and gross domestic product (GDP), although at the cost of delaying the deregulation of labour markets.

A less tangible, but highly relevant, policy of post-1983 governments has been a growing emphasis on Asia in Australia's international relations and a complementary stress on multiculturalism in domestic affairs. Australia's participation in the Vietnam War and, more decisively, the end of the "white Australia" policy in the late 1960s were early signs of the shift away from Europe, but this shift has been more actively promoted since the late 1980s. The influential report, *Australia and the Northeast Asian Ascendancy* (Garnaut, 1989), emphasized the need for attitude changes, the promotion of language teaching, and other pro-Asian measures if Australia were to take advantage of rapid growth in East Asia. The East Asia Analytical Unit, located in the Department of Foreign Affairs and Trade, has continued in the tradition of the Garnaut Report by actively promoting official research on Asian economies. In foreign policy, Australia has increasingly exerted leadership in Asian matters, notably in the administration of the 1993 United Nations-sponsored Cambodian elections, which was led by Australia and Japan, and in the promotion of APEC. The current prime minister's goal of establishing a republic by the time of the 2000 Olympic Games in Sydney is simultaneously an assertion of sovereignty and of the anachronism of having a European head of state.

Explaining Australian Export Success in Asia

AUSTRALIA'S EXPORT PERFORMANCE over the past four decades has depended primarily on the expansion of a handful of primary industries. The direction of trade was influenced by U.K. accession to the European Community (now the European Union) and the high levels of European protection for agriculture, which blocked the expansion of Australia's agricultural exports to Europe apart from wool and feed grains. Some exports were redirected from other traditional markets, such as petroleum exports from the United States to East Asia. The general point made above is relevant here: with more or less homogeneous primary products being sold at more or less competitively determined world prices, the ultimate destination matters little to the exporting country. East Asia has taken a rapidly increasing share of Australia's primary product exports both because its demand has been booming and because of some trade diversion. The important determinants of Australian export success, however, have been on the supply side: Australia has a resource endowment that has enabled it to take advantage of the growing demand for specific goods and services.

Australia's agricultural sector underwent substantial changes during the 1950s and 1960s. Technical change was accompanied by substitution of capital for labour. Although real farm output grew by 2 and 3 percent per annum during the 1950s and 1960s, respectively, agriculture's share of national income fell from 25 percent in 1950 to 8 percent in 1970; during the same period, employment in agriculture declined by about 15 percent from the Korean War boom peak of 480,000 (Freebairn, 1987, p. 140). Exports became relatively more important to the rural sector, but its share of national exports dropped sharply (Table 1). Agricultural exports were hurt by Britain's accession to the European Community, but probably more important was the impact of the EC's Common Agricultural Policy on world markets; as the Community accumulated surpluses in grains, dairy products, sugar, and so on, it sought ways to dispose of them, which depressed world prices, hurting more efficient producers such as Australia.[20] Constant-market-share analysis of Australian exports from the early 1960s to the mid-1970s has shown that specialization in wool, meat, and grains was the major source of slow export growth (McColl and Nicol, 1980), and this decline in established export items accompanied a decline in importance to Australia of established export markets.

Meanwhile, the mining sector, which had been in decline during the first half of the century, experienced a boom. Taxation incentives favouring investment in oil and mineral exploration were implemented in 1946, and modern techniques were applied to the search for and exploitation of large operations. Major projects in the 1950s included the Mount Isa copper mine, uranium mining to support the growing atomic power industry overseas, mining of beach sands for the aerospace industry, and the King Island tungsten operation.

Export revenues grew more rapidly during the 1960s as new mining projects were developed, especially in response to the demands for iron ore and coking coal from the fast-growing Japanese steel industry. Before the 1960s a policy goal was to ensure availability to meet growing domestic demand, and some exports were banned (notably iron ore between 1938 and 1960), which deterred exploration (Smith, 1983). After the lifting of such restrictions, the development of very large-scale equipment and 100,000 tonne-plus capacity bulk carriers facilitated exporting from the open-cut iron ore mines of Western Australia and coal mines of Queensland. Similar developments based on technical advances and very large-scale capital-intensive operations characterized the growth of bauxite mining and alumina refining. New scientific exploration techniques also led to the discovery of the Bass Strait oilfield and of new sources of nickel, manganese, and phosphate rock.

The 1970s were a difficult decade for the primary producers. After a boom in the early years of the decade, prices leveled off for most Australian exports and demand was depressed by the decline of world growth after 1973. The main exception was coal, which benefited from the oil crisis and from Australia's low-cost production: Australian coal production doubled between 1970 and 1980. Bauxite and alumina exports also grew rapidly from a low base; Australia produced no bauxite before 1963, but by 1980 was the third largest exporter of bauxite and the number one exporter of alumina (Helliwell, 1984). After a minerals-led investment boom in the early 1980s, export of energy minerals grew rapidly during the decade. As pointed out earlier, coal, oil, and natural gas now play a prominent role in Australia's exports; even aluminum is, in part, an energy export since the energy-intensive transformation of bauxite into aluminum is a means of indirectly exporting Australian coal.

Freebairn (1987, p. 149) has highlighted four positive influences on world mineral demand, all of which are exogenous to Australia but have worked very much to the country's advantage during the second half of this century – first, the rapid growth of the Japanese economy in the postwar years and, later, of other resource-poor East Asian economies; second, the advent of large bulk carriers and large-scale port handling facilities, which have reduced transport costs for low-value bulky cargoes; third, technological changes that have created demand for materials such as uranium, mineral sands, bauxite, and special metal-hardening additives; and fourth, the oil price increases of the 1970s, which stimulated demand for alternative energy sources such as coal and natural gas. The Australian policy response to such opportunities was largely passive, removing obstacles such as export bans.

The supply response of Australia's primary producers was sharpened in the 1980s by the trade liberalization described earlier. Under the old trade regime, mining was heavily discriminated against and agriculture was also less advantaged than manufacturing (Table 6). Reducing trade restrictions increased the relative attractiveness of allocating resources to those two sectors, and currency depreciation after the dollar floated in 1983 added to the attractiveness of exporting.

Recent Australian sectoral policies have had a negative but probably minor impact on exports. A ban on opening new uranium mines was introduced in the early 1980s. Under the Mabo legislation, passed in late 1993, new mining projects or renewals of leases will involve negotiation with Aboriginal title holders, which may create some uncertainty about property rights where traditional claims might be involved and deter exploration (Anderson and Findlay, 1995, p. 88). The relatively slow growth of previously dominant wool exports also has something to do with domestic policies. When export prices dropped in the late 1980s, the Australian wool board borrowed to purchase any wool not fetching a minimum floor price at auction. This attempt to support prices in a world market in which Australia still appeared to have market power seems to have been completely ineffective: the

TABLE 6			
EFFECTIVE RATES OF ASSISTANCE IN AUSTRALIA			
	MINING[1]	AGRICULTURE	MANUFACTURING
1970-74	n.a.	22	33
1974-78	n.a.	11	26
1978-82	n.a.	9	23
1982-86	−4	13	21
1986-90	−3	12	17
1990-91	−3	15	14

1 n.a. = not available, but "much more negative" than in the 1980s; the 1982-86 entry for mining is for 1983-84 only.
Source: Anderson and Findlay (1995), p. 77, based on Industry Commission, *Annual Reports*.

price of wool remained below its 1987-88 peak through the early 1990s, and stocks mounted.

It is important to bear in mind, however, that the reorientation of Australian exports towards East Asia has occurred over four decades. Australian policies may have boosted the process, but the underlying reasons must be found elsewhere. These reasons are straightforward. As the resource-poor Japanese economy grew, demand for resource-based imports grew too, and this was exactly where Australia's comparative advantage lay.

THE ELABORATELY TRANSFORMED MANUFACTURES DEBATE

WHILE PRIMARY PRODUCTS ACCOUNT FOR MOST of Australia's incremental exports, the domestic debate over export performance has been mainly concerned with manufactured exports. Since such exports were not a large part of Australia's export bundle before the 1980s, the debate has focused on the need for aggressive targeting of the needs of specific industries if they are to become internationally competitive.

The debate over Australia's recent performance in diversifying its exports and the policy implications was fuelled by *The Rebirth of Australian Industry* (Sheehan, Pappas and Cheng, 1994). The "executive summary" of the study identified "a historic change in Australian trade performance in elaborately transformed manufactures since 1985," which was due to "four factors – a sharp and sustained reduction in costs, a positive change in Australian attitudes to competitiveness and to exporting, the success of a range of industry specific policies targeting export growth, and the expansion of world import demand, especially in Asia." The emphasis on manufactured exports and arguments for industrial policy provided the basis for controversy.

The definition of elaborately transformed manufactures (ETM) is rather broad, including 27 two-digit categories from SITC classes 5 to 8. Exports of ETM in 1993 amounted to $13.3 billion, about one-fifth of all commodity exports and 3 percent of GDP. Among the ETM some are resource-based activities (the largest two-digit exporter is iron and steel, with $1,387 million in 1993), some of which belie the hi-tech image of ETM; for example, textiles, footwear, clothing, cork and wood products, leather manufactures, and travel goods are among the two-digit categories included. The high growth rate of ETM exports is heavily influenced by a single category – computing equipment, exports of which increased from $184 million in 1985 to $1,169 in 1993 (and even faster at the constant prices in which most of Sheehan, Pappas and Cheng's analysis is conducted).

The industry policy conclusions are based primarily on a sub-group of six "policy ETM" – pharmaceuticals, computing equipment, telecommunications equipment, road vehicles, other transport equipment, and clothing. As the most striking evidence of the effectiveness of industrial policy in promoting manufactured exports, Sheehan, Pappas and Cheng (1994, pp. 24-6) focus on passenger road vehicles, a cosseted industry whose exports grew from $67 million in 1980-81 to $510 million in 1992-93. In our earlier discussion of the automobile industry, it was pointed out that the Export Facilitation Scheme did promote exports and may have increased welfare but that the first best policy would have been to remove the protective trade barriers, in which case the EFS would have been harmful.

As mentioned previously, the policy debate over Australian industrial policy and export promotion has been vituperative, and not very illuminating in view of the shortage of empirical work. The Sheehan, Pappas and Cheng study is the fullest statement of the case for supporting "sunrise industries," but it still fails to distinguish between before-and-after comparisons and causal relationships, and to separate an industry's gross exports from its net exports. Earlier, on the basis of empirical work on 10 fast-growing manufactured exports, Krause (1984, pp. 292-3) argued that Australian manufactured export success was negatively related to government assistance. Neither of the conflicting hypotheses can be considered proven, but even if Sheehan, Pappas and Cheng are correct in stating that interventionist government policies have stimulated exports of elaborately transformed manufactures, the case would still have to be made that the net social effects were beneficial.

THE ROLE OF REGIONAL TRADING ARRANGEMENTS

UNTIL THE EARLY 1980S AUSTRALIA PURSUED a unilateral trade policy, paying little attention to either multilateral or bilateral trade diplomacy. During that decade, however, Australia adopted a more active international stance as the loose free trade arrangement with New Zealand was deepened and broadened in the 1983 Closer Economic Relations (CER) treaty (McLean, 1995; Scollay, 1996). In the Uruguay Round of multilateral trade negotiations, which began in 1986, Australia not only participated actively in a GATT Round for the first time but played a major role in creating and leading the Cairns Group, which brought agriculture back onto the negotiating table (Higgott and Cooper, 1990).

At the same time as Australia was displaying its multilateral credentials, there was growing concern about the threat of regionalism, exemplified by the European Union's 1992 program and the North American Free Trade Agreement (NAFTA) negotiations. With rumours of an East Asian regional trading arrangement being floated, especially by Malaysia, the Australian government feared being left as the only kid without a bloc and preemptively launched Asia Pacific Economic Cooperation (Pomfret, 1996). The initiative was supported by Japan, which feared that an exclusive Asian trading bloc would alienate the United States, as well as by the United States itself, and it received high-profile support at the first APEC summit, held in Seattle in 1993. Malaysia was the most reluctant Asian participant, as it promoted the alternative East Asian Economic Caucus (EAEC).

Australia's current policy is one of firm support for the World Trade Organization (WTO) and multilateralism, but statements about regionalism are not always consistent. The current Labour Party government has a certain paternal interest in APEC, which is promoted as a forum for open regionalism consistent with the WTO. There is also concern about the evolution of the ASEAN into a free-trade area and about the possibility of articulating CER with the ASEAN to forestall discrimination against Australia in the fast-growing South-East Asian markets. The importance of these debates lies in the potential challenge to the WTO-based system of multilateralism if discriminatory regional trading arrangements were to emerge in East Asia. Note, however, that discriminatory trade policies have played no part in Australia's trade with East Asia so far.

Lessons for Canada

DESPITE THEIR APPARENT SIMILARITY as primary-product exporters, Australia and Canada differ substantially in their export composition.[21] Australian exports are far more primary-product oriented, as shown by Table 7. While the exports of both Australia and Canada are more oriented than the world average towards primary products, the orientation is more pronounced in Australia for both minerals and agricultural products. And while both countries' primary-product orientation shifted from agricultural exports towards mineral exports during the 1980s, that bias has become much greater in Australia; by 1993, the share of mineral-intensive items in Australia's exports was two and a half times the corresponding ratio for Canada.

TABLE 7

REVEALED COMPARATIVE ADVANTAGE FOR AUSTRALIA AND CANADA[1]

	AUSTRALIA		CANADA	
	1980	1993	1980	1993
Mineral-intensive goods	0.98	3.53	0.96	1.40
Agricultural-intensive goods	3.05	2.43	1.61	1.39
Technology-intensive goods	0.49	0.41	0.73	0.57
Labour-intensive goods	0.23	0.18	0.14	0.31
Human-capital-intensive goods	0.32	0.30	1.46	1.52
Capital-intensive goods	0.41	0.36	1.06	0.96

1 The RCA index is the share in a country's total exports of products from a given classification divided by the same ratio for the world.
Source: Bora (forthcoming), Table 2.

The mix of primary products also differs between the two countries. The weight of coal, iron ore, bauxite, wool, and gold in Australia's exports is not matched in Canada's export bundle. Australia may also be better placed for food exports because of its proximity to East Asian markets and of growing Asian demand for dairy products and shellfish; wheat is the only major overlap in export crops. Canada could export aluminum and has forest products and some minerals that Australia does not have, but the implications require product-specific analysis rather than the drawing of simple lessons.

Comparing the two countries' manufactured exports also reveals important differences. While both are high-income countries with a comparative disadvantage in labour-intensive goods, Australia's comparative disadvantage carries across the whole range of broadly defined manufactured categories listed in Table 7. Canada, by contrast, has a fairly strong comparative advantage in human-capital-intensive goods and its export share of capital-intensive goods is close to the global average.[22] If Australia is to substantially diversify its export mix, then a sustained investment in human capital will be necessary, but it will take time to catch up with Canada, whose enrolment ratios in secondary and tertiary education have been substantially higher than Australia's for decades, with a cumulative impact on the stock of human capital.[23]

With respect to service exports, Canadian tourism and education suppliers can compete with Australia. Both countries have benefited from real depreciations relative to the United States over the past 15 years. Both offer space-intensive vacations, although the package clearly differs between Cairns and Banff. In the education sector, Australia and Canada are more obviously in direct competition; together with New Zealand, the United Kingdom, and the United States, they share the advantage of a well-established English-language higher-education system. However, Canadian education suppliers appear to be discouraged by a reward system in which the benefits of export earnings do not filter down sufficiently to the primary providers (e.g., university departments); they are also victims of significant non-price disadvantages in Asian markets, given Asian perceptions of the threats to personal safety in North America and given Canada's distance from the fast-growing markets of South-East Asia.

Import liberalization and domestic deregulation were probably necessary conditions for Australia's accelerated export growth of the 1980s and 1990s, but some of this was catch-up of opportunities previously forgone. Canada's postwar trade regime has been more liberal than Australia's, and that remains true in the mid-1990s even after Australia's trade reforms. Australia's complex array of export promotion measures may have stimulated some export categories, but there is no convincing evidence that they had any significant impact on the overall rate of export growth; moreover, there has been little attempt to assess their net benefits. My impression is that industrial policies have yielded little if any net benefits, and as part of the government overinvolvement that continues to characterize the Australian economy, they are part of the problem rather than of the solution.

CONCLUSIONS

ONE THEME OF THIS REVIEW of Australian export performance has been the gap between reality and the perceptions of policymakers and their advisers. Australia's impressive performance in exporting to Asia over recent decades has been mainly based on primary-product exports. The dramatic increase in dependence on mineral-related exports is the single most important aspect of Australian exports since 1960.

A large part of the Australian policy debate over exports, however, has concerned the role of manufactured exports and the desirability of shaping industrial policy so as to promote such exports. Some industrial policies have been associated with increased exports, but there has been an unwillingness to ask what would have happened in the policies' absence rather than make a simple before-and-after comparison. In assessing the desirability of industrial policies, there has also been a blindness to the distinction between gross and net benefits and to the possibility that export growth is not an end in itself.

What can Canada learn from Australia's experience? The disappointing answer is: not much that Canada did not know already. Reducing trade barriers has favoured the non-industrial sector and has benefited export activities, and wider economic deregulation has helped to make Australian producers more internationally competitive; this is true for established export giants like BHP and for newcomers like the University of Adelaide. Broader changes in attitude have played an important role as producers have become more internationally aware, and the government has fostered greater consciousness of Australia's proximity to Asia. Specific export promotion policies, however, have played almost no role; government-sponsored reports claiming the opposite tend to be conceptually flawed and, in some cases, self-serving when they are produced by agencies whose very *raison d'être* is to implement such policies.

APPENDIX 1

SELECTED THREE-DIGIT SITC EXPORT CATEGORIES

TABLE A-1	
OTHER AUSTRALIAN RESOURCE-BASED MERCHANDISE EXPORTS IN 1994[1]	
	A$ MILLIONS
Crops	
263 Cotton	671
081 Feed for animals	403
054 Vegetables	390
042 Rice	366
043 Barley	340
057 Fruits and nuts	328
048 Cereal preparations	243
Alcoholic beverages (112)	471
Dairy products (022 milk $855m; 024 cheese $386m; 023 butter $180m)	1,420
Other livestock products	
012 Non-bovine meat	771
211 Hides and skins	448
611 Leather	396
001 Live animals	359
411 Animal oils and fats	184
Crustaceans and molluscs (036)	851
Wood chips (246)	478
Pearls and precious stones (667)	493

Table A-1 (Cont'd)

	A$ MILLIONS
Mineral related products	
Copper (682+283)	1,124
Iron and steel (672+673+674+675)	1,265
Zinc (686)	384
Nickel (284+683)	589
Lead (685)	278
Gold coin (951)	247
Uranium and thorium ores (286)	170
Non-ferrous base metal waste (288)	181
Confidentialized mineral ores (280)	452
Misc. (278 crude minerals and 699 metals n.e.s.)	524
Total of listed items	14,026

1 Classified by SITC categories (from rank 14 to 75).
Source: Australia. Department of Foreign Affairs and Trade (1995a), pp. 22-3.

Endnotes

1 In this paper "dollar" and the sign $ refer to the Australian dollar, which was close to parity with the Canadian dollar in 1995, but was worth less than the Canadian dollar previously. Australian data reported in the paper are often expressed for the financial year, which runs from July 1st until June 30th. For helpful comments on a preliminary draft, I am grateful to Bijit Bora, Laura Brewer, and Christopher Findlay. Thanks also to the conference participants for their comments, especially the official discussant, Peter J. Sagar.

2 One problem in evaluating Australian economic performance is the lack of any suitable comparator; Catherine Mann makes this point during the discussion in Lowe and Dwyer (1994, p. 107).

3 The same point applies to Canada before 1992. Australia consistently ran a trade deficit with Canada throughout the 1980s; in 1984 Australia's imports from Canada were more than double its exports to Canada.

4 All figures are from the Australian Department of Foreign Affairs and Trade's publication, *Composition of Trade, Australia, 1994*; "East Asia" refers to Japan, South Korea, China, Taiwan, Hong Kong, and the six pre-1995 ASEAN members (Brunei, Indonesia, Malaysia, Philippines, Singapore, and Thailand).

5 This pattern has influenced the political economy of trade liberalization, as the declining industrial sectors that typically demand protection in

the OECD countries have been competing with exporting sectors that have important stakes in Asian markets and lobby for Australia to take a lead in regional trade liberalization. The textile, clothing and footwear industry successfully fought for protection in the 1970s (Capling and Galligan, 1992, pp. 235-37), and the automobile industry has historically received substantial government assistance, but the ability of either sector to protect the domestic market has declined with the growth of Australia's exports to Asia.

6 This statistic may be misleading in a way that is typical of much analysis of Australian exports. The leading manufactured export to Japan is aluminum, which with just over $1 billion accounted for over one-fourth of manufactured exports in 1990. In 1994 aluminum exports had fallen to $767 million, accounting for most of the aggregate decline in manufactured exports. Thus the statistic is heavily influenced by a single commodity, and aluminum may not be the kind of item conjured up by the term "manufactured exports."

7 Interview surveys in Hong Kong and Singapore prepared for the Murdoch University Asia Research Centre identified four significant non-price reasons for selecting Australian higher-education institutions: climate, proximity, safety, and the presence of friends and relatives (reported in Lewis, 1995).

8 An important change in the Australian higher-education system has been the increased share of foreign student fees being retained by university departments in the 1990s (almost 50 percent at the University of Adelaide, compared to close to zero in Ontario universities). This has led to departments being more active in recruiting overseas students directly and being willing to contribute to university-wide marketing initiatives.

9 Pincus (1995) and Corden (1995) analyze the history of Australian trade policies. Pomfret (1993, pp. 90-123) summarizes Canadian tariff history.

10 The other major exception to the liberalization process has been Australia's proclivity towards the use of anti-dumping duties. Although the anti-dumping actions in force in Australia declined from a high of 200 in 1984-85 to less than 50 by the end of 1980s, initiations have increased again in the 1990s and the 1994 GATT Trade Policy Review of Australia observed that "Australia accounted for close to one-third of all initiations reported under the GATT Anti-Dumping Code in 1991-92" (GATT, *Trade Policy Review: Australia*. Volume 1. Geneva: GATT, May 1994, p. 57). Feaver and Wilson (1995) argue that Australian anti-dumping laws have not been used as a discretionary tool of protection.

11 In 1991 Singapore, Hong Kong, Taiwan, and South Korea were graduated from the Australian System of Tariff Preferences (ASTP); in 1993, textiles, clothing and footwear, chemicals, sugar, canned tuna, and vegetable and fruit preparations were excluded from the ASTP. In July 1994, ASTP rates were frozen so that they would be reduced or eliminated as general tariff rates fall to 5 percent in July 1996 and zero by the end of the century.

12   The relative decline in Australia's trade position was highlighted by its drop from being the world's 11th biggest exporter in 1973 to 22nd in 1983 (where it remains in the 1990s). Canada, by contrast, retained an almost constant rank – 6th or 7th – in 1973, 1983 and 1993.

13   AusAID's defensiveness probably reflects Australia's lack of a strong tradition of granting aid on humanitarian grounds. The share of aid to GDP is one of the lowest among OECD countries, and studies of the distribution of aid according to the recipient countries' income levels also give Australia poor marks – that is, Australian aid goes to countries with higher income levels than does the aid of most European donors or Canada (White and McGillivray, 1995).

14   Gruen (1993) is an example, which also illustrates the tendency to justify industry-specific policies in terms of "strategic" trade theory. Gruen was an adviser to the Treasurer and is currently with the Industry Commission. Pomfret (1992) reviews the applicability of strategic trade policy to Australia, with a negative conclusion.

15   Trade policies towards the automotive industry are discussed in Bora and Pomfret (1995, pp. 98-104). Grdosic (1994) analyses the EFS in greater detail.

16   Krueger and Duncan (1993) have analyzed the pressures towards increasing complexity in any discretionary policy, as well as the reasons that complexity will be fostered by administrators and by interest groups who benefit.

17   Forsyth (1992) provides an introduction to the microeconomic reforms. The collection of papers in Lowe and Dwyer (1994) assesses the reform package's role in integrating Australia into the global economy.

18   The domestic duopoly, consisting of Qantas and Ansett Airlines, was temporarily broken by a new entrant, Southern Cross, but the new company quickly went bankrupt. The web of international agreements designed to protect the profitability of Qantas could have been undermined by a liberal air-transport clause associated with the Closer Economic Relations agreement with New Zealand, but the Australian government unilaterally disavowed the clause a week before it was due to take effect, either out of fear of Air New Zealand's competitive impact or to enforce its own view of how Australasian suppliers should be organized. Findlay (1985) argues the benefits of air transport liberalization for Australia, while the Centre for International Economics Tourism Report (1988, p. 74) warns of the danger of losing rent.

19   The debate is surprisingly strident. Stanford (1993, p. 53) refers to the polarisation and "the peculiar argot it has generated." Advocates of strategic trade and industrial policies define any activity whose domestic production they wish to expand as "strategic," while opponents of intervention are castigated as "economic rationalists" – a term of abuse in Australia!

20   Agricultural protectionism and subsidized exports are rampant not only in the European Union. The United States has also participated in

export subsidy wars, and the high-performing North-East Asian countries have protected their farmers.
21 There may be more similarity between western Canada and Australia. Regional economic differences are less pronounced in Australia than in Canada, although in the 1990s a differentiation is emerging between the more dynamic states of Queensland and Western Australia, whose growth is led by primary-product exports, and the "rust belt" states of Victoria and South Australia, where the automobile and textiles, clothing and footwear industries are concentrated.
22 Some of Canada's manufactured export success has been due to the Auto Pact, but even excluding automobiles, Canadian manufactured exports are larger and more diversified (that is, include more human-capital-intensive items) than Australia's. This interpretation of Canadian trade by factor-content is similar to that of Hejazi and Trefler (in this volume) in which they identify Canada as a net exporter of secondary and college education.
23 The OECD (1977) reported that, in 1975, 44 percent of the 15-19 age group were enrolled in full-time education in Australia, compared with 66 percent in Canada; in the 20-24 age group, enrolment rates were 5.4 and 14.5 percent, respectively. According to the World Bank's *World Development Report* for 1994, 1991 secondary enrolment rates were 82 percent in Australia and 104 percent in Canada; tertiary enrolment rates were 39 percent in Australia and 99 percent (!!!) in Canada; these rates are constructed as the enrolment in secondary or tertiary institutions, divided by the number of 12-17 and 20-24 year-olds, respectively.

BIBLIOGRAPHY

Albon, Robert. "Export Subsidisation Through Export Facilitation: The Case of Passenger Motor Vehicle." Paper presented at the Bureau of Industrial Economics Annual Conference, Canberra, July 1992.

Anderson, Kym. "Australia's Changing Trade Pattern and Growth Performance," in *Australia's Trade Policies*. Edited by Richard Pomfret. Melbourne: Oxford University Press, 1995, pp. 29-52.

Anderson, Kym and Christopher Findlay. "Policies Affecting Australia's Primary Sectors," in *Australia's Trade Policies*. Edited by Richard Pomfret. Melbourne: Oxford University Press, 1995, pp. 74-90.

Anderson, Kym and Ross Garnaut. *Australian Protection: Extent, Causes and Effects*. Sydney: Allen & Unwin, 1987.

Australia. Department of Foreign Affairs and Trade. *Composition of Trade, Australia, 1994*. Canberra: Commonwealth of Australia, 1995a.

---------. *Trade in Services Australia 1993-94.* Canberra: Commonwealth of Australia, 1995b.
Bora, Bijit. "Trade and Investment in the APEC Region, 1980-1993," in *International Trade and Migration in the APEC Region.* Edited by Peter Lloyd and L. Williams. Melbourne: Oxford University Press, forthcoming.
Bora, Bijit and Richard Pomfret. "Policies Affecting Manufacturing," in *Australia's Trade Policies.* Edited by Richard Pomfret. Melbourne: Oxford University Press, 1995, pp. 91-111.
Capling, Ann and Brian Galligan. *Beyond the Protective State.* Cambridge, U.K.: Cambridge University Press, 1992.
Centre for International Economics Tourism Report. *Economic Effects of International Tourism.* Canberra: Centre for International Economics, 1988.
Corden, Max. The 1995 Joseph Fisher lecture, delivered at the University of Adelaide, 26 October 1995.
Feaver, Donald and Kenneth Wilson. "An Evaluation of Australia's Anti-Dumping and Countervailing Law and Policy." Victoria University of Technology, Department of Applied Economics Working Paper 4/95, May 1995. Forthcoming in the *Journal of World Trade.*
Findlay, Christopher. *The Flying Kangaroo: An Endangered Species?* Sydney: Allen and Unwin, 1985.
---------. "Service Sector Policies," in *Australia's Trade Policies.* Edited by Richard Pomfret. Melbourne: Oxford University Press, 1995, pp. 112-34.
Forsyth, Peter (ed.). *Microeconomic Reform in Australia.* Sydney: Allen and Unwin, 1992.
Freebairn, John. "Natural Resource Industries," in *The Australian Economy in the Long Run.* Edited by Rodney Maddock and Ian McLean. Cambridge, U.K.: Cambridge University Press, 1987, pp. 133-64.
Grdosic, Mike. "Sunk Costs, Hysteresis, and Export Facilitation: A Conceptual Synthesis Used to Analyse the Automotive Component Producers 'Export Success'." Economics honours thesis, University of Adelaide, November 1994.
Gruen, N. "Export Assistance, Trade Liberalisation, Strategic Trade Theory and the New Development Consensus." Centre for Economic Policy Research, Australian National University, 1993.
Helliwell, John. "Natural Resources and the Australian Economy," in *The Australian Economy: A View from the North.* Edited by Richard Caves and Lawrence Krause. Sydney: Allen and Unwin, 1984, pp. 81-126.
Higgott, R. and A. Cooper. "Middle Power Leadership and Coalition Building: Australia, the Cairns Group, and the Uruguay Round of Trade Negotiations." *International Organisation* 44 (1990):589-632.
Krause, Lawrence. "Australia's Comparative Advantage in International Trade," in *The Australian Economy: A View from the North.* Edited by Richard Caves and Lawrence Krause. Sydney: Allen and Unwin, 1984, pp. 275-311.
Krueger, Anne and Roderick Duncan. "The Political Economy of Controls: Complexity." NBER Working Paper 4351. Cambridge, Mass.: National Bureau of Economic Research, April 1993.
Lewis, Philip. "Indonesian Demand for Higher Education in Australia." Paper presented to the 24th Conference of Economists, Adelaide, 25-27 September 1995.
Lowe, Philip and Jacqueline Dwyer. *International Integration of the Australian Economy.* Sydney: Reserve Bank of Australia, 1994.
McColl, G. and R. Nicol. "An Analysis of Australian Exports to its Major Trading Partners, Mid-1960s to Late 1970s." *The Economic Record* 56 (1980):145-157.

McLean, Ian. "Trans-Tasman Trade Relations: Decline and Rise," in *Australia's Trade Policies*. Edited by Richard Pomfret. Melbourne: Oxford University Press, 1995, pp. 171-89.

McMullan, Bob. *Winning Markets: Australia's Future in the Global Economy*. A statement by the Minister of Trade, International Public Affairs Branch, Department of Foreign Affairs and Trade, Canberra, June 1995.

Organisation for Co-operation and Development. *Australia: Transition from School to Work or Further Study*. Paris, 1977.

Pincus, Jonathan. "Evolution and Political Economy of Australian Trade Policies," in *Australia's Trade Policies*. Edited by Richard Pomfret. Melbourne: Oxford University Press, 1995, pp. 53-73.

Pomfret, Richard. *Unequal Trade: The Economics of Discriminatory International Trade Policies*. Oxford, U.K.: Basil Blackwell, 1988.

_____. "Imperfect Competition and Trade Policy in Small Open Economies," in Bureau of Industry Economics, *1992 Conference of Industry Economics, Papers and Proceedings*. Canberra: Australian Government Publishing Services, 1992, pp. 25-39.

_____. *The Economic Development of Canada*. Second edition. Scarborough, Ont.: Nelson Canada, 1993.

_____ (ed.). *Australia's Trade Policies*. Melbourne: Oxford University Press, 1995.

_____. "Blocs: The Threat to the System and Asian Reactions," in *Regional Integration and the Asia Pacific*. Edited by Bijit Bora and Christopher Findlay. Melbourne: Oxford University Press, 1996, pp. 13-24.

Scollay, Robert. "Australia-New Zealand Closer Economic Relations Agreement," in *Regional Integration and the Asia-Pacific*. Edited by Bijit Bora and Christopher Findlay. Melbourne: Oxford University Press, 1996.

Sheehan, Peter, Nick Pappas and Enjiang Cheng. *The Rebirth of Australian Industry; Australian Trade in Elaborately Transformed Manufactures, 1979-1993*. Centre for Strategic Economic Studies, Victoria University of Technology, Melbourne, 1994.

Smith, Ben. "Resources and Australian Economic Development," in *Surveys of Australian Economics*. Volume 3. Edited by Fred Gruen. Sydney: Allen and Unwin, 1983, pp. 77-123.

Stanford, Jon. "Industrial Policy in Australia," in *Industrial Policy in Australia and Europe*. Edited by Jon Stanford. Canberra: Australian Government Printing Service, 1993, pp. 36-86.

White, Howard and Mark McGillivray. "How Well Is Aid Allocated? Descriptive Measures of Aid Allocation: A Survey of Methodology and Results," *Development and Change* 26 (1995):163-83.

World Bank. *World Development Report, 1994*. New York: Oxford University Press, 1994.

Comment

Peter J. Sagar
Entrepreneurship and Small Business Office
Industry Canada

IS THE GLASS HALF FULL, OR HALF EMPTY? Dr. Pomfret has looked into the glass containing Australia's experiences in exporting to APEC and declared it at least half-empty. From this perspective, it is virtually devoid of lessons for Canada. But perhaps this is as much a question of perspective as of reality.

In essence, Professor Pomfret examines the Australian experience and comes away with the following observations:

- Australia is a very resource-intensive country.
- Much of Australia's experience in exporting to APEC has been based on its natural resources.
- Manufacturing exports lag behind.
- There is little evidence that at a macro level, Australian industrial policy has supported manufacturing trade.
- Canada has had relatively greater success in manufacturing exports.
- Therefore, Australia holds no lessons for Canada.

All of these points are supportable, and indeed are supported in Dr. Pomfret's paper. He is obviously right. Except, and it is an important except, he has only looked at the top half of the glass, and not surprisingly, he has found it empty.

But given that the topic is Australia, perhaps looking at the under side of the glass is not only necessary, but appropriate.

Let us begin with the irrefutable facts:

- Australia is resource rich, but relatively short of human resources (*in quantitative terms only*).
- Australia is a major and highly successful resource producer and successful exporter.
- Australian manufacturing is thin in depth and scope.

But what about Canada? Well, despite Canada's apparently greater manufacturing base than Australia's, the reality is that much of our manufacturing base owes its presence to a combination of proximity to the U.S. market (and some farsighted policy decisions ranging from the Auto Pact to medicare), plus a relative abundance of raw materials. Strip away Canada's exports that are not related to these factors, that is precisely the criteria against which Australia is being held to account, and the story is not dissimilar.

- Canada is resource rich but relatively short of human resources (*in quantitative terms only*).
- Canada is a major and highly successful resource exporter.
- Canada's manufacturing base is relatively thin in depth and scope.

Given these similarities, surely there must be some lessons to be learned from one country to the other. And there are.

The starting point for this look at the bottom half of the Australian glass is the overall trade performance of Australia in APEC. A recent study by the Australian Parliamentary Research Service notes the shifting patterns of Australian trade. In 1985, 48.2 percent of Australia's exports went to Asian APEC countries. In just seven years, this share has risen to 57.2 percent.[1] While this market is a rapidly growing opportunity indeed, Australia has had to take major steps to stay in the game against fierce international rivals. If not completely successful as demonstrated by Dr. Pomfret, these steps appear to have enabled Australia to expand its relative focus on the fiercely contested Asian APEC market.

What is the basis of this performance? Is it really just a resource-driven comparative advantage or is something else at work?

The answer is of course a bit of both. Yes, Australia has some advantages in the resources area. But resource markets globally are highly contested with relatively few areas of country-specific monopolies (if any) existing. Australia has had to compete fiercely for these markets, including with Canada, and to the extent that it has won, it has not done so on the basis of resource riches alone, but through a very highly developed strategy of improvements to shipping infrastructure, to workplace practices, to quality in product and delivery terms, and to vigorously seeking out new technologies and approaches. This is as true for coal exports as for broccoli and fresh dairy.

In these areas, at the micro if not macro level there are lessons for Canada. These lessons do not appy only to the resource sector but, in fact, have more in common with the lingo of the manufacturing companies: targeted, client-focussed marketing; adroit application of technology; quality and just-in-time delivery; workforce flexibility and participation; total quality; and so on.

On the manufacturing side, Australia has undergone an extensive transformation, from a massively protected, inwardly focussed economy to one where the public level of understanding of globalization and internationalisation would put Canada's awareness of these issues to shame. Obviously, the Australian manufacturing base remains small and truncated compared to many countries – a legacy of its past colonial evolution and present day global pressures – but within that base reside a host of small, dynamic, and rapidly emerging companies.

Can Canada learn anything here? Perhaps not on the trade side, although this is more a function of similarities than of differences between the two countries. Both have similar trade support programs and structures. Both compete head-to-head in certain sectors. And both have experienced success and failure in meeting the challenges of the developed and developing countries.

But in terms of domestic policies, the two countries have followed different routes, reflecting different histories and realities as well as different development paradigms. It may be here that the lessons lie. But unfortunately, Professor Pomfret's paper dwells more on finding weaknesses in Australia's approach than in seeking out the successes and in comparing these to the Canadian approach.

This is not the place to try to do that job, but areas where Australian interventions should be looked at for lessons would include the targeted efforts under the National Industrial Extension Service to develop a cadre of skilled management consultants and management tools, and to support their use by small- and medium-sized businesses. The promotion of best practices, networking, globalization and trade awareness are also areas worth investigation. A comparison of the two countries' relative successes in managing the major restructuring of the past two decades would also be interesting.

Returning to the slightly tortured and wholly clichéd analogy of the half-full glass for the last time, the lessons for Canada and Australia can indeed be imputed from the half-empty portion of the glass – those areas where our efforts have not been successful. However the more refreshing and informative lessons are likely to be found by looking at the half-full part – the areas where each country can point to success. Such examinations are unlikely to detect dramatic reversals in Australia's performance at the macro level. But they should yield lessons for both countries in terms of building competitive firms able, and equally important, willing to seize the rich opportunities of the ASEAN region. That is a challenge for another paper, another time.

ENDNOTE

1   Patricia Sagar. "Growth of Intra Asian APEC Trade and Some Implications for Australia." Parliamentary Research Service, Background Paper No. 19, Table 6, Australian Parliament, 1994.

Sylvia Ostry
Centre for International Studies
University of Toronto

The New Trading System: Global or Regional?

ASIA PACIFIC ECONOMIC COOPERATION (APEC) has been described as "four adjectives in search of a noun." It is an apt description in a much broader sense than that intended by the Australian foreign minister who coined it because it captures the combination of uncertainty and groping that underlies not just the future of APEC but, more importantly, the emergence of a new global economy and the new trade and investment rules that may govern it in the 21st century. There is, indeed, a profound irony in the current situation. After the completion of the most ambitious and comprehensive trade negotiation in history – the Uruguay Round of the General Agreement on Tariffs and Trade (GATT) – as well as an unprecedented wave of unilateral liberalization in countries in Asia and Latin America and a worldwide move to the market system following the demise of the former Soviet empire, the future of the global trading system is shrouded in a thick fog of uncertainty. The purpose of this paper is to spell out the main reasons for this curious conjuncture and then try to assess the relationship between the multilateral rules-based system and regional integration initiatives such as APEC.

To tackle these issues, first let me sketch out the main impact of two major structural changes that are shaping the economic and policy environment. Layered on these transformative forces is a third causal factor, and arguably the most important, which is the fundamental shift in U.S. trade policy that took place during the 1980s. While no longer the sole economic superpower, the United States remains key to the evolution of the global system in the foreseeable future. In that sense, we are in a unipolar world, but given the amazing changes that have been occurring in the Pacific region, this unipolarity can only be viewed as temporary. So the evolution of APEC – the search for the "noun" – could be of major significance.

Structural Changes

There are two pervasive and ongoing transformations in the international economy, separate but closely interrelated, that are reshaping the global and domestic policy agendas. One has been termed globalization – a *terme du jour* that is widely used and rarely defined – and the other is the accelerating revolution in the converging information, computer, and communications technologies, which I will call the ICT revolution for the sake of convenience. Not surprisingly, governments and international institutions, which tend to make policy in a rearview mirror, have found it increasingly difficult to adapt to this emerging new techno-economic paradigm. It is, indeed, the mismatch between accelerating change and lagging policy response that feeds the climate of uncertainty. But more on that below. First I highlight the main features of these changes before turning to closer-to-home questions about the shift in American policy and its consequences.

Globalization

This much-overused word has no agreed definition but one common theme does emerge from the discourse – that of increasing interdependence among countries. Increasing interdependence or linkage began after the Second World War as the tariff barriers erected during the 1930s were dismantled through successive multilateral negotiations led by the GATT. During the 1970s and early 1980s, the deregulation of financial markets and the increasing use of information and communication technologies spurred an enormous increase in financial linkage so that today financial flows vastly outweigh trade flows among countries. But in the second half of the 1980s, an unprecedented surge in foreign direct investment (FDI), which grew at three times the rate of trade and four times the rate of output, ushered in a new phase of interdependence that is *qualitatively* different, in its policy implications, from previous linkages. After the "bulge" of 1985-90 there was a slowdown in FDI outflows, but now we see a new pick-up and a significant shift to host countries in East Asia, especially China and the so-called Chinese Economic Area in this region. Most of the FDI is in technology-intensive industry and in services. Less tied to the location of natural resources or to supplying protected markets than in the past, this FDI is more "mobile."

While the bulge of the 1980s was partly due to one-off factors (new forms of protectionist pressures in Europe and the United States, for example), there were and are underlying forces generating globalization. The ICT revolution is both an enabling factor and a driver, fostering innovation in products and production processes and also in organization at the level of the firm and of the industry. Greatly intensified international rivalry, especially in technology-intensive industries, spurs the transnational corporations (TNC) to capture economies of scale and scope, customize products to satisfy consumer tastes, gain access to inter- and intra-firm networks and to knowledge, both technological and "tacit." Hence the transnational corporations are the dominant global actors. And the TNC are increasingly the main channels for trade, finance, and technology – the sources of growth.

For example, it is estimated that one-third of world exports of goods and services today takes place within firms. Among the member states of the Organisation for Economic Co-operation and Development (OECD) and, to a growing degree, the newly industrialized economies (NIE) of East Asia, trade in manufacturing is predominantly within the same industry and, increasingly, in capital- and technology-intensive components of complex systems products such as autos, aircraft, machinery, telecommunications equipment, and so on, reflecting the development of transnational production processes. Moreover, the global impact of the multinational enterprises (MNE) is understated by only looking at their role in trade. Once sales of foreign affiliates, licensing and royalty payments for technology, and franchising fees are taken into account, a recent estimate for the United States suggests that *three-quarters* of all its international transactions are linked to the activities of American MNE. Since the U.S. firms are the most transnational, the comparable figures for Europe and certainly for Japan would be smaller, but moving in the same direction.

The policy consequences of this new phase of interdependence are now becoming visible. Globalization has pushed international policy within national borders. For the TNC, border barriers are less important than domestic "structural impediments," which are barriers to effective market access by trade or effective presence by investment – another two-way funnel for technology flows. These impediments can arise, often unintentionally, from regulatory policies, legal cultures, the behaviour of private actors – in other words, from differences in *systems* among countries. The "shallow integration" of the GATT, which centred mainly on the removal of border barriers to trade, is being replaced by a policy agenda of "deeper integration," which centres on domestic policies and institutions, involves both trade and investment as complementary modes of entry, and also involves an intrinsic pressure for *harmonization*. The GATT agenda, in profound contrast, implied a primacy of trade and a preservation of system *diversity*.

In sum, the agenda of deeper integration covers trade, investment, and technology and is far more intrusive and erosive of national sovereignty because the pressure for harmonization will, at the limit, generate continuing system friction. The threat of delocalization by TNC adds to this harmonization pressure as they seek to reduce the transaction costs associated with different regulations and legal regimes. The Uruguay Round, especially in broaching "new" issues such as services and intellectual property, marked a transition to the new agenda. All future negotiations, whether multilateral, regional, or bilateral, will involve a move inside national borders because the main forces of globalization – FDI and the ICT revolution – will not abate.

Finally, while globalization is transforming the international policy agenda and is closely interrelated to the ICT revolution, it is worth spelling out briefly the impact of technological change on domestic policies, which in turn have international spillovers. I now turn to the second structural transformation – the ICT revolution.

The ICT Revolution

Current and ongoing changes in technology are so pervasive that they are seen as a new techno-economic paradigm, comparable to the Industrial Revolution. However, I want simply to cite one aspect of that new paradigm – namely, locational competition.

Just as rivalry among TNC has become far more intense and is a major cause of the vast increase in FDI, and because FDI is the major source of the three engines of growth – trade, finance and technology – countries are competing for foreign investment. Locational competition has a number of consequences. Fiscal and other financial incentives, especially those directly linked to technology, have proliferated in both OECD and non-OECD countries since the mid-1980s. A new form of protectionism in the shape of conditional national treatment is emerging in high-tech sectors that are eligible for government R&D funding. More broadly, the significance of locational competition is to shift comparative advantage to the architecture of the knowledge infrastructure.

While this shift is not yet apparent as a coherent policy focus in any of the OECD countries, and while there has been much debate about the nature of effective innovation policies, since there are no international rules to define the interface between domestic high-tech policies and their international spillover, high-tech battles have continued to flare, most visibly between the United States and Japan. But, more importantly in the context of this conference, many countries in East Asia – and most of all China – see technology access as the key to convergence in living standards. The most effective funnel of technology flows is, as stated previously, the TNC, which brings both coded and tacit knowledge, including managerial and marketing skills and sophisticated intra- and inter-firm networks. This will be discussed further when considering APEC but first I briefly deal with the issue of American trade policy.

U.S. Trade Policy

The role played by the United States in building the postwar architecture of international economic cooperation was largely a product of the cold war. The postwar architecture was a magnificent achievement that created unprecedented prosperity and stability for the Western world. But it had some unexpected consequences. One of these was the creation of the "convergence club" of the major OECD countries which, by the 1970s, had by and large converged in technological capabilities, capital per worker, education levels, and managerial capacity. The United States, the undisputed economic and technological superpower in 1950, never expected that several decades later Europe and Japan would catch up with American dominance or even challenge it in some respects. And there was another surprise. While all of the OECD countries chose capitalism and rejected communism, it became clearer over time that capitalism came in a variety of brands. There were not just two systems – capitalism and communism – but a range of variants, with structural differences that affected openness to trade, investment, and technology.

These two unexpected outcomes have shaped the evolution of U.S. trade policy. First of all, from an exclusive focus on multilateralism and the GATT in the three decades following the end of the war, by the end of the 1980s U.S. policy had shifted to a multitrack – multilateral, regional, and unilateral – approach. Unilateralism was applied through a variety of so-called Section 301 or "Super 301" trade-remedy regulations, whereby the U.S. government could judge a trade practice to be "unfair" or "unreasonable" and could apply sanctions if it deemed them necessary. ("Super 301" is a regulation that enables the U.S. government to declare that an *entire country* is "unfair" in its trade practices.) Indeed, because of structural differences in ease of access and of other system differences defined as unfair or unreasonable by the United States, the focus of American trade policy increasingly shifted to what were once considered domestic policies and practices, and also expanded beyond trade to include investment and technology. This was most apparent in the high-tech battles with Japan and, to a lesser extent, with Europe during the 1980s, culminating in the paradigmatic U.S.-Japanese structural impediments initiative (SII) at the end of the decade, in which the Americans listed dozens of practices, ranging from land-use regulations to competition policy enforcement, as unfairly or unreasonably impeding access to the Japanese market.

With these developments, the agenda of deeper integration was launched. These fundamental changes in American policy purpose and ambience amplified and highlighted the inadequacies of the GATT. They were manifested, in part, by the inclusion of services and intellectual property in the agenda of the Uruguay Round launched under U.S. leadership. Indeed, it was the long delay in the launch of that Round that led to the first major U.S. *bilateral negotiation*: the Free Trade Agreement (FTA) with Canada was designed as a "wake-up call" for the Europeans and for some developing countries, whose foot-dragging and opposition had stalled the multilateral negotiations for several years.

By the time the Uruguay Round was concluded at the end of 1993 (after much brinkmanship involving missed deadlines), American multitrack policy was firmly entrenched. On the regional track, the FTA was followed by the

North American Free Trade Agreement (NAFTA) and various initiatives in Latin America and East Asia. Unilateralism has continued in Section 301 disputes with Japan and other countries, and indeed the Clinton administration has twice renewed Super 301 since it lapsed in 1990. If unilateralist actions continue, they could seriously, and perhaps irrevocably, undermine the credibility of the GATT's successor, the World Trade Organization (WTO), because its much strengthened dispute-settlement mechanism was designed to eliminate or, at least, seriously constrain the use of such trade instruments by member states. The most recent "close call" was the dispute this year with Japan over autos, where a unilateral application of trade sanctions would have been clearly illegal under the new WTO rules. It has been argued by some that the Americans never intended to apply sanctions because they "knew" that at the end of the day, the Japanese would agree to a deal. Be that as it may, it hardly needs saying that this kind of brinkmanship is a dangerous game to play when outcomes are intrinsically uncertain, as was unquestionably the case in the auto dispute. And uncertainty is indeed the essential element in the evolution of the trading system, as I have repeatedly stressed.

That brings me to a broader issue. An extremely important contextual factor, which is changing and will continue to profoundly change American policy, is the end of the cold war. The cold war context was fundamental, not only in fostering the creation of the postwar trading system, but also in acting as a constraint on trade disputes through what has been called a spillover from high policy to low policy. That constraint is gone and, increasingly, the spillover will be in the opposite direction. This, too, greatly adds to the uncertainty of policy impact since it makes it nearly impossible to calibrate the repercussions of a trade dispute on foreign-policy and security issues, especially in a region as diverse and complex as East Asia, where the role of the world's next superpower, China, is a puzzle to all (including, I expect, its own current leaders). The U.S.-Japanese auto dispute was carefully studied in most Asian countries, but it is not clear what lessons were drawn. Many American trade policy officials believe that it underlines the "success" of Super 301 – in other words, that threats will get results *without* the use of sanctions. That may well be so, if you like high-stakes poker. But the international trading system should not be a poker game!

Finally, the end of the cold war will significantly lessen the power of the U.S. presidency, especially in trade matters, as Congressional deference continues to decline rapidly. The first signs of this are clear in the current debate over fast-track renewal, which has less to do with side agreements than with a shift in the structure of power in the world's most diffuse governance system. This rise in Congressional power will also make American trade policy much more unpredictable. So where does this climate of uncertainty leave us in discussing the theme of this paper?

Globalism, Regionalism and APEC

JACOB VINER'S CLASSIC WORK on regionalism and multilateralism established the terms of the debate on this issue: do regional agreements, on balance, divert trade or create trade? An updated but fundamentally similar version that has emerged over the past few years is couched in jazzier terms: Are regional agreements building blocks or stumbling blocks to global liberalization? More recently, however, especially since the difficulties over the Uruguay Round, a renewed interest in the subject is generating a more comprehensive approach – a most welcome approach, since the Vinerian paradigm is clearly inadequate in today's mutating global environment.

In a recent study of regionalism, the newly established World Trade Organization noted that in the 1990s there was a surge of new regional integration agreements (RIA) – 33 in all between 1990 and 1994. This brought the total at the end of 1994 to 109, and more accords of this type are coming on stream. A friend has remarked that in the year 2021, when the European Union (EU) and the United States discover that they have to attend 287 dispute panels and re-estimate 20,416 rules of origin (ROO), a brilliant policy analyst will write a memo suggesting a new global institution based on a near-forgotten agreement called the GATT!

Jokes aside, it is partly this vast proliferation of RIA – the pursuit of new initiatives such as TAFTA, FTAA, CEFTA, and so on – that has provoked the new debate. The proponents of RIA tend to argue that they are additive – they all promote more liberalization – or, indeed, that they are symbiotic: RIA spur other RIA, which spur global liberalization. Opponents counter that the emerging crazy quilt of arrangements may actually multiply new sorts of impediments – for example, in extremely complex ROO, which divert both trade and investment by providing easier access for protectionist lobbies than would be the case in a multilateral negotiation. On the other hand, one strong argument of proponents is that RIA are easier and faster than negotiations in the WTO. Thus, the argument goes, it is better to tackle new issues in an RIA first and then move to the WTO later. This synergy argument was the rationale for the American-led push to launch the Multilateral Agreement on Investment (MAI) in the OECD rather than the WTO. The strong attack on the OECD/MAI from non-OECD countries, expressed most recently at a joint seminar of the WTO and the United Nations Conference on Trade and Development (UNCTAD), is an early warning signal that the symbiosis argument may be a trifle simplistic – especially in the case of investment, since the most attractive host countries are increasingly outside the OECD club, particularly in East Asia.

Which brings us to APEC. The current debate about RIA and the WTO will not provide adequate policy guidance unless it clearly recognizes a number of fundamental issues that go beyond the traditional trade policy framework. I enumerate some of these because they are especially relevant in the APEC context.

The first issue is that trade policy is less and less focused on trade and more and more "modal-neutral" – to use the newest "OECD-ese" to define the complementarity of trade and investment. In APEC, the engine of regional integration has been investment, especially investment by the Japanese, the Americans, the overseas Chinese, and now the European and especially the Germans. In this form of "natural integration," the private sector stimulated unilateral liberalization as countries sought FDI, the engine of growth. Locational competition assisted the regional integration led by investment. Any realistic agenda for successful deeper integration in East Asia must deliver on a comprehensive investment code that, above all, includes meaningful commitments on transparency, performance requirements, and dispute settlement. This, of course, enters the terrain of domestic impediments, including regulatory policies, technology transfer, and legal systems. Technology transfer is especially important since so much of the East Asian "miracle" has come from increased factor inputs rather than increasing total factor productivity. More broadly, no region in the world is as diverse as APEC and hence so fertile a terrain for the generation of system friction. Given this fact, the view – strongly espoused by many American policy experts – that deeper integration is more feasible in regional negotiations than in the WTO seems a trifle hyperbolic. However, the resistance of a number of East Asian countries to an institutionalization of APEC stems from precisely the same perception! Indeed, in the case of the Chinese and perhaps others, the absence of transparency and of institutional arrangements seems an invaluable asset when foreign firms (and governments) compete for deals. As to governments, at the September 1995 EU-East Asia Summit, Asian leaders pushed for the creation of a new forum, to be called ASEM (Asia-Europe Meeting), in parallel with APEC.

Secondly, this example of the difficulties of pursuing deeper integration in APEC underlines another aspect of the RIA-versus-WTO debate – namely, that regionalism is not regionalism is not regionalism. The deepest integration in a regional arrangement is, of course, the European Union. But one could hardly pose a more striking contrast than that between European and East Asian integration. All arrangements are path-dependent, and the origins of the EU were political: economics was a means to an end. The natural integration of East Asia is private sector led, *facilitated* by international non-governmental organizations (INGO), with governments *following*. The contrast between the institutional surplus of Europe and the institutional deficit of East Asia reflects these different launch forces. There is no "one size fits all" model of RIA, and the policy debate must reflect the importance of path dependency. So what is the right model for APEC? The answer, obviously, is very murky, and the Osaka meeting of APEC leaders last November did little to change that.

In the context of APEC, the disagreements among member countries – most significantly between the Americans and the Australians, on the one hand, and the Japanese, on the other – are, as mentioned earlier, well captured in the aphorism, "four adjectives in search of a noun." Of the three tracks of APEC – trade facilitation, liberalization, and cooperation – the Americans and the Australians have placed the highest priority on liberalization; and they have charged the Japanese with foot-dragging over the agriculture issue. The Japanese have led on cooperation, generating widespread suspicion that they are using official development assistance (ODA) funds to improve the production base for Japanese firms and to stymie the Americans, who are cutting back on international aid because of pressures from a hostile Congress. Trade facilitation garners the support of most countries, being the safest territory. But in effect, Japan can play three of the policy objectives and the United States, scarcely one. That makes the game much more complex – chess, rather than high-stakes poker. Layered on top of these different views of the future direction of APEC's evolution are profound geopolitical and security concerns – most of all, the role of China, the emerging regional hegemon.

This brings us back to the issue of the crucially important role of American policy. The absence of President Clinton from the Osaka meeting because of the budget gridlock is emblematic: "It's economics and the 1996 election, not foreign policy, stupid!" In any country, the allocation of political effort will be decided on the basis of domestic considerations: a fight with Congress over fast-track negotiations or most-favoured-nation (MFN) treatment for China, or over the substitution of competition policy for anti-dumping, for example, does not look like much of a "winner" in Washington until, *at least*, after the 1996 election. Hence APEC's emphasis on trade policy liberalization rings a bit hollow at present. And the synergy argument for regionalism looks a bit thin until 1997.

In conclusion, then, the search for the noun will continue. Meanwhile, what about the WTO – the Mercedes without gas? What will happen at next year's APEC meeting in Singapore? No country should take American commitment to a multilateral rules-based system as a given. All the time and effort expended on the Osaka meeting might have been better spent on planning for the Singapore conference. Coalitions of middle powers played a key role in the launch and ongoing negotiations of the Uruguay Round; they could do so again, in my view. At the very least, while regional efforts are now part of the game, each should be carefully assessed in terms of risks and benefits, and of their linkages to multilateralism.

About the Contributors

Ashfaq Ahmad is an investment analyst with the Strategic Investment Analysis group at Industry Canada. His research focus centres on global direct investment and merger and acquisition activity. He is co-author of Industry Canada papers on formal and informal investment barriers in the G-7 countries and on the activities and performance of Canadian-based multinationals.

Michael Baker is an assistant professor at the University of Toronto's Department of Economics. He has completed several papers on the economic outcomes of immigrants to Canada. His other research interests include unemployment dynamics, the Unemployment Insurance system, the effects of minimum wages on youth employment, sex- and ethnic-based differentials in labour market outcomes, and family labour supply.

Dwayne Benjamin is an associate professor at the Department of Economics of the University of Toronto. He received his B.Sc. from that university and a Ph.D. from Princeton University. His main area of research is the study of labour markets, both in a Canadian context and in developing countries. He has written several papers on immigrant economic outcomes in Canada, as well as on male/female wage differentials, minimum wages, and the economics of aging in Canada. In the development economics area, he has published papers on the changing role of women and models of household behaviour in rural areas, as well as evaluations of the role of labour and land markets for household welfare.

John F. Chant is professor of economics at Simon Fraser University, where he specializes in monetary economics. He has also taught at the University of Edinburgh, Duke University, Queen's University, and Carleton University, and he served as research director at the Economic Council of Canada, producing the study *Efficiency and Regulation*, which served as background for the 1980 *Bank Act*. He is author of several books and, as an expert on the regulation of financial institutions in Canada, has served as a consultant to the Bank of Canada, the World Bank, the Royal Commission on the Economic Union and Development Prospects of Canada, and other government departments and agencies.

Francis X. Colaço is senior advisor to the vice-president, East Asia and Pacific Region, at the World Bank in Washington. He attended Bombay University and received his doctorate from the University of California at Berkeley. While at the World Bank, he has served as principal economic advisor to the vice-president, Sector Policy and Research, and director of the Eastern Africa Department. He has written a number of articles and books on international capital flows and economic development.

Brian R. Copeland is professor of economics at the University of British Columbia. He is a graduate of that university and obtained his Ph.D. in economics from Stanford University. His teaching and research interests have covered international trade theory and policy as well as environmental economics, including the links between trade and transboundary pollution.

John W. Craig is a lawyer in private practice in Toronto. He is a graduate of the University of Toronto and of the Osgoode Hall Law School. He is a former partner at McMillan Binch, with which he has an ongoing consulting relationship. While a partner at McMillan Binch, he maintained an active business law practice, with special emphasis on foreign investment in Canada. He managed the firm's Japanese practice, assisting Japanese companies investing in Canada. Mr. Craig is a frequent writer and speaker in Canada and Japan on a variety of business, trade, and investment issues.

Wendy Dobson is a professor at the University of Toronto and director of its Centre for International Business. She was educated at the University of British Columbia, and at Harvard University. She holds a Ph.D. in economics from Princeton University. Dr. Dobson's background combines research and practice. Prior to joining the University of Toronto in 1990, she served as president of the C.D. Howe Institute and later as Associate Deputy Minister of Finance in the federal government. Dr. Dobson is author and editor of numerous books and articles on international economic issues.

Donald J. DeVoretz has been with Simon Fraser University since obtaining his doctorate in economics from the University of Wisconsin. His main research interests include the economics of immigration, with special emphasis on the employment, income, and savings effects of Canadian immigration flows.

Stephen T. Easton is a professor of economics at Simon Fraser University. He received his Ph.D. in economics from the University of Chicago. He has published in the areas of international trade, economic history, and the economics of education. Current interests include trade and education.

ABOUT THE CONTRIBUTORS

Frank Flatters is professor of economics at Queen's University and associate director of the John Deutsch Institute for the Study of Economic Policy. He works on public policy in the fields of international trade and public sector economics, has published in academic journals in these areas, and has been a consultant to a number of governments in the developed and developing world. From 1983 to 1986 he was resident advisor and coordinator of a Harvard University advisory project on fiscal and commercial policy reform in Indonesia. In recent years, he has been the director of collaboration projects at Queen's with the Malaysian Institute of Economic Research and the Thailand Development Research Institute.

Reuven Glick is a vice-president at the Federal Reserve Bank of San Francisco, where he serves as chief of the International Studies Section of the Economics Research Department and also as director of the Bank's Center for Pacific Basin Monetary and Economic Studies. He has taught economics and international business at the Graduate School of Business of New York University and in the Economics Department of the University of California at Berkeley. He has also served as a consultant to the World Bank on international debt issues. He received a Ph.D. degree in economics from Princeton University. He is the author of many journal articles and other professional writings, and his current research interests include international macroeconomic policy and Pacific Basin economic developments.

Edward M. Graham is senior fellow at the Institute for International Economics (IIE) in Washington. Previously he served on the faculties of the Massachusetts Institute of Technology (MIT) and the University of North Carolina and as an international economist at the U.S. Treasury. He has written extensively on international direct investment and multinational enterprises. His best-known work, co-authored with Professor Paul Krugman of MIT, is an analysis of the effects of foreign investment on the U.S. economy. He has co-authored or co-edited books on these and other topics, and is currently co-directing with Professor J. David Richardson a new IIE project on competition policy and the world economy.

Richard G. Harris is professor of economics at Simon Fraser University and an associate of the Program on Economic Growth and Policy of the Canadian Institute for Advanced Research (CIAR). His area of specialization is international economics, especially the economics of integration. He is currently involved in research on the North American Free Trade Area, the North American currency union, and the globalization of labour markets. He is the editor of the present volume.

Tim Hazledine has been professor of economics at the University of Auckland since 1992. Before that, he taught at Queen's University and at the University of British Columbia, where he was a member of the Department of Agricultural

Economics. His Ph.D. is from the University of Warwick. Professor Hazledine's main research interest is the performance of small open-market economies. He is currently working on the impacts on Canada of free trade with the United States and on the consequences of New Zealand's massive program of economic liberalization, begun in 1984.

Keith Head is an assistant professor of policy analysis in the Faculty of Commerce and Business Administration, University of British Columbia, where he teaches courses on international business management and public policy analysis. He has a Ph.D. from the Massachusetts Institute of Technology. Dr. Head's research interests include foreign direct investment, international trade policy, and economic geography. His current research focuses on the immigrants' impact on trade and the effects of trade liberalization on North American manufacturing.

Walid Hejazi is assistant professor at Scarborough College, University of Toronto, and a research associate at the University's Centre for International Studies. He is a graduate of the University of Western Ontario and obtained his Ph.D. in economics from the University of Toronto. His research interests are in the areas of microeconomics and international economics. Papers in progress deal with investment, output, and exchange rate volatility (with Peter Pauly) and with productivity, innovation, and economic growth (with Melvyn Fuss and Len Waverman).

John F. Helliwell is a professor of economics at the University of British Columbia. He obtained a D.Phil. from Oxford. His main interests in economics are quantitative macroeconomics (national and international), comparative empirical studies of economic growth, international finance, taxation and monetary policy, and natural resources, energy, and environmental issues. He is also the W.L. Mackenzie King Visiting Professor of Canadian Studies at Harvard University.

Richard G. Lipsey is a professor of economics at Simon Fraser University and the founding director of the Program of Economic Growth and Policy of the Canadian Institute for Advanced Research. He has worked over three decades in reformulating neoclassical theory. His current research covers such areas as economic growth theory and public policy; Canadian trade policy; and spatial economics and oligopoly theory. Dr. Lipsey has contributed to pioneering theories, including the "general theory of second best," opening up new directions of research on economic policy. He is well known to university students for his introductory text book in economics, available in 14 languages.

James McRae is the director of, and a professor at, the University of Victoria's School of Public Administration. He is a graduate of that university and completed his Ph.D. in economics at the University of Western Ontario. His numerous research articles and monographs have dealt with regional economic development, regulated industries, taxation, international trade, and anticombines policy. Currently, he is an expert witness to the federal Department of Justice on anticombines proceedings in the health care industry and in the area of commercial solid waste collection and disposal. He has served as economic advisor to the British Columbia Task Force on Small Business Programs and Services.

Robert N. McRae obtained his Ph.D. from the University of British Columbia. He is a professor in, and the head of, the Department of Economics at the University of Calgary. His area of specialization is energy economics; his research has focused on estimating systems of fuel consumption equations and analyzing the implications of various energy policy initiatives. Over the past several years, he has been examining the energy consumption behaviour of developing countries of Asia. He has published extensively in the area of energy economics, and has presented his research to numerous national and international conferences, university seminars, business organizations, and government agencies.

Masao Nakamura is a professor in the faculties of Commerce and Business Administration and of Applied Science at the University of British Columbia, where he teaches international business and technology management. He is also Japan Research Chair professor at the UBC Institute of Asian Research. He obtained his Ph.D. in operations research and industrial engineering from Johns Hopkins University. He has published research on economics and managerial decision processes underlying firm and household behaviour in North America and Japan, among other countries. His current research topics include technology and new product management, international joint ventures, internationalization of Japanese firms, and econometrics.

Marcus Noland was educated at Swarthmore College (B.A.) and Johns Hopkins University (Ph.D.). He is currently a senior fellow at the Institute for International Economics and a visiting associate professor at Johns Hopkins University. Dr. Noland is the author or co-author of several books, all published by the Institute for International Economics. He has also written numerous scholarly articles on international economics and the economies of the Asia Pacific region, and served as a consultant to the World Bank, the New York Stock Exchange, and the Advisory Committee on Trade Policy and Negotiations.

Sylvia Ostry is chairman of the Centre for International Studies at the University of Toronto. She has a Ph.D. in economics from McGill University and Cambridge University. She has held a number of senior positions with the federal government and was latterly the Deputy Minister of International Trade, ambassador for multilateral trade negotiations, and the prime minister's personal representative for the economic summit. Dr. Ostry is a frequent guest lecturer and has published many articles on governance and on international trade and finance.

Tae H. Oum is Van Dusen Foundation professor of management at the Faculty of Commerce and Business Administration, University of British Columbia. Dr. Oum's primary areas of research are regulatory and industry policy analysis, demand modeling and forecasting, and cost and productivity analysis. His recent research has focused on airline and telecommunications policy issues at the national and international level, including the "open skies" policy, globalization of the airline networks, airline corporate strategies, etc. He has advised various Canadian and other government agencies on air policy matters, as well as five major airlines in Canada, the United States, and Asia.

Richard Pomfret is professor of economics at the University of Adelaide. Previously he was professor of economics at the Johns Hopkins University School of Advanced International Studies in Washington, Bologna, and Nanjing. He has also worked at Concordia University in Montreal and at the Institut für Weltwirtschaft at the University of Kiel. In 1993 he was seconded to the United Nations Economic Commission for Asia and the Pacific for a year, acting as advisor on macroeconomic policy to the Asian republics of the former Soviet Union. He has been consultant to the World Bank, the Arab Monetary Fund, and the European Union. He has written numerous scholarly books and journal articles, as well as textbooks on international trade and economic development.

Someshwar Rao is the director of the Strategic Investment Analysis Directorate, Industry and Science Policy Sector, at Industry Canada. He directs activities related to the department's research publications program, including an analysis of the impact of investment (domestic as well as foreign), trade and technology, and Canada's corporate governance structure on the microeconomic growth potential of Canadian industry. Prior to joining Industry Canada, he was with the Economic Council of Canada, where he played a key role in the preparation of major reports on Canada's competitiveness. He also served as acting director of the group that was responsible for the development of CANDIDE, a disaggregated model of Canadian industry. Dr. Rao obtained his Ph.D. from Queen's University.

John Ries is assistant professor at the University of British Columbia's Faculty of Commerce and Business Administration, where he teaches courses on international business, government and business, and the Japanese business environment. He has a B.A. from the University of California at Berkeley and a Ph.D.

in economics from the University of Michigan. Dr. Ries' primary research interests are international trade and business and the Japanese economy. He spent one year at Japan's Ministry of International Trade and Industry. He has been a consultant to the Canadian government and has contributed articles to numerous academic conferences and journals. His current research includes analysis of the factors that underlie a multinational firm's choice of where to locate foreign investment, an assessment of the Canada-U.S. Free Trade Agreement on North American manufacturing, and a study of Canadian trade and immigration.

Peter J. Sagar is director general, Entrepreneurship and Small Business Office, Industry Canada. He has extensive experience in the area of economic and industrial development within the federal government, including the departments of Finance and Industry, the Economic Council of Canada, the Senate Committee on Economic and Regional Development, and the Privy Council Office. He has just returned from a two-year exchange with the government of Australia, where he assisted in the development of AusIndustry, an organization that specializes in business development programs and services. Mr. Sagar has a Bachelor of Mathematics (Statistics) from the University of Waterloo and an M.A. in economics from Queen's University.

Lawrence L. Schembri is associate professor of economics at Carleton University, where he has taught courses in international trade and finance, statistics, advanced macroeconomics, and business finance. His primary research interests are international trade and finance, with an emphasis on empirically oriented research. He has published papers on exchange rate pass-through; trade hysteresis; the international transmission of macroeconomic shocks; exchange rate policy; international R&D spillovers and economic growth; the impact of trade liberalization on employment, productivity growth, and competitiveness; and international trade in services. Professor Schembri graduated from Trinity College, University of Toronto; he holds an M.Sc. in economics from the London School of Economics and a Ph.D. in economics from the Massachusetts Institute of Technology.

Louise Séguin-Dulude is a professor at the Institute of Applied Economics at the École des Hautes Études Commerciales (HEC) in Montreal. Previously, she was director of HEC's Centre for International Business Studies and of the Montreal-China Management Education Program. She holds a Ph.D. in economics from the University of Montreal and an M.A. from the University of Toronto. Her main areas of teaching are international business, international economics and technology management. Her research interests are the economic and managerial aspects of multinational enterprises with a focus on technology management and R&D strategy. She is the author of numerous articles and papers on these topics.

ABOUT THE CONTRIBUTORS

Murray G. Smith is the director of the Centre for Trade Policy and Law, Carleton University and the University of Ottawa. Previously, he was director of the International Economics Program of the Institute for Research on Public Policy. Before that, he was with the C.D. Howe Institute, where he served as a senior policy analyst and Canadian research director for the Canadian-American Committee. During the Tokyo Round of multilateral negotiations, he was the director of International Economic Relations in the British Columbia government. He is the co-author or co-editor of several books on Canada's trade relations. He serves on the roster of panelists for disputes under Chapter 18 of the Canada-U.S. Free Trade Agreement.

Daniel Trefler is a faculty member at the Harris School of Public Policy Studies, University of Chicago and the University of Toronto's Department of Economics. He is a research affiliate at the National Bureau of Economic Research (Boston), as well as at the Institute for Policy Analysis and the Centre for International Studies (both at the University of Toronto). His current research examines the relationship between international trade and labour market outcomes such as earnings, employment, and wage inequality across industrial sectors and skill groups. Recent work includes an assessment of the wage and employment effects of the Canada-U.S. Free Trade Agreement. He is also involved in basic research into the appropriate modeling of trade flows and the effects of trade policy.

Russel M. Wills is president of Cognetics, a Vancouver-based research and consulting firm engaged in industrial technology development; environmental and social impact assessments of projects; analyses of forest-based industries; and technology and environmental policy. He has been principal investigator in a variety of research projects throughout Asia and Africa for the World Bank, the Academy for Educational Development, the Canadian International Development Agency, the International Development Research Centre (IDRC), the U.S. Agency for International Development (USAID), and the MacArthur Foundation, as well as a number of national governments and the private sector. He is a fellow at the Centre for Asian Legal Studies at the University of British Columbia and recently worked as principal consultant in an IDRC-sponsored project to help the office of the prime minister of Laos to research and formulate a science and technology policy. He has taught at Stanford University and received a Ph.D. from the University of Washington.